Materials Science and Technologies

nova
science publishers
www.novapublishers.com

Materials Science and Technologies

Physics and Mechanics of New Materials and Their Applications, 2021 – 2022
Ivan A. Parinov, DrSc, Shun-Hsyung Chang, PhD
and Arkady N. Soloviev, DrSc (Editors)
2023. ISBN: 979-8-88697-542-0 (eBook)

Recycling of Discarded Carpets for Structural Polymer Composites
Rajesh Kumar Verma, PhD, Jogendra Kumar, PhD
and Shivi Kesarwani (Editors)
2023. ISBN: 979-8-88697-538-3 (Softcover)
2023. ISBN: 979-8-88697-588-8 (eBook)

Magnesium Alloys: Advances in Research and Applications
Catalin Iulian Pruncu, PhD
and Kavian Omar Cooke, PhD (Editors)
2022. ISBN: 978-1-68507-975-8 (Softcover)
2022. ISBN: 979-8-88697-237-5 (eBook)

Materials Science and Technologies
Catalin Iulian Pruncu, PhD
and Kavian Omar Cooke, PhD (Editors)
2022. ISBN: 978-1-68507-975-8 (Softcover)
2022. ISBN: 979-8-88697-237-5 (eBook)

Photosensitizers and Their Applications
Davor Margetic, PhD and Renjith Thomas, PhD (Editors)
2022. ISBN: 978-1-68507-880-5 (Hardcover)
2022. ISBN: 979-8-88697-039-5 (eBook)

More information about this series can be found at https://novapublishers.com/product-category/series/materials-science-and-technologies/

Noureddine Abidi
and Julia L. Shamshina
Editors

Properties and Applications of Ionic Liquids

nova
science publishers
www.novapublishers.com

DOI: https://doi.org/10.52305/HFEO4188.

Library of Congress Cataloging-in-Publication Data

ISBN: 979-8-88697-635-9

Published by Nova Science Publishers, Inc. † New York

This book is dedicated to my wife Amal, and my kids Merwan and Nabil, and in memory of my late father, mother, and sister. You are always with me.

Noureddine Abidi, Ph.D.

To my late father, Leonid Fyodorov.

J. L. Shamshina, Ph.D.

Contents

Preface

This book reviews recent developments in the field of ionic liquid (IL). The ILs are composed of ions chosen to achieve maximum destabilization of the crystalline solid state of the substance. In Walden's original paper (1914) which examined the electrical conductivity and the molecular size, of mono-, di- and tri-alkylammonium nitrates, he described this class of substances as "salts that exist in a liquid state under standard conditions due to the peculiarities of their chemical structure", and mentioned that "the behavior of these organic salts at temperatures below or around 100°C resembles experiences with inorganic molten salts".

There are many ions that might be selected for the IL formation using "anti-crystal engineering" approach, by balancing of ions symmetry and interionic interactions. ILs are structurally organized on different length scales, from supramolecular (ion clusters) to mesoscopic (hydrogen-bonding networks). Using appropriate combinations of the component ions, it is possible to prepare ILs with essentially any structure and targeted physicochemical properties. The presence of ions results in a number of unique properties, such as wide liquid ranges, non-volatility, and high decomposition temperatures. However, there are many claims accepted as true by the members of the wider scientific communities, – ILs are green, ILs do not burn, ILs are non-toxic, – that eventually were found to be speculative due to limited scientific findings available at the time, and erroneous impressions about the degree of certainty. The overview of ILs is covered in Chapters 1 and 2.

The field has gained a substantial growth from 1990 onwards, due to efforts by the United States Air Force (AFOSR) in applying ILs as electrolytes. Indeed, ILs with high electrical conductivity, high dielectric constants, large electrochemical windows, and redox-stability have been utilized in a number of electrochemical processes, in batteries, capacitors, photovoltaic devices, and electrochemical sensors. Electrical conductivity and dielectric properties of ILs and their solutions are discussed in Chapter 3.

Researchers have been able to knowledgeably prepare ILs based on desired properties for specified purposes, by selecting the ions that are able to impart needed characteristics, at the same time maintaining the core desirable features of the liquid state of matter. Investigations of the various ions and their combinations at the molecular level provided the understanding for the rational development of tailored ILs, so-called Task Specific Ionic Liquids (TSILs), in which a functional group is covalently attached to one (or both ions), for a specific task. This included TSILs for the replacement of volatile amines in reversible capture of CO_2, covered in Chapter 4. The book also covers another type of TSILs, surface-active ILs (“SAILs”), that possess both polar hydrophilic and hydrophobic group in their structure, whose properties and applications are covered in Chapter 5. While initially SAILs found applications in synthesis and catalysis, their usage was later expanded to other areas, such as petroleum industry (covered in Chapter 6).

Another important application of ILs is their extensive utilization in pharmacology, biomedicine, and, lately, biotechnology. Historically, ILs been used as active pharmaceutical ingredients (API-ILs) and as solvents in pharmaceutical industry, for the synthesis of drugs or drug intermediates, pharmaceutical crystallization, continuous manufacturing, and drug delivery of poorly water-soluble drugs. Since recently, they have been employed as relevant alternative solvents to conventional media for stabilization, processing, handling, and storage of nucleic acids. This relatively recent area of investigations is covered in Chapter 7.

Perhaps, the most important application of ILs is their utilization as new solvent systems with enhanced, or even unique, performance. This included IL-assisted graphite exfoliation, the method that has high potential to improve commercial availability of graphene. In this process, the cation–anion coordination, particularly π-stacking in the case of aromatic cations, contributed significantly to the success of the exfoliation process (Chapter 8). ILs provide suitable replacements for volatile organic compounds (VOCs) identified as significant environmental pollutants thus being extensively used for a wide range of separations. Heavy metal ions extraction is covered in Chapter 9, whereas separation of pollutants is covered in Chapter 10.

Importantly, ILs are suitable for the pre-treatment of various biomasses such as wood (hardwood and softwood), grasses, crop residues, and agricultural biproducts (e.g., crustacean biomass). There are numerous IL-based technologies now available for the separation of biopolymers from biomasses which offer the opportunity to use the clean separated biomass components as a reproducible feedstock for further valorization into chemical building blocks, or for the preparation of advanced materials from them. Experimental insight into cellulose dissolution mechanism in ILs/cosolvent mixed systems is covered in Chapter 11. These unique solvent properties of the ILs opened up scope for new applications and improvements in existing technologies. Thus, a process called "natural fiber welding" allows for transformation of loose cellulosic (or protein) fibers into congealed network using an IL solvent, to increase the performance of biopolymer-based (cotton, silk) textiles. The state of natural fiber welding technology is covered in Chapter 12. In summary, the book strongly indicates the significant activity currently taking place in the IL field, and the future holds many new discoveries.

Chapter 1

What Are Ionic Liquids? Embracing the Immensity

Paula Berton[1]
Noureddine Abidi[2]
and Julia L. Shamshina[2]
[1] Chemical and Petroleum Engineering Department,
University of Calgary, Calgary, Alberta, Canada
[2] Fiber and Biopolymer Research Institute,
Department of Plant and Soil Science, Texas Tech University, Lubbock, Texas, USA

Abstract

Considering the rapid development of this class of compounds and exponential growth in applications, summarizing what ionic liquids (ILs) are is "embracing the immensity". The perception of ILs has changed and keeps changing, or rather evolving, over time. It is a dynamic, ongoing process with a capacity for self-correction, like any scientific learning. The needs of users of ILs change, and new technologies and new products require new and efficient systems.

Keywords: ionic liquids, ionicity, applications of ionic liquids, ionic liquid processes, ionic liquid types

Introduction

A brief and, at the same time, up-to-date and original introduction is a challenge when it comes to summarizing the ionic liquid's (ILs) field progress, even when the active development of the field of ILs has taken place for only the last four decades. One could think that trying to condense what ILs are in a few pages feels like "embracing the immensity," considering the rapid development of this class of compounds and exponential growth in applications.

Out of the numerous reviews available on the topic, three stand out, since they are written by some of the founders of the field – John Wilkes (2002) [1], Austen Angell (2012) [2], and Tom Welton (2018) [3]. While citations of the original research papers are typically preferred over citations of review articles, these works highlight the early chronicles of the field and are

In: Properties and Applications of Ionic Liquids
Editors: Noureddine Abidi and Julia L. Shamshina
ISBN: 979-8-88697-635-9

well spread out in time, allowing interested readers to closely follow the development of the discipline. There is also an interesting perspective that looks at the ILs through the lens of patenting field [4]. This work reviewed patents and claims and provided an overview of IL-based technologies from the point of their commercial potential, elucidating trends toward new applications for ILs. Finally, a recent book chapter by Roland Kalb on the industrialization of ILs summarized the opinions of academic and industrial leaders in the field and gave an overview of both pilot and fully commercialized IL-based technologies [5]. It also identified opportunities and future research areas.

The field as we know it originated from catalytic and electrochemical studies conducted using room-temperature molten salts – substituted alkylpyridinium and imidazolium chloroaluminates [6-12]. It was Professor Ken Seddon, who, when became fascinated with these "neoteric solvents", co-founded the Queen's University Ionic Liquids Laboratory Research Centre (QUILL) [3]. This was the first research center that partnered with industrial companies: BP Chemicals, Shell, Proctor and Gamble, Petronas, Unilever, Imperial Chemical Industries (ICI), etc., to work on ILs for the research on clean technologies and catalysis [13]. In 2000, Ken Seddon, Robin Rogers, and Sergey Volkov organized a NATO Advanced Research Workshop "Green Industrial Applications of Ionic Liquids", the first international meeting dedicated to the area of ILs [14], which laid the foundation for a new era in using ILs for green technologies. Soon after that, multiple conferences and symposia dedicated to ILs followed, including the 1st International Congress on Ionic Liquids (COIL) in Salzburg, in 2005. Now, the list of conferences on ILs is exhaustively long and includes the Ionic Liquid Conference, Ionic Liquids in Chemical and Biochemical Systems Conference, Ionic Liquids in Purification Technologies Conference, Ionic Liquids in Separation and Purification Technologies Conference, Chemistry of Ionic Liquids Conference, Gordon Research Conference on Ionic Liquids, not to mention numerous IL symposia [15].

"All Our Knowledge Has Its Origins in Our Perceptions" (Leonardo Da Vinci)

The perception of ILs has changed and keeps changing, or rather evolving, over time. First of all, it is a dynamic, ongoing process with a capacity for self-correction, like any scientific learning. Secondly, the needs of users of ILs change, and new technologies and new products require new and efficient systems. Historically, ILs were defined as ionic compounds with a melting point below the boiling point of water (100°C), and multiple research articles claimed it was Tom Welton who introduced this temperature limitation in his 1999 review [16]. According to Tom Welton himself, this was most certainly not the case [3]. Instead, he recommended that the term "ionic liquid" assumes low-melting salt. Yet, as Isiah Warner put it when talking about GUMBOS (Group-of-Uniform-Materials-Based-on-Organic-Salts): "There is no logical reason why we cannot extend our melting point range to 250°C while maintaining the tunability which is inherent to ILs" [17]. Nonetheless, right now the term IL assumes an ionic compound with a melting point below 100°C, often below room temperature (room-temperature ILs or RTILs).

It was also initially declared that "ILs are composed solely of ions" [18, 19]. The question of this restrictive nomenclature is looked at as a limitation in the field after the development of multicomponent systems of intermediate ionicity [20]. In 2010, the term "confused" or oligomeric ILs was introduced, where ILs contained oligomeric hydrogen-bonded cations or anions [21]. Investigations at the molecular level revealed that a large number of liquid forms exhibited what was called "intermediate ionicity". These are hydrogen-bonded complexes [19,22] of ionized and unionized forms that share a delocalized proton [20]. In simplified terms, these systems are somewhere in between fully ionized (i.e., ILs) and non-ionized (i.e., deep eutectic solvents) structures [23-26]. As an example, a simple, protic IL triethylammonium acetate is classified as a complex system with oligomeric ions [27]. This said, we agree that the term "ionic liquids" for this type of liquid salt is somewhat controversial.

That is not to mention that ILs were defined to be composed of an "organic cation and either an organic or inorganic anion". However, systems containing more than two ions, so-called "double salt ionic liquids" (DSILs [28-30]), can be prepared by simply mixing two (or more) ILs. Such mixing allowed manipulating physicochemical properties of the systems through the additional degree of freedom, and often exhibited nonlinear changes in comparison with "original" ILs [31], due to differences in ion interaction energies and formation of hydrogen bonds.

Despite all the controversies about the definition and number of compounds that should be included under the term "ionic liquids", they are game changers: They can be prepared with specific physicochemical properties targeted for a specific process. A sign of their importance and potential is signed by the fact that ILs won the Great British Innovation Vote in 2013 [32] as the innovation presumably able "to shape the 21st century."

A Boom in Applications

It became evident by 2000, that the low vapor pressure of these liquids allows using them as safe solvents [33] for transition metal-catalyzed reactions [34-37], enzymatically-catalyzed reactions [38], and other chemical transformations [39]. Concurrently, Jim Davis introduced the term Task-Specific Ionic Liquids (TSILs), in which a functional group was covalently attached to one or both of the ions of an otherwise "ordinary" IL [40,41], with a particular application in mind: as catalysts [42], chiral solvents [43], magnetically susceptible materials [44], for CO_2-capture [45], and so on (the list of TSILs' is endless!).

The discovery that ILs are suitable for the dissolution of biopolymers and biomasses [46] opened a whole new avenue of research. Initially, biopolymer-dissolving halide-containing ILs were used for the purpose [47,48]. This was followed by the current state-of-the-art acetate-based ILs [49,50], produced according to the patented CBILS® halide-free process. Later on, super-base ILs with high dissolution power were developed [51,52]. Out of the relatively recent processes, it is worth mentioning the HighPerCellCarbon® process which involves the wet-jet spinning of cellulose fibers from ILs followed by transformation into carbon fibers by carbonization [53,54].

Around 2007, the study of ILs in the pharmaceutical industry was initiated [55]. Initially, ILs were used as solvents in the synthesis of drugs or drug intermediates, solvents for extraction of drugs from natural products, and as solvents, co-solvents, anti-solvents and emulsifiers in pharmaceutical separations and continuous pharma manufacturing. As the field developed, ILs were used as pharmaceutical ingredients, as a tool to reformulate APIs into more bioavailable forms [56,57] where liquefaction was a result of "anti-crystal engineering" [58] or combining an IL strategy with the prodrug strategy to improve the delivery of solid APIs, representing an enabling technology platform. Even though the concept of API-ILs provides integrated drug design, development, and manufacturing, and the numerous academic studies, there are limited examples of commercialization, such as the case of a reverse transcriptase inhibitor for the treatment of HIV from GlaxoSmithKline (GSK), GSK2838232 [59-61] or CAGE [62].

Conclusion

In these transitioning times, when the world is looking for alternative, more sustainable technologies, ionic liquids can, once again, be part of innovative, revolutionary solutions, for example, in battery-related and energy storage applications [63-67], to solve obstacles in solar cells-based photovoltaics technology [68,69] and to reduce the human impact on the environment by capturing CO_2 or toxic gases such as mercury from gaseous effluents [70-76] or by treating industry-generated tailings [77], to name a few.

The list of applications and innovations unlocked by ILs goes on and on, and some of the latest advances are already turning into processes and innovations to answer real-world problems in different fields [78]. Nowadays, it is not uncommon to read in the news about scaling up IL-based technologies. For example, Ioniqa Technologies is a Dutch polyethylene terephthalate (PET) recycler that uses ILs to convert colored PET waste into a monomer for producing new PET [79]. Ioniqa's demonstration plant can digest 10,000 tonnes of PET annually, and Koch Industries recently announced a $30 million investment to help scale up the process [80]. Although many of these technologies are well-developed and de-risked, technological innovations based on ILs continue to emerge each year, and ILs are still demonstrating that their tunability can be used in the large domain of cutting-edge fields. For example, the tunable properties of ILs were recently used, for the first time, in physical reservoir computing (PRS) systems, allowing a versatility that was unthinkable before. This new IL-based PRS design could lead to Artificial Intelligence devices that can directly learn at the time scales of environmentally generated signals, in real-time [81].

This book presents recent trends in different areas of ILs chemistry and technologies. The readers will find eleven chapters on recent applications of ILs in the fields of the petroleum industry, stabilization of nucleic acids, separations, and biopolymers' processing, including trends and critical perspectives.

References

[1] Wilkes, J. S. A short history of ionic liquids – from molten salts to neoteric solvents. *Green Chem.* 2002, *4*, 73–80. https://doi.org/10.1039/B110838G.

[2] Angell, C. A.; Ansari, Y.; Zhao, Z. Ionic Liquids: Past, present and future. *Faraday Discuss.* 2012, *154*, 9–27. https://doi.org/10.1039/C1FD00112D.

[3] Welton, T. Ionic liquids: a brief history. *Biophys. Rev.* 2018, *10*, 691–706. https://doi.org/10.1007/s12551-018-0419-2.

[4] Morton, M. D.; Hamer C. K. Ionic liquids – The beginning of the end or the end of the beginning? – A look at the life of ionic liquids through patent claims. *Separation Purification Tech.* 2018, *196*, 3–9. https://doi.org/10.1016/j.seppur.2017.11.023.

[5] Kalb, R. S. Toward Industrialization of Ionic Liquids. In: Shiflett, M. (eds) Commercial Applications of Ionic Liquids. Green Chemistry and Sustainable Technology. Springer Cham, 2020. https://doi.org/10.1007/978-3-030-35245-5_11.

[6] Ubbelohde, A. R. Research on Molten Salts: Introduction. In: *Ionic Liquids*, Inman, D.; Lovering, D. G. (Eds.), Plenum Press, New York (1981). https://doi.org/10.1007/978-1-4757-0920-9_1.

[7] Wilkes, J. S.; Levisky, J. A.; Wilson, R. A.; Hussey, C. L. Dialkylimidazolium chloroaluminate melts: a new class of room-temperature ionic liquids for electrochemistry, spectroscopy and synthesis. *Inorganic Chem.* 1982, *21(3)*, 1263–1264. https://doi.org/10.1021/ic00133a078.

[8] Osteryoung, R. A.; Osteryoung, J. G. Studies in ambient temperature ionic liquids. AFOSR Report (1987), (AFOSR-TR-87-0879; Order No. AD-A182723).

[9] Zawodzinski, T. A., Jr.; Osteryoung, R. A. 1-Methyl-3-ethylimidazolium hydrogen dichloride: synthesis and application to the study of protons in ambient-temperature chloroaluminate ionic liquids. *Inorganic Chem.* 1988, *27(24)*, 4383–4384. https://doi.org/10.1021/ic00297a011.

[10] Osteryoung, R. A. AFOSR Report (1990). AFOSR-TR-90-0084; Order No. AD-A217742: Chemical and electrochemical studies in ionic liquids.

[11] Avent, A. G.; Chaloner, P. A.; Day, M. P.; Seddon, K. R.; Welton, T. Evidence for hydrogen-bonding in solutions of 1-methyl-3-ethylimidazolium halides, as determined by 1H, 35Cl and 127I NMR spectroscopy, and its implications for room-temperature halogenaluminate (III) ionic liquids. *Proceedings - Electrochemical Society* 1990, 90-17 (Proc. Int. Symp. Molten Salts, 7th, 1990), 98–133.

[12] Dent, A. J.; Seddon, K. R.; Welton, T. The structure of halometallate complexes dissolved in both basic and acidic room-temperature haloaluminate(III) ionic liquids, as determined by EXAFS. *J. Chem. Soc. Chem. Commun.* 1990, *4*, 315–316. https://doi.org/10.1039/C39900000315.

[13] Seddon, K. R. Ionic liquids for clean technology. *J. Chem. Technol. Biotechnol.* 1997, 68, 351–356. https://doi.org/10.1002/(SICI)1097-4660(199704)68:4<351::AID-JCTB613>3.0.CO;2-4.

[14] *Green Industrial Applications of Ionic Liquids*. NATO Science Series. Rogers, R. D.; Seddon, K. R.; Volkov, S. (Eds.). Kluwer Academic Publishers (2002).

[15] Ionic Liquids Conferences. Available at: https://waset.org/ionic-liquids-conferences. Last assessed 09-20-22.

[16] Welton T. Room-temperature ionic liquids. Solvents for synthesis and catalysis. *Chem. Rev.* 1999, *99*, 2071–2084. https://doi.org/10.1021/cr980032t.

[17] Warner Research Group, Louisiana State University. Available at: https://faculty.lsu.edu/iwarner/our-research.php. Last accessed 09-20-22.

[18] Lei, Z.; Chen, B.; Koo, Y.-M.; MacFarlane, D. R. Introduction: Ionic Liquids. *Chem. Rev.* 2017, *117*, 10, 6633–6635. https://doi.org/10.1021/acs.chemrev.7b00246.

[19] Ohno, H. Ionic Liquids. In: *Functional Organic Liquids*. Nakanishi, T. (Ed.). Wiley-VCH Verlag GmbH & Co. KGaA (2019). https://doi.org/10.1002/9783527804948.ch12.

[20] Kelley, S. P.; Narita, A.; Holbrey, J. D.; Green, K. D.; Reichert, W. M.; Rogers, R. D. Understanding the effects of ionicity in salts, solvates, co-crystals, ionic co-crystals, and ionic liquids, rather than nomenclature, is critical to understanding their behavior. *Cryst. Growth Des.* 2013, *13*, 965–975. https://doi.org/10.1021/cg4000439.

[21] Bica, K.; Rogers, R. D. Confused ionic liquid ions—a "liquification" and dosage strategy for pharmaceutically active salts. *Chem. Commun.* 2010, *46*, 1215–1217. https://doi.org/10.1039/B925147B.

[22] Bica, K.; Shamshina, J. L.; Hough, W. L.; MacFarlane, D. R.; Rogers, R. D. Liquid forms of pharmaceutical co-crystals: exploring the boundaries of salt formation. *Chem. Commun.* 2011, *47*, 2267–2269. https://doi.org/10.1039/C0CC04485G.

[23] Ławniczak Ł, Syguda A, Borkowski A, Cyplik P, Marcinkowska K, Wolko Ł, Praczyk T, Chrzanowski Ł, Pernak J. Influence of oligomeric herbicidal ionic liquids with MCPA and Dicamba anions on the community structure of autochthonic bacteria present in agricultural soil. *Sci. Total Environ.* 2016, *563-564*, 247–255. https://doi.org/10.1016/j.scitotenv.2016.04.109.

[24] Berton, P.; Kelley, S. P.; Wang, H.; Myerson, A. S.; Rogers, R. D. Separate mechanisms of ion oligomerization tune the physicochemical properties of n-butylammonium acetate: cation-base clusters vs. anion-acid dimers. *Phys. Chem. Chem. Phys.* 2017, *19*, 25544. https://doi.org/10.1039/C7CP04078.

[25] Frizzo, C. P.; Bender, C. R.; Salbego, P. R. S.; Farias, C. A. A.; Villetti, M. A.; Martins, M. A. P. Heteroassembly ability of dicationic ionic liquids and neutral active pharmaceutical ingredients. *ACS Omega* 2018, *3(2)*, 2282–2291. https://doi.org/10.1021/acsomega.7b02097.

[26] Stoimenovski, J.; Dean, P. M.; Izgorodina, E. I.; MacFarlane, D. R. Protic pharmaceutical ionic liquids and solids: Aspects of protonics. *Faraday Discuss.* 2012, *154*, 335. https://doi.org/10.1039/C1FD00071C.

[27] Berton, P.; Kelley, S. P.; Wang, H.; Rogers, R. D. Elucidating the triethylammonium acetate system: Is it molecular or is it ionic? *J. Mol. Liq.* 2018, *269*, 126–131. https://doi.org/10.1016/j.molliq.2018.08.006.

[28] Wang, H.; Kelley, S. P.; Brantley III, J. W.; Chatel, G.; Shamshina, J. L.; Pereira, J. F. B.; Debbeti, V.; Myerson, A. S.; Rogers, R. D. Ionic fluids containing both strongly and weakly interacting ions of the same charge have unique ionic and chemical environments as a function of ion concentration. *ChemPhysChem* 2015, *16*, 993–1002. https://doi.org/10.1002/cphc.201402894.

[29] Wang, H.; Pereira, J. F. B.; Myerson, A. S.; Rogers, R. D. Double salt ionic liquids prepared by mixing partially miscible ionic liquids: Tuning the solubility of lipophilic molecules. *ECS Transactions* 2014, *64*, 33–44.

[30] Wang, H.; Pereira, J. F. B.; Berton, P.; Rogers, R. D. Double salt ionic liquids containing the trihexyl(tetradecyl)phosphonium cation: The ability to tune the solubility of aromatics, ethers, and lipophilic compounds, *ECS Transactions* 2016, *75*, 451–465.

[31] Chatel, G.; Pereira, J. F. B.; Debbeti, V.; Wang, H.; Rogers, R. D. Mixing ionic liquids – "simple mixtures" or "double salts"? *Green Chem.* 2014, *16*, 2051–2083.

[32] Stoye, E. Ionic liquids win Great British Innovation Vote. *Chemistry World*, March 27, 2013. Available at: https://www.chemistryworld.com/news/ionic-liquids-win-great-british-innovation-vote/6018. Last assessed 09-20-22.

[33] Earle, M. J.; Seddon, K. R. Ionic liquids. Green solvents for the future. *Pure Applied Chem.* 2000, *72*, 1391–1398. https://doi.org/10.1351/pac200072071391.

[34] Wasserscheid, K. Ionic liquids - new "solutions" for transition metal catalysis. *Angewandte Chemie* (International ed. in English) 2000, *39(21)*, 3772–3789. https://doi.org/10.1002/1521-3773(20001103)39:21<3772::AID-ANIE3772>3.0.CO;2-5.

[35] Sirieix, J.; Ossberger, M.; Betzemeier, B.; Knochel, P. Palladium catalyzed cross-couplings of organozincs in ionic liquids. *Synlett* 2000, *2000(11)*, 1613–1615. https://doi.org/10.1055/s-2000-7913.

[36] Wasserscheid, P.; Waffenschmidt, H. Ionic liquids in regioselective platinum-catalyzed hydroformylation. *J. Molecular Catalysis A: Chemical* 2000, *164(1-2)*, 61–67. https://doi.org/10.1016/S1381-1169(00)00259-4.

[37] McLachlan, F.; Mathews, C. J.; Smith, P. J.; Welton, T. Palladium catalyzed Suzuki cross-coupling reactions in ambient temperature ionic liquids: Evidence for the Importance of Palladium imidazolylidene complexes. *Organometallics* 2003, *22*(25), 5350–5357. https://doi.org/10.1021/om034075y.

[38] Madeira Lau, R.; van Rantwijk, F.; Seddon, K. R.; Sheldon, R. A. Lipase-Catalyzed Reactions in Ionic Liquids. *Org. Lett.* 2000, *2*(26), 4189–4191. https://doi.org/10.1021/ol006732d.

[39] Dubreuil, J. F.; Bazureau, J. P. Rate accelerations of 1,3-dipolar cycloaddition reactions in ionic liquids. *Tetrahedron Lett.* 2000, *41*(38), 7351–735. https://doi.org/10.1016/S0040-4039(00)01237-5.

[40] Wierzbicki, A.; Davis, J. H. Jr. Envisioning the second generation of ionic liquid technology: design and synthesis of Task-specific Ionic Liquids (TSILs). Proceedings of the symposium on advances in solvent selection and substitution for extraction. AIChE 2000, New York.

[41] Davis, J. H. Jr. Task-specific ionic liquids. *Chem. Lett.* 2004, *33(9)*, 1072–1077. https://doi.org/10.1246/cl.2004.1072.

[42] Cole, A. C.; Jensen, J. L.; Ntai, I.; Tran, K. L. T.; Weaver, K. J.; Forbes, D. C.; Davis, J. H. Novel Brønsted-acidic ionic liquids and their use as dual solvent−catalysts. *J. Am. Chem. Soc.* 2002, *124*, 5962–5963. https://doi.org/10.1021/ja026290w.

[43] Schulz, P. S.; Müller, N.; Bösmann, A.; Wasserscheid, P. Effective chirality transfer in ionic liquids through ion-pairing effects *Angew. Chem. Int. Ed.* 2007, *46*, 1293–1295. https://doi.org/10.1002/anie.200604406.

[44] Okuno, M.; Hamaguchi, H.; Hayashi, S. Magnetic manipulation of materials in a magnetic ionic liquid. *Appl. Phys. Lett.* 2006, *89*, 132506. https://doi.org/10.1063/1.2353808.

[45] Yokozeki, A.; Shiflett, M. B. Separation of carbon dioxide and sulfur dioxide gases using room-temperature ionic liquid [hmim][Tf2N]. *Energy Fuels* 2009, *23*, 4701–4708. https://doi.org/10.1021/ef900649c.

[46] Swatloski, R. P.; Spear, S. K.; Holbrey, J. D.; Rogers, R. D. Dissolution of cellulose with ionic liquids. *J. Am. Chem. Soc.* 2002, *124*, 4974–4975. https://doi.org/10.1021/ja025790m.

[47] Turner, M. B.; Spear, S. K.; Holbrey, J. D.; Daly, D. T.; Rogers, R. D. Ionic liquid-reconstituted cellulose composites as solid support matrices for biocatalyst immobilization. *Biomacromolecules* 2005, *6*, 2497–2502. https://doi.org/10.1021/bm050199d.

[48] Turner, M. B.; Spear, S. K.; Holbrey, J. D.; Rogers, R. D. Production of bioactive cellulose films reconstituted from ionic liquids. *Biomacromolecules* 2004, *5*, 1379–1384. https://doi.org/10.1021/bm049748q.

[49] Zavgorodnya, O.; Shamshina, J. L.; Berton, P.; Rogers, R. D. Translational Research from Academia to Industry: Following the Pathway of George Washington Carver, In: ACS Symposium Series, Ionic Liquids: Current State and Future Directions, American Chemical Society, 2017; Vol. 1250, pp. 17-33. https://doi.org/10.1021/bk-2017-1250.ch002.

[50] Hermanutz, F.; Gähr, F.; Uerdingen, E.; Meister, F.; Kosan, B. New developments in dissolving and processing of cellulose in ionic liquids. *Macromol. Symp.* 2008, *262*, 23–27. https://doi.org/10.1002/masy.200850203.

[51] Sixta, H.; Michud, A.; Hauru, L.; Asaadi, S.; Ma, Y.; King, A. W. T.; Kilpelainen, I.; Hummel, M. Ioncell-F: A High-strength regenerated cellulose fibre. *Nordic Pulp & Paper Res. J.* 2015, *30(1)*, 43–57.

[52] Elsayed, S.; Hummel, M.; Sawada, D.; Guizani, C.; Rissanen, M.; Sixta, H. Superbase-based protic ionic liquids for cellulose filament spinning. *Cellulose* 2021, *28*, 533–547. https://doi.org/10.1007/s10570-020-03505-y.

[53] The "Cellulose Fibre Innovation of The Year 2022 Award" Goes To DITF Denkendorf. Available at: https://www.textileworld.com/textile-world/fiber-world/2022/02/the-cellulose-fibre-innovation-of-the-year-2022-award-goes-to-ditf-denkendorf/. Last assessed 09-19-22.

[54] Vocht, M. P.; Beyer, R.; Thomasic, P.; Muller, A.; Ota, A.; Hermanutz, F.; Buchmeiser, M. A. High-performance cellulosic filament fibers prepared via dry-jet wet spinning from ionic liquids. *Cellulose* 2021, *28*, 3055–3067. https://doi.org/10.1007/s10570-021-03697-x.

[55] Hough, W. L.; Smiglak, M.; Rodríguez, H.; Swatloski, R. P.; Spear, S. K.; Daly, D. T.; Pernak, J.; Grisel, J. E.; Carliss, R. D.; Soutullo, M. D.; Davis, J. H. Jr.; Rogers, R. D. The third evolution of ionic liquids: Active pharmaceutical ingredients. *New J. Chem.* 2007, 31, 1429. https://doi.org/10.1039/B706677P.

[56] Ferraz, R.; Branco, L. C.; Prudêncio, C.; Noronha, J. P.; Petrovski, Ž. Ionic liquids as active pharmaceutical ingredients. *ChemMedChem* 2011, *6*, 975–985. https://doi.org/10.1002/cmdc.201100082.

[57] Rogers, R. D.; Daly, D. T.; Swatloski, R. P.; Hough-Troutman, W. L.; Hough-Troutman, J. J. H. L.; Marcin, S.; Juliusz, P.; Spear, S. K. Multifunctional Ionic Liquid Compositions for Overcoming Polymorphism and Imparting Improved Properties for Active Pharmaceutical, Biological, Nutritional and Energetic Ingredients. U.S. Patent No. 8,232,265; PCT/US2006/039454; WO 2007/044693 A2 (2007), 18 October 2012.

[58] Dean, P. M.; Turanjanin, J.; Yoshizawa-fujita, M.; Macfarlane, D. R.; Scott, J. L. Exploring an anti-crystal engineering approach to the preparation of pharmaceutically active ionic liquids. *Cryst. Growth Des.* 2009, *9*, 1137–1145. https://doi.org/10.1021/cg8009496.

[59] Hatcher, M. A.; Johns, B. A.; Martin, M.T.; Tabet, E. A.; Tang, J. Preparation of betulin derivatives for the treatment of HIV. WIPO (PCT) Patent WO2013090664A1, 20 June 2013.

[60] DeJesus, E.; Harward, S.; Jewell, R. C.; Johnson, M.; Dumont, E.; Wilches, V.; Halliday, F.; Talarico, C. L.; Jeffrey, J.; Gan, J.; Xu, J.; Felizarta, F.; Scribner, A.; Ramgopal, M.; Benson, P.; Johns, B. A. A Phase IIa study evaluating safety, pharmacokinetics, and antiviral activity of GSK2838232, a novel, second-generation maturation inhibitor, in participants with human immunodeficiency virus type 1 infection. *Clin. Infect. Dis.* 2019, Pii: ciz938. https://doi.org/10.1093/cid/ciz938.

[61] Johns, B. A. Derivatives of betulin. U.S. Patent US 9,102,685 B2, 11 August 2015.

[62] Banerjee, A.; Ibsen, K.; Brown, T.; Chen, R.; Agatemor, C.; Mitragotri, S. Reply to Rogers and Gurau: Definitions of ionic liquids and deep eutectic solvents. *Proc Natl Acad Sci USA* 2018, *115*, E11000–E11001. https://doi.org/10.1073/pnas.1815526115.

[63] Vogl, T.; Menne, S.; Kühnel, R.-S.; Balducci, A. The Beneficial Effect of Protic Ionic Liquids on the Lithium Environment in Electrolytes for Battery Applications. *J Mater Chem A* 2014, *2*, 8258–8265. https://doi.org/10.1039/c3ta15224c.

[64] Lin, M. C.; Gong, M.; Lu, B.; Wu, Y.; Wang, D. Y.; Guan, M.; Angell, M.; Chen, C.; Yang, J.; Hwang, B. J. An ultrafast rechargeable aluminium-ion battery. *Nature* 2015, *520*, 324–328. https://doi.org/10.1038/nature14340.

[65] Cheng, E.J.; Shoji, M.; Abe, T.; Kanamura, K. Ionic liquid-containing cathodes empowering ceramic solid electrolytes. iScience, 10 February 2022. https://doi.org/10.1016/j.isci.2022.103896.

[66] Lewandowski, A.; Świderska-Mocek, A. Ionic liquids as electrolytes for Li-ion batteries - An overview of electrochemical studies. *J. Power Sources* 2009, *194*, 601–609. https://doi.org/10.1016/j.jpowsour.2009.06.089.

[67] Watanabe, M.; Thomas, M. L.; Zhang, S.; Ueno, K.; Yasuda, T.; Dokko, K. Application of ionic liquids to energy storage and conversion materials and devices. *Chemical Reviews* 2017, *117*, 7190–7239. https://doi.org/10.1021/acs.chemrev.6b00504.

[68] Inomata, T.; Matsunaga, A.; Jin, G.; Kitagawa, T.; Muramatsu, M.; Ozawa, T.; Masuda, H. Improvements in photoelectric performance of dye-sensitised solar cells using ionic liquid-modified TiO_2 electrodes. *RSC Adv.* 2022, *12*, 19624–19631. https://doi.org/10.1039/D2RA03230A

[69] Gao, X.-X.; Ding, B.; Zhang, Y.; Zhang, S.; Turnell-Ritson, R. C.; Kanda, H.; Abuhelaiqa, M.; Shibayama, N.; Luo, W.; Li, M.; Tirani, F.; F.; Scopelliti, R.; Kinge, S.; Züttel, A.; Zhu, D.; Zhang, B.; Feng, Y.; Fei, Z.; Nazeeruddin, M. K.; Dyson, P. J. Halide exchange in the passivation of perovskite solar cells with functionalized ionic liquids. *Cell Rep. Phys. Sci.* 2022, *3(4)*, 100848. https://doi.org/10.1016/j.xcrp.2022.100848.

[70] Abai, M.; Atkins, M. P.; Hassan, A.; Holbrey, J. D.; Kuah, Y.; Nockemann, P.; Oliferenko, A. A.; Plechkova, N. V.; Rafeen, S.; Rahman, A. A.; Ramli, R.; Shariff, S. M.; Seddon, K. R.; Srinivasan, G.; Zoub, Y. An ionic liquid process for mercury removal from natural gas. *Dalton Trans.* 2015, *44*, 8617–8624. https://doi.org/10.1039/C4DT03273J.

[71] Bara, J. E.; Camper, D. E.; Gin, D. L.; Noble, R. D. Room-temperature ionic liquids and composite materials: Platform technologies for CO_2 capture *Acc. Chem. Res.* 2010, *43*, 152–159. https://doi.org/10.1021/ar9001747.

[72] Bates, E. D.; Mayton, R. D.; Ntai, I.; Davis, J. H. Jr. CO_2 Capture by a Task-Specific Ionic Liquid. *J. Am. Chem. Soc.* 2002, *124*, 926–927. https://doi.org/10.1021/ja017593d.

[73] Blanchard, L. A.; Gu, Z. Y.; Brennecke, J. F. High-Pressure Phase Behavior of Ionic Liquid/CO_2 Systems. *J. Phys. Chem. B* 2001, *105*, 2437–2444. https://doi.org/10.1021/jp003309d.

[74] Zhang, J.; Zhang, S.; Dong, K.; Zhang, Y.; Shen, Y.; Lv, X. Supported absorption of CO_2 by tetrabutylphosphonium aminoacid ionic liquids. *Chem. – Eur. J.* 2006, *12*, 4021–4026. https://doi.org/10.1002/chem.200501015.

[75] Wang, Y.; Lu, Y.; Wang, C.; Zhang, Y.; Huo, F.; He, H.; Zhang, S. Two-dimensional ionic liquids with an anomalous stepwise melting process and ultrahigh CO_2 adsorption capacity. *Cell Rep. Phys. Sci.* 2022, *3(7)*, 100979. https://doi.org/10.1016/j.xcrp.2022.100979.

[76] Recker, E. A.; Green, M.; Soltani, M.; Paull, D. H.; McManus, G. J.; Davis, J. H. Jr.; Mirjafari, A. Direct air capture of CO_2 via ionic liquids derived from "waste" amino acids. *ACS Sustainable Chem. Eng*. 2022, *10*, 11885–11890. https://doi.org/10.1021/acssuschemeng.2c02883.

[77] Berton, P.; Chen, Y. N.; Manouchehr, S.; Wong, K.; Ahmadi, Z.; Bryant, S. L.; Rogers, R. D. *AIChE J.* 2022, *68(2)*, e17494. https://doi.org/10.1002/aic.17494.

[78] Shiflett, M.B. Commercial Applications of Ionic Liquids. Green Chemistry and Sustainable Technology. Springer Cham, 2020.

[79] Ioniqa, https://ioniqa.com/ Last accessed 09-21-22.

[80] Blois, M. Koch invests in lithium extraction and PET recycler. *C&E News*, September 16, 2022, *100*, 33. https://cen.acs.org/environment/sustainability/Koch-invests-lithium-extraction-PET/100/i33. Last accessed 09-21-22.

[81] Koh, S.-G.; Shima, H.; Naitoh, Y.; Akinaga, H.; Kinoshita, K. Reservoir computing with dielectric relaxation at an electrode–ionic liquid interface. *Sci. Reports* 2022, *12(1)*. https://doi.org/10.1038/s41598-022-10152-9.

Chapter 2

Ionic Liquids: From Laboratory Research to Industrial Applications

George G. Tamas*
Bradley University, Peoria, Illinois, USA

Abstract

Since their discovery at the beginning of the 20th century, ionic liquids have gained much attention in the academic and industrial worlds. Their structure, based on a virtually unlimited combination of cations and anions, is responsible for their high tunability. As organic salts – with remarkable properties such as wide liquid range at relatively low temperatures, negligible vapor pressure, good thermal stability, and broad electrochemical windows – ionic liquids have exceptional solvation properties adjustable to target a wide range of applications, including, but not limited to, chemical catalysis, textiles, energy storage, sustainable lubricants, energy storage, bioactive compound extraction, membrane separation, and pharmaceuticals. Although the research on ionic liquids grew rapidly starting in the early 1990's, the technological transfer from the laboratory to industrial applications has not been as fast paced. This chapter summarizes some remarkable scientific breakthroughs that proved to be successful industrial transitions.

Keywords: ionic liquids, industry, commercial applications

Introduction

In 1914, Paul Walden published his research on ethylammonium nitrate ([$EtNH_3$][NO_3]) and reported some unique physical properties of the new chemical formed by the neutralization of ethylamine with concentrated nitric acid [1]. This compound had an unexpected feature; its melting point was below room temperature, at around 12ºC. Other salts become liquids only at much higher temperatures (e.g., NaCl m.p. = 801°C, $CaCl_2$ m.p. = 772°C) [2]. Although Walden's paper did not reach a large audience at the time, a century later, his work is considered the starting point of a new dynamic chapter of chemistry: ionic liquids.

---

* Corresponding Author's Email: gtamas@fsmail.bradley.edu.

In: Properties and Applications of Ionic Liquids
Editors: Noureddine Abidi and Julia L. Shamshina
ISBN: 979-8-88697-635-9

The name ionic liquids (ILs) acknowledges the uniqueness of these molten organic salts. Those in liquid state below 25°C are called room-temperature ionic liquids (RTIL). A more flexible definition of ILs comprises all the organic salts with melting points under 100°C [3]. Ionic liquids are built of organic cations and organic or inorganic anions, and the type of ions combined dictate their properties that makes the ILs highly tunable chemicals, which is why some scholars consider them *designer solvents*.

Several comprehensive reviews on ILs properties, synthetic protocols, purification methods, and fields of applicability have been published [4, 5, 6, 7, 8, 9, 10]. The ILs field has had a fast-paced growth with the start of the 21st century, becoming a fully mature and stable area of research, with over eight thousand papers published yearly for the past decade (Figure 1).

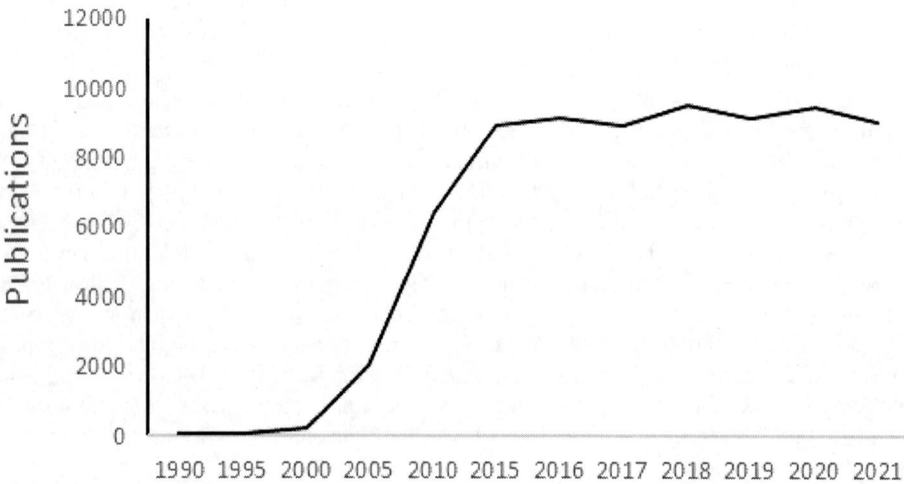

Figure 1. SciFinder™ publications search results for the keywords "ionic liquid" up to 2021.

In the ILs field, commercial fruition of the new technologies has been delayed, with an apparent disconnection between the research publications trend and the investment in production facilities. It has been argued that the ILs field followed a hype curve where the peak of high technological expectancy was followed by a trough of disillusionment. However, the current status continues on a steady growth [11].

This chapter summarizes the IL-based technologies that are already well established commercial applications as well as the ones at the start-up stage and it is structured according to the role of the ILs in the fields of application: solvents and catalysis, electrochemistry, functional fluids, and analytical chemistry.

Solvents and Catalysis

Ionic Liquids and Cellulose Dissolution

Cellulose is a ubiquitous biopolymer provided by nature at a billion tons scale [12, 13]. As the main structural constituent of plants, cellulose is a regenerable resource with numerous potential applications. Still, its lack of solubility in water or other common solvents was for a while the main impediment in its processability [14]. Solutions to this problem were proposed and two procedures were well established and scaled up, viscose [15] and lyocell [16]. Viscose fibers (also known as rayon) are obtained mainly from wood pulp, after the cellulose is chemically modified by treatment with sodium hydroxide, followed by derivatization with carbon disulfide. Then dopes with low viscosity undergo wet spinning through a spinneret immersed in an aqueous bath of sulfuric acid. This procedure has a heavy environmental load since carbon disulfide is toxic and it is not fully recovered [17]. Conversely, lyocell fibers are produced through a direct dissolution technique rather than a chemical protocol and benignly replaces hazardous compounds such as carbon disulfide with the less harmful N-methyl morpholine-N-oxide (NMMO) [18]. Unfortunately, NMMO is a thermally labile compound and, as an oxidant, is sensitive to catalytic impurities, requiring the addition of stabilizers to the dope at an early stage of dissolution [19]. This procedure is environmentally friendly, with more than 99% of solvent recovery per production cycle, although it still involves a significant amount of energy in solvent recycling [20].

As early as 1934, Graenacher patented N-alkylpyridinium organic salts as direct solvents for cellulose, both pure and mixed with suitable diluents [21]. However, it took more than half a century until Swatloski discovered that 1-butyl-3-methylimidazolium chloride IL could dissolve cellulose without activation or pretreatment, triggering an increase of research on this subject [22, 23].

Based on these findings, two novel approaches have been successfully developed and are currently going through the early start-up commercial stages. In the United States, Natural Fiber Welding Inc. patented the *fiber welding* technology, which employs 1-ethyl-3-methylimidzolium acetate [EMIm][OAc] IL and a molecular solvent to transform cellulose into a *congealed network* while still conserving the native biopolymer structure [24]. The raw cellulose fibers are briefly exposed to the appropriate formulation of the *welding solvent* without allowing a complete dissolution. The outer fibrous layer is swollen and solubilized by the IL, disrupting the hydrogen bonding network. The gained mobility of the cellulose permits interaction with the surfaces of the contact fibers and consequently the reshape of the material. Reaction quenching by adding an antisolvent and removing the IL generates a yarn with a completely new morphology (Figure 2) [25]. The new morphology (Figure 3) generates fabric with improved durability, abrasion resistance, reduced pilling and moisture management properties [25]. Furthermore, it allows the upcycling of shorter fiber yarns which are usually considered not desirable by the textile industry [26].

The second approach in development occurs in Finland, where researchers from Aalto University and the University of Helsinki developed Ioncell® [27], a dry-jet wet spinning technology, employing 1,5-diazabicyclo[4.3.0]non-5-ene acetate [DBNH][OAc] as cellulose dissolution agent (Figure 4a) [28].

Similar to the welding protocol, the acetate anion disrupts the hydrogen bonding network, but in this case, a full dissolution of the cellulosic wood pulp occurs at moderate temperatures. Once the homogenous, viscous, dope is formed, a piston extruder is used to spin it, via a multi-hole spinneret. Continuous filaments are generated, subsequently exposed to an air gap, and then coagulated in a cold water bath (Figure 4b) [29, 30]. Highly oriented cellulose fibers are produced with tenacities consistently above 50 cN/tex and high initial moduli of up to 34 GPa [29].

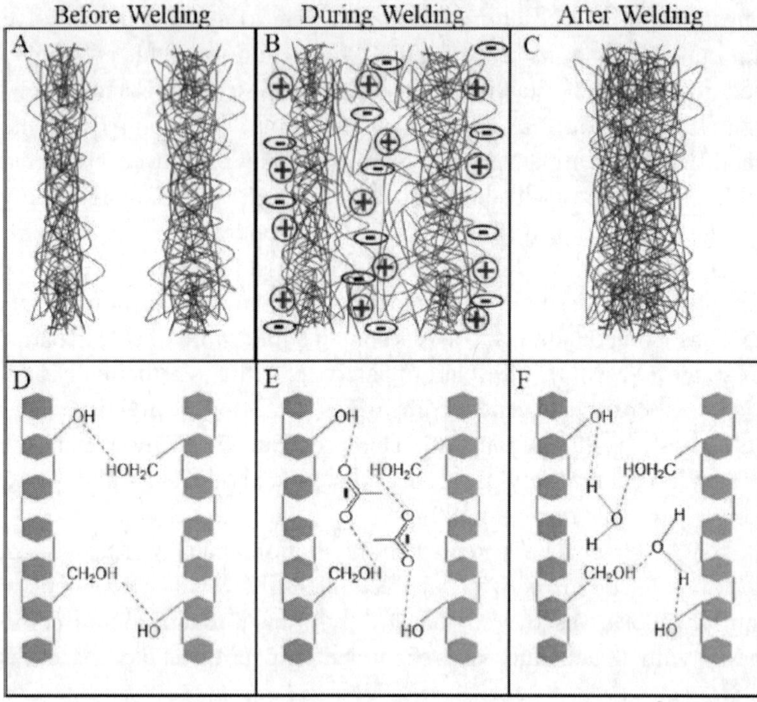

Figure 2. Representation of the welding protocol. It starts with two separate cellulosic yarns (A) that are briefly exposed to the IL-organic solvent formulation promoting the rearrangement of the outer shell fibers (B), leading to the reconstituted single yarn post-reaction quenching (C). At the polymer level, the strong H-bonding network (D) is disrupted by the insertion of the ionic liquid ions (E), which subsequently are removed (F). Reprinted with permission from ref [25] Copyright ©2021 American Chemical Society.

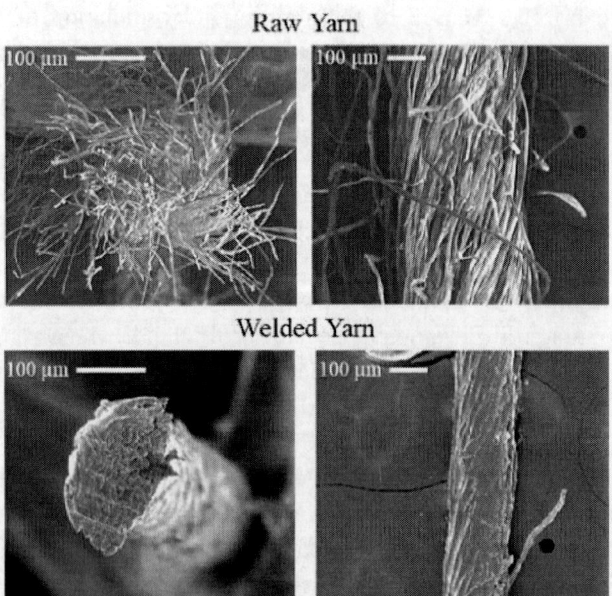

Figure 3. SEM cross (left column) and longitudinal (right column) sections of a cotton yarn. The upper row depicts a raw yarn, whereas the lower row shows the same material after applying the welding treatment – *welded yarn*. Reprinted with permission from ref [25] Copyright ©2021 American Chemical Society.

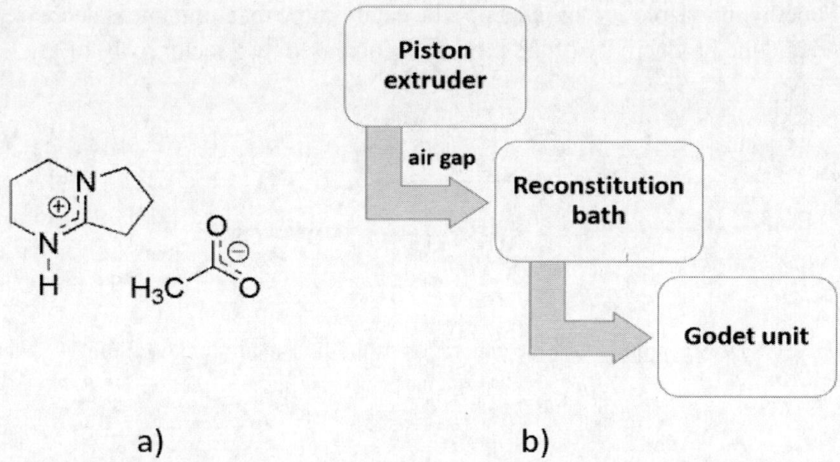

Figure 4. a) Structure of [DBNH][OAc]; b) Schematic of Ioncell® dry-jet wet spinning process.

While showing remarkable cellulose dissolution capabilities, both IL-based techniques have to overcome two major disadvantages: the cost of ILs (not a commodity chemical yet) and the need to quantitatively remove the water from the operating fluids. Water removal could be achieved by developing a highly efficient solvent recycling system, although it comes with the adjacent capital expense.

Biphasic Acid Scavenging Utilizing Ionic Liquids (BASIL)

One of the first multi-ton scale commercial application involving ILs was introduced in 2002 by BASF under the trade name BASIL™ (i.e., Biphasic Acid Scavenging utilizing Ionic Liquids) at its production plant in Ludwigshafen, Germany. This new procedure improved the synthesis of alkoxyphenylphosphines, precursors of Lucerine® photoinitiators used in UV curing of coatings and printing inks [31–34]. The original chemical protocol generated hydrochloric acid as by-product that was trapped with trietilamine (Figure 5). The tertiary ammonium chloride formed separates as a solid non-stirrable paste, which becomes relatively difficult to manage in large scale production [32].

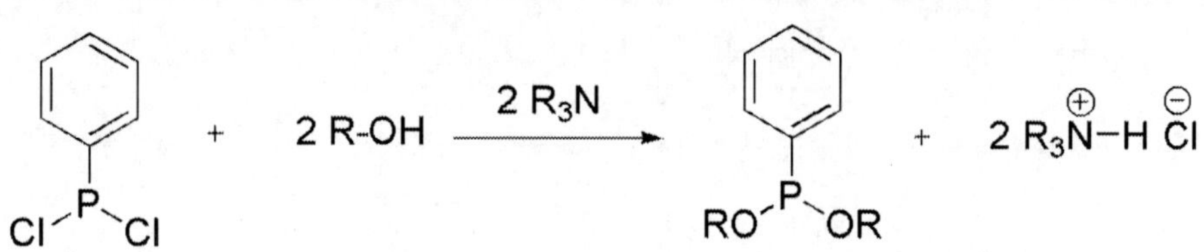

Figure 5. The original synthesis of the alkoxyphenylphosphines using tertiary amines as acid scavengers.

By replacing the triethylamine base with 1-methylimidazole, an IL is generated: 1-methylimidazolium chloride ([HMIm][Cl]) (Figure 6). Since the organic salt is liquid at the reaction temperature, a clear biphasic system is formed, allowing a facile separation of the phosphine derivative target molecule and decreasing the volume of the reactor [35]. Also, because 1-methylimidazole acts as nucleophilic catalyst, the reaction time is decreased to under one second and the productivity of the process is increased by a factor of 104 [36].

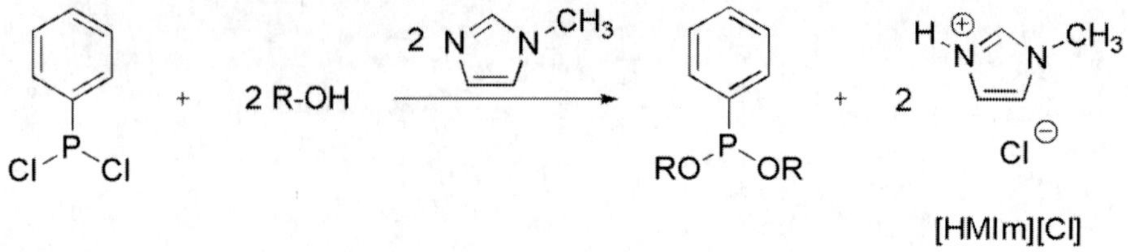

Figure 6. BASF's BASIL protocol for the generation of the IL 1-methylimidazolium chloride ([HMIm][Cl]).

Replacing Toxic Phosgene with Nucleophilic HCl

The toxicity of various chlorinating agents such as phosgene or thionyl chloride is well known, and it is a real hazard when the processes occur at an industrial scale [37]. The chlorination of aromatic alcohols in the presence of hydrogen chloride was reported by Earle in 2004 [38]. However, attempts to chlorinate 1,4-butanediol (BDO) in similar conditions generated large amounts of secondary products (i.e., tetrahydrofuran (THF), 4-chloro-1-butanol (CBO)) and only a small quantity of the target product: 1,4-dichlorobutane (DCB) (Figure 7a) [39]. A new alcohol chlorination protocol (patented by BASF, 2005) solved this problem by replacing the toxic phosgene with *nucleophilic HCl* as a halogenating agent in the presence of ILs. Under the activating effect of the organic salt, more than 98% selectivity towards DCB product was achieved (Figure 7b) [40].

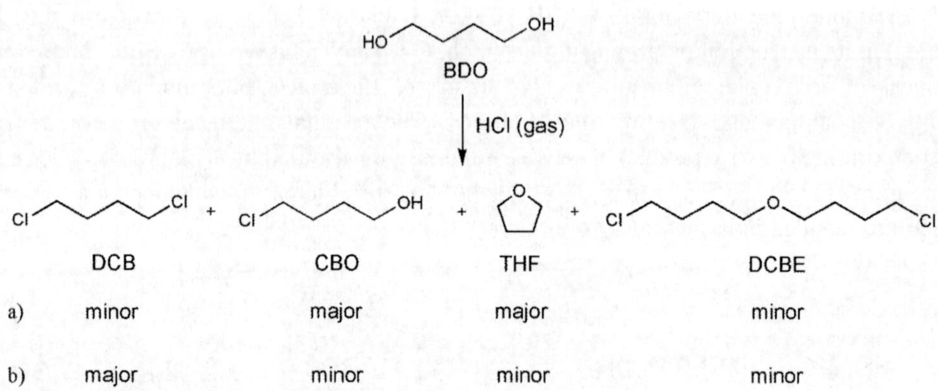

Figure 7. Reaction of BDO with nucleophilic HCl. Product amount distribution a) in absence of the ionic liquid; b) in presence of the ionic liquid.

DIFASOL™ Dimerization

In the oil refining industry, the fluid and steam cracking processes are a significant source of light olefins (C2-C5 fractions). Oligomerization of these olefines by homogenous catalysis adds value to lower grade chemicals. In 1977, the Institut Français du Pétrole (IFP) developed and commercialized an alkene dimerization protocol under the trade name Dimersol [41, 42] (Figure 8). The catalytic system involved in the process was based on a cationic complex that generates an activated nickel hydride species (Ni-H)+ [41]. Homogenous organometallic catalysis, under right metal-ligand and metal-activator selections, provides better control of the final product. Depending on the feed, Dimersol GTM and Dimersol XTM were licensed out for dimerization of propene and butenes into more desirable hexenes and octenes [43].

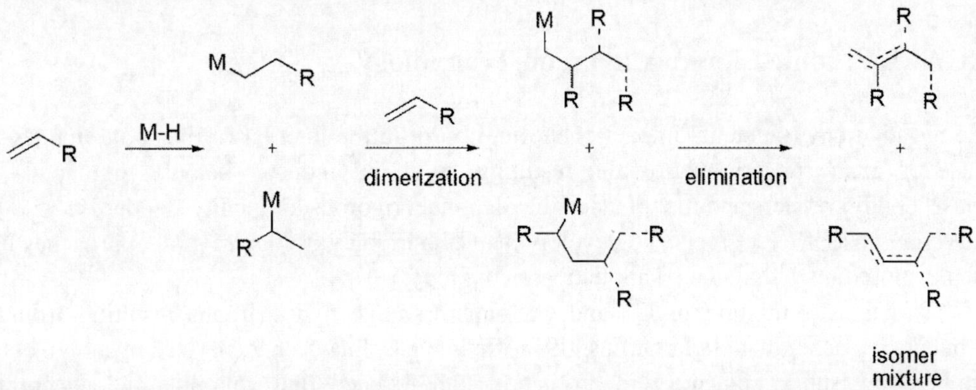

Figure 8. Dimerization of a terminal alkene generating a mixture of isomers (dashed lines representing the points of variability).

Specifically, a soluble nickel (II) metal and a chloroaluminium derivative co-catalyst (i.e., Ni(II) + $AlCl_{3-x}Et_x$) generate *in situ* the active species used in the nonregioselective industrial dimerization of low cut alkenes (C2-C4), with an observed decreasing olefine reactivity going from ethylene to n-butenes [44, 45]. Unfortunately, the catalyst recyclability requires a water wash step that has negative effects on the overall yields and production costs [46].

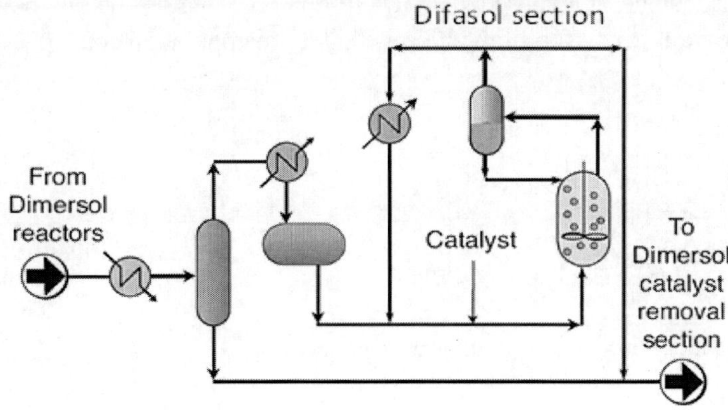

Figure 9. DIFASOL dimerization. Reprinted with permission from ref [41]. Copyright ©2021 American Chemical Society.

To surpass the Dimersol™ monophasic protocol limitations, IFP further developed Difasol™, as an IL-based biphasic system [44, 47]. The technique exploits the immiscibility of the dimerization product into the IL-catalyst phase, where the reaction takes place. This difference in miscibility enables a facile separation by simple phase decantation. A decrease in the Ni(II) catalyst consumption and better dimer selectivity were reported [47]. Also, since both phases are liquids, there is an improved mixing and heat transfer, with a direct effect on the volume of the reaction vessel. Furthermore, the new approach handles more efficiently the diluted monomer feedstocks [43]. The Difasol process can be retrofitted to the current Dimersol production units reducing the capital expense (see Figure 9).

ISOALKY: Ionic Liquid Alkylation Technology

The negative environmental impact of the fuel burnt in the internal combustion engines is a cause of great concern worldwide, resulting in an oil industry heavily regulated [48]. Consequently, refiners and fuel blenders have to meet rigorous air quality standards while still delivering cost-efficient fuels. To develop cleaner-burning gasoline, several approaches have been explored, including the IL alkylation technology.

Alkylate is a mixture of C7 and C8 naphtha-range hydrocarbons resulting from the isobutane's reaction with light olefins [49]. Alkylate gasoline is characterized by a high octane number, low sulfur content, zero aromatics, and reduced fuel volatility and evaporative emission (Reid vapor pressure) [50]. According to the American Fuel and Petrochemical Manufacturers association, alkylate gasoline is essential in complying with a series of recent regulations such as California demand for reformulated gasoline, Clean Air Act Reid Vapor Pressure-compliant fuel, and the Corporate Average Fuel Economy targets for mileage efficiency and carbon reduction [48].

The traditional alkylation technologies employ either hydrofluoric or sulfuric acid catalysts to produce alkylated gasoline, with a relatively equal distribution of the two chemical routes among the existing refining plants in the US [51]. The reaction is based on acid-promoted generation of carbocations provided by the C3-C5 Fluid catalytic cracking (FCC) olefins, which react with the isobutane in a subsequent step (Figure 10). Unfortunately, both acid catalysts have a heavy environmental footprint and raise safety manipulation concerns.

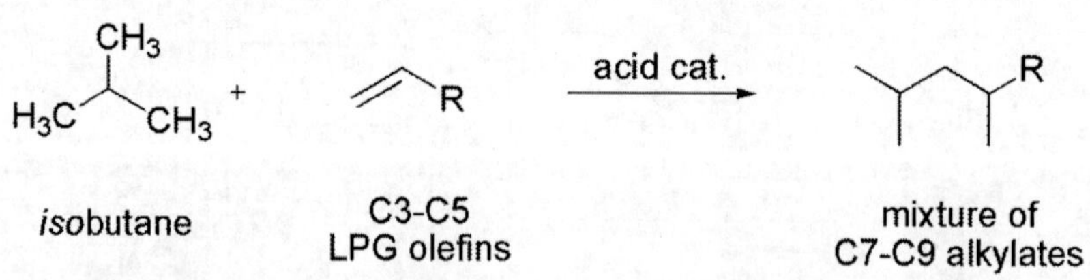

Figure 10. Conventional acid-catalyzed alkylation protocol.

Chevron (United States) and Honeywell UOP (United States) developed ISOALKY technology for the production of alkylate gasoline using non-volatile and environmentally friendly ILs [52, 53]. Specifically, ISOALKY protocol uses a chloroaluminate IL catalyst and employs anhydrous hydrochloric acid as co-catalyst [50].

The superacid properties of the chloroaluminate organic salts were previously investigated by Ma and collaborators. A 2:1 ratio of aluminum chloride: trimethylsulfonium chloride (TMSuCl) in the presence of 1 atm hydrochloric acid formed several aryl and arylalkyl carbocations that could be analyzed by NMR and UV-Vis spectroscopy (Figure 11). The protonation of the liquid hydrocarbons takes a couple of minutes, whereas the protonation of the solid compounds can take around 15 min [54].

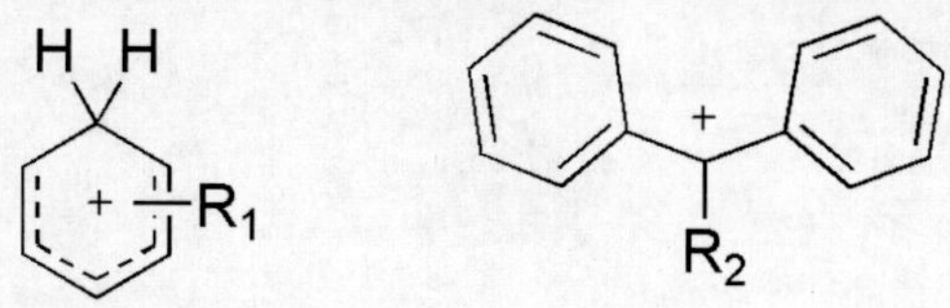

Figure 11. Various carbocations produced by superacid chloroaluminate organic salts in the presence of HX (R_1 = Me, R_2 = Ph, Me, H).

The ISOALKY system is described as an "anhydrous $AlCl_3$ - based Lewis acid catalyst promoted with anhydrous hydrochloric acid, Brønsted acid and exhibiting the characteristics of a superacid." This strong acidity has an immediate beneficial impact on the size of the reactor and the amount of catalyst needed in the total reaction mass to promote alkylation, which is decreased from over 50% in the case of conventional HF/H_2SO_4 to 3-6% for the new IL approach [50].

The organic salts claimed in the Chevron ISOALKY patents are hydrocarbyl substituted pyridinium or imidazolium chloroaluminate compounds (Figure 12), where the anion is $Al_2Cl_7(-)$. All of them are liquid salts and, as catalysts, show the highest Lewis acidity when the mole fraction of $AlCl_3$ in the IL is $x(AlCl_3) = 0.67$ [50]. The new procedure can alkylate feedstocks spanning from ethylene to amylenes, while using only trace amounts of hydrochloric acid co-catalyst recycled *in situ* [55].

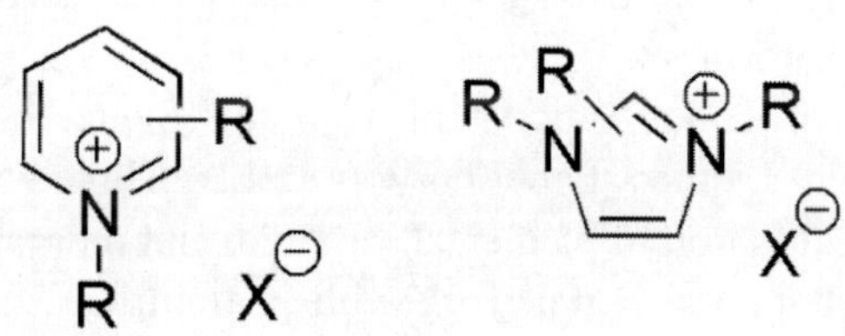

Figure 12. Structures of the ILs present in the ISOALKY catalyst.

Chevron completed commissioning the first commercial IL–based alkylation unit at its Salt Lake City refinery in 2021. Honeywell UOP announced that Big West Oil (United States) and Tüpraş (Turkey) refineries will revamp their hydrofluoric acid units to the ISOALKY technology [56, 57].

SILP Catalysis

The Supported Ionic Liquid Phase (SILP) catalysis refers to a methodology of converting a purely homogeneous catalyst system into a heterogeneous one [58, 59, 60, 61]. A SILP catalyst structure has three key components: a solid supporting material, an IL, and the active catalyst (Figure 13).

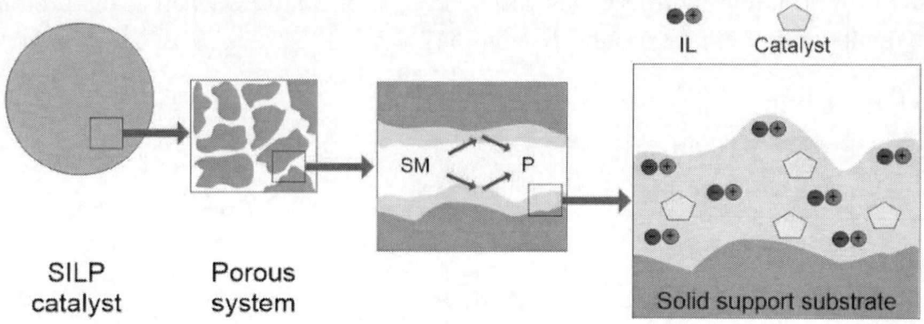

Figure 13. A SILP catalyst is a heterogeneous system providing a homogeneous IL-catalyst layer where the starting material (SM) is converted into product (P).

Traditionally, the solid support material is an inert porous substrate with a large internal surface area that fulfills only a mechanical role. This surface is subsequently coated with a homogeneous mixture of an IL and the catalyst. The IL is a non-reactive carrier, ensuring a high dispersion of the contained molecular species [62]. Although, the SILP catalyst is in the form of solid pellets or powder, at a microscopic level, the chemistry happens in homogeneous phase, allowing for high selectivity and mild reaction conditions. This setup is especially beneficial for industrially scaled-up processes where an easy separation from the reaction mixtures is always desirable [63].

Natural gas reservoirs are usually contaminated with elemental and oxidized forms of mercury in concentrations ranging from 0.1 – 5000 $\mu g\ m^{-3}$ [64]. All these forms of mercury (i.e., Hg(0), Hg(I), and Hg(II)) are known to be toxic [65] and corrosive, impacting the personnel and causing irreversible damages to the aluminum-based equipment via a liquid-metal-embrittlement mechanism [66].

Petronas (Malaysia), in collaboration with Clariant (Switzerland) and the scientists at Queen's University Ionic Liquids Laboratory (QUILL, United Kingdom), developed a scrubbing agent for the extraction of elemental, organic, and inorganic mercury from natural gas [67]. Abai and collaborators' initial work with [C4mim]Cl-CuCl2 and $[N_{4441}]Cl\text{-}CuCl_2$ couples, at a $\chi CuCl_2 = 0.50$ ratio, supported a reaction with mercury following the mechanism described by Eq. (1):

$$2[Cu_2Cl_6]_2^- + 2Hg(0) \rightarrow [Hg_2Cl_6]_2^- + 2[CuCl_2]^- + 2CuCl(s) \quad (1)$$

The equation suggests that a liquid phase IL-based mercury scrubbing from natural gas is possible. Still, a scale-up would be less sustainable at the industrial level, considering the viscosity of the organic salts. Employing the SILP methodology, QUILL researchers proved that mercury vapors from a natural gas effluent could be efficiently trapped in the presence of chlorocuprate(II) ILs impregnated on highly porous solid supports [64]. The heterogeneous approach avoids the slow mass transfer generated by IL's high viscosity while using more efficiently only thin films of IL deposited on the scrubber pellets [62].

In 2014, Clariant and Petronas introduced the results of this research to the market under the trade name HycaPure™ Hg technology, which claimed to have a lifetime three times longer than sulfur-impregnated activated carbon [67]. Practically, thirty-five years after De Munck and Hjortkjaer (1981) introduced the heterogenization concept of supported liquid phase (SLP) systems, Malaysian oil and gas company Petronas made the mercury removal methodology a fully commercial application [68, 69, 70].

Ionic Liquid Cooling Technology (ILTEC)

Considering the high temperatures required by the various metallurgical processes, the need to cool the aggregates is permanently present and currently fulfilled by water. As a conventional coolant, water has high thermal conductivity and is an available low-cost resource. However, if leaking occurs, and water enters in contact with molten metal, catastrophic explosions may occur. Such event is mainly caused by the rapid volume expansion along with hydrogen gas generation. Therefore, as a precautionary measure, the water cooling systems are restricted to an operating temperature of maximum 60°C and are positioned at higher levels than the metal melt baths [11].

To improve the safety of the personnel and the equipment in metallurgical plants, Mettop GmbH (Austria) and Proionic (Austria) developed ILTEC technology, as an IL-based cooling alternative of the metallurgical equipment involved in producing non-ferrous metals, ferroalloys, and steel [71, 72]. The new working fluid (i.e., IL-B2001) is based on 1-ethyl-3-methylimidazolium tetrafluoroborate (97%) and 1-ethyl-3-methylimidazoliumfluoride (1.5%) organic salts. This non-water cooling medium has several advantages [71, 73]:

- It has excellent long-term stability at operating temperature between 50-200°C, and a short-term stability up to 250°C. This extended working range makes the recovery of the dissipated heat more efficient.
- In case of leaking, IL-B2001 decomposes into elements without a rapid volume expansion or hydrogen gas generation, therefore mitigating the risk of explosion.
- It is produced under a chlorine-free technology, thus it is non-corrosive for the equipment.
- It undergoes crystallization under -30°C, preventing piping damage by freezing volume expansion in case of system failure.

These factors support the transition from a water-cooling operating system placed in a red safety zone to a more desirable green area of operational awareness (Figure 14).

The ILTEC technology could also be retrofitted to the existing water cooling setup since it can remove a comparable amount of heat through the piping and pumping system already in place [71].

Construction-wise, the equipment has three major components (Figure 15):

- A storage tank where IL-B2001 is kept under nitrogen to prevent moisture absorption due to organic salts hygroscopicity.
- A dual redundant pumping system to move the coolant through the piping lines.
- Two heat exchangers for heat release to a secondary cooling circuit.

Mettop's novel technology is already in operation at ArcelorMittal, Luxembourg (2015 – tap hole cooling); Nyrstar, Belgium (2015 – side wall cooling); and Voestalpine Stahl Donawitz, Austria (2017 – flange cooling) production facilities [71].

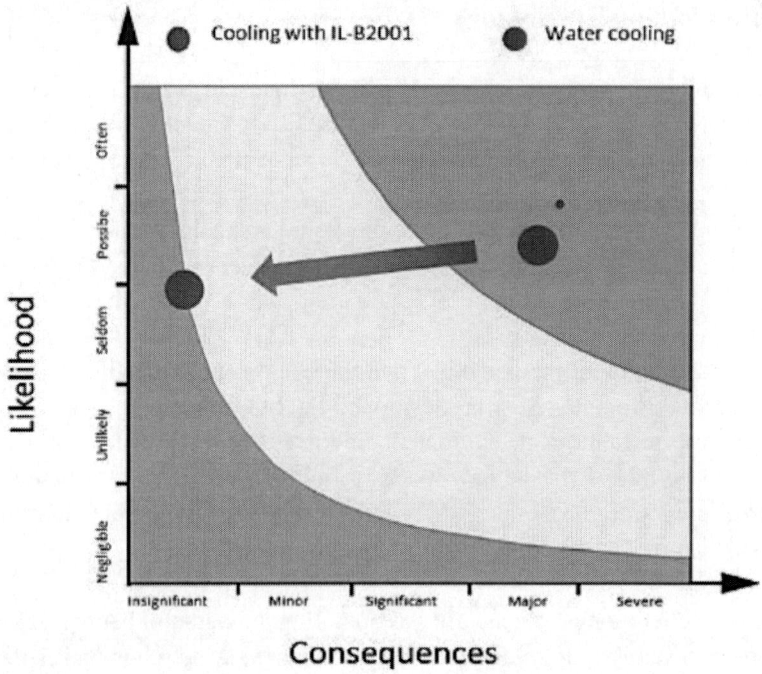

Reprinted with permission from ref [71]. Copyright Mettop GmbH.

Figure 14. Consequences vs. likelihood of an incident event, comparing the water-cooling operating system and the IL-based cooling operating system.

Reprinted with permission from ref [71]. Copyright Mettop GmbH.

Figure 15. Basic design of an ILTEC equipment.

Electrochemistry

Due to their large electrochemical windows, high ionic conductivities, good thermal stabilities, non-flammability, and non-volatility features, ILs are excellent candidates for the development of electrochemical applications [74]. The two areas of progress where ILs electrolytes showed promising results are metal electroplating and energy storage.

Chromium Electroplating Process

Scionix (Netherlands) developed and patented an IL-analog system designed for chromium electroplating, which can deposit crack-free coatings on metals at more than 90% current efficiency. This is remarkable considering that the traditional aqueous acid-based protocols use only 20% of the overall energy consumed for actual chromium electrodeposition, negatively impacting the economic efficiency of the process [75, 76].

Scionix's formulation is a 1:2 mixture of choline chloride and chromium(III) chloride hexahydrate forming a deep eutectic solvent under the appearance of a dark green viscous liquid at room temperature [77]. Notably, this electrochemical pair eliminates the highly toxic chromium(VI) from its composition. Furthermore, the absence of bulk water reduces hydrogen evolution during electrochemical process and concomitantly mitigates the embrittlement of the coating layer [77].

Energy Storage

Almost fifty years ago, after Oseteryoung and collaborators published a comprehensive study on 1-butylpyridinium chloride and aluminum chloride [78], Nardy, with the United States Air Force support, patented the use of aluminum chloride and the more generic 1-Alkyl pyridinium chlorides room-temperature ILs mixtures as electrolytes for primary and secondary batteries [79]. However, these systems had a high moisture sensitivity problem, where even traces of water would decompose the ILs [80]. Chemical stability issues, along with the ILs' known hygroscopicity, were later largely overcome with the introduction of hydrophobic anions. Specifically, Grätzel synthesized a multitude of hydrophobic imidazolium IL-based on anions such as $[OTf]^-$ (trifluoromethanesulfonate) and $[NTf_2]^-$ (bis((trifluoromethyl)sulfonyl)amide) [81]. These new room-temperature liquid salts showed much larger electrochemical windows (5-6 V) than water (1.23 V) [82]. Once reliable electrolytes were available, their applicability to energy storage devices such as batteries and supercapacitors was the expected progression.

Lithium-ion batteries are ubiquitous in all lightweight mobile devices. Although their specific energy density is high, an improvement in terms of power density is always desirable [83]. Better electrolytes are an essential channel for lithium-ion batteries advancement. IoLiTec (Germany) offers a suite of battery dedicated electrolytes based on $[NTf2]^-$ and $[BF4]^-$ anions. They are, in general, combined with relatively small size cations to reduce ILs viscosity and promote good ion mobility [84]. NOHMs Technologies (United States) offers NanoLyte battery electrolytes based on phosphate ILs. The company claims a non-flammable and non-volatile product with a liquid temperature ranging from -80 to 300°C and good lithium-ion conductivity [85].

Supercapacitors are operationally safe and have a very long lifespan, usually higher by one to two orders of magnitude than lithium-ion batteries [83]. They can deliver bursts of energy under rapid charge/discharge cycles due to their very high energy per unit volume or mass. Two types of supercapacitors are commercially available:

- Electrostatic double-layer capacitors (EDLC) have the highest energy density and capacity per volume, with a high-rate charge-discharge capability. Furthermore, the absence of any chemical reactions warrants a long lifespan. EDLCs store energy through electrostatic adsorption of the ionic species at the electrode-electrolyte interface, where the separation of charge takes place through a Helmholtz double-layer formation [86, 87].
- Electrochemical pseudocapacitors have redox-active material electrodes, such as metal oxides, where reversible faradic reactions occur. The pseudocapacitance is mainly achieved through faradic processes, including electrosorption, redox reactions, and intercalation, and only minorly through EDCL mechanism [88, 89].

Several companies are key players in the supercapacitor market (for both EDLC and ultracapacitors) [90], including Maxwell Technologies (United States) [91], Kyocera AVX (United States) [92], Panasonic (Japan) [93], Nippon Chemi-Con (Japan) [94], Nichicon (Japan) [95], and Tokin (Japan) [96] to mention some.

Functional Fluids

Ionic liquids are excellent candidates as operating fluids in various industrial applications due to their thermal stability, low vapor pressure and compressibility, good lubrication, moderate viscosity, and non-hazardous nature.

Lubricants

The potential of using ILs as specialty lubricants either under their pure form or as oil additives has been addressed since the early 2000s [97]. Their thermal stability and conductivity, low vapor pressure, low flammability, and dipolar structure, make them suitable lubricants. Additionally, certain atoms present in the IL composition (e.g., N, P, F, and B) are active reaction sites for the metals resulting during the friction process, potentially generating antiwear compounds [98].

Currently, two classes of ILs on the market are sold as lubricants: the electro-conductive lubricant from Klüber (Germany) and the high-pressure gas compression lubricant from Proionic.

Klüber Lubrication developed electro-conducting lubricants designed to protect the electric motors' rolling bearings from premature surface wearing caused by electrical arcing [99]. With the automotive industry transitioning from the 12 V to 48 V standard, stronger electric fields are expected to be experienced by the moving parts of the alternators and electric motors of the cars [100]. At high rotation speeds, the grease around the ball bearings, while minimizing the frictions, also act as an insulator. Therefore, the ball bearings are transformed into capacitors when the lubricating film is in a contiguous state. They can charge until the voltage is high enough to penetrate the lubricating grease. Small discharge sparks will form at the raceway level, generating hotspots capable of locally melting the bearing surface. The phenomenon is called electrical pitting and causes vibrations and ambient noise [101].

Two well-established approaches to prevent electrical erosion are bearing insulation and bearing conduction. Insulated bearings have the outer ring coated in ceramic or polyphenylene sulfide, both options adding to the cost of the product. Bearing conduction prevents the formation of stray voltages by using electrically conductive *black greases* based on carbonaceous materials such as carbon nanotubes, graphene, carbon blacks, or fullerenes [101]. The addition of solid particulates affects smooth running, though, and usually, this type of lubricant's conductivity decreases over time [99].

Klüber Lubrication patented the use of ILs as either base-oil or additive lubricants. The Klüberlectric BQ 72-72 rolling bearing grease exploits the intrinsic electric conductivity, thermal stability, and low vapor pressure of the ILs, achieving speed factors exceeding 1 million n x dm, classifying it as a high-speed, long-life grease [102].

Ionic Compressor – Linde PLC

An interesting engineering implementation of ILs superior characteristics is the *ionic compressor*, developed and patented by Linde PLC (Germany) and Proionic in 2005 [103, 104, 105]. In the ionic compressor (Figure 16), an IL seal system replaces the conventional compressor metal piston, exploiting the sparing solubility of the hydrogen in the organic salts [106]. This equipment has been succesfully commercialized and currently used in over 180 hydrogen fueling stations worldwide [104].

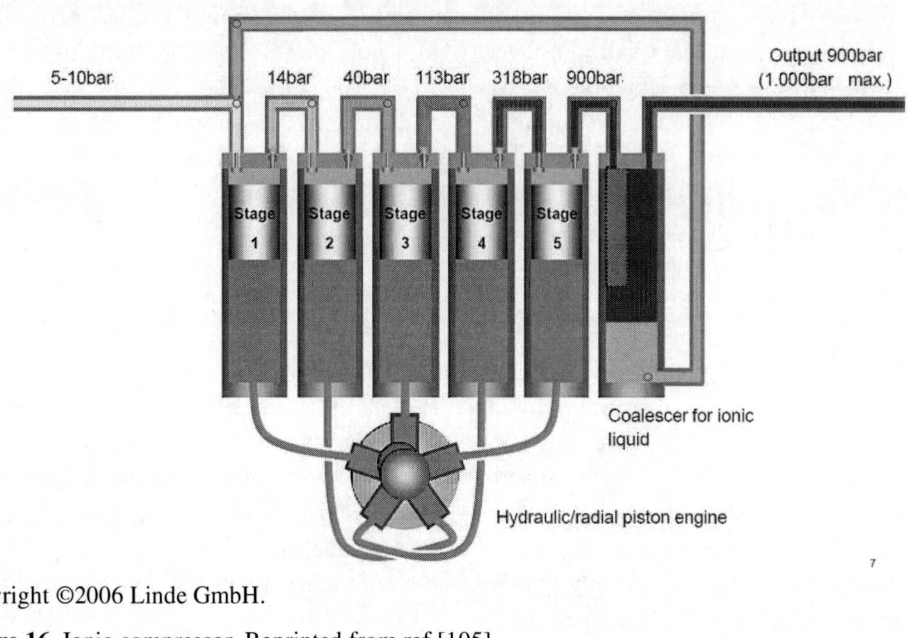

Figure 16. Ionic compressor. Reprinted from ref [105].

A five-stage system can compress the hydrogen gas up to 90 MPa (*e.g.*, IC90 model) while reducing the number of moving parts of the equipment down to eight [107], enhancing the service life and reducing the operating costs. Also, due to the ILs' good thermal stability and conductivity, the heat exchangers conventionally installed on the outside of the pistons are not needed as the heat transfer is achieved through the IL, supporting a controlled isothermal process [105].

Analytical Chemistry

Supelco™ (United States) employed polycationic ILs as stationary phase in gas chromatogram columns and started commercialization of the Supelco SLB®-IL series in 2008. The new columns could be operated at higher temperatures and are less affected by moisture and oxygen contamination [108]. An interesting development of the concept is the Watercol™ columns, which can quantitatively analyze water in various matrices. In tandem with a thermal conductivity detector, water could be reproducibly measured in the 0.5 ppm – 99% range [109]. SLB®-ILPAH series of GC columns claim to separate the polycyclic aromatic hydrocarbons (PAH) such as phenanthrene, anthracene, benzo[a]anthracene based on dicationic IL 1,12-di(tripropylphosphonium)dodecane bis(trifluoromethanesulfonyl)imide phase [110].

Conclusion

The impressive R&D effort in the ILs field, both at the academic and industrial level, has revealed a plethora of essential information about these novel organic salts and pointed out various possible industrial applications. In the last three decades, though, the conversion of the new knowledge into commercially viable applications has lagged behind the discoveries' trend. This lag is not related to the quality of the processes but rather to the possibility of achieving economically sustainable production cycles.

As the quantities of the ILs produced by the industry start passing only now the thousand tons barrier, the prices of some ILs are still prohibitive. It is a critical loop where a decrease in price is conditioned by a significant market demand while the same market demand could not be easily increased if the costs of the inputs are too high. In an ideal world, all the participants should share the burden of exiting this loop fairly.

As a consequence of this status quo, there is an acute need for post-production recycling. Fortunately, one of the main reasons the ILs are considered green is precisely their readiness to be recycled. Their recyclability has beneficial long-term financial results for the companies, promoting the circularity of the chemistries and the planet's health. However, it does come with an associated capital expense, especially for cases where a high purity recycled IL is requested. As the ILs community knows, any quantitative recycling targeting over 97% purities and returns becomes laborious and sometimes involves expensive specialized equipment. Such an initial investment in a recycling plant could be a major financial burden for the startups, directly undermining the ILs market growth. One plausible solution is for the ILs producers to offer inventory recycling/maintenance services.

Given the versatility of these designer solvents, their potential applications are virtually endless. The amount of research dedicated to ILs is promising; companies developing and using ILs continue to multiply; thus, the flourishing of industrial applications is inevitable in the near future.

References

[1] Walden, P. Ueber die Molekulargrösse und elektrische Leitfähigkeit einiger geschmolzenen Salze [*On the molecular size and electrical conductivity of some molten salts*]. *Bulletin* de *l'Académie impériale* des sciences de *St.-Pétersbourg*, 1914, *8*, 405–422.

[2] Janz, G. J. Atomic and ionic radii, melting points, and boiling points of single salts and eutectics." In G. J. Janz (Ed.), *Molten Salts Handbook,* 1967 (pp. 1–38). Academic Press. https://doi.org/10.1016/B978-0-12-395642-2.50006-X.

[3] Bockris, J. O., Mackenzie, J. D.; Kitchener, J. A.. Viscous flow in silica and binary liquid silicates. *Transactions of the Faraday Society*, (1955), *51*(0), 1734–1748. https://doi.org/10.1039/TF9555101734.

[4] Plechkova, N. V., Seddon, K. R. "Applications of ionic liquids in the chemical industry." *Chemical Society Reviews*, 2008, *37*(1), 123–150. https://doi.org/10.1039/B006677J.

[5] Gutowski, K. E. Industrial uses and applications of ionic liquids. *Physical Sciences Reviews*, 2018, *3*(5), 20170191. https://doi.org/doi:10.1515/psr-2017-0191.

[6] Warner, I. M., El-Zahab, B.; Siraj, N. Perspectives on moving ionic liquid chemistry into the solid phase. *Analytical Chemistry*, 2014, *86*(15), 7184–7191. https://doi.org/10.1021/ac501529m.

[7] Greer, A. J., Jacquemin, J; Hardacre, C. Industrial applications of ionic liquids. *Molecules*, 2020, *25*(21). https://doi.org/10.3390/molecules25215207.

[8] Ventura, S. P. M., e Silva, F. A., Quental, M. V., Mondal, D., Freire, M. G.; Coutinho, J. A. P. Ionic-liquid-mediated extraction and separation processes for bioactive compounds: Past, present, and future trends. *Chemical Reviews*, 2017, *117*(10), 6984–7052. https://doi.org/10.1021/acs.chemrev.6b00550.

[9] Meksi, N.; Moussa, A. A review of progress in the ecological application of ionic liquids in textile processes. *Journal of Cleaner Production*, 2017,*161*, 105–126. https://doi.org/10.1016/j.jclepro.2017.05.066.

[10] Toledo Hijo, A. A. C., Maximo, G. J., Costa, M. C., Batista, E. A. C.; Meirelles, A. J. A.. Applications of ionic liquids in the food and bioproducts industries. *ACS Sustainable Chemistry & Engineering*, 2016, *4*(10), 5347–5369. https://doi.org/10.1021/acssuschemeng.6b00560.

[11] Kalb, R. S. Toward industrialization of ionic liquids. In M. B. Shiflett (Ed.), *Commercial Applications of Ionic Liquids,* 2020, (pp. 261–282). Springer International Publishing. https://doi.org/10.1007/978-3-030-35245-5_11.

[12] Olsson, C.; Westman, G. Direct dissolution of cellulose: background, means and applications. In T. a. de Ven & L. Godbout (Eds.), *Cellulose - Fundamental Aspects*. IntechOpen. 2013, https://doi.org/10.5772/52144.

[13] Klemm, D., Heublein, B., Fink, H.-P.; Bohn, A.. "Cellulose: Fascinating biopolymer and sustainable raw material." *Angewandte Chemie International Edition*, 2005, *44*(22), 3358–3393. https://doi.org/10.1002/anie.200460587.

[14] Hauru, L. K. J., Hummel, M., King, A. W. T., Kilpeläinen, I.; Sixta, H. Role of solvent parameters in the regeneration of cellulose from ionic liquid solutions. *Biomacromolecules*, 2012, *13*(9), 2896–2905. https://doi.org/10.1021/bm300912y.

[15] Liebert, T. Cellulose solvents – Remarkable history, bright future." In T. F. Liebert, T. J. Heinze & K. J. Edgar (Eds.) *Cellulose Solvents: For Analysis, Shaping and Chemical Modification,* 2010, (Vol. 1033, pp. 3–54). American Chemical Society. https://doi.org/10.1021/bk-2010-1033.ch001.

[16] Luo, M., Roscelli, V. A., Neogi, A. N., Sealey, I. J. E.; Jewell, R. A.. *Lyocell fibers, and compositions for making same* (U.S. Patent No. US6210801B1). U.S. Patent and Trademark Office, 2001. https://patents.google.com/patent/US6210801B1/en.

[17] Sayyed, A. J., Deshmukh, N. A.; Pinjari, D. V.. A critical review of manufacturing processes used in regenerated cellulosic fibres: Viscose, cellulose acetate, cuprammonium, LiCl/DMAc, ionic liquids, and NMMO based lyocell." *Cellulose*, 2019, *26*(5), 2913–2940. https://doi.org/10.1007/s10570-019-02318-y.

[18] Goswami, P., Blackburn, R. S., El-Dessouky, H. M., Taylor, J.; White, P.. Effect of sodium hydroxide pre-treatment on the optical and structural properties of lyocell." *European Polymer J.*, 2009, *45*(2), 455–465. https://doi.org/10.1016/j.eurpolymj.2008.10.030.

[19] Rosenau, T., Potthast, A., Hofinger, A., Sixta, H.; Kosma, P. Instabilities in the system NMMO/water/cellulose (lyocell process) caused by Polonowski type reactions. *Holzforschung*, 2002, *56*(2), 199–208. https://doi.org/10.1515/HF.2002.033.

[20] Zhang, S., Chen, C., Duan, C., Hu, H., Li, H., Li, J., Liu, Y., Ma, X., Stavik, J.; Ni, Y. Regenerated cellulose by the lyocell process, a brief review of the process and properties." *BioResources*, 2018, *13*(2), 4577–4592.

[21] Graenacher, C. *Cellulose solution* (U.S. Patent No. US1943176A). U.S. Patent and Trademark Office, 1934. https://patents.google.com/patent/US1943176A/en.

[22] Swatloski, R. P., Spear, S. K., Holbrey, J. D.; Rogers, R. D. Dissolution of cellulose with ionic liquids. *J. American Chemical Society*, 2002,*124*(18), 4974–4975. https://doi.org/10.1021/ja025790m.

[23] Swatloski, R. P., Rogers, R. D.; Holbrey, J. D. *Dissolution and processing of cellulose using ionic liquids* (U.S. Patent No. US6824599B2). U.S. Patent and Trademark Office, 2004. https://patents.google.com/patent/US6824599/en/en11..

[24] DeLong, H. C., Trulove, P. C., Haverhals, L. M.; Reichert, W. M. *Natural fiber welding* (U.S. Patent No. US8202379B1). U.S. Patent and Trademark Office, 2012. https://patents.google.com/patent/US8202379B1/en.

[25] Horvath, A., Curry, J., Haverhals, L. M.; Shaw, S. K. Practical online monitoring of ionic liquid fiber welding solvent." *ACS Omega*, 2021, *6*(34), 22367–22373. https://doi.org/10.1021/acsomega.1c03122.

[26] Haverhals, L. M., Durkin, D. P.; Trulove, P. C. Natural fiber welding." In M. B. Shiflett (Ed.), *Commercial applications of ionic liquids,* 2020, (pp. 211–226). Springer International Publishing. https://doi.org/10.1007/978-3-030-35245-5_9.

[27] *Ioncell.* (n.d.). Ioncell. Retrieved February 12, 2022, from https://ioncell.fi/.

[28] Michud, A., Tanttu, M., Asaadi, S., Ma, Y., Netti, E., Kääriainen, P., Persson, A., Berntsson, A., Hummel, M.; Sixta, H. Ioncell-F: ionic liquid-based cellulosic textile fibers as an alternative to viscose and Lyocell." *Textile Res. J.*, 2016, *86*(5), 543–552. https://doi.org/10.1177/0040517515591774.

[29] Sixta, H., Michud, A., Hauru, L., Asaadi, S., Ma, Y., King, A. W. T., Kilpeläinen, I.; Hummel, M. Ioncell-F: A High-strength regenerated cellulose fibre." *Nordic Pulp & Paper Research J.* 2015, *30*(1), 43–57. https://doi.org/10.3183/npprj-2015-30-01-p043-057.

[30] Parviainen, A., Wahlström, R., Liimatainen, U., Liitiä, T., Rovio, S., Helminen, J. K. J., Hyväkkö, U., King, A. W. T., Suurnäkki, A.; Kilpeläinen, I. Sustainability of cellulose dissolution and regeneration in 1,5-diazabicyclo[4.3.0]non-5-enium acetate: A batch simulation of the IONCELL-F process. *RSC Advances*, 2015, *5*(85), 69728–69737. https://doi.org/10.1039/C5RA12386K.

[31] Seddon, K. R. A taste of the future. *Nature Materials*, 2003,*2*(6), 363–365. https://doi.org/10.1038/nmat907.

[32] Freemantle, M. BASF's smart ionic liquid. *Chemical & Engineering News Archive*, 2003, *81*(13), 9. https://doi.org/10.1021/cen-v081n013.p009.

[33] Rogers Robin D.; Seddon Kenneth R. Ionic liquids—solvents of the future?" *Science*, 2003, *302*(5646), 792–793. https://doi.org/10.1126/science.1090313.

[34] Maase, M.; Huttenloch, O. *Procede de separation d'acides contenus dans des melanges reactionnels chimiques, a l'aide de 1-alkylimidazoles* [*Process for the separation of acids contained in chemical reaction mixtures, using 1-alkylimidazole*]. (Patent No. WO2005061416), 2005. Weltorganisation für geistiges Eigentum Internationales Büro. https://patentscope.wipo.int/search/fr/detail.jsf?docId=WO2005061416.

[35] Volland, M., Seitz, V., Maase, M., Flores, M., Papp, R., Massonne, K., Stegmann, V., Halbritter, K., Noe, R., Bartsch, M., Siegel, W., Becker, M.; Huttenloch, O. *Method for the separation of acids from*

chemical reaction mixtures by means of ionic fluids. (U.S. Patent No. US7351339B2). U.S. Patent and TrademarkOffice, 2008. https://patents.google.com/patent/US7351339B2/en.

[36] Chojnowski, J., Cypryk, M.; Fortuniak, W. The extension of the mechanistic concept of the nucleophilic catalysis in the silicon chemistry to some reactions of the P(III) center: Analogies between silylation and phosphorylation." *Heteroatom Chemistry*, 1991, *2*(1), 63–70. https://doi.org/10.1002/hc.520020109.

[37] Gad, S. C. Phosgene. In P. Wexler (Ed.), *Encyclopedia of Toxicology (Third Edition),* 2014, (pp. 904–906). Academic Press. https://doi.org/10.1016/B978-0-12-386454-3.00903-9

[38] Earle, M. J., Katdare, S. P.; Seddon, K. R.. "Paradigm confirmed: The first use of ionic liquids to dramatically influence the outcome of chemical reactions." *Organic Letters*, 2004, *6*(5), 707–710.

[39] Ryan, T. A., Seddon, E., Seddon, K.; Ryan, C. *Phosgene: And related carbonyl halides,* 1996. Elsevier.

[40] Stegmann, V.; Massonne, K. Method for producing haloalkanes from alcohols (U.S. Patent No. US20070055084A1). U.S. Patent and Trademark Office, 2007. https://patents.google.com/patent/US20070055084/en?oq=Method+for+Producing+Haloalkanes+From+Alcohols.

[41] Andrews, J.; Bonnifay, P. The IFP dimersol process for dimerization of propylene into isohexenes." In L. F. Albright & A. R. Goldsby (Eds.) *Industrial and Laboratory Alkylations,* 1977. (Vol. 55, pp. 328–340). American Chemical Society. https://doi.org/10.1021/bk-1977-0055.ch020.

[42] IFPEN | Accueil. (n.d.). *IFPEN*. Retrieved February 23, 2022, from https://www.ifpenergiesnouvelles.fr/home.

[43] Ionic liquids in synthesis, 2 Volume Set, 2nd Edition | Wiley. (n.d.). *Wiley.Com.* Retrieved February 23, 2022, from https://www.wiley.com/en-us/Ionic+Liquids+in+Synthesis%2C+2+Volume+Set%2C+2nd+Edition-p-9783527312399.

[44] Olivier-Bourbigou, H.; Magna, L. Ionic liquids: Perspectives for organic and catalytic reactions. *J. Molecular Catalysis A: Chemical*, 2002, *182–183*, 419–437. https://doi.org/10.1016/S1381-1169(01)00465-4.

[45] Forestière, A., Olivier-Bourbigou, H.; Saussine, L. Oligomerization of monoolefins by homogeneous catalysts." *Oil & Gas Science and Technology - Revue de l'IFP*, 2009, *64*(6), 649–667. https://doi.org/10.2516/ogst/2009027.

[46] Pârvulescu, V. I. & Hardacre, C. Catalysis in ionic liquids. *Chemical Reviews*, 2007, *107*(6), 2615–2665. https://doi.org/10.1021/cr050948h.

[47] Chauvin, Y., Einloft, S.; Olivier, H. "Catalytic dimerization of propene by nickel-phosphine complexes in 1-butyl-3-methylimidazolium chloride/AlEtxCl3-x (x = 0, 1) ionic liquids." *Industrial & Engineering Chemistry Research*, 1995; *34*(4), 1149–1155. https://doi.org/10.1021/ie00043a017.

[48] *Alkylate: Understanding a Key Component of Cleaner Gasoline*. (n.d.). AFPM. Retrieved March 5, 2022, from https://www.afpm.org/newsroom/blog/alkylate-understanding-key-component-cleaner-gasoline.

[49] Fitzgibbon, T. (n.d.). *Alkylation unit*. Energy insights by McKinsey. Retrieved March 5, 2022, from http://www.mckinseyenergyinsights.com/resources/refinery-reference-desk/alkylation-unit/.

[50] Timken, H. K., Luo, H., Chang, B.-K., Carter, E.; Cole, M. ("ISOALKY™ technology: Next-generation alkylate gasoline manufacturing process technology using ionic liquid catalyst." In M. B. Shiflett (Ed.), *Commercial Applications of Ionic Liquids,* 2020, (pp. 33–47). Springer International Publishing. https://doi.org/10.1007/978-3-030-35245-5_2.

[51] Tullo, A.H. The time is now for ionic liquids. *Chemical & Engineering News* (2020, February 3). https://cen.acs.org/materials/ionic-liquids/time-ionic-liquids/98/i5.

[52] Elomari, S., Trumbull, S., Timken, H. K. C.; Cleverdon, R. *Alkylation Process using Chloroaluminate Ionic Liquid Catalysts* (U.S. Patent No. US7432409B2). U.S. Patent and Trademark Office, 2008, https://patents.google.com/patent/US7432409B2/en?oq=Patent+7%2c432%2c409.

[53] Timken, H. K. C., Elomari, S., Trumbull, S.; Cleverdon, R. *Integrated Alkylation Process using Ionic Liquid Catalysts* (U.S. Patent No. US7432408B2). U.S. Patent and Trademark Office, 2008. https://patents.google.com/patent/US7432408B2/en?oq=Patent+7%2c432%2c408.

[54] Ma, M. & Johnson, K. E. Carbocation formation by selected hydrocarbons in trimethylsulfonium bromide-AlCl3/AlBr3-HBr ambient temperature molten salts. *J. American Chemical Society*, 1995, *117*(5), 1508–1513. https://doi.org/10.1021/ja00110a007.

[55] *Isoalky^{TM} Alkylation Process Solution*. (n.d.). Honeywell UOP. Retrieved March 2, 2022, from https://uop.honeywell.com/content/uop/en/us/home/industry-solutions/refining/gasoline/alkylation.html.

[56] Manochio, T. (2021, November 11). "Big west oil proceeds with Honeywell to revamp alkylation unit to ISOALKY^{TM} Technology." *Honeywell UOP*. https://uop.honeywell.com/content/uop/en/us/home/news-events/2021/november/big-west-oil-proceeds-with-honeywell-to-revamp-alkylation-unit-to-isoalky-technology.html.

[57] Frett, N. (2022, February 9). "Tupras selects Isoalky^{TM} technology for its refineries to produce high octane alkylate." *Honeywell UOP*. https://uop.honeywell.com/content/uop/en/us/home/news-events/2022/february/tupras-selects-isoalky-technology-for-its-refineries-to-produce-high-octane-alkylate.html.

[58] Riisager, A., Eriksen, K. M., Wasserscheid, P. & Fehrmann, R. (2003). "Propene and 1-octene hydroformylation with silica-supported, ionic liquid-phase (SILP) Rh-phosphine catalysts in continuous fixed-bed mode." *Catalysis Letters*, *90*(3), 149–153. https://doi.org/10.1023/B:CATL.0000004109.46005.be.

[59] Haumann, M. & Riisager, A. (2008). "Hydroformylation in room temperature ionic liquids (RTILs): Catalyst and process developments." *Chemical Reviews*, *108*(4), 1474–1497. https://doi.org/10.1021/cr078374z.

[60] Riisager, A., Fehrmann, R., Flicker, S., van Hal, R., Haumann, M. & Wasserscheid, P. (2005). "Very stable and highly regioselective supported ionic-liquid-phase (SILP) catalysis: Continuous-flow fixed-bed hydroformylation of propene." *Angewandte Chemie International Edition*, *44*(5), 815–819. https://doi.org/10.1002/anie.200461534.

[61] *Supported Ionic Liquid Phase (SILP) Technology*. (n.d.). SILP. Retrieved March 12, 2022, from http://www.silp-technology.de/.

[62] Qureshi, Z. S., Deshmukh, K. M. & Bhanage, B. M. (2014). "Applications of ionic liquids in organic synthesis and catalysis." *Clean Technologies and Environmental Policy*, *16*(8), 1487–1513. https://doi.org/10.1007/s10098-013-0660-0.

[63] Haumann, M. (2020). "Continuous catalytic processes with supported ionic liquid phase (SILP) materials." In M. B. Shiflett (Ed.), *Commercial Applications of Ionic Liquids* (pp. 49–67). Springer International Publishing. https://doi.org/10.1007/978-3-030-35245-5_3.

[64] Abai, M., Atkins, M. P., Hassan, A., Holbrey, J. D., Kuah, Y., Nockemann, P., Oliferenko, A. A., Plechkova, N. V., Rafeen, S., Rahman, A. A., Ramli, R., Shariff, S. M., Seddon, K. R., Srinivasan, G. & Zou, Y. (2015). "An ionic liquid process for mercury removal from natural gas." *Dalton Transactions*, *44*(18), 8617–8624. https://doi.org/10.1039/C4DT03273J.

[65] Liu, G., Cai, Y., O'Driscoll, N., Feng, X. & Jiang, G. (2012). "Overview of mercury in the environment." In G. Liu, Y. Cai & N. O'Driscoll (Eds.), *Environmental Chemistry and Toxicology of Mercury* (pp. 1–12). John Wiley & Sons, Inc.

[66] Wilhelm, S. M. (2009). "Risk analysis for operation of aluminum heat exchangers contaminated by mercury." *Process Safety Progress*, *28*(3), 259–266. https://doi.org/10.1002/prs.10322.

[67] Neil, S. (2015, March 11). "Mercury-grabbing ionic liquids hit the gas." *Chemistry World*. https://www.chemistryworld.com/news/mercury-grabbing-ionic-liquids-hit-the-gas/8363.article.

[68] De Munck, N. A., Verbruggen, M. W. & Scholten, J. J. F. (1981). "Gas phase hydroformylation of propylene with porous resin anchored rhodium complexes part I. Methods of catalyst preparation and characterization." *Journal of Molecular Catalysis*, *10*(3), 313–330. https://doi.org/10.1016/0304-5102(81)85052-3.

[69] Hjortkjaer, J., Heinrich, B. & Čapka, M. (1990). "Influence of the ligand structure on SLP-catalysed hydroformylation of propene." *Applied Organometallic Chemistry*, *4*(4), 369–374. https://doi.org/10.1002/aoc.590040409.

[70] Harris, P. (2014, March 27). "Petronas' mercury removal technology enters the international market." *Oil and Gas Technology*. http://www.oilandgastechnology.net/downstream-news/petronas%E2%80%99-mercury-removal-technology-enters-international-market.

[71] *ILTEC Technology. Ionic Liquid Cooling Technology-Description and Application.* (n.d.). Mettop GmbH. Retrieved March 19, 2022, from https://mettop.com/api/cdn/uploads/1526032491_k7timr64.pdf.

[72] Mettop GmbH. (2010). *Verfahren zur kühlung eines metallurgischen ofens sowie kühlkreislaufsystem für metallurgischeöfen* [*Method of cooling a metallurgical furnace and cooling circuit system for metallurgical furnaces*]. (Patent No. AT508292A1). Osterreichische Patentanmeldung https://patents.google.com/patent/AT508292A1/en?oq=508292.

[73] Filzwieser, I., Filzwieser, A., Hanel, M. & Ruhs, S. (2016, January 1). "ILTEC - METTOP´S revolutionary cooling solution for the steel industry." *METTOP*. https://www.mettop.com/api/cdn/uploads/1476433948_bf6nnc6i.pdf.

[74] Karuppasamy, K., Theerthagiri, J., Vikraman, D., Yim, C.-J., Hussain, S., Sharma, R., Maiyalagan, T., Qin, J. & Kim, H.-S. (2020). "Ionic liquid-based electrolytes for energy storage devices: A brief review on their limits and applications." *Polymers*, *12*(4), 918. https://doi.org/10.3390/polym12040918.

[75] Abbott, A. P., Davies, D. L., Capper, G., Rasheed, R. K. & Tambyrajah, V. (2002). *Ionic liquids and their use as solvents* (World Intellectual Property Organization Patent No. WO2002026701A2). World Intellectual Property Organization International Bureau. https://patents.google.com/patent/WO2002026701A2/en?oq=WO2002026701A2.

[76] *OQEMA*. (n.d.). OQEMA. Retrieved February 21, 2022, from http://oqema.com//en/.

[77] Abbott, A. P., Capper, G., Davies, D. L. & Rasheed, R. K. (2004). "Ionic liquid analogues formed from hydrated metal salts." *Chemistry – A European Journal*, *10*(15), 3769–3774. https://doi.org/10.1002/chem.200400127.

[78] Chum, H. L., Koch, V. R., Miller, L. L. & Osteryoung, R. A. (1975). "Electrochemical scrutiny of organometallic iron complexes and hexamethylbenzene in a room temperature molten salt." *Journal of the American Chemical Society*, *97*(11), 3264–3265. https://doi.org/10.1021/ja00844a081.

[79] Nardi, J. C., Hussey, C. L. & King, L. A. (1978). *AlCl3 /1-alkyl pyridinium chloride room temperature electrolytes* (U.S. Patent No. US4122245A). U.S. Patent and Trademark Office. https://patents.google.com/patent/US4122245A/en.

[80] Abdul-Sada, A. K., Avent, A. G., Parkington, M. J., Ryan, T. A., Seddon, K. R. & Welton, T. (1993). "Removal of oxide contamination from ambient-temperature chloroaluminate(III) ionic liquids." *Journal of the Chemical Society, Dalton Transactions*, *22*, 3283–3286. https://doi.org/10.1039/DT9930003283.

[81] Bonhôte, P., Dias, A.-P., Papageorgiou, N., Kalyanasundaram, K. & Grätzel, M. (1996). "Hydrophobic, highly conductive ambient-temperature molten salts." *Inorganic Chemistry*, *35*(5), 1168–1178. https://doi.org/10.1021/ic951325x.

[82] Ong, S. P., Andreussi, O., Wu, Y., Marzari, N. & Ceder, G. (2011). "Electrochemical windows of room-temperature ionic liquids from molecular dynamics and density functional theory calculations." *Chemistry of Materials*, *23*(11), 2979–2986. https://doi.org/10.1021/cm200679y.

[83] Karuppasamy, K., Theerthagiri, J., Vikraman, D., Yim, C.-J., Hussain, S., Sharma, R., Maiyalagan, T., Qin, J. & Kim, H.-S. (2020). "Ionic liquid-based electrolytes for energy storage devices: a brief review on their limits and applications." *Polymers*, *12*(4). https://doi.org/10.3390/polym12040918.

[84] *Battery Electrolytes* (n.d.). IoLiTec. Retrieved March 27, 2022, from https://iolitec.de/en/products/energy-cleantech/battery-electrolytes.

[85] *Safety Solvents, Lithium-ion Batteries* (n.d.). NOHMs Technologies Inc. Retrieved March 28, 2022, from https://www.nohms.com/products/safetysolvents/.

[86] Inagaki, M. & Kang, F. (2014). Chapter 3—Engineering and applications of carbon materials. In M. Inagaki & F. Kang (Eds.), *Materials Science and Engineering of Carbon: Fundamentals (*2nd ed., pp. 219–525). Butterworth-Heinemann. https://doi.org/10.1016/B978-0-12-800858-4.00003-6.

[87] *Electrolytes for Supercapacitors* (n.d.). IoLiTec. Retrieved March 27, 2022, from https://iolitec.de/en/products/energy-cleantech/supercap-electrolytes.

[88] Viswanathan, B. (2017). Chapter 13—Supercapacitors. In B. Viswanathan (Ed.), *Energy Sources: Fundamentals of Chemical Conversion Processes and Applications* (pp. 315–328). Elsevier. https://doi.org/10.1016/B978-0-444-56353-8.00013-7.

[89] Sarno, M. (2020). Chapter 22 - Nanotechnology in energy storage: The supercapacitors. In A. Basile, G. Centi, M. D. Falco & G. Iaquaniello (Eds.), *Studies in Surface Science and Catalysis* (Vol. 179, pp. 431–458). Elsevier. https://doi.org/10.1016/B978-0-444-64337-7.00022-7.

[90] Kailas (2022, March 23). "How to handle symmetric supercapacitor market challenge with ease using these tips | Key players are Maxwell, Panasonic, NEC TOKIN, Nesscap, AVXELNA, Korchip, Nippon Chemi-Con, IoxusLS Mtron, Nichicon." *ChattTenn Sports*. https://chatttennsports.com/120736/business/how-to-handle-symmetric-supercapacitor-market-challenge-with-ease-using-these-tips-key-players-are-maxwell-panasonic-nec-tokin-nesscap-avxelna-korchip-nippon-chemi-con-ioxusls-mtron-nichicon/.

[91] *Maxwell Technologies—Products*. (n.d.). Maxwell. Retrieved March 29, 2022, from https://maxwell.com/products/.

[92] *SuperCapacitors*. (n.d.). KYOCERA AVX. Retrieved March 27, 2022, from https://www.kyocera-avx.com/products/supercapacitors/.

[93] *Backup Power Supply Modules with Electric Double Layer Capacitors*. (n.d.). Panasonic Industry. Retrieved March 29, 2022, from https://industrial.panasonic.com/ww/products/pt/edlc-backup-power-supply-modules.

[94] *Supercapacitors*. (n.d.) Nippon Chemi-Con Corporation. Retrieved March 27, 2022, from https://www.chemi-con.co.jp/en/products/edlc/.

[95] *Electric double layer capacitors "EverCAP®."* (n.d.) Nichicon Corporation. Retrieved March 27, 2022, from https://www.nichicon.co.jp/english/products/evercap.html.

[96] *Super capacitor*. (n.d.). TOKIN Corporation. Retrieved March 29, 2022, from https://www.tokin.com/products/capacitor/super-capacitor/.

[97] Ye, C., Liu, W., Chen, Y. & Yu, L. (2001). "Room-temperature ionic liquids: A novel versatile lubricant." *Chemical Communications*, *21*, 2244–2245. https://doi.org/10.1039/B106935G.

[98] Cai, M., Yu, Q., Liu, W. & Zhou, F. (2020). "Ionic liquid lubricants: When chemistry meets tribology." *Chemical Society Reviews*, *49*(21), 7753–7818. https://doi.org/10.1039/D0CS00126K.

[99] *Ionic Liquids – Innovative Lightning Conductor in e-Mobility*. (n.d.). Kluber Lubrication. Retrieved March 20, 2022, from https://www.klueber.com/global/en/company/newsroom/news/ionic-liquids-innovative-lightning-conductor-in-e-mobility/.

[100] Nicolas, R. (n.d.). "Conductive lubricants for reduced bearing wear." *Car Engineer*. Retrieved March 20, 2022, from https://www.car-engineer.com/conductive-lubricants-bearings/.

[101] Suzumura, J. (2016). "Prevention of electrical pitting on rolling bearings by electrically conductive grease." *Quarterly Report of RTRI*, *57*(1), 42–47. https://doi.org/10.2219/rtriqr.57.1_42.

[102] Schmidt-Amelunxen, M., Sohn, D., Grundei, S. & Höpke, A. (2013). *Lubricating Grease Composition Based On Ionic Liquids* (U.S. Patent No. US8455407B2). U.S. Patent and Trademark Office. https://patents.google.com/patent/US8455407B2/en?q=kluber+ionic+liquids&oq=kluber+ionic+liquids.

[103] Adler, R., Pfandl, M. & Siebert, G. (2015). *Multistage piston compressor* (U.S. Patent No. US9109587B2). U.S. Patent and Trademark Office. https://patents.google.com/patent/US9109587B2/en?oq=Patent+9%2c109%2c587.

[104] *Hydrogen refueling technologies*. (n.d.). Linde Engineering. Retrieved March 18, 2022, from https://www.linde-engineering.com/en/plant-components/hydrogen-refueling-technologies/index.html.

[105] Kömpf, M. (2006). "Mobility under high pressure." *Linde Technology Report*, *2006*, 24–29.

[106] Lei, Z., Dai, C. & Chen, B. (2014). "Gas solubility in ionic liquids." *Chemical Reviews*, *114*(2), 1289–1326. https://doi.org/10.1021/cr300497a.

[107] Gutowski, K. E. (2018). "Industrial uses and applications of ionic liquids." *Physical Sciences Reviews*, *3*(5). https://doi.org/10.1515/psr-2017-0191.

[108] *Ionic liquid GC capillary columns*. (n.d.). Merck. Retrieved March 29, 2022, from https://www.sigmaaldrich.com/US/en/technical-documents/protocol/analytical-chemistry/gas-chromatography/ionic-liquid-gc-capillary-columns.

[109] *Watercol^{TM} series columns*. (n.d.). Merck. Retrieved March 29, 2022, from https://www.sigmaaldrich.com/US/en/technical-documents/protocol/analytical-chemistry/gas-chromatography/gc-measurement-of-water-using-watercol.
[110] *SLB®-ILPAH Capillary GC Column Sigma-Aldrich*. (n.d.). Retrieved March 29, 2022, from https://www.sigmaaldrich.com/US/en/product/supelco/29799u.

Chapter 3

Electrical Conductivities and Dielectric Properties of Ionic Liquids and Their Solutions

Natalia V. Plechkova[1,*]
Umme Salma[2]
Vladimir V. Shcherbakov[3,†]
and Yulia M. Artemkina[3]

[1] Wellcome-Wolfson Institute for Experimental Medicine,
The Queen's University of Belfast, Belfast, UK
[2] Istituto di Nanotecnologia (NANOTEC) – CNR,
c/o Campus Ecotekne, Lecce, Italy
[3] Department of General and Inorganic Chemistry,
Mendeleev University of Chemical Technology, Moscow, Russia

Abstract

This chapter considers the electrical conductivities of ionic liquids (ILs) and their mixtures with molecular solvents. The electrical conductivity of pure ILs decreases with increased alkyl chain length on the cation (as a consequence of the increased viscosity) and conversely increases by enlarging the anion size. This electrical conductivity increase is due to hydrogen bonding and charge delocalization. The dependence on the solvent composition (IL: solvent ratio) and temperature is additionally discussed.

ILs behave like strong electrolytes in dilute solutions with not large association constants Ka and molar conductivities that decrease proportionally to the square root of the IL concentration. The association of the ILs depends on the solvation nature of their cations and anions, solvent permittivity and temperature. Various equations for the molar electrical conductivity concentration dependence are currently applied which results in an inconsistency in the association constant Ka. As for strong electrolytes, the specific electrical conductivity of concentrated solutions of ILs in polar solvents passes a maximum in concentration growth due to an increase in the association of the IL ions in concentrated solutions and therefore an increase in their viscosity.

The dielectric characteristics of ILs determine the ionic and molecular processes that occur in them - the dissociation and solvation of ions and molecules. The electrical

* Corresponding Author's Email: n.plechkova@qub.ac.uk.
† Corresponding Author's Email: shcherb@muctr.ru.

In: Properties and Applications of Ionic Liquids
Editors: Noureddine Abidi and Julia L. Shamshina
ISBN: 979-8-88697-635-9

conductivity (EC) κ and the static dielectric permittivity (DP) εs of ILs determine the ratio of conduction currents to bias currents in an alternating electromagnetic field. Provided that these currents are equal, electrical (ionic) relaxation takes place in ILs. The ionic relaxation time θ equals to the ratio of absolute DP ($\varepsilon s \varepsilon o$) to specific EC κ. A significant value of direct current (DC) conductivity for ILs, makes it difficult to determine their dielectric characteristics and to interpret the dielectric relaxation spectra (DRS).

As a result of the interfering effect of DC conductivity, the results of determining the dielectric characteristics of the ILs cannot always be reliable, as evidenced by the variation in the values of the static DP εs and the dipole dielectric relaxation time τ, as published in literature. The influence of the electromagnetic field frequency and the DC conductivity on the DRS is also considered. A rather complex type of DR and OKE spectra for ILs is due to the different polarization processes occurring over a wide frequency range. These processes include the polarization of the ILs molecules (α-relaxation), their ions and more complex structures formed in ILs and in their solutions.

The DRS of protic and aprotic ILs and currently used options for explaining the nature and mechanisms of different absorption regions were analyzed. It was shown that as the frequency gets larger, the high-frequency (HF) EC ILs increases, which in the microwave range becomes commensurate with their DC conductivity. Equations describing the dispersion of HF EC for ILs are presented and analyzed. Under the condition of $\omega\tau >> 1$ (where ω is angular frequency), the active HF EC κ' reaches its limit value of $\kappa\infty$, which does not depend on the frequency. The limiting HF EC $\kappa\infty$ is determined by the ratio of the absolute dielectric constant $\varepsilon s \varepsilon 0$ to the dielectric relaxation time τ. The limiting HF EC $\kappa\infty$ determines the temperature dependence of DC conductivity in ILs and their solutions in polar solvents: as the temperature of DC conductivity increases, the κ increases directly in proportion to their limiting HF EC $\kappa\infty$. The proximity of the activation energies for the dielectric relaxation processes, viscosity η and DC conductivity κ indicate the commonality of the polarization processes and the translational movement of the particles (molecules and ions) in ILs.

The time for dipolar relaxation τ of ILs is defined with lower accuracy, than the viscosity η. Therefore, for the analysis and generalization of the temperature dependences of DC conductivity κ in ILs, it is more preferable to use the dependences $\kappa - \varepsilon s T/\eta$ instead of the dependences $\kappa - \kappa\infty$. In the analysis of conductivity-viscosity correlations of ILs, in addition to the Walden rule, it is proposed to use equations that take into account the effect on DC conductivity not only of viscosity, but also of the permittivity of εs and temperature T. The conditions for the formation of ion pairs different in structure are considered in IL solutions. It has been suggested that the maximum on the dependence of the specific EC on the concentration of an IL is associated with the formation of contact ion pairs in the solution.

This chapter includes proposed equations which allow the distance d between the molecules and ions of ILs in solutions to be estimated, as well as estimating the concentration of the solution c at which there is a certain value of distance d between ions in IL solutions.

Keywords: ionic liquids, permittivity, low frequency and high frequency electrical conductivity, relaxation processes

Abbreviations

AAE	Arrhenius activation energy
AC	Alternating current

AN	Acetonitrile
BAF	Butylammonium formate
CC	Cole-Cole
c_{IL}	Concentration of an IL in solution, mol/L
$c(\kappa_{\text{max}})$	Concentration corresponding to the maximum specific Electrical conductivity
c_{max}	Maximum specific electrical concentration at a given temperature
CIP	Contact ion pairs
c_{SIP}	Concentration corresponding to separated by solvent Molecules formation
$[C_2mim]^+$	1-ethyl-3-methylimidazolium cation
$[C_4mim][NTf_2]$	1-butyl-3-methylimidazolium bis{(trifluoromethyl)sulfonyl}amide
$[C_4mim][OTf]$	1-butyl-3-methylimidazolium trifluoromethanesulfonate triflate
$[C_4m_\beta py][NTf_2]$	1-butyl-3-methylpyridinium bis{(trifluoromethyl)sulfonyl}amide
$[C_5mim]^+$	1-pentyl-3-methylimidazolium cation
$[C_2mim][CF_3SO_3]$	1-ethyl-3-methylimidazolium triflate
$[C_8mim][OTf]$	1-octyl-3-methylimidazolium triflate
$[N_{2\,4\,4\,4}][N(CN)_2]$	Ethyltributylammonium dicyanamide
$[N_{4\,4\,4\,8}][N(CN)_2]$	Octyltributylammonium dicyanamide
$[P_{6\,6\,6\,14}]Cl$	Trihexyltetradecylphosphonium chloride
C153	Coumarin 153
$d_{\text{(ILs)}}$	Molecular size of ionic liquids
D	Debye
DC	Direct current
DCM	Dichloromethane
d	Density
DF	Debye and Falkenhagen
DMF	Dimethylformamide
DMSO	Dimethyl sulfoxide
DP	Dielectric permittivity
DR	Dielectric relaxation
DRS	Dielectric relaxation spectra
EAF	Ethylammonium formate
EAN	Ethylammonium nitrate
EC or κ	Electrical conductivity
ε	Dielectric constant
ε_0	Absolute DP of the vacuum
ε_s	Static dielectric permittivity
ε^*	Complex dielectric constant
ε' and ε''_d	Dielectric characteristics
ε''_{ion}	Contribution of direct current conductivity
θ	Ionic relaxation time
FTIR	Fourier transform far-infrared
F_o	Frequency
ΔG_A^o	Gibbs-free energies
ΔH^o_A	Change in the enthalpy of the association

HF	High-frequency
ILs	Ionic liquids
IR	Infrared
K_a	Association constants
κ_{max}	Maximum specific electrical conductivity at a given temperature
κ_∞	Limiting high-frequency (HF) electrical conductivity
κ^*	Complex electrical conductivity
κ''_{opt}	Optical component of reactive conductivity
KCl	Potassium chloride
KDE	Kernel Density Estimation
Λ_0	Limiting molar conductivities
Λ	Molar conductivities
M	Molar mass
MAF	Methylammonium formate
MeOH	Methanol
η	Viscosity
N	Number of solvent molecules in solution
N_{K+}	Coordination numbers of the cation
N_{An-}	Coordination numbers of the anion
$[NTf_2]$	Bis{(trifluoromethyl)sulfonyl}amide anion
OKE	Optical Kerr-effect
PAF	Pentylammonium formate
PAN	Propylammonium nitrate
PCA	Principal component analysis
PrOH	Propanol
QSPR	Quantitative structure-property relationship
R	Desired resistance
R_a	Measured equivalent active resistance of ionic liquids
$\delta(\varepsilon''_d)$	Error of definition of a dipolar component
ΔS^o_A	Change in the entropy of the association
SED	Stokes-Einstein-Debye
SIP	Separated by solvent molecules
τ	Dielectric relaxation time
T	Temperature
TAF	Triethylamine + formic acid
THz	Terahertz
τ_{i-d}	Time of ion-dipole relaxation
τ_{ILs}	Dielectric relaxation time of ionic liquids
VFT	Vogel-Fulcher-Tammann
V_m	Molar volume
$V_{m,IL}$	Molar volume of an IL, L/mol
VOCs	Volatile organic compounds
$V_{slip\ eff,cal}$	Volumes for boundary conditions of slipping
$V_{stick\ eff,cal}$	Volumes for boundary conditions of sticking
ω	Frequency of the electromagnetic field
X	Molar confraction

X_{im}	Reactance of ionic liquids
2-HEAB	2-hydroxyethylammonium butanoate
2-HEAF	2-hydroxyethylammonium formate
2-HEAP	2-hydroxyethylammonium propanoate

Introduction

Ionic liquids (ILs) are a class of materials that possess a unique combination of physicochemical properties: negligible vapor pressure at room temperature, low melting points (conventionally lower than 100°C), thermal and chemical stabilities and the ability to dissolve both polar/nonpolar and organic/inorganic substances [1]. Hence, they can be alternatives for some volatile organic compounds (VOCs) in extraction, synthesis, catalysis and separations. The chemical and physical properties of ionic liquids are determined by the choice of cations and anions. By carefully selecting the cations and anions, the physical properties of the IL, such as its density, melting point, viscosity, electrical conductivity, dielectric permittivity, miscibility, hydrophilicity/hydrophobicity, can be adapted for a particular application; therefore, ionic liquids were coined as "designer solvents" [1].

ILs are often used not in their pure form, but as co-solvents or additives. In these capacities, they demonstrate improved and enhanced thermodynamic and transport properties. Hence, knowledge of the properties of binary IL mixtures is crucial for industrial applications. For electrical conductivity, most commonly used electrolytes are aqueous solutions of salts or salt solutions with volatile organic solvents (VOCs). Due to the volatile nature of water and organic solvents, there is a limited use of IL solutions especially at elevated temperatures. One notable property of ionic liquids is their conduction at temperatures higher than 100°C, which makes them beneficial materials for electrical conductivity applications even in their neat state.

The conductivity of ionic liquids emerges from the inherent motion of the cations and anions in ionic liquids when an electric potential difference is applied, so they are prospective *electrolyte materials for electrochemical devices* [2]. Generally, ionic liquids exhibit conductivities over a wide range e.g., 0.01-2 S m^{-1}, but high conductivities in the order of 10 S m^{-1} can be observed in imidazolium cation-based ionic liquids [3]. The conductivity values increase in binary mixtures of ILs with molecular solvents and are comparable to those of organic electrolyte solutions in VOCs [4]. For these reasons, ionic liquids and their solutions are considered as promising candidates as electrolytes.

A distinctive feature of ILs is the presence in the liquid phase of both undissociated molecules, which can be considered as a solvent, and ions into which these molecules dissociate. The degree of dissociation in ILs is unknown, but it can be assumed that it is not high, since the DC conductivity of ILs is much lower than the electrical conductivity of simple molten salts. Some considerations about the possible values of the degree of dissociation of ILs are considered in this work when analyzing the distance between ions and molecules.

When analyzing the properties of pure ILs, in particular when interpreting their dielectric spectra, the electrical conductivity κ is usually used. By multiplying this value by the molar volume IL V_m, the molar conductivity λ of the ILs can also be obtained. The values of λ for ILs are significantly lower than the molar conductivities of aqueous and non-aqueous solutions of simple electrolytes and, as a rule, do not exceed 5.0 S cm^2 mol^{-1} at 25°C [5].

When analyzing the electrical conductivity of ILs solutions in polar solvents, both molar and specific conductivity are used. The molar conductivity of dilute solutions of ILs decreases with increasing concentration. Based on the molar conductivity of dilute solutions of ILs, the values of the limiting molar conductivity λ_0 and association constants K_A are calculated. In the case of concentrated solutions of ILs, the electrical conductivity κ is usually analyzed. In these solutions, as the concentration increases, there is a maximum in the dependence of κ on the concentration *c*.

The association of ions in dilute solutions of ILs and the nature of the maximum in the concentration dependence of the electrical conductivity of concentrated solutions of ILs in polar solvents have been considered.

The dielectric characteristics of ILs that determine the ionic and molecular processes that occur in them - the dissociation and solvation of ions and molecules have been considered next.

The time for dipolar relaxation τ of ILs is defined with lower accuracy, than the viscosity η. In the analysis of conductivity-viscosity correlations of ILs, in addition to the Walden rule, it is proposed to use equations that take into account the effect on DC conductivity not only of viscosity, but also of the permittivity of ε_s and temperature *T*. The conditions for the formation of ion pairs different in structure are considered in IL solutions. It has been suggested that the maximum on the dependence of the specific EC on the concentration of an IL is associated with the formation of contact ion pairs in the solution. This Chapter includes proposed equations which allow the distance d between the molecules and ions of ILs in solutions to be estimated, as well as estimating the concentration of the solution c at which there is a certain value of distance d between ions in IL solutions.

Electrical Conductivity of Ionic Liquids: Dependence on Composition and Temperature

Dependence of the Ionic Liquid Electrical Conductivity on the Composition: The Nature and Size of the Cation and Anion

The *nature of cations* has a significant influence on the specific conductivities κ of ionic liquids. The conductivities κ of four dicyanamide based ionic liquids have been compared [6]. At 298 K, the value of conductivity for 1-butyl-3-methylimidazolium dicyanamide, $[C_4mim][N(CN)_2]$, reaches 1.09 [7].

The length of the alkyl chains on ions also plays an important role in determining the electrical conductivity κ of various ionic liquids and has been observed to decrease with the increased length of alkyl chains. Table 1 summarizes the measured values of electrical conductivities (κ) for tetrafluoroborate ionic liquids with varying alkyl chain lengths (where $[C_n mim]$ is 1-alkyl-3-methylimidazoluim cation; $n = 2$ (ethyl), $n = 4$ (butyl), $n = 6$ (hexyl) and $n = 8$ (octyl)) [6]. The studied ILs showed a decrease of κ with increasing alkyl chain length as the consequence of increasing viscosity [6]. Hence, for applications of ionic liquids as electrolytes in electrochemical devices and to choose the ionic liquids with higher electrical conductivity among their homologues, longer alkyl chains should be avoided [8].

Table 1. Electrical conductivities (κ) of the investigated ionic liquids[a]; taken from [6]

T/°C	κ/S·m^{-1}			
	[C_2mim][BF_4]	[C_4mim][BF_4]	[C_6mim][BF_4]	[C_8mim][BF_4]
-35	-	0.00212	0.000405	0.0001524
-25	0.1059	0.00778	0.001685	0.000650
-15	0.228	0.0225	0.00544	0.00218
-5	0.423	0.0542	0.01451	0.00605
5	0.703	0.1132	0.0332	0.01441
15	1.076	0.211	0.0672	0.0304
25	1.546	0.352	0.1229	0.0579
35	2.11	0.556	0.207	0.1012
45	2.77	0.821	0.326	0.1647
55	3.51	1.150	0.483	0.252
65	4.34	1.545	0.683	0.368
75	5.23	2.00	0.926	0.512
85	6.20	2.52	1.214	0.688
95	7.22	3.10	1.545	0.895
105	8.29	3.73	1.92	1.134
115	9.41	4.41	2.33	1.403
125	10.56	5.13	2.78	1.701
135	11.75	5.88	3.26	2.03
145	12.97	6.64	3.77	2.38
155	14.22	7.45	4.31	2.75
165	15.48	8.29	4.88	3.15
175	16.75	9.15	5.46	3.56
185	18.03	10.04	6.07	3.99
195	19.3	10.94	6.69	4.44

[a]The relative standard uncertainty of κ is 0.5%.

The Effect of the Anion

A set of ionic liquids was studied by keeping the same cation (1-butyl-3-methylimidazolium one) and by changing the anions (Cl^-, Br^-, I^-, $[NCS]^-$, $[N(CN)_2]^-$ and $[BF_4]^-$) to observe changes in the electrical conductivities [9]. With an increase in the size of the anion, a decrease in the viscosity of the ILs occurs, which leads to a natural increase in their electrical conductivity (Table 2). At room temperature, an increase in electrical conductivity was observed with increasing the anion size for the halide anions [9]. Hydrogen bonding and charge delocalization are the two primary interactions that overcome the predicted decrease in molecular mobility as ionic size increases.

Firstly, these selected ILs are considered based on their hydrogen bonding ability. Hydrogen bonds are formed between the hydrogen atoms on the imidazolium ring and a halide ion [10, 11]. If the halide ion gets larger, the strength of this hydrogen bond weakens due to increasing charge delocalization. The decrease in hydrogen bond strength can be interpreted as a decrease in the charge carrier's interaction with its surroundings, resulting in an increase in molecular mobility and conductivity.

The second factor affecting conductivity is that as the ion gets larger, the charge distributes out more across its surface, lowering the charge surface density and hence lowering the Coulombic interaction. Similarly to the hydrogen bond, a decrease in interionic interaction leads to an increase in conductivity.

Table 2. Conductivities and viscosities of 1-butyl-3-methylimidazolium chloride ([C4mim]Cl), bromide ([C4mim]Br), iodide ([C4mim]I), thiocyanate ([C4mim][NCS]), dicyanamide ([C4mim][N(CN)2]) and tetrafluoroborate ionic liquids. Adapted from [9]

Ionic liquid	Temperature (°C)	κ (S/m)	Ref.	Viscosity (mPa s)	Ref
[C4mim]Cl	20	0.0023	[9]		
	-20	8.4×10^{-7}	[9]		
	-40	4.1×10^{-10}	[9]		
[C4mim]Br	20	0.0077	[9]		
[C4mim]I	20	0.028	[9]		
	75	0.71	[9]		
	120	2.5	[9]		
[C4mim][NCS]	20	0.42	[9]		
	25			35.9	[12]
[C4mim][N(CN)2]	20	0.94	[9]	29.3	[13]
	25	1.09	[13]	6.9	[13]
	70	2.85	[13]		
[C4mim][BF4]	20	0.32	[9]	180	[14]
	25	0.35	[14]	104.2	[15]

The Influence of Temperature on the Conductivity of Ionic Liquids: Activation Energy of the Specific Electrical Conductivity of Ionic Liquids

As the electrical conductivity depends on the temperature of the sample and the concentration of the ions, the higher the temperature, the lower the viscosity and the higher the kinetic energy of the ions in the ionic liquids and therefore, the higher the electrical conductivity.

There are several equations used to explain the effect of temperature on the electrical conductivity. One of these is the Arrhenius equation which was introduced by the Swedish chemist Svante Arrhenius. He combined the concepts of the activation energy and the Boltzmann distribution law into one of the most important relationships in physical chemistry, which can be expressed by the following formula:

$$\kappa = \kappa_0 \exp(-\frac{E_a}{k_B T}). \tag{1}$$

Here, κ is the electrical conductivity (S/m), κ_0 is the pre-exponential constant (which can be thought of as the maximum electrical conductivity which is at infinite temperature), E_a is the activation energy of the electrical conductivity (the threshold energy that ions must acquire in order to move from one position to another under the action of an external field), k_B is Boltzmann constant and T is the temperature in Kelvin. The term "k_BT" represents the average kinetic energy of the ions and it is obvious that with higher temperature, the kinetic energy will be higher and thus speed up the conductivity.

The viscosity of ionic liquids is generally high and decreases with the increases in temperature. The viscosity of ionic liquids mainly depends on the intermolecular interactions (hydrogen bonds and van der Waals interactions). With a rise in temperature, these interactions become weaker which causes a drop in the dynamic viscosity of the ionic liquids. Generally, the relation between viscosity (η) and temperature (T) is stated exponentially in the Arrhenius-type equation:

$$\eta = \eta_0 \exp(-\frac{E_a}{k_B T}). \quad (2)$$

In the above equation, E_a is the activation energy, η_0 is a pre-exponential constant, and k_B is the Boltzmann constant or the universal gas constant ($k_B = 8.314$ J·mol^{-1}·K^{-1}).

Another ratio is the Vogel-Fulcher-Tammann (VFT) equation, which was proposed to explain the dependence of the electrical conductivity on the temperature. The VTF equation can be written in different forms, but the popular form is:

$$\kappa = A \exp(-\frac{B}{T - T_g}). \quad (3)$$

where A, B and T_g are fitting parameters. The modified version of the VTF- type equation would be written as:

$$\kappa = \kappa_0 \exp(-\frac{E_a}{k_B (T - T_g)}). \quad (4)$$

where κ_0, E_a and T_g can be obtained from the best fit of the equation to experimental data. Overall, the electrical conductivity of ILs has been found to be influenced by external effects such as temperature, electric field, magnetic field, etc. As an *important factor, temperature* variations (according to the above equations) considerably affect the electrical conductivity of the ILs. The electrical conductivity of the ionic liquids significantly increases with an increase in temperature (Figure 1) [16].

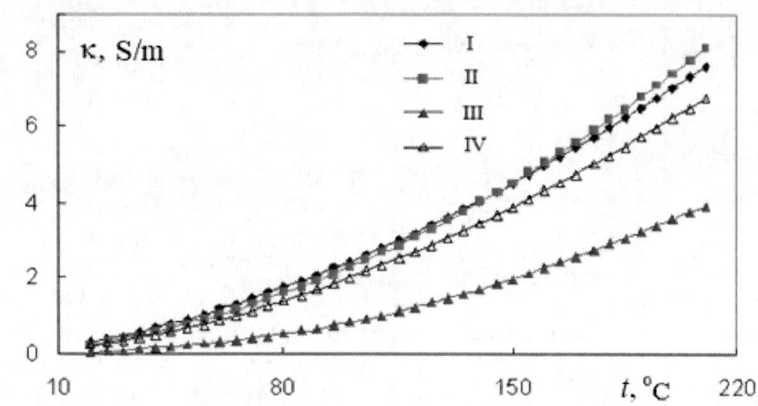

Figure 1. Electrical conductivity (κ) of 1-butyl-3-methylimidazolium bis{(trifluoromethyl) sulfonyl}amide ([C_4mim][NTf_2]) (I); 1-butyl-3-methylimidazolium trifluoromethane sulfonate (triflate) ([C_4mim][OTf]) (II); 1-octyl-3-methylimidazolium triflate ([C_8mim][OTf]) (III) and 1-butyl-3-methylpyridinium bis{(trifluoromethyl) sulfonyl}amide ([C_4mpy][NTf_2]) (IV) ionic liquids at different temperatures (T/K) [16], reprinted with permission from [16].

It is worth paying attention to the nature of the curves: for ILs they are concave, for solutions they are linear and convex (in a wide temperature range). The mobility of ions in ionic liquids is considerably different from that in non-aqueous solvents and there have been very few attempts to model the conductivity of ionic liquids. The conductivity and viscosity of an IL is often combined into what is termed as Walden's rule (Equation 5) [17]:

$$\lambda\eta = \text{constant} \quad (5)$$

wherein λ is the molar conductivity and η is the viscosity of the solvent, and it is given by Equation (6):

$$\Lambda = \kappa M/d \quad (6)$$

where M is the molecular weight and d is the density of the IL. Ideally, the Walden product (Λη) remains constant for a given IL regardless of temperature. The magnitude of the Walden product for different ILs has been shown to vary inversely with ion size. This inverse relationship between ion size and the magnitude of Λη is generally followed for the cations. Figure 2 demonstrates the Walden plot of log(molar conductivity, Λ) against log(reciprocal viscosity η^{-1}) graphs for some dicyanamide ionic liquids [18].

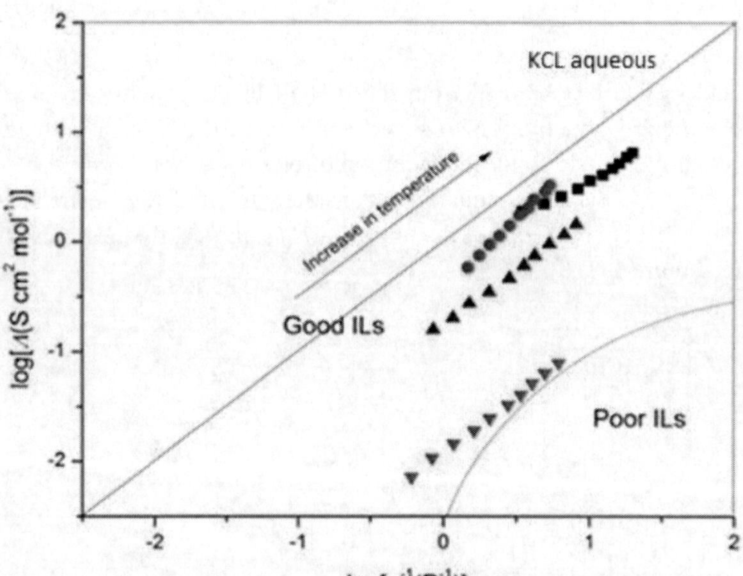

Figure 2. Walden plots of log(molar conductivity, Λ) against log(reciprocal viscosity η^{-1}), 1-butyl-3-methylimidazolium dicyanamide [C_4mim][$N(CN)_2$] (■), 1-butyl-2,3-dimethylimidazolium dicyanamide [C_4dmim][$N(CN)_2$] (●), ethyltributylammonium dicyanamide [$N_{2\,4\,4\,4}$][$N(CN)_2$] (▲), and octyltributylammonium dicyanamide [$N_{4\,4\,4\,8}$][$N(CN)_2$] (▼) [18], reprinted with permission from [18].

Some combinations of anions and cations might result in minimum values of Tg (glass temperature). The identification of such combinations would provide a useful guide to likely "good" ionic liquids, provided that the melting points and fragility factors were also agreeable. Some ILs have conductivity that is almost an order of magnitude lower than ideal. Such a liquid must have a high degree of correlation in the motion of its anions and cations to justify the poor conductivity at a given fluidity (so called a "poor IL"), and would also be anticipated to have a much higher vapor pressure than the salts with ideal Walden behavior, Figure 2.

Deviations from the Walden rule observed in proton transfer ionic liquids have previously been applied to explain ionic association. This rule does not fully extend to ionic liquids as it is only justifiable for ions at infinite dilution where ion-ion interactions can be disregarded, which is obviously not the case in ionic liquids.

In 2021, variations in the conductivities of a large number of ionic liquids with respect to temperature were reported [19]. The authors presented the Kernel Density Estimation (KDE) contour plot showing the distribution of conductivities data and temperature for the database of 3784 points comprising 285 individual ionic liquids (Figure 3). The majority of ionic liquids in this study represented the range of conductivities from 7×10^{-3} S m^{-1} to 7 S m^{-1}.

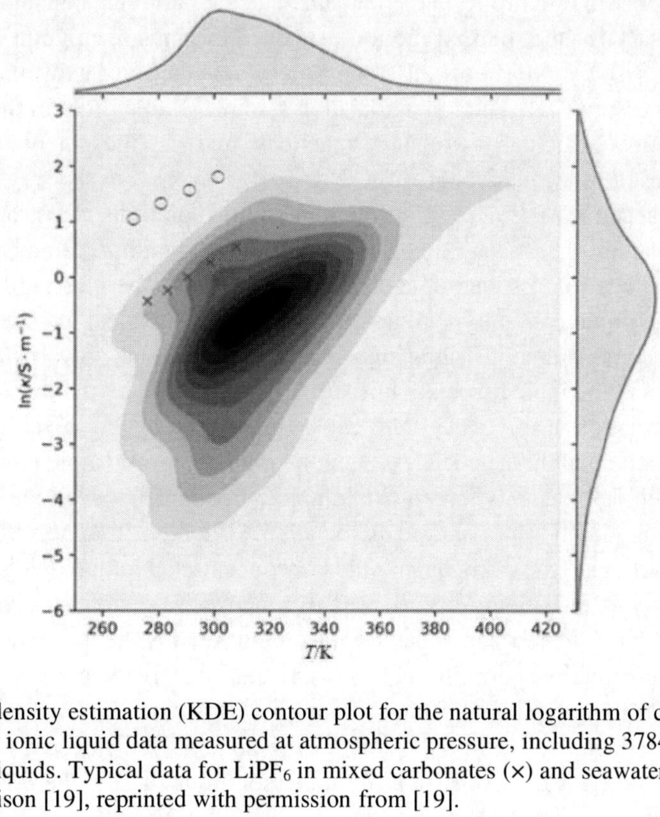

Figure 3. Kernel density estimation (KDE) contour plot for the natural logarithm of conductivity versus temperature for all ionic liquid data measured at atmospheric pressure, including 3784 data points and 285 unique ionic liquids. Typical data for $LiPF_6$ in mixed carbonates (×) and seawater (o) are also shown for comparison [19], reprinted with permission from [19].

From these data, the generic trend of increasing conductivity with increases in temperature is noticed due to improved ion mobility. Moreover, diffusivity is also increased with the rise in temperature, which has been shown to boost the conductivity of ionic liquids. The elevation of conductivities at higher temperatures have many applications in the energy sector. One notable property of ionic liquids is the conduction at temperatures higher than 100°C, which makes them beneficial, for example, as electrolytes for electrical conductivity applications even in their neat state. Additionally, at higher temperatures water and organic solvents evaporate, so it is beneficial to use neat ionic liquids at higher temperatures. As the temperature rises, the high viscosity of ionic liquids decreases, which leads to an increase in the mobility of ionic particles. As a result, the areas of practical use of electrolytes with ionic liquid are expanding.

Electrical Conductivity of Ionic Liquid Solutions and the Association of Ions

Dilute Solutions of Ionic Liquids

Ion association is a typical phenomenon in non-aqueous or mixed solutions, as well as for electrolytes in water that are categorized as strong. The partial connection of oppositely charged ions in electrolyte solutions tends to produce different chemical species known as ion pairs. These ions must be near enough together for the Coulombic attraction energy to overcome the repulsive heat energy in order to form the ion pair. The ion pair must be long-lived, longer than the period necessary for the thermal motion to induce the ions to migrate around as distinct species [20]. Ion association affects the ionic liquids' conductivity as well as other variables such as vapour pressure [21]. The ion association also plays a key role in the mechanism of a variety of reactions, particularly exchange reactions, and has the potential to influence the stereochemical result of a chemical process [22].

The conductance measurements of dilute ionic liquid solutions across a wide temperature range allow the values of association constants and thermodynamic parameters of association to be determined. The ion association constants are known to be affected by the alkyl chain length of the ionic liquid cation, type of anion, temperature, physical features of the molecular solvents and interactions amongst ions and solvent. The solvents solvate the ionic liquids to varying degrees, and the ionic association is strongly influenced by the ion solvation.

The ion dissociation in solvent systems can be influenced by the solvent properties such as polarity and H-bonding ability (which is especially critical in protic ionic liquids). Polar liquids having high dielectric constants screen Coulombic attractions between the counterions more effectively, resulting in an improved ion dissociation. H-bonding can occur between the anion and the cation (particularly in pure ionic liquids or concentrated ionic liquid solutions). In most ionic liquid solutions, H-bonding occurs primarily between the solvent and the anion and is mainly responsible for the ion pair separation, resulting in an increased number of "free" ions [23]. Due to the high relative permittivity ($\varepsilon \sim 78.4$) and extensive H-bonding, it is unsurprising that dilute aqueous ionic liquids are almost completely dissociated.

The molar conductivity λ of the ILs solutions in polar solvents decreases with an increase in their concentration. At the same time, in dilute solutions of ILs, there is an almost linear decrease in λ from the square root of the concentration $c^{1/2}$, Figure 4a.

Figure 4 compares the $\lambda - c^{1/2}$ dependencies of ILs in different polar solvents. The following pattern of λ change in polar solvents has been established [10, 24, 25]:

$\lambda(\text{AN}) > \lambda(\text{Water}) > \lambda(\text{DMF}) > \lambda(\text{DMSO})$

The dependences of the molar conductivity λ on the concentration of the strong and weak electrolytes differ significantly, Figure 4b. This figure compares the concentration dependences of λ in aqueous solutions of strong (5) and weak (6) inorganic acids. In the first case, there is a slight decrease in λ with increasing concentration. In this case, a practically linear dependence of λ on the square root of the concentration is observed, as is the case in IL solutions, Figure 4a. The proportionality of the molar conductivity on the square root of the concentration is called Kohlrausch's law [26]. For a weak acid, there is a sharp decrease in λ with increasing concentration, which is due to a decrease in the degree of dissociation with an increase in the content of a weak electrolyte in the solution. As noted above, the DC conductivity measurements are used to calculate the values of the limiting molar conductivity λ_0 and association constants K_a of ionic liquids. Before considering the results of determining these values, it is necessary to briefly analyze the methodology for their calculation based on the measurements of DC conductivity of dilute solutions.

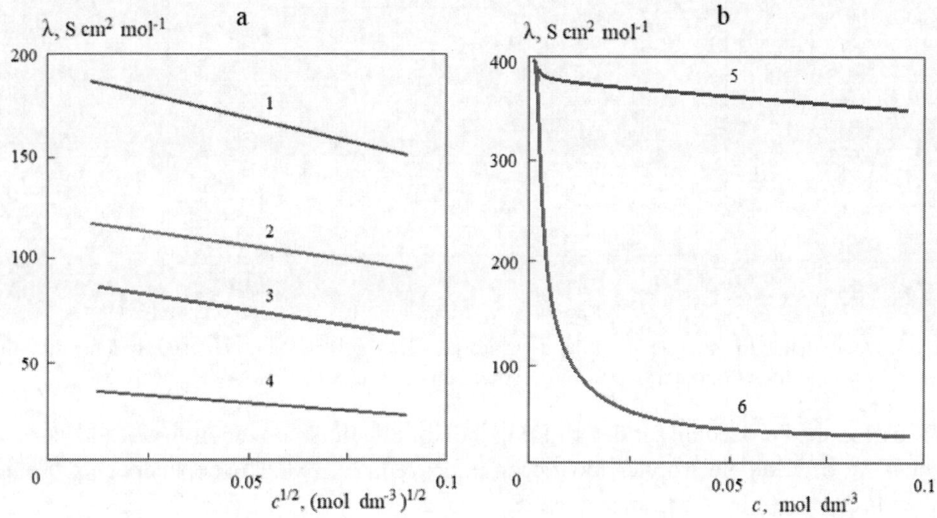

Figure 4. Relationship between molar conductivities of ILs in acetonitrile (AN) (1), water (2), dimethylformamide (DMF) (3), dimethylsulfoxide (DMSO) (4) (a); the dependence of the molar conductivity of a strong (5) and weak (6) electrolyte on the concentration (b).

A relatively simple calculation procedure of the limiting molar conductivity λ_0 and association constants K_a is used for solutions of weak (strongly associated) electrolytes. If we assume that there is practically no electrostatic interaction between ions in the dilute solutions of these electrolytes, then the degree of dissociation α is determined by the ratio:

$$\alpha = \frac{\lambda}{\lambda_0}. \qquad (7)$$

Substituting this expression into the Ostwald equation for the dependence of the degree of dissociation on concentration, it is obtained after transformation:

$$\frac{1}{\lambda} = \frac{1}{\lambda_0} + \frac{K_a}{\lambda_0^2}\lambda c. \qquad (8)$$

Thus, the dependence of the inverse molar conductivity $1/\lambda$ on the product λc is the equation of a straight line, the intercept of which is $1/\lambda_0$, and the slope is the ratio $K_a/(\lambda_0)^2$, Figure 5.

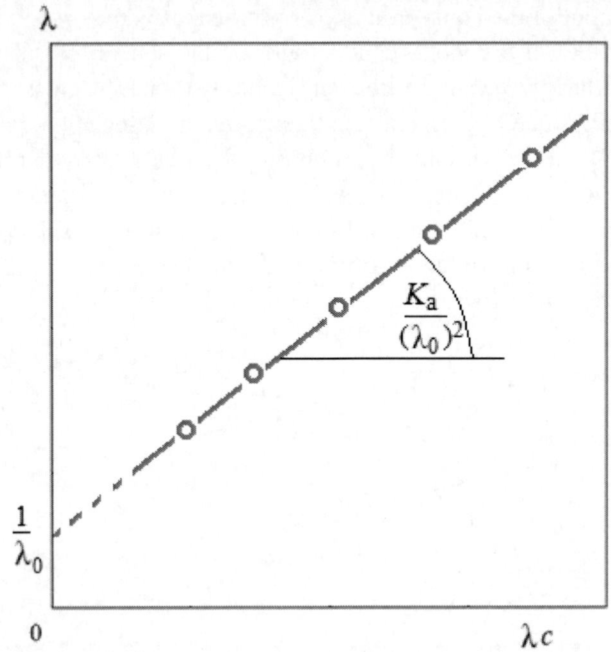

Figure 5. Dependence of the inverse molar conductivity $1/\lambda$ on the product λc used to determine the values of limiting molar conductivity λ_0 and association constants K_a of weak electrolytes.

The described procedure for determining the limiting molar conductivity λ_0 and association constants K_A of weak electrolytes have been improved [27, 28]. The dependence presented in Figure 5 is described by the Equation 9 [28]:

$$\frac{1}{\lambda\, S(Z)} = \frac{1}{\lambda_0} + \frac{\lambda c \gamma_{\pm}^2 K_a S(Z)}{(\lambda_0)^2}. \tag{9}$$

In the case of strong electrolyte solutions, various modifications of the Onsager equation obtained on the basis of the Debye-Hückel theory are used. For I-I electrolytes, and hence for solutions of ILs, this Equation 10 is written as [26]:

$$\lambda = \lambda_0 - (A + B\lambda_0)\sqrt{c}. \tag{10}$$

In this equation, A and B describe the contribution to the decrease in molar conductivity due to electrophoretic and relaxation effects. The coefficients A and B depend on temperature T, permittivity ε and viscosity η of the solvent [26]:

$$A = \frac{8.248 \cdot 10^{-4}}{\sqrt{\varepsilon T}\eta},\ B = \frac{8.204 \cdot 10^{5}}{\sqrt{(\varepsilon T)^3}}. \tag{11}$$

The Onsager equation is practically not used for the analysis of conductometric data. To determine the limiting molar conductivity λ_0 and association constants K_a, several equations were proposed in the second half of the last century that describe the dependence of molar conductivity on concentration. The description of these equations and the scope of their application for the calculation of the limiting molar conductivity λ_0 and association constants K_a are considered in the reviews [29-34]. It should be noted that the limiting molar conductivity values λ_0 of the strong electrolytes solutions obtained using different equations quantitatively agree with each other, which cannot be said about the values of association constants K_a, which can differ significantly [33]. At the same time, the degree of dissociation $\alpha = 0.90 \pm 0.04$ are typical for solutions of ILs in polar solvents; however, the patterns of changes in the association constants depending on the nature of cations and anions have not been fully elucidated [5].

Various electrical conductivity equations are currently used to interpret the results of conductometric measurements of dilute solutions of ILs in polar solvents [34-41]. A review of the works devoted to the study of the electrical conductivity and association of ILs in VOCs solutions has been provided [25]. Particular attention in this review was given to 1-ethyl-3-methylimidazolium tetrafluoroborate, $[C_2mim][BF_4]$, and 1-butyl-3-methylimidazolium tetrafluoroborate, $[C_4mim][BF_4]$. The solvent properties and association parameters of these ILs are shown in Table 3.

Table 3. Comparison of limiting molar conductivities, Λo, association constants, K_A, and Gibbs-free energies ΔG_A^o for solutions of $[C_2mim][BF_4]$ and $[C_4mim][BF_4]$ in various molecular solvents (dichloromethane, DCM; propanol, PrOH; methanol, MeOH; acetonitrile, ACN; dimethylformamide, DMF; and dimethylsulfoxide, DMSO) of viscosity, η, and relative permittivity, ε_r, at 298.15 K [25]

Solvent	Ionic liquid	η/m Pa s	ε_r	Λo/S cm^{2}mol^{-1}	K_A/ dm^3 mol^{-1}	ΔG_A^o /kJ mol^{-1}	Ref.
DCM	$[C_2mim][BF_4]$	0.415	8.82	181.07	595.110	-33.0	[42]
	$[C_4mim][BF_4]$	---	---	161.08	478.920	-32.4	[42]
PrOH	$[C_2mim][BF_4]$	1.957	20.45	31.43	943	-17.0	[43]
	$[C_4mim][BF_4]$	---	---	30.29	987	-17.1	[43]
MeOH	$[C_2mim][BF_4]$	0.5438	32.63	134.0	66.1	-10.4	[44]
	$[C_4mim][BF_4]$	---	---	121.84	37.7	-9.0	[45]
ACN	$[C_2mim][BF_4]$	0.3413	35.96	197.10	15.9	-6.9	[46]
	$[C_4mim][BF_4]$	---	---	189.29	15.7	-6.8	[47]
DMF	$[C_2mim][BF_4]$	0.8455	36.81	93.47	11.30	-6.0	[48]
	$[C_4mim][BF_4]$	---	---	89.83	10.57	-5.9	[48]
DMSO	$[C_4mim][BF_4]$	1.9946	46.52	41.51	4.6	-3.8	[45]
Water	$[C_2mim][BF_4]$	0.8903	78.36	114.27	0.89	---	[49]
	$[C_4mim][BF_4]$	---	---	107.39	0.16	---	[49]

It was shown that the magnitude of the electrical conductivity and the degree of association of the ILs depend on the properties of the solvent, such as its permittivity and viscosity [25]. The conductivity value and the degree of association of the ILs in VOCs are also affected by the structure of the solvent, in particular, its ability to form hydrogen bonds. In addition, ion transport is affected by the alkyl chain length in the imidazolium cation, the structure and size of the ILs anion, and, of course, temperature. As a result of the studies carried out in this work, a clear dependence of the association on DP ε_s of VOCs was established: $[C_2mim][BF_4]$ and $[C_4mim][BF_4]$ are practically not associated in water ($\varepsilon_s \sim 80$) and are weakly associated in acetonitrile, N,N-dimethylacetamide, N,N-dimethylformamide and dimethylsulfoxide ($30 \leq \varepsilon_s \leq 50$). As the DP of the solvent increases, the solvation of ions in the solution increases. Strong association occurs in dichloromethane, butan-1-ol, propan-2-ol and propan-1-ol ($\varepsilon_s \leq 20$).

Thus, strong solvation of the IL ions weakens their association, while weak solvation enhances it. It has also been shown that the ion association of the ILs in non-aqueous solvents increases with increasing temperature.

The question of the effect of temperature on the association of the ILs in aqueous solutions remains not fully elucidated. In the temperature range of 10–45°C, the association constant of $[C_2mim][BF_4]$ and $[C_4mim][BF_4]$ passes through a minimum, which is observed at 20–25°C temperatures [49]. In the temperature range of 0–40°C, the association constant of $[C_2mim][BF_4]$ decreases from 7.2 to 4.5 [24]. The mentioned above difference in the association constant values K_a, obtained using various equations of electrical conductivity should be considered. In particular, for $[C_4mim][BF_4]$ in methanol at 25°C [24] using Barthel's low-concentration chemical model [34] K_a=37.7 was obtained, and in [44] by the method [40, 41, 50] K_a=53.7. An even greater discrepancy for this IL occurs in water: 0.16 according to [25], Barthel's low-concentration chemical model [34] and 105 according to [51], method of [40, 41, 50]. In this regard, [52] noting the existence of numerous equations for calculating λ_0 and K_a, the authors express the opinion that comparing the K_a values for different ILs in polar solvents should be done using the same electrical conductivity equation.

Somewhat surprising is the overestimated accuracy of the values of the association constants (K_a) given in [25], obtained by the author himself: 595110 and 478920, respectively, for $[C_2mim][BF_4]$ and $[C_4mim][BF_4]$ in dichloromethane and Eyring activation enthalpy of charge transport ($\Delta H_\lambda^\ddagger$) – accurate to four to five significant figures for the same ILs in the solvents under consideration. A representation with an accuracy of three, and in exceptional cases – four digits is optimal for these values.

Dichloromethane has larger values of molar conductivities (Λ) for ionic liquids than dimethylformamide, water, and 1-propanol, which change very little with temperature. As a result, limiting molar conductivities, Λ_0, values are significantly greater than those seen in 1-propanol, dimethylformamide, and water. The results of the thermodynamic functions of association show that the association process is spontaneous. The positive values of enthalpy $\Delta H^o{}_A$ indicate that the ion-pairing process is endothermic, as the Gibbs-free energy is negative, entropic rather than enthalpic factors appear to dominate, and so ion pair production of ionic liquids in dichloromethane occurs spontaneously in both circumstances. The positive value of the change in the entropy of the association ($\Delta S^o{}_A$) and the fact of $T\Delta S^o{}_A > \Delta H^o{}_A$ should be noted. This happens not only in dichloromethane, but in other solvents too. Various findings indicate that the cations and anions are less associated in the solvent, which has a high relative permittivity due to ion solvation. Relative permittivity is the factor by which the electric field between the charges is decreased relative to vacuum. The strongest association is seen in dichloromethane, a low-permittivity solvent with low solvation strength for the ions [43, 46, 53].

Concentrated Solutions of Ionic Liquids

In ionic liquid solutions with molecular solvents, the general trend of electrical conductivity is an increase in conductivities compared to pure ionic liquids, but the increase in electrical conductivity is not linear and depends on the added amount of solvent in the solution. Since the ions of ILs in solutions are separated by solvent molecules, dilution of ionic liquids by solvents reduces medium viscosity and boosts ionic conductivity. This is because of the number of free ions (charges) available in the solutions, which start aggregating after a certain limit of rise in conductivity. At high ionic concentrations, however, all solvent molecules may be engaged in the solvation shell, resulting in a system that is more similar to pure ionic liquids than solutions. With rising concentrations of ionic liquids, ionic conductivity levels approach a limit, after which they reduce as a general trend is shown in Figure 6. The specific electrical conductivity κ reaches its maximum value at a concentration *c* whose value depends on the nature of the ionic liquid and the polar solvent. This decrease is due to the lack of solvent molecules for solvation and the ongoing association of ions not separated by solvent molecules. In contrast, after the maximum for the same reason (lack of solvent for solvation of ions), the viscosity increases sharply.

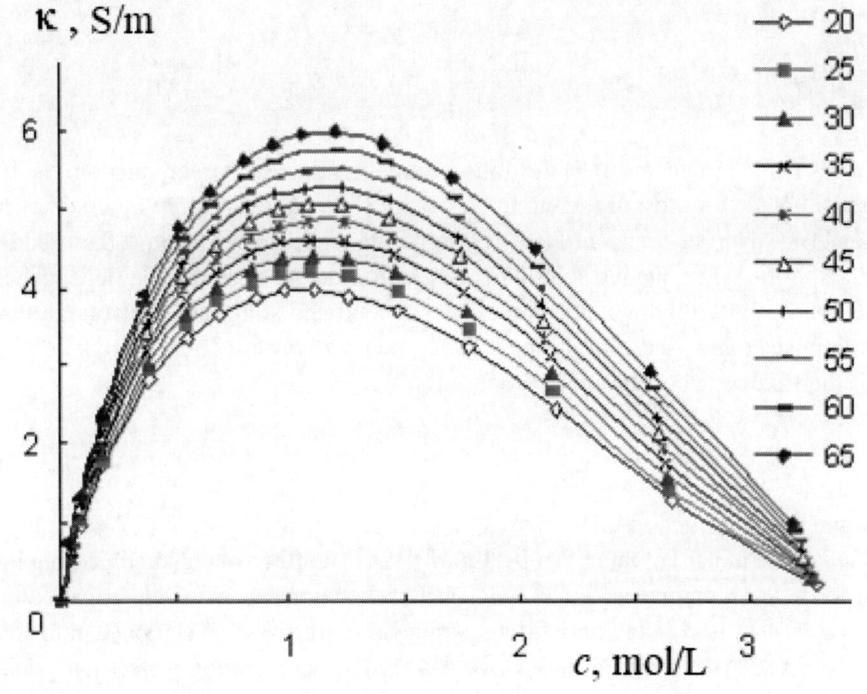

Figure 6. Electrical conductivity (*κ*), corresponding to molar confraction (*x*) of 1-butyl-3-methylpyridinium bis{(trifluoromethyl)sulfonyl}amide in acetonitrile; temperature values (°C) are shown in the graph [54], reprinted with permission from [54].

The dependence of the specific electrical conductivity of the IL solutions in polar solvents is similar to this dependence in electrolyte solutions. The position of the maximum electrical conductivity on the concentration axis in electrolyte solutions depends on the degree of solvation of ions, permittivity of the solvent, and temperature. It is believed that an increase in the solvation of ions, an increase in the dielectric constant of the solvent, and a decrease in temperature lead to a shift in the maximum electrical conductivity of electrolyte solutions towards less concentrated solutions [33]. A similar dependence of the position of the maximum on the concentration axis on solvation, solvent permittivity, and temperature should also be expected for solutions of ILs in polar solvents.

To describe the dependence of the electrical conductivity κ of concentrated electrolyte solutions on the molality m (mol/kg solvent), the Casteel-Amis equation [55] is widely used:

$$\frac{\kappa}{\kappa_{max}} = (\frac{m}{m_{max}})^a \exp\left[b(m - m_{max})^2 - a\frac{m - m_{max}}{m_{max}}\right]. \tag{12}$$

In this equation, m_{max} is the molality corresponding to the maximum electrical conductivity κ_{max} at a given temperature, a and b are fitting coefficients (shape parameters).

The maximum electrical conductivity of the ILs in VOCs is observed at x_{ILs} 0.1 – 0.25 mole fractions of the ILs [56]. In this work, the authors used the Casteel-Amis equation, in which the molality was replaced by the mole fraction of x_{ILs}:

$$\frac{\kappa}{\kappa_{max}} = (\frac{x}{x_{max}})^a \exp\left[b(x - x_{max})^2 - a\frac{x - x_{max}}{x_{max}}\right]. \tag{13}$$

As a result, it turned out that the four-parametric Casteel-Amis equation "is excellent, yielding $r^2 > 0.999$" for most of the mixtures studied [56]. An even better approximation of the conductometric data of concentrated IL solutions is obtained using the Padé (n, m) approximation [56]. In aqueous solutions of ILs, in order to describe the dependence of the electrical conductivity on the concentration of concentrated solutions, the volume fraction of ILs was proposed to be used [57]. Using a lattice-hole theory for conductivity κ in ionic liquid mixtures, the following equation was proposed [57]:

$$\kappa = (1 - \varphi)\, \varphi\, \kappa_0 + \varphi\, \kappa_{IL} \tag{14}$$

in which the volume fraction of IL $\varphi \approx c_{IL}V_{m,IL}$ (c_{IL} is the concentration of an IL in solution, mol/L, $V_{m,IL}$ is the molar volume of an IL, L/mol), κ_0 is an effective conductivity that implicitly accounts for hydrodynamic and ionic atmosphere relaxation, κ_{IL} is the electrical conductivity of a pure IL. κ_0 was used as the only fitting parameter in this work [57]. For [C_2mim][BF_4], for example, $\kappa_0 = 32.34$, $\kappa_{IL}=1.553$ S/m, $V_{m,IL}=0.154$ L/mol. The maximum electrical conductivity in aqueous solutions of ILs occurs at a volume fraction of IL $\varphi \approx 0.5$ [57].

To calculate the electrical conductivity values of the IL concentrated solutions in a wide range of concentrations and temperatures, one can use the parameters of the Casteel-Amis equation in the form κ/κ_{max} and c_{IL}/c_{max}. At the same time, in the $\kappa/\kappa_{max} - c_{IL}/c_{max}$ coordinates, the κ/κ_{max} values of strong concentrated solutions [58] and weak [59] electrolytes, as well as IL solutions in polar solvents [60]. The generalized $\kappa/\kappa_{max} - c_{IL}/c_{max}$ relationships shown in Figure 19, together with the other three ILs, are described in Table 8 below (Part 6). The advantage of presenting the electrical conductivity values of the IL solutions is that there is the possibility of estimating their κ values based on the values of κ_{max} and c_{max} over a wide range of concentrations and temperatures.

Some additional information on the temperature and concentration dependences of the electrical conductivity of ILs and their solutions is provided in the subsequent sections of this chapter.

Based on the electrical conductivity consideration of dilute and concentrated IL solutions in polar solvents described above, the following can be deducted:

- The molar conductivity for most solutions of ILs in polar solvents decreases in proportion to the square root of the concentration (Figure 4a), the association constants K_a in dilute solutions of ILs are not large; these two factors indicate that ILs behave like strong electrolytes in dilute solutions;
- The association of ILs depends on the solvation nature of its ions, the permittivity of the solvent, and temperature; strong solvation reduces the association of ions; with an increase in the permittivity of the solvent, the association also decreases; the association constants K_a of ILs in polar solvents increase with temperature, which indicates an increase in the association of IL ions upon heating;
- To determine the thermodynamic characteristics of ion association, various equations of the molar electrical conductivity concentration dependence are currently used, which can give inconsistent values of the association constants K_a. To compare the K_a values of different ILs in polar solvents of different nature, it is advisable to use the data obtained using the same electrical conductivity equation. This is to say that as the concentration c increases, the specific electrical conductivity κ of the concentrated solutions of the ILs in polar solvents passes through a maximum. Taking the case of solutions of strong electrolytes for example, the existence of a maximum in the $\kappa - c$ dependences is due to an increase in the association of the IL ions in concentrated solutions and an increase in their viscosity. The maximum electrical conductivity κ_{max} at a given temperature and the corresponding concentration c_{max} can be used to generalize the concentration and electrical conductivity temperature dependences of concentrated solutions of ILs in polar solvents.

Dielectric Properties of Ionic Liquids

The most important characteristics of ionic liquids (ILs) are the static dielectric permittivity (DP) ε_s and the dielectric relaxation (DR) time τ. These determine the processes enabled by the ionic liquids such as the dissociation into ions, the solvatation, as well as the mobility of the ions and molecules. The dissociation into ions is closely related to the structural characteristics of the liquid.

Aprotic ILs are characterized by relatively low (10-20) ε_s values [5, 61, 62]. The high DP values of protic ILs [63-67], as well as polar solvents such as water, formamide and N-methylformamide [68, 69] (ε_s of formamide and N-methylformamide are 110 and 190, respectively) are caused by a rather strong and branched network of hydrogen bonds formed in the liquid.

If the DP is a structural characteristic of ILs, then DR τ can be considered their kinetic characteristic. The DR time characterizes the mobility of the solvent molecules, in particular, the rate of dielectric relaxation V ($V \sim 1/\tau$) in an alternating electromagnetic field. The DR time increases with increasing viscosity of the liquid and with the growth of the size of its molecules. Sufficiently viscous ILs are characterized by relatively high DR times (tens and hundreds of ps) [64, 65, 67, 70]. The DR time of acetone and acetonitrile at room temperature is about 3 ps, and water is about 8.3 ps (25°C) [68, 69].

To describe the transport properties of the ILs and electrolyte solutions in polar solvents, it is useful to consider a third κ_∞ value combining the structural and kinetic characteristics - limiting high-frequency (HF) electrical conductivity (EC). It is the ratio of absolute DP ($\varepsilon_s\varepsilon_0$) to the

DR time (τ), Equation (15):

$$\kappa_\infty = \frac{\varepsilon_s\varepsilon_0}{\tau} \tag{15}$$

In this expression, ε_s is the static dielectric permittivity, and ε_0 is the absolute DP of the vacuum ($\varepsilon_0 = 8.854 \cdot 10^{-12}$ F/m).

The value of κ_∞ is an important parameter of ILs and polar solvents, as it defines the specific electrical conductivity and its temperature dependence [60]. The maximum κ_∞ value occurs for solvents with high DP (water, formamide) and low DR times (acetone, acetonitrile) [60]. Therefore, it is no coincidence that aqueous solutions as well as solutions in acetone and acetonitrile and mixtures thereof are characterized by maximum electrical conductivity [10, 25, 71]. The physical meaning of the κ_∞ value is discussed below.

Due to the particular importance of the dielectric characteristics of ILs' influence on the processes carried out in them, increased attention has been paid to these characteristics. Several dozen papers have been published in literature over the years, which provide the results of theoretical and experimental studies on the dielectric characteristics of ILs. An analysis of the results of these studies was provided [61-64].

Unlike conventional polar organic compounds, ILs are both a solvent and an electrolyte. Therefore, as with solutions of electrolytes in polar solvents, the experimental determination of the dielectric characteristics of ILs and their interpretation are associated with certain difficulties caused by direct current (DC) electrical conductivity. As a result, the microwave frequency domain was required to measure the DP of ILs.

Measurements of dielectric characteristics of conductive solutions have been actively carried out since the middle of the last century [72, 73], when the results of the measurements of ε_s and τ of electrolyte solutions in polar solvents in the centimeter and millimeter wavelength range appeared in the literature [68].

These measurements were carried out at fixed frequencies. Since then, the measuring equipment has changed significantly and allows almost the entire dielectric spectrum of ILs to be obtained, starting from several MHz and up to tens and hundreds of THz, that is, almost to the optical wavelength range [61-64].

Given the importance of the DC electrical conductivity influence on the results of the dielectric measurements and their interpretation, modern thinking on the dependence of the measured electrical conductivity and dielectric properties on alternating current (AC) frequency will be considered in this chapter.

Relaxation Processes in Ionic Liquids

Relaxation of Electrical Conductivity (Ionic Relaxation) in Ionic Liquids

To estimate the region of electromagnetic field frequencies in which it is possible to measure the DP of ILs, an electrical equivalent circuit of ILs (Figure 7a) in the form of a parallel connected electrical resistance *R* and electric capacitance *C* is considered [74].

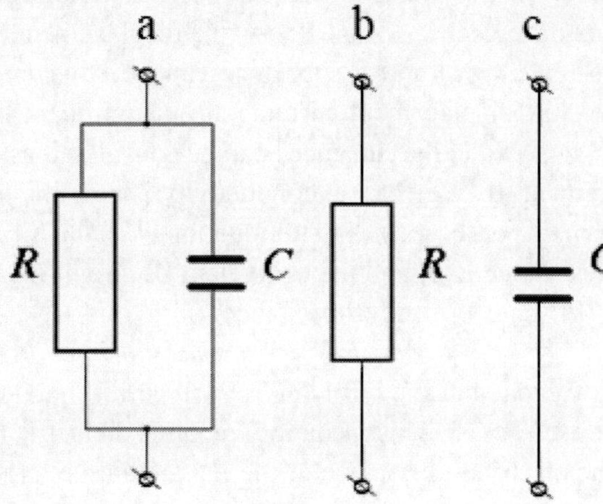

Figure 7. Representation of the electric properties of ILs in the form of a resistance *R* and capacitance *C* connected in parallel over a wide frequency range (a), as well as under the conditions $\omega CR \ll 1$ (b) and $\omega CR \gg 1$ (c).

In this scheme, the electrical resistance *R* characterizes the conducting IL properties ($\kappa \sim 1/R$), and electrical capacity *C* characterizes dielectric permittivity ($\varepsilon_s \sim C$).

When an alternating electric field of the angular frequency ω ($\omega = 2\pi F$, *F* is linear frequency) is applied, the resistance of the *RC* circuit (Figure 7a) becomes a complex value. The complex resistance (impedance) of the *Z* chain consists of its active (real) R_a and reactive (imaginary) X_{im} components, Equation (16):

$$Z = R_a + iX_{im} \tag{16}$$

where $i = \sqrt{-1}$. In this case, the dependence of R_a and X_{im} on frequency is described by the Equations (17) and (18):

$$R_a = \frac{R}{1 + (\omega CR)^2} \tag{17}$$

$$X_{im} = \frac{1}{\omega C} \cdot \frac{(\omega CR)^2}{1 + (\omega CR)^2}. \tag{18}$$

The value of ωCR is the ratio of the displacement currents to conduction currents, Equation (19) [74]:

$$\omega CR = \frac{\omega \varepsilon_s \varepsilon_o}{\kappa} \tag{19}$$

The numerical value of the product ωCR can serve as a criterion for assigning a substance to the class of conductors or dielectrics. If $\omega CR >> 1$, the displacement currents greatly exceed the conduction currents ($\omega \varepsilon_s \varepsilon_o >> \kappa$) and a substance can be considered dielectric. Under a condition $\omega CR << 1$, the shift displacement currents are insignificant in comparison with the conduction currents ($\omega \varepsilon_s \varepsilon_o << \kappa$) and the substance can be considered as a conductor.

The correct measurement of IL electrical conductivity is possible at low frequencies, i.e., provided that $\omega CR << 1$. In this case, according to Equation (17), the measured active resistance R_a will be equal to the resistance IL R and the equivalent IL circuit (Figure 7a) is transformed into a circuit (Figure 7b). The electric capacitance C in this case does not affect the R measurement results.

The measured electric capacitance $C_{mes} = 1/(\omega X_{im})$ will correspond to the dielectric constant of ILs under the condition $\omega CR >> 1$. This inequality requires dielectric measurements of ILs at high frequencies, and the electrical resistance of the ILs should be large enough. In this case, the equivalent IL circuit (Figure 7a) is converted to the circuit, Figure 7c.

Since the Equations (17) and (18) for the active R_a and reactive X_{im} components of the complex resistance Z include the value of $(\omega CR)^2$, the inequalities $\omega CR << 1$ and $\omega CR >> 1$ are fulfilled, respectively, under the condition of $\omega CR < 0.1$ and $\omega CR > 10$. Thus, when the frequency of the electromagnetic field ω is increased in the frequency region in which the inequality (20) is fulfilled:

$$0.1 \le \omega CR \le 10 \tag{20}$$

There is a decrease in the measured values of electrical resistance R_a and X_{im}, despite the fact that the real values of resistance R and electric capacitance C (Figure 7a) remain constant.

Since the ωCR is the ratio of displacement currents to conduction currents, as shown in Equation (19), there is an apparent relaxation process (electrical or ionic relaxation) in the inequality (20) frequency interval, which must be considered when interpreting the measurement results of the dielectric IL characteristics.

In a real experiment, an IL is located in a contact or non-contact measuring cell. Therefore, the electrical circuit (Figure 7a) is complicated as a result of the addition of an electric capacity with a double layer of contact cell or capacitance in the walls of the contactless measuring cell [74]. At low frequencies, in the case of using contact measuring cells, the measurement results are influenced by the polarization electric capacitance C_p and resistance R_p [75] As a result, the measured IL resistance R_a exceeds its true value R. Circular diagrams of the electrical relaxation in the absence of and in the presence of the polarization effects are shown in Figure 8.

In order to eliminate the effect of the polarization on the electrodes during precision conductometric measurements, the measured resistance R_a is extrapolated to an infinite frequency in the $R_a - 1/F$ coordinates [75].

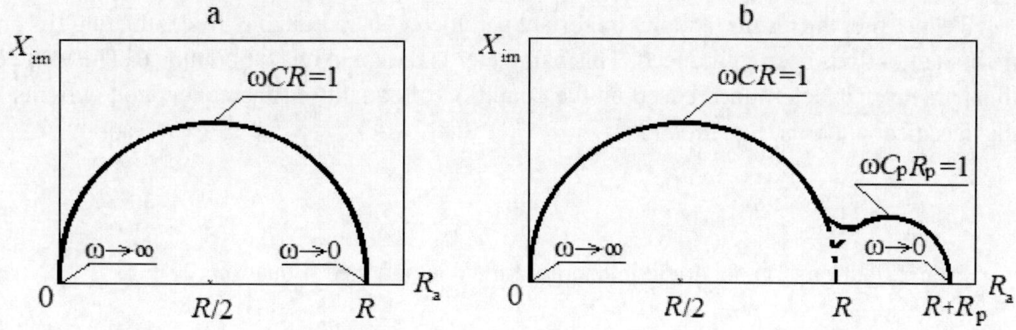

Figure 8. Pie diagrams of the electrical relaxation in the absence of polarization effects (a) and in the presence of polarization effects (b).

An analogous situation occurs in electrolyte solutions over a wide frequency range. The dependence of the measured resistance and capacitance of the conductive solution on frequency has been discussed in literature [60, 74]. As a result, it was proposed to interpret the changes in resistance and capacitance considered above as an apparent (fictitious) electrical relaxation process (ionic relaxation). Ionic relaxation does not lead to a change in the real electrical conductivity and permittivity of the solution under study. Ionic relaxation is considered when performing precision conductometric measurements by extrapolating the inverse measured resistance $1/R$ to zero frequency at the coordinates $1/R_a - F^2$ [60].

In the middle of the last century, the results of dielectric measurements of electrolyte solutions were obtained, which were used as a confirmation of the Debye–Falkenhagen (DF) effect (increase in permittivity with increasing concentration) [76]. In fact, the ionic relaxation discussed above was not considered when processing the results of these measurements. At the same time, the measured apparent (effective) DP of the various electrolyte solutions increased directly in proportion to the square of their specific conductivity [60].

It should also be noted that for dilute aqueous electrolyte solutions, the ionic relaxation time coincides with the relaxation time of the ionic atmosphere in the Debye-Hückel theory [60]. Naturally, the question arises as to whether there is a dispersion of electroconductivity and permittivity of electrolyte solutions predicted theoretically by Debye and Falkenhagen [77] or whether it is a simple redistribution of the conduction currents and displacement currents in the scheme shown in Figure 7a. This issue remains unresolved for ILs as well.

In dilute electrolyte solutions, the DF effect is very small. Possibly the dispersion of EC and DP occurs in the water-dioxane and methanol-toluene mixtures. Both systems exhibit an impedance dispersion fully characterized by volumetric permittivity and volumetric conductivity [77]. The EC dispersion of polar solvents will be discussed in the next section.

Ionic relaxation should also be considered when performing dielectric measurements of ILs and interpreting the results. To do this, for each IL, it is necessary to preset the frequency region defined by the inequality (20) in which it occurs. To this end, an estimate of the frequency F_0 ($F_0=\omega_0/2\pi$) at which the conduction currents are equal to the displacement currents ($\omega CR=1$ или $\omega\varepsilon\varepsilon_0=\kappa$) for some protic and aprotic liquids has been made, and the values obtained have been compared with the values for the diluted and concentrated aqueous KCl solutions.

To interpret the results, another important parameter characterizing the conducting ILs and their solutions has been introduced. This parameter is the ionic relaxation time θ. The value of this parameter is determined based on the condition of equality of the conduction currents to displacement currents, Equation (21):

$$\omega_o CR = \omega_o\theta = \omega_o\varepsilon_s\varepsilon_o/\kappa = 1 \qquad (21)$$

From Equation (21), the ionic relaxation time θ is obtained, Equation (22):

$$\theta = \varepsilon_s\varepsilon_o/\kappa \qquad (22)$$

The frequency F_o, at which the conduction currents ILs are equal to the displacement currents, is also obtained from Equation (21), considering that $\omega_o=2\pi F_o$, Equation (23):

$$F_o = \kappa/(2\pi\varepsilon_s\varepsilon_o)=17.975\ \kappa/\varepsilon_s \qquad (23)$$

By way of example, for some ILs, Table 4 shows the specific electrical conductivity (κ), dielectric constant (ε), times of dipole τ, ionic relaxation θ and frequency F_o corresponding to the equality of the conduction current to displacement currents, and the limiting high-frequency electrical conductivity κ_∞. For comparison, Table 4 also shows similar values for aqueous KCl solutions in the concentration range of 10^{-3} to 1 mol/l.

Table 4. DC conductivity κ, static dielectric permittivity ε_s, dipole τ and ion θ relaxation times, frequency F_o, corresponding to the equality of conduction current to displacement currents and limiting high-frequency electrical conductivity κ_∞ of methylammonium formate (MAF), ethylammonium formate (EAF), butylammonium formate (BAF), ethylammonium nitrate (EAN), 2-hydroxyethylammonium formate (2-HEAF), 2-hydroxyethylammonium propanoate (2-HEAP), 2-hydroxyethylammonium butanoate (2-HEAB) and also of some aqueous KCl solutions; t = 25°C

ILs, c_{KCl}, M	κ, S/m	ε_s	τ, ps	θ, ps	F_o, GHz	κ_∞, S/m
MAF	4.38[a]	40.3[a]	86[a]	81.5	1.95	4.15
EAF	1.216[a]	30.3[a]	392[a]	221	0.721	0.68
EAF	1.361[b]	34.14[b]	582[b]	222	0.717	0.519
BAF	0.31[a]	29.2[a]	940[a]	834	0.191	0.27
BAF	0.2161[b]	24.77[b]	2620[b]	1015	0.157	0.0837
EAN	2.24[c]	27.3[c]	185[c]	108	1.47	1.30
2-HEAF	0.30[d]	47.1-61[d]	1100[d]	1390[d]	0.114-0.088	0.38-0.29
2-HEAP	0.06[d]	19.1[d]	2200[d]	2818[d]	0.056	0.077
2-HEAB	0.04[d]	15.3[d]	1900[d]	3387[d]	0.047-0.021	0.07
0.001 M KCl	0.0147	78.3[e]	8.3[e]	47200	0.00337	83.5
0.01 M KCl	0.141	78.3[e]	8.3[e]	4917	0.0324	83.5
0.1 M KCl	1.29	77.3[e]	8.2[e]	531	0.300	83.5
1 M KCl	11.8	70.5[e]	7.6[e]	52.9	3.01	77.7

[a][64]; [b][67]; [c][65]; [d][70]; [e][68]

The first thing to note by comparing the values given in Table 4 is the significant discrepancy between the τ values obtained in the results for EAF [64] and BAF [67]. With EAF, for example, these values differ by a magnitude of 1.5, for BAF they differ by 2.8. The differences noted are primarily related to the interfering effect of the DC conductivity in determining the dielectric characteristics. This is discussed in the next section in more detail. At the same time, it should be noted that the proximity of the the DC conductivity κ values and the limiting high-frequency electrical conductivity κ_∞, as well as the times of dipole τ and ionic relaxation θ of MAF, BAF, 2-HEAF and 2-HEAP respectively. In the case of MAF, for example, the discrepancy of these values does not exceed 5-6%. The values of EC κ and κ_∞ are considered, as well as the times τ и θ for EAF and 2-HEAB, which differ by almost 1.8 times. The reason for this difference may also be an error in the determination of the τ values, as indicated by the authors [64, 70].

In aqueous electrolyte solutions at higher concentrations, the values of EC κ and κ_∞, as well as the relaxation times τ and θ, also converge [60]. With the transition from a 0.001 M KCl solution to a 1 M solution of this electrolyte, the ionic relaxation time θ decreases by almost 900 times (Table 4) and approaches the dipole dielectric relaxation time τ of water (8.3 ps) [78].

As the concentration of the aqueous electrolyte solution increases, its DC EC also approaches the HF EC of water (83.5 Sm/m at 25°C). The maximum DC EC is observed for concentrated solutions of strong inorganic acids (HCl, HNO_3, H_2SO_4). At 25°C, it does not exceed 84 S/m [79], so it practically coincides at this temperature with the κ_∞ value of water [60].

In ILs, as in the case of concentrated aqueous electrolyte solutions [80], ionic and dipole relaxation can be combined into a unified ion-dipole relaxation process. In this case, the time of ion-dipole relaxation ($\tau_{i\text{-}d}$) is determined on the basis of the τ_{ILs} and θ values using Equation (24):

$$\frac{1}{\tau_{i\text{-}d}} = \frac{1}{\tau_{ILs}} + \frac{1}{\theta}; \quad \tau_{i\text{-}d} = \frac{\tau_{ILs} \cdot \theta}{\tau_{ILs} + \theta}. \tag{24}$$

Equation (24) predicts a reduction in the dielectric relaxation time in aqueous electrolyte solutions with an increase in their concentration, which has previously been observed experimentally [68]. Using this equation, based on the dielectric relaxation time of water (8.3 ps) and the ionic relaxation times of the θ solutions (Table 4), it is possible to estimate the values of $\tau_{i\text{-}d}$ for aqueous KCl solutions. For a 1M aqueous KCl solution, for example, using the values τ of the water (8.3 ps) and θ of the solution (52.9 ps), $\tau_{i\text{-}d}$ is equal to $8.3 \cdot 52.9/(8.3+52.9) = 7.2$ ps. The obtained value is close to the τ for a 1M KCl solution (7.6 ps, Table 4).

Based on the above, the summary of this part can be drawn:

- The specific electrical conductivity (EC) κ and static dielectric permittivity (DP) ε_s of ILs define the relationship of the conductivity currents to the displacement currents in an alternating electromagnetic field; provided these currents are equal in ILs, an electrical (ionic) relaxation occurs, the time of which θ is equal to the ratio of absolute DP ($\varepsilon_s\varepsilon_o$) to the specific EC κ, Equation (21);

- As a result of ionic relaxation, the measured equivalent active resistance of the ILs R_a becomes less than the desired resistance R, which is perceived as an apparent increase in the electrical conductivity with an increase in the frequency of the electromagnetic field ω; similarly, the reactance of the ILs $X_{im}=1/(\omega C'_{mes})$ decreases with increasing frequency, which is perceived as an apparent increase in the dielectric constant of ILs with increasing frequency; in fact, the specific EC κ and DP ε_s of the ILs retain their constant value in the region of frequencies corresponding to ionic relaxation, which is independent of the frequency of the electromagnetic field;
- Similar ionic relaxation is observed in dilute aqueous solutions of electrolytes, which is considered in the theory of these solutions as relaxation of ionic atmospheres; wherein, the ionic relaxation time of the diluted electrolyte solution is substantially shorter than the dipole dielectric relaxation time of water; in concentrated aqueous electrolyte solutions, as well as in ILs, the ionic relaxation time θ becomes commensurate with the dipole dielectric relaxation time τ of the polar liquid, Table 4;
- The DC conductivity of the ILs is closely related to their dielectric characteristics, in particular, to the ratio of the absolute DP ($\varepsilon_s\varepsilon_o$) to the time DR ($\tau$), which is equal to the limiting HF EC of the ILs, Equation (15); on the basis of the values specified in Table 4, it is possible to suggest that the DC conductivity κ is close to the limiting HF EC κ_∞, and the time of ionic relaxation θ is close to the time of the dipolar relaxation τ in ILs.

The above analysis is valid provided that the electrical conductivity and dielectric constant of ILs are independent of the frequency of the electromagnetic field. This independence occurs in the low frequency domain. Therefore, it is necessary to take this relationship into account and consider the phenomena of permittivity dispersion and the high-frequency conductivity dispersion of ILs.

Dispersion of the Dielectric Permittivity of Ionic Liquids

Despite a thorough analysis of the effect of ionic conductivity on the results of ILs' dielectric measurements [62, 63], this analysis requires some additions, taking into account ionic and dipole relaxations, Equation (23), as well as the HF EC dispersion of the ILs themselves. We will carry out this analysis considering our approach on the unity of the electrical relaxation processes in conducting liquids and their solutions. The dependence of the electromagnetic properties of the ILs, as well as organic solvents, on the frequency of the electromagnetic field can be described both in terms of the complex dielectric constant ($\varepsilon^* = \varepsilon' - i\varepsilon''$) and in terms of the complex electrical conductivity ($\kappa^* = \kappa' + i\kappa''$) [81].

In general, the relationship between the values of κ^* and ε^* and their frequency dependence is described by Maxwell's equations and the Kramers-Kronig relations [82].

In the dielectric constant dispersion region for complex DP and EC, the following ratios can be written, Equations (25) and (26):

$$\varepsilon^* = \kappa^*/(i\omega\varepsilon_o) = \kappa''/(\omega\varepsilon_o) + \kappa'/(i\omega\varepsilon_o) = \varepsilon' - i\,\varepsilon'' \tag{25}$$

$$\kappa^* = i\omega\varepsilon_o\varepsilon^* = \omega\varepsilon_o\varepsilon'' + i\omega\varepsilon_o\varepsilon' = \kappa' + i\,\kappa'' \tag{26}$$

Equations (25) and (26) are two equivalent methods for describing the electromagnetic properties of a substance. For polar solvents and ILs, it is customary to consider the electromagnetic properties in terms of the complex DP [61, 63-65, 69]. In the simplest case, the dependence of the complex DP on frequency is described by the Debye equation [83]:

$$\varepsilon' = \varepsilon_\infty + \frac{\varepsilon_s - \varepsilon_\infty}{1 + i\omega\tau} \tag{27}$$

By separating the active and reactive components, for the active ε′, Equation (28), and a reactive ε″, Equation (29), components of the complex DP equation are derived:

$$\varepsilon' = \varepsilon_\infty + \frac{\varepsilon_s - \varepsilon_\infty}{1 + (\omega\tau)^2} \tag{28}$$

$$\varepsilon'' = \frac{\varepsilon_s - \varepsilon_\infty}{1 + (\omega\tau)^2}(\omega\tau) \tag{29}$$

which are obtained by using an equivalent electrical circuit of the dielectric relaxation process, Figure 9a.

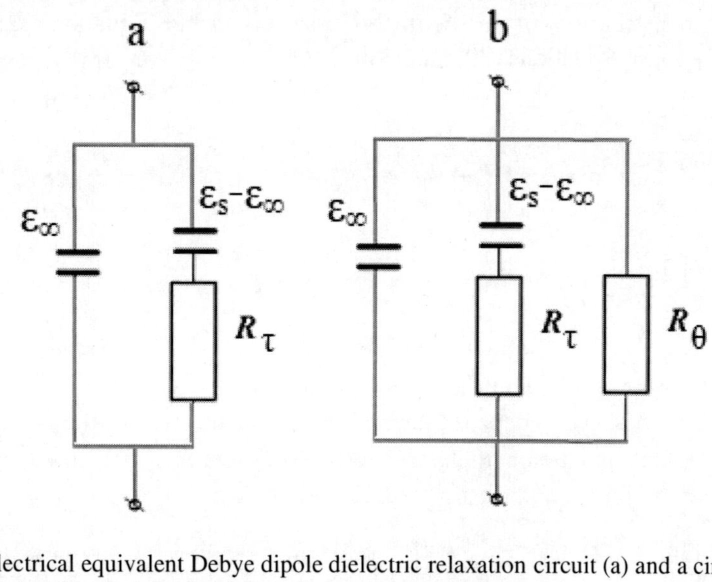

Figure 9. An electrical equivalent Debye dipole dielectric relaxation circuit (a) and a circuit supplemented by an electrical conductivity relaxation process (b).

The dependences ε' and ε" from frequency for a model IL for which ε_s =15.0, ε_∞ =2.0, τ=100 ps, and DC conductivity is equal to zero are given in Figure 10.

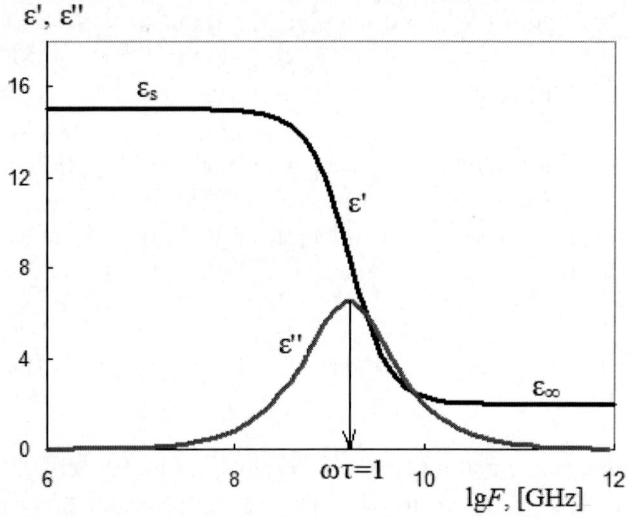

Figure 10. The frequency dependence of an active ε' and reactive ε" component of a complex DP for a model IL (ε_s=15.0, ε_∞=2.0, τ=100 ps, and κ_{DC}=0).

The DP of an IL remains constant at lower frequencies of the electromagnetic field (ωτ<<1) which is called the static dielectric permittivity ε_s. In the 0.1 ≤ ωτ ≤ 10 frequency region (DP dispersion region), the dipolar IL molecules do not have time to follow the change in the external field. In this case, the active component of the complex DP of an IL ε' decreases, reaching a value of ε_∞, called the optical component of the complex DP, in the limit where ωτ>>1. The reactive component of the complex DP (dielectric loss coefficient) ε" with an increase in frequency *F* passes through the maximum that occurs under the condition of ωτ=1, Figure 10. The condition of ωτ=1 can be considered as the equality of the conduction currents to the displacement currents of an IL. In this case, the Debye Equations (28) and (29) are converted to the form, Equations (30) and (31):

$$\varepsilon'_{(\omega\tau=1)} = \frac{\varepsilon_s + \varepsilon_\infty}{2} \tag{30}$$

$$\varepsilon''_{(\omega\tau=1)} = \frac{\varepsilon_s - \varepsilon_\infty}{2} \tag{31}$$

For the model IL (ε_s =15.0, ε_∞ =2.0) under consideration, using Equations (30) and (31), the following values are obtained: $\varepsilon'_{(\omega\tau=1)}$=8.5 and $\varepsilon''_{(\omega\tau=1)}$=6.5. When analyzing the frequency (dielectric relaxation spectra) dependencies ε′ and ε″ of the ILs by the frequency value F_0, corresponding to the maximum dielectric loss coefficient ε″, the numerical value of the dielectric relaxation time τ can be estimated, Equation (32):

$$\tau = \frac{1}{\omega} = \frac{0.159}{F_0} \tag{32}$$

In this equation the frequency is expressed in Hz.

The maximum ε″ of the model IL occurs at the frequency F_0=1.59 GHz. Using Equation (32), we obtain the relaxation time: $\tau = 0.159/(1.59\cdot10^9) = 1.0\cdot10^{-10}$ s = 100 ps. This method of estimating the dielectric relaxation time may be useful where there are difficulties in analyzing the dielectric relaxation spectra due to the DC electrical conductivity of ILs, which is assumed to be zero for the model IL in question.

Strictly speaking, Debye's theory is true for molecules in the form of a ball that do not interact with each other and, of course, do not form intermolecular chemical bonds. However, Debye Equations (27), (28) and (29) are more widely used when considering the dielectric relaxation processes of water and other polar solvents with a developed system of hydrogen bonds, as well as ionic liquids. The relationship of the dipole relaxation time τ with the viscosity of the liquid η is subject to the Stokes-Einstein-Debye equation (SED) [84], Equation (33):

$$\tau = \frac{4\pi a^3}{3}\cdot\frac{\eta}{k_B T} \tag{33}$$

in which a is the radius of the molecule, k_B is the Boltzmann constant and T is the absolute temperature. The first term in Equation (33) is the effective volume of a molecule (in this case, an IL molecule).

The analysis of the dielectric relaxation time τ dependence on the viscosity η and temperature T ($\tau=f(\eta/T)$) confirms the validity of SED equation for various protic (water, alcohols) and aprotic (acetonitrile, acetone, DMF and DMSO) solvents. It is important to emphasize that determining the time of the dipole relaxation of ILs is associated with significant methodological difficulties and the fact that with a change in temperature, the τ value increases directly in proportion to the ratio η/T and allows estimation the DR time of solvents and ILs at various temperatures. For water, i.e., up to a critical temperature, the relaxation time is estimated by the equation $\tau = (2.77 \pm 0.09)10^{-9}\eta/T$ [60].

The Debye Equations (27) and (28) can be combined into equation of a circle, Equation (34):

$$(\varepsilon' - \frac{\varepsilon_s - \varepsilon_\infty}{2})^2 + (\varepsilon'')^2 = (\frac{\varepsilon_s + \varepsilon_\infty}{2})^2 \tag{34}$$

It has been proposed to use the dependencies $\varepsilon''=f(\varepsilon')$ in the analysis of dielectric relaxation processes in polar solvents and in their solutions. These dependencies are called the Cole-Cole diagrams [85]. They are semi-circular arcs centered on the abscissa axis at the point $(\varepsilon_s - \varepsilon_\infty)/2$, Figure 11a. For solutions of electrolytes and conducting ILs, these dependencies make it relatively easy to determine the value of static DP by extrapolating an arc to zero frequency in the $\varepsilon''=f(\varepsilon')$ coordinates. When conducting IL studies in a wide temperature range, it is convenient to use the generalized coordinates $\varepsilon''/(\varepsilon_s - \varepsilon_\infty) - (\varepsilon' - \varepsilon_\infty)/(\varepsilon_s - \varepsilon_\infty)$, Figure 11b. In this case, the experimental values fit on a single curve in the entire investigated temperature range, Figure 11.

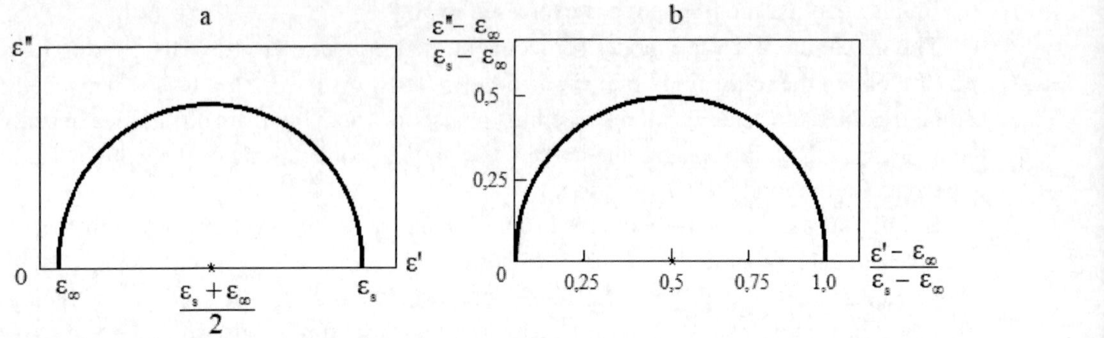

Figure 11. Pie diagrams of dipole relaxation of ILs in the ε″ - ε′ coordinates (a) and in generalized coordinates (b).

The Debye Equations (27), (28), and (29) are valid for systems with a single relaxation time. The studies carried out in the middle of the last century on the dielectric relaxation processes in polar solvents and electrolyte solutions in these solvents showed that for some polar solvents and their solutions there is a distribution of relaxation times when the center of the semi-circle is below the abscissa axis, Figure 12a, or the dependence ε"=*f*(ε') is a beveled arc, Figure 12b.

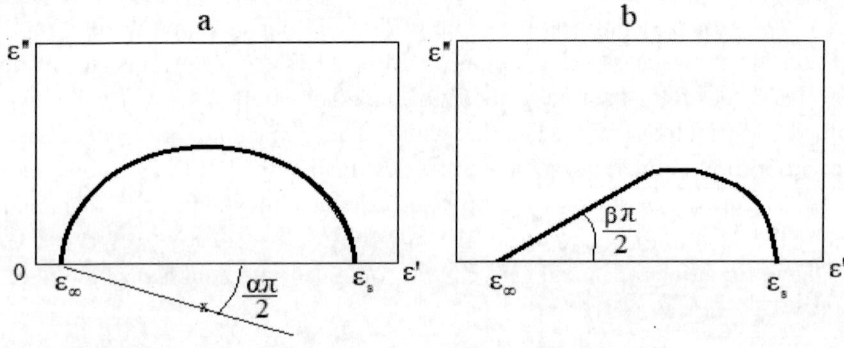

Figure 12. Pie chart with relaxation time distribution by the Cole-Cole model (a) and the Cole-Davidson model (b).

In the first case, the distribution of the relaxation times according to the Cole-Cole model [85] is shown. The DP dependence on frequency in this case is described by Equation (35):

$$\varepsilon' - \varepsilon_{\infty} = \frac{\varepsilon_s - \varepsilon_{\infty}}{1+(i\omega\tau)^{1-\alpha}} \tag{35}$$

In the second case, the distribution of the relaxation times according to the Davidson-Cole model [86] is shown, which is described by Equation (36):

$$\varepsilon' - \varepsilon_{\infty} = \frac{\varepsilon_s - \varepsilon_{\infty}}{(1+i\omega\tau)^{\beta}} \tag{36}$$

It should be noted that one of the authors of the work [85], Robert Cole in his publication with David Hall [87] refused to use the idea of the distribution of the relaxation times in solutions and considered the dielectric spectrum as the sum of two Debye dispersion regions [87]. As well as in conductive solutions of electrolytes, the measured reactive component of the complex DP of ILs ε''_{mes} is composed of its ionic ε''_{ion} and dipole ε''_{d} components. The ionic component ε''_{ion} is determined by the DC conductivity value, Equation (37):

$$\varepsilon''_{ion} = \frac{\kappa_{DC}}{\omega\varepsilon_0}. \tag{37}$$

The frequency dependence of the dipole component ε" is described by Equation (29). As a result, for the measured value ε''_{mes} of ILs, Equation (38) is obtained:

$$\varepsilon''_{meas} = \varepsilon''_{d} + \varepsilon''_{ion} = \frac{\varepsilon_s - \varepsilon_\infty}{1+(\omega\tau)^2}(\omega\tau) + \frac{\kappa_{DC}}{\omega\varepsilon_0}. \tag{38}$$

At high frequencies, the contribution of the ionic component (DC conductivity) is not large and does not affect the appearance of the diagram, Figure 11a. In the low frequency region, the contribution of the ionic conductivity becomes significant. As a result, in the low-frequency area, as the appearance of the circular chart changes (Figure 13), the value of the static DP ε_s of a liquid is not possible to determine by extrapolation in the $\varepsilon''=f(\varepsilon')$ coordinates.

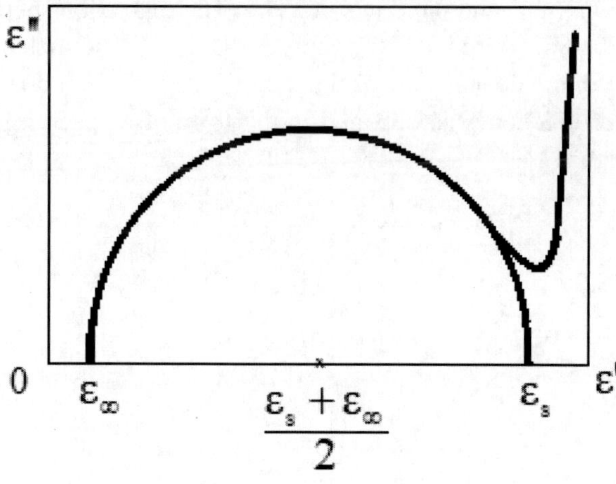

Figure 13. The increase in the dielectric loss factor ε''_{mes} in the low frequency domain due to DC conductivity.

In this case, when analyzing the results of IL dielectric studies, as well as when constructing the pie diagrams for dielectric relaxation, it is not the measured value ε''_{meas} that is used, but only its dipole component ε''_{d}:

$$\varepsilon''_{d} = \varepsilon''_{meas} - \frac{\kappa_{DC}}{\omega\varepsilon_0}. \tag{39}$$

Here, an important clarification should be made in the entire frequency range under investigation (currently the equipment allows dielectric measurements at frequencies of tens and hundreds of THz). it is believed that DC-conductivity remains constant and does not depend on frequency. Thus, the dispersion of the electrical conductivity for electrolyte solutions, predicted by Debye and Falkenhagen (DF) at the beginning of the last century [88], is not taken into account. According to DF theory, as a result of reducing the contribution of the relaxation process of ionic atmospheres, the electrical conductivity of solutions should increase with increasing frequency. No justification for the independence of DC-conductivity from frequency is currently available and, starting from the middle of the last century, the analysis of the dielectric measurements of conductive solutions was corrected for DC-conductivity according to Equation (39) [72].

As some justification for the notion of DC-conductivity independence from frequency, two reasons can be considered. Firstly, the mobility of ions in the external electric field is significantly higher than the dipole mobility of the polar ILs molecules. The polarization of ions occurs at frequencies that are several orders of magnitude higher than the dielectric polarization region. Secondly, as has been shown above, the ionic relaxation time θ described by Equation (22) defines the redistribution region of the conduction currents and displacement currents, Equation (20). At the same time, there is no increase as the frequency of the DC-conductivity gets larger. However, there is a dispersion of electrical conductivity in the microwave region, which will be discussed below after analyzing some features which take into account the DC-conductivity in dielectric measurements.

Due to the rapid development of measurement technology, in addition to dielectric relaxation spectra, it became possible to significantly expand the field of relaxation research into the terahertz region (tens and hundreds of THz). The relaxation spectra of ILs are usually analyzed in the ε',ε" = *f*(*F*) coordinates. For aprotic ILs, the static dielectric permittivity ε_s of which is 8-16 units [5], the dependence of the measured active ε' and reactive ε" components of the complex DP on frequency is obtained in the form of a spectrum, the general view of which is shown in Figure 14.

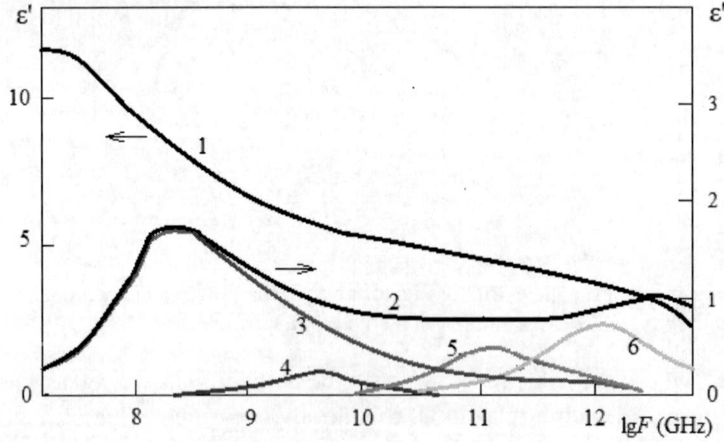

Figure 14. A general view of the active ε' (1) and reactive ε" (2) component frequency dependence of the complex DP of aprotic ILs; 3, 4, 5, 6 are the contributions of various relaxation processes to the measured value ε".

As a result of these studies, it was found that in addition to the Debye relaxation spectrum, due to the reorientation of ILs molecules (curves 1 and 3), dielectric losses are observed in the frequency range F > 1 GHz up to frequencies 1000 GHz (curve 2) indicating the presence of other, faster relaxation processes that may be associated with the polarization of ILs cations and anions, as well as other types of polarization (curves 4, 5, 6), Figure 14. These relaxation processes are discussed in the next section of this chapter. Here the effect of the DC electrical conductivity on the Debye dielectric relaxation spectrum ILs should also be considered.

As shown by the analysis of the literature data, the relative error of the ε' and ε'' measurements, as a rule, does not exceed 2-3%. We will estimate an error of calculation of size ε''_d for protic methylammonium formate (MAF) IL at 25 °C (Table 4) in the 0.1 - 1000 GHz range of the frequencies, considering that $\varepsilon_s \approx 40.3$, $\tau \approx 86$ ps, $\varepsilon_\infty = 2.59$, $\kappa = 4.38$ S/m [64], and $\delta(\varepsilon'') = 3\%$ and $\delta(\kappa) = 0.5\%$. To do this, Equations (28), (29) and (39) will be used, as it was performed for electrolyte solutions [60]. Table 5 shows the calculated values of ε', ε'', ε''_d, ε''_{ion} and an error $\delta(\varepsilon''_d)$. It should be noted that the error $\delta(\varepsilon''_d)$ was calculated as the difference error, Equation (40):

$$\delta_{(A\text{-}B)} = \frac{A}{A-B}\delta_{(A)} + \frac{B}{A-B}\delta_{(B)} \qquad (40)$$

where $A = \varepsilon''_{meas}$, $B = \varepsilon''_{ion} = \kappa/(\omega\varepsilon_0)$, Equation (33).

Table 5. Calculated using Debye Equations (28) and (29) dielectric characteristics (ε' and ε''_d) MAF, a contribution of DC conductivity (ε''_{ion}), total value of ε''_{meas} and an error of definition of a dipolar component $\delta(\varepsilon''_d)$

Frequency, GHz	ε'	ε''_d	ε''_{ion}	ε''_{meas}	$\delta(\varepsilon''_d)$, %
0.1	40.2	2.00	787.3	789.3	1362
0.5	37.7	9.50	157.5	167.0	63.4
1	31.8	15.8	78.7	94.5	22.6
1.85	21.5	18.7	42.7	61.4	12.6
5	7.13	12.3	15.7	28.0	8.9
25	2.80	2.78	3.15	5.93	8.3
50	2.64	1.40	1.57	2.97	8.3
100	2.60	0.69	0.79	1.48	8.3
1000	2.59	0.069	0.079	0.148	8.3

From the data presented in Table 5, it follows that the error in determining the dipole component of the dielectric loss coefficient $\delta(\varepsilon''_d)$, even at a frequency corresponding to the maximum ε''_d (1.85 GHz) exceeds 10% and increases sharply with a decrease in frequency. This increase is caused by the proximity of the ε''_{meas} and ε''_{ion} values, Table 5. At frequencies F < 1.85 GHz, $\delta(\varepsilon''_d)$ can reach several tens and hundreds of percent. As a result, it is not possible to detect and identify low-frequency relaxation processes in the IL (MAF) in question, whose relaxation time exceeds 86 ps (Table 4) to the left of the maximum. With frequencies F>> 1.85 GHz, a definition error ε''_d remains practically a constant (8.3%) that is connected with the fact that under the condition $\omega\tau >> 1$, ionic and dipolar components ε'' equally depend on frequency - they decrease proportionally to $1/\omega$.

The effect of DC conductivity on the dielectric relaxation spectra of protic (a) and aprotic (b) ILs is considered, Figure 15. For protic IL, the values $\varepsilon_s = 30.0$ $\tau = 500$ ps (a) are adopted. For aprotic ILs, the values are ε_s=15, τ=100 ps (b). For both ILs ε_∞=2.0. The calculation was performed according to Equations (28), (29) and (38) for DC-conductivity values: 2 - 0; 3 – 0.01; 4 – 0.1; 5 – 1.0; 6 - 10 S/m, Figure 15.

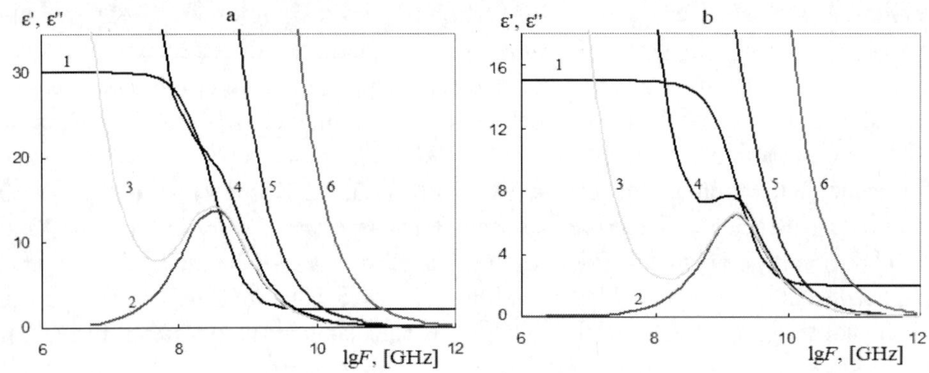

Figure 15. The frequency dependence of active ε' (1) and reactive ε" (2, 3, 4, 5, 6) components of complex DP of two model ILs, for which ε_s=30.0, τ=500 ps (a) and ε_s=15.0, τ=100 ps (b); DC-conductivity: 2 – 0; 3 – 0.01; 4 – 0.1; 5 – 1.0; 6 – 10 S/m.

A maximum of ε" occurs at frequencies of 0.318 GHz at τ = 100 ps (a) and 1.59 GHz at τ=100 ps. For protic IL (ε_s=30, τ=500 ps), the value ε" calculated by Equation (38) does not practically allow the isolation of the dipole component ε''_d already at DC-conductivity of 0.1 S/m, curve 4, Figure 15a. For the aprotic IL at this DC-conductivity value, it is still possible to isolate the dipole component ε''_d from the value ε", curve 4, Figure 15b. Since the DC conductivity of the protic ILs is typically greater than 0.1 S/m (Table 4), this conductivity substantially shields the low frequency (Debye) dielectric relaxation region of ILs. As a result, it is not possible to identify the molecular processes occurring in ILs at frequencies lower than 1 GHz. Due to the large errors in the definition of ε''_d (Table 5), large errors in the determination of IL dielectric characteristics may occur. This is especially true for the dielectric relaxation times, Table 4. This raises the question of the validity of using the relaxation time distribution model in the analysis of the DRS (dielectric relaxation spectra) described by Equations (35) and (36).

To summarize the above:

- When analyzing the dielectric relaxation spectra of conducting ILs, correction for the DC conductivity is necessary using Equation (39);
- At comparatively high DC conductivity values of ILs defined by Equation (24), the ionic losses ε''_{ion} shield the dipole loss ε''_d, preventing identification of the molecular processes occurring in the ILs;
- When analyzing the dielectric relaxation spectra and determining the parameters of the relaxation time distribution described by Equations (35) and (36), it is necessary to take into account the calculation errors of Equation (27) of the dipole component of the dielectric loss coefficient ε''_d, which increases sharply with a decrease in frequency, Table 5;

- The numerical value of the dielectric relaxation time τ can be estimated from the frequency corresponding to the maximum ε''_d using Equation (32).

High Frequency Conductivity Dispersion of Ionic Liquids and Organic Solvents

Description of the electromagnetic properties of ILs and volatile organic compounds (VOCs), as well as electrolyte solutions in these solvents [60, 80] at high frequencies of the electromagnetic field is advisable to carry out not only in terms of the complex DP ε^*, but also in terms of the components of the complex EC κ^*, Equations (12) and (13) [81]. The basis of the phenomenological description of the HF EC IL and VOC dispersion is the DP dispersion and the Debye Equations (27), (28) describing it.

Substituting the Debye expressions (28) and (29) for ε' and ε'' in Equation (26), for the active κ' and reactive κ'' components of the complex HF EC ILs and VOCs, the following Equations (41) and (42) are obtained [60]:

$$\kappa'=\frac{\kappa_\infty}{1+(\omega\tau)^2}\times(\omega\tau)^2 \tag{41}$$

$$\kappa''=\kappa''_{opt}+\frac{\kappa_\infty}{1+(\omega\tau)^2}\times\omega\tau \tag{42}$$

In Equations (41) and (42), κ_∞ is the limiting HF EC, Equation (15), κ''_{opt} is the optical component of reactive conductivity ($\kappa''_{opt}=\omega\varepsilon_0\varepsilon_\infty$).

When these equations were derived, it was necessary that in the field of DP dispersion $\varepsilon_s >> \varepsilon_\infty$. Therefore, in Equation (15) with κ_∞ as the numerator instead of ($\varepsilon_s-\varepsilon_\infty$), one is offered to use only ε_s [80].

Equations (41) and (42) describe the dispersion of the high frequency conductivity of the ILs and VOCs. As frequency is increased, there is an increase in an active component of conductivity κ' from zero under a condition $\omega\tau<<1$, up to its extreme value κ_∞ which does not depend on frequency under a condition $\omega\tau>> 1$. According to current understanding, the limiting HF EC κ_∞ is a fundamental characteristic of a substance and it is equal to the ratio of absolute DP $\varepsilon_s\varepsilon_0$ to dipole relaxation time τ. The value of κ_∞ limits the DC conductivity of aqueous electrolyte solutions and determines their temperature dependence [60]. As will be shown below, κ_∞ also determines the temperature dependence of DC EC of ILs and their solutions in VOCs [60]. The idea of the existence of the limiting HF EC VOCs was first formulated in 1977 [89, 90]. In the works [91-94], it was clarified and significantly supplemented. Since the limiting HF EC is used to describe the temperature dependence of the DC conductivity of the ILs, it seems necessary to elaborate on some characteristics of this feature of ILs and VOCs. At low frequencies, the condition $\omega\tau<<1$ is fulfilled, and Equation (41) is converted to Equation (43):

$$\kappa'=\varepsilon_s\varepsilon_0\tau\omega^2 \tag{43}$$

That is, the active EC κ' of the ILs and VOCs increases in direct proportion to the square of the frequency. In this frequency region, the κ' of the ILs is significantly lower than their DC conductivity. For methylammonium formate (MAF), for example, (ε=40.3, τ=86 ps [64] at a frequency of 1 MHz using Equation (43), $\kappa'=1.21\cdot10^{-6}$ S/m is obtained, which is significantly less than DC-conductivity of MAF (4.38 S/m). At a frequency of 1 GHz for the calculation of κ', it is already necessary to use Equation (41), having substituted the values ω, τ and κ_∞ (Table 4), in which κ'=0.95 S/m is obtained. This HF EC value is already commensurate with the DC-conductivity of MAF.

At high frequencies, when the condition ωτ>>1 is met, ε''_d decreases with frequency increase according to Equation (35), and the active conductivity of the ILs and VOCs κ' is independent of the frequency and reaches its limit value κ_∞, Equation (15).

Figure 16 shows, as an example, the HF EC ethylammonium nitrate (EAN) and ethylammonium formate (EAF) dispersion curves were built according to κ' и κ'' calculated using Equations (42) and (43), shown in Figure 16. The values ε_s and τ of EAN and EAF, which are necessary for the calculation are taken from the works [67] and [65].

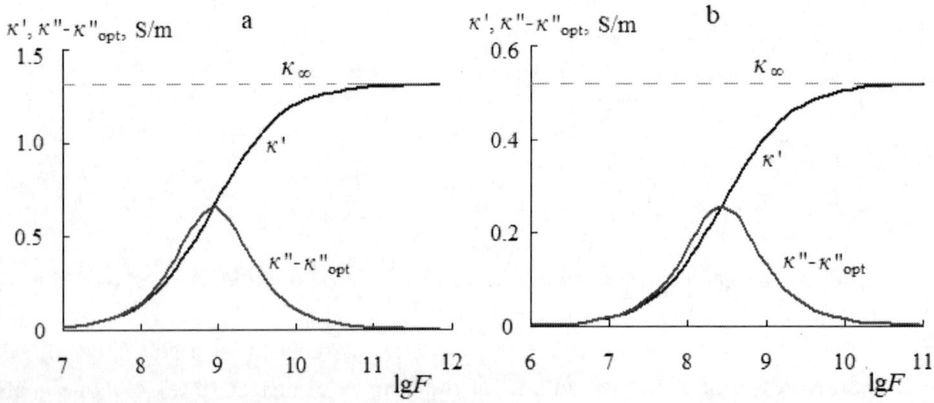

Figure 16. HF EC dispersion curves of ethylammonium nitrate (a) and ethylammonium formate (b), t=25°C.

The limiting HF EC values of some ILs at 25°C are shown in Table 3. Table 6 shows the κ_∞ values of some protic ILs in the 278.15 - 338.15 K temperature range. These values were calculated according to Equation (15) according to the results of the works [67] and [65].

Table 6. Limiting HF EC κ_∞ (S/m) of some ILs in the 278.15 - 338.15 K temperature range

T, K	Ethylammonium formate (EAF), butylammonium formate (BAF), and pentylammonium formate[a] (PAF)			Ethylammonium nitrate (EAN) and propylammonium nitrate[b] (PAN)	
	EAF[a]	BAF[a]	PAF[a]	EAN[b]	PAN[b]
278.15	0.277	0.0351	0.0272	0.750	0.0272
288.15	0.405	0.0543	0.0425	1.031	0.0425
298.15	0.519	0.0837	0.0575	1.307	0.0575
308.15	0.667	0.1018	0.0823	1.628	0.0822
318.15	0.861	0.1663	0.1280	2.027	0.1280
328.15	1.007	0.1899	0.1629	2.508	0.1629
338.15	1.260	0.2407	0.2145	2.803	0.2145

[a][67]; [b][65];

Since ILs are characterized by high viscosities η, and therefore large dielectric relaxation times, the limiting HF EC of ILs is significantly lower than that of VOCs [60]. Based on the analysis of the dipole relaxation processes in ILs, the following can be noted:

- In alternating electromagnetic fields of high frequency, with frequency growth, the increase in conductivity κ' of the active component is observed for ILs; at comparatively low frequencies (for which $\omega\tau<<1$), the increase of κ' occurs in proportion to the square of the frequency, Equation (26), wherein the high frequency conductivity of ILs is substantially lower than their DC conductivity;
- In the range of electromagnetic field frequencies for which inequality (20) is true:

$$0.1\leq\omega\tau\leq10, \qquad (20)$$

The HF EC of ILs becomes commensurable with their DC conductivity, and under a $\omega\tau>>1$ condition, HF EC does not depend on frequency and reaches the extreme value equal to κ_∞ which is defined by the $\varepsilon_s\varepsilon_0/\tau$ relation, Equation (15):

- Because the IL viscosity is 1 to 2 orders of magnitude higher than the VOC viscosity in the tested temperature range, the frequency region in which DP relaxation is observed and HF EC shifts in frequency lower by one to two orders of magnitude;
- When considering relaxation processes in ILs, in addition to dipole relaxation processes, it is also necessary to consider the relaxation of HF EC.

Dielectric Characteristics of Ionic Liquids

Static Dielectric Constant

The static dielectric permittivity (DP) ε_s plays an important role in the processes associated with the dissociation of ions and their solvation in ILs. As a result of the studies carried out in the last 20 years, it was found that the static DP of aprotic ILs averages 10-15 units [5], while the ε_s of protic ILs can reach several dozen units [62, 65-67]. This fact has previously been confirmed [95] using a theoretical assessment of the ε_s of 1500 ILs based on the QSPR model (quantitative structure-property relationship).

The equations for this model use information on the polarizability, size and shape of the ILs molecules, but also consider the interactions between ions, i.e., the contribution of ionic polarization to the ILs dielectric constant. For 52 ILs with imidazolium, pyridinium, pyrrolidinium, ammonium and sulfonium cations, the obtained calculated values of ε_s are compared with their experimental values. In addition, the calculated values of ε_s were compared in the work [95] with the DP of acetone, DMSO and water.

As a result, the DP of most ILs is shown to be lower than that of acetone (ε_s = 20.5) and averages 15.5 units. However, there is a group consisting of 218 ILs, whose DP is located between the ε_s of DMSO (ε_s = 46.8) and water (ε_s = 78.4). This group primarily includes ammonium ionic liquids with 2-hydroxyethyl)dimethylammonium cation, as well as several imidazolium, sulfonium and pyrrolidinium ILs containing oxygen in the structure of the cation.

The QSPR method used in the work [95] could allow, according to the authors, a promising way for selecting ILs for further study.

Before comparing the quantitative values of the ε_s of ILs, it is necessary to take into account an important observation [63] that "the static dielectric constants of ionic liquids measured by DRS are at least an order of magnitude less accurate than the data for VOCs obtained by conventional methods." A significant decrease in the accuracy of the definition of the ε_s of ILs is explained, as shown above, by the presence of significant DC conductivity. Therefore, the comparison of the DP of the ILs below and the consideration of the temperature effect on the ε_s should be treated with a certain degree of caution, but the resulting patterns could be considered as possible options.

The static DP of 1 alkyl-3-methylimidazolium aprotic ILs with a poor polar bis{(trifluoromethyl)sulfonyl}amide ([NTf$_2$]) anion increases with the growth of the alkyl chain length from 12.0 ([C$_2$mim]$^+$) to 15.0 ([C$_5$mim]$^+$) at 25°C [63]. This fact is not consistent with the effect of the hydrocarbon chain length on the ε_s of VOCs, Figure 17.

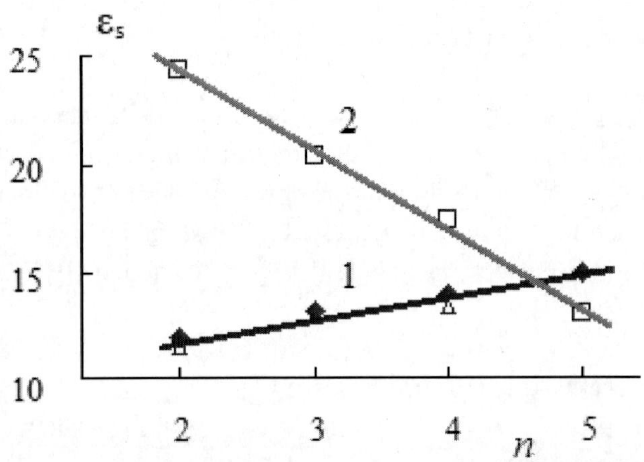

Figure 17. The DP (ε_s) dependence of 1-alkyl-3-methylimidazolium bis{(trifluoromethyl)sulfonyl} amides, [C$_n$mim][NTf$_2$], (1) and alcohols (2) on the number of carbon atoms (n).

The DP of alcohols, for example, at the same temperature decreases from 24.5 (ethanol, C_2H_5OH) to 13.7 (propanol, $C_5H_{11}OH$) [69]. The increase of the ε_s in the range of 1 alkyl-3-methylimidazolium ILs is difficult to explain by the presence of a weakly polar [NTf$_2$]$^-$-anion. In this connection, the DP values of the four 1-alkyl-3-methylimidazolium salts can be considered more reliable: the values of the ε_s at 298.15 K lie between 12.8 and 8.9, and decrease with increasing chain length of the alkyl residue of the cation for non-polar [BF$_4$]$^-$ and [PF$_6$]$^-$ anions [96]. At the same time, an increase in the DP of 1-ethyl-3-methylimidazolium-based, [C$_2$mim]$^+$ ILs, along with an increase in the polarity and polarizability of the anion is consistent with existing thinking: the maximum DP value (35.0-35.5) occurs in the case of the ethylsulfonate anion, [C$_2$H$_5$SO$_3$]$^-$ [63]. The high DP of protic ILs is due to the presence, for example, of high polarity anions such as ethylsulfate, [C$_2$H$_5$SO$_4$]$^-$ or hydroxyl-functionalization of alkyl chains [63]. As in the case of VOCs, abnormally high values of ε_s in this case are due to the formation of sufficiently strong hydrogen bonds between the IL molecules.

Of particular interest is the temperature dependence of the static DP of ILs. However, increasing the DC conductivity of the ILs with raising the temperature significantly enlarges the error when determining the ε_s value. The static DP of the ILs decreases with increasing temperature [65, 67], with an almost linear dependence of the DP on the temperature for ethylammonium nitrate (EAN), propylammonium nitrate (PAN) [65], ethylammonium formate (EAF), and butylammonium formate (BAF). With rising temperature, the thermal motion increases, which reduces the dipole-dipole interactions. As a result, a decrease in ε_s with an increase in temperature ($-\partial\varepsilon_s/\partial T$) is much higher for protic ILs than for aprotic ones. This effect is enhanced by replacing the nitrate anion in protic ILs with a bipolar formate anion [67]. In general, the temperature coefficients DP ($-\partial\varepsilon_s/\partial T$) of the considered protic ILs are similar to those of hydrogen-bonded molecular liquids such as, for example, alcohols [65].

The EAF has high values of static permittivity and a more pronounced temperature dependence compared to an EAN [67]. The likely reason for this difference between the EAF and EAN, is because of the different nature of the hydrogen bond formed between the formate anion and the ethylammonium cation compared to nitrate, as this results in a higher ionic packing factor [67].

Relaxation Processes in Ionic Liquids

The review [63] summarized the results of ILs studies by dielectric relaxation, magnetic relaxation, Kerr optical effect spectra, and molecular dynamics simulations published prior to 2010-2011. This work analyzed in detail the possible relaxation processes that occur in ILs over a wide frequency range, ranging from the megahertz to the terahertz range. It has been shown that in ILs, the dipole relaxation process (similar to that of VOCs) is superimposed by ionic relaxation processes, caused by the movement of the IL ions in an alternating electric field of high frequency. In the case of simple ions in a variable high-frequency field, translational movement occurs. For complex ions with an electric dipole moment, both rotational and translational motion in an alternating field is possible [63].

Accordingly, in the dielectric relaxation spectrum (DRS), several maxima are observed due to the dependence of ε'' on the frequency, Figure 14. The first, most pronounced maximum ε'' of DRS is due to the polarization of the ILs molecules. The maximum value of the dielectric relaxation time τ_1 corresponds to this process. Since the viscosity of the ILs is one to two orders of magnitude higher than the viscosity of the VOCs, the absorption caused by the polarization of the IL molecules shifts towards the lower frequencies. In this frequency domain, as shown in the previous section, a significant contribution to the measured value ε''_{mes} is made by the ion loss ε''_{ion}, Equation (37). As a result, for ILs with high viscosity, it is not possible to identify the relaxation processes occurring at low frequencies, since they are shielded by the electrical conductivity (DC conductivity), Figure 15.

There is an agreement among scientists that the dominant mode in the megahertz/gigahertz range mainly reflects the cooperative reorientation of the dipolar cations, as supported by a comparison with the magnetic relaxation data, Kerr optical effect spectra, and the molecular dynamics modeling [63].

Simple anions and complex anions with high molecular symmetry (for example, tetrafluoroborate, $[BF_4]^-$, and hexafluorophosphate, $[PF_6]^-$, are dielectrically inactive. The spectral contribution of dipolar anions is difficult to estimate, since for most of these anions, the uptake is small or absent at all. An exception is alkyl sulfates (ILs containing, for example, the ethylsulfate anion $[C_2H_5SO_4]^-$), which are characterized by higher values of static dielectric permittivity [63].

Comparatively long-lived electron pairs can also contribute to dipole relaxation, since their formation is accompanied by very high electrical dipole moments. All of the contributions listed are used to interpret the spectra by comparing the dependencies of theoretically calculated frequency and the experimental frequency.

Before analyzing the results of the spectral studies of the ILs performed after 2010-2011, it should be noted that the dielectric relaxation times of the main (Debye) regions of the DP dispersion τ_1 are determined with greater accuracy compared to the relaxation times (τ_2, τ_3 and others) of the faster processes. This is due to the greater blurring of the high-frequency dispersion regions (4, 5, 6, Figure 14), their smaller amplitude, as well as the asymmetric nature of the spectrum when used to calculate the τ_2 and τ_3 of the Davidson-Cole model [86], Figure 9b. Since both models of relaxation time distribution (Davidson-Cole and Cole-Cole, Figure 9) are empirical, their application has no theoretical justification, and the shape of the spectrum does not provide grounds for molecular interpretation [63]. The values of τ_1 together with static DP of ILs are used in this work to calculate the limiting HF EC in the final section and to establish the correlations between the DC conductivity and the dielectric properties of the ILs.

Particular attention was paid to the analysis of the spectra in the giga- and terahertz ranges, for works published in the last decade, in order to clarify the nature of the processes occurring in the ILs. Before starting to analyze these works, it should be noted that in very viscous ILs, as is in crystalline substances, when applying an alternating electromagnetic field, there is no physical reorientation of molecules and ions. At the frequencies of the field commensurate with the reverse relaxation time, absorption of the energy of the external field occurs, which is recorded by research equipment and further analyzed in order to split a rather complex absorption spectrum into various dipole, ionic and other components. Some aspects of microwave energy absorption by substances containing polar molecules are discussed [60].

The broad-range dielectric spectra of a homologous series of protic ionic liquids based on alkylammonium formates, combining microwave dielectric, THz time domain, and broadband Fourier transform far-infrared (FTIR) spectroscopy were investigated in the study [64]. The combination of these methods allowed the analysis of the complex dielectric permittivity in an interval covering more than four orders of frequencies. To interpret the obtained results based on the spectrum in the terahertz (THz) range, a principal component analysis (PCA) was carried out yielding a clear visualization of correlations in the data sets. As a result, it was found that due to the first two principal components, almost 99% of the whole spectral variance is covered.

For comparison, the dielectric spectra of other ILs, as well as VOCs (alcohols) and water obtained in the same frequency region, were also included in a number of analyzed spectra. It has been shown that the observed processes relate to the low-frequency reorientational relaxation of the dipolar ions and the intermolecular vibrations between cations and anions. The spectral behavior of ethylammonium formate (EAF), butylammonium formate (BAF), and 2-hydroxyethylammonium formate (2-HEAF) seems very similar to that of water, which, in turn, exhibits the largest values by far. At the same time, there is a strong dependence of the dielectric spectra in both the microwave and the THz regions on the alkyl chain length, polarity and choice of anion) [64].

The calculated and experimentally obtained frequency-dependent dielectric spectra of 1-ethyl-3-methylimidazolium triflate ($[C_2mim][CF_3SO_3]$) have been compared [97]. When performing calculations, the complete spectrum of the generalized dielectric constant was considered as consisting of the dielectric permittivity spectrum and the dielectric conductivity spectrum. In this case, the latter is divided into its translational and non-translational components. For 1-ethyl-3-methylimidazolium triflate over the entire frequency range, both spectral components are shown to contribute with approximately equal weight and they overlap. At the same time, the low-frequency region of the dielectric conductivity also undergoes changes, while its high-frequency part remains practically unchanged.

Inductive effects appear not only at high frequencies, but also contribute in the sub gigahertz range. The calculated peak obtained in this (sub gigahertz) region correlates well with the experimental OKE (Optical Kerr-effect) spectroscopy spectrum. The authors of the work [97] explain this peak by the correlation between the induced dipole moment of the cations and the local electric field created by the anion framework. In the frequency interval of 500 MHz – 20 GHz, the authors of this work state almost quantitative agreement of the calculated and experimental spectra. Good agreement of the spectra is also obtained up to a frequency of 10 THz. However, the experimental value of the optical component of DP $\varepsilon_{\infty,exp}$ for 1-ethyl-3-methylimidazolium triflate, $[C_2mim][CF_3SO_3]$, ($\varepsilon_\infty = 2.3$) is 1.21 times higher than its calculated $\varepsilon_{\infty,calc}$ value. The lower value of the $\varepsilon_{\infty,calc}$ is caused, according to the authors of the work [97], by the lack of accounting for the theoretical calculation of the polarization spectrum of hydrogen atoms of the 1-ethyl-3-methylimidazolium cation $[C_2mim]^+$ cation. As a result of the studies, it was concluded that "induced dipole moments have to be added to the collective non-translational dipole moment and thus become part of the dielectric permittivity."

Terahertz dynamics and the slower relaxation of protic alkylammonium nitrate ILs were studied using femtosecond optical spectroscopy, dielectric relaxation spectroscopy and terahertz spectroscopy in the temperature range of 5-65°C [61]. As a result of the studies, it was found that the Debye spectrum associated with reorientation is symmetrically expanded due to an additional high-frequency contribution. This contribution fades by the power law. Due to the higher symmetry of the polarizability tensor, relaxation in the high-frequency region occurs three times faster than relaxation in the Debye region. It was also found that when the temperature rises, as in the case of aqueous electrolyte solutions [68], the Debye component of the spectrum shifts towards the higher frequencies. It has been suggested that the relaxation time in each case is determined by the rate of hydrogen bond breakage rather than the diffusion process. The difference in libration frequencies, as well as the abnormal relaxation rates, is explained by the weak polarization of the nitrate ion. Therefore, the dielectric spectrum is sensitive to cation motions, whereas the OKE spectrum is sensitive to anions.

A comparison of the dielectric and OKE spectra obtained at 25°C for ethylammonium nitrate (EAN) showed that in both cases a peak associated with librations (hindered rotational movements) prevails at higher frequencies. At the same time, it turned out to be strange that in EAN, the libration band turns out to be 2-3 times higher in frequency for the dielectric spectrum compared to the OKE spectrum.

The differences in libration frequencies and abnormal relaxation rates are explained in the work [61] by the behavior of the nitrate ion. This delocalized-moment flat ion has a strongly aromatic-free bound lean charge system and is therefore weakly polarized. As a result, the dielectric spectrum is sensitive to cation motions, whereas the OKE spectrum is sensitive to anions. Two intense bands at frequencies of 2 and 6 THz are usually cause of bending and stretching of the hydrogen bonds (i.e., water). However, since the dielectric and OKE signals are primarily sensitive to rotations, the dominant contribution in this region is likely to be librational. In this case, the shoulder observed just below 1 THz would be a likely candidate for the hydrogen bond stretch.

As a result of using the Stokes-Einstein-Debye (SED) Equation (33) in the temperature range of 5-65°C, the effective molecular volume $V_{eff.}$ of EAN was estimated. The values of *ca.* $V_{eff.}$ = 15 Å^3 were found by the means of OKE and DRS, but for the latter, the additional factor of 3 gives 5 Å^3.

Corrections may be applied for the molecular aspect ratio and slip or stick conditions [98]. These corrections show that the $V_{eff.}$ for the ethylammonium ion is underestimated. The active motion for the nitrate ion (in OKE) is an out-of-plane rotation, and for the alkylammonium ion (in DRS) the active motion is an end-over-end rotation, which does not suggest that the $V_{eff.}$ for the ammonium ion should be particularly small. As an explanation for the observed effect in the work under consideration [61], it is proposed that the multiple hydrogen bonds lead to relaxation by the molecular jump mechanism. In this case, the relaxation time in each case is determined by the rate of hydrogen bond breaking rather than the diffusional process intrinsic to the SED relation (19).

It should be added that the resulting value of the $V_{eff.}$ of the EAN (15 Å^3) means that the effective radius of the EAN (1.53 Å) is substantially smaller than the actual radius of it. For water (molecule radius 1.38 Å) the $V_{eff.}$ = 11 Å^3. The dipole dielectric relaxation time of the water is significantly less - 8.3 ps [69]. According to the SED Equation (33), the relaxation time is proportional to the volume, so for the EAN the τ should be of the order of 11-12 ps, which is significantly lower than the obtained τ value in the work [61].

The dynamics of ILs have been studied in the work [99] using dielectric relaxation (DR) and Optical Kerr-effect spectroscopy (OKE) with a time resolution in the frequency range of ~ 10 MHz to ~ 20 THz. This work shows that the OKE and DR spectra are dominated by relaxation in the gigahertz range. It extends to a relatively sharp band at a frequency of about 10 THz. Unlike other ILs, imidazolium salts show an additional sub-α relaxation that dominates the OKE signal. The proximity of the Arrhenius activation energy (AAE) values of DR and OKE α-relaxation, viscosity and DC-conductivity AAE for EAN indicates the presence of an extensive hydrogen bond, leading to strong cooperativity in the movements of cations and anions. For aprotic ILs that do not form an expanded network of hydrogen bonds, the activation energy for α-relaxation is significantly lower than that of EAN and PAN.

According to the results of this work, in the terahertz range, intermolecular dynamics may be overlapped by intramolecular fluctuations. The wide bandwidth covered by the ILs spectra indicates that for these compounds, the intermolecular motion timeline ranges from ~100 femtoseconds to nanoseconds. The rotational and translational movements of the anions and cations are closely related and these movements may be modulated by the hydrogen bond interactions. As a rule, DR and OKE are simultaneously sensitive to anions and cations, although with different masses, so information on specific ions from the obtained spectra is difficult to extract. An example of this is the aprotic sulfonium and pyrrolidinium ILs, which were previously studied [99]. The situation is different for the protic EAN and PAN ILs, as the DR has the characteristic of cation reorientation, while the α-relaxation of OKE controls the anion. Similar activation energies for the rotation of the anion and cation indicate cooperativity in their dynamics, and a comparison of the total reorientation of the ethylammonium ion (DR spectroscopy) with reorientation of N-H (IR spectroscopy with time resolution) indicates a jump reorientation due to the hydrogen-bond network of this IL.

It was also shown that for the studied imidazolium ILs in α-relaxation, both the DR and OKE spectra are dominated by the reorientation of cations through jumps at high angles [99]. Dipolar anions give an additional DR mode at a frequency ~ 5-10 GHz, which is not large in amplitude, while the α-relaxation mode increases compared to the ILs containing non-polar anions. It has also been shown that in mixtures of imidazolium salts with polar solvents, these ILs retain their spectrum up to unexpectedly low mole fractions ($x \approx 0.2 - 0.4$) before the transition to the dynamics characteristic of the electrolyte solutions occurs, in which cation solvation prevails, as well as the formation (at higher x) of contact ion pairs. The degree of ion pairing varied depending on the dielectric permittivity of the solvent but was always similar to that of salts with organic cations.

The dielectric relaxation spectra of the three members of the protic ILs family: ethylammonium (EAF), butylammonium (BAF) and pentylammonium (PAF) formates, as well as pseudo-IL triethylamine + formic acid (1:2 molar ratio; TAF) were studied in the work [67] over a wide frequency range (from 50 MHz to 89 GHz) in the temperature range of 5-65°C. In parallel, these studies were supplemented by measurements of density, viscosity and conductivity of these ILs. As a result of these studies, it was found that the dominant relaxation of the EAF, BAF and PAF occurs both at the expense of cations and the reorientation of anions, which are synchronized in their dynamics due to hydrogen bonds. In addition to their charges, all alkylammonium cations, as well as the formate anion, have a constant dipole moment. As a result, the dielectric spectra of the studied ILs, including TAF, are a superposition of contributions from ion translation and dipole rotation.

Compared to the nitrate, the amplitude and relaxation times indicate a different nature of H-binding between the formate anion and ethylammonium cation. For ethylammonium nitrate in α-relaxation observed in the DR spectra, cationic reorientation clearly prevails, since the anion has no dipole moment. Comparison with the time-resolved spectra of the Optical Kerr-effect sensitive to nitrates showed that rotations of anions and cations occur on the same time scale. In the case of alkylammonium formates, both the cation and the anion contribute to the α-relaxation. Assuming that dipole rotation is the source of the dominant mode, the experimentally determined collective dielectric relaxation time τ_1 of the studied formates was transformed into the corresponding time of rotation correlation of the individual τ'_j dipoles. The obtained correlation time τ'_j according to the SED Equation (33) increased linearly with η/T growth, cutting off some τ^0_j value corresponding to the $\eta/T \rightarrow 0$ value on the τ'_j axis.

Comparison of the V_m experimental values of the rotation volumes obtained for α-relaxation with the calculated values of $V_{eff}(\tau_1)$ and volumes for boundary conditions of sticking ($V_{stick\ eff,cal}$) and slipping ($V_{slip\ eff,cal}$), showed that for all alkylammonium formats, the $V_{eff}(\tau_1)$ experimental data are significantly lower than theoretical limit values of $V_{stick\ eff,cal}$, but are comparable for the $V_{slip\ eff,cal}$ value of their cations. For instance, for the ethylammonium cation, the $V_{slip\ eff,cal}$ value of 6.9 Å^3 is close to the experimental rotation volume of 9.8 Å^3. As a result, it has been suggested that reorientation of cations prevails for ethylammonium formate (EAN) in the τ_1. However, the anion also affects τ_1. For EAN with its symmetric nitrate ion, $V_{eff}(\tau_1)$=5 Å^3 [61]. Thus, reorientation of ethylammonium cations experiences greater friction in EAF compared to EAN, which is due to the different way hydrogen bonds are formed in these two (EAF and EAN) ionic liquids [67].

Using the Arrhenius equation, $E_a(\tau_1)$ and $E_a(\tau_2)$ activation energies were estimated from the dependencies of relaxation times $\tau_1=f(T)$ and $\tau_2=f(T)$ [67]. These values were compared with the viscosity and conductivity $E_a(\kappa)$ activation energies $E_a(\eta)$ calculated in the same temperature range. As a result, it was found that for EAF, DRS $E_a(\tau_1)$ and $E_a(\tau_2)$ agree well with $E_a(\kappa)$, while $E_a(\eta)$ is somewhat higher. For BAF and PAF, $E_a(\tau_1) \approx E_a(\eta) \approx E_a(\kappa)$, whereas $E_a(\tau_2)$ is significantly lower.

The study [67] also analyzed the dependence of logλ on logη (Walden plot). The deviation of this dependence from the "ideal" KCl line is usually taken as a measure of the ionic nature of the sample, reflecting correlations in the translational motion of cations and anions [100]. The deviation from the ideal KCl line is established for all ILs studied, with the degree of ionicity decreasing with the transition from EAN to EAF, then to PAF and to BAF. According to the previous discussion, in the considered deviation from the ideal KCl line, the different nature of the H bond for nitrate and formate is reflected and the influence of the alkyl chain length is seen. The possible reasons for the deviation of logλ on logη dependencies from the ideal KCl line are discussed in the next section in the analysis of the electrical conductivity and dielectric characteristics of ILs. The activation energy of DC conductivity of ILs is also considered there.

When analyzing the dielectric relaxation spectra of electrolyte solutions in the work [101], it was also proposed to consider the relaxation of the resulting ion pairs of various structures - separated by solvent molecules (SIP), and contact ion pairs (CIP). Such (SIP and CIP) ion pairs were found in concentrated solutions of magnesium, aluminum and scandium sulfates [101]. In conjunction with how it is indicated in the work [101], numerous DR studies have shown the complete absence of discrete ion pairs in ILs, at least on the DR timescale [102, 103].

The possibility of the existence of SIP and CIP ion pairs in ILs can be investigated by analyzing the distances between molecules and ions. This analysis is performed in the last section of this chapter. Based on the above discussion of static DP and polarization processes occurring in ILs, the following can be formulated:

- Due to the interfering effect of DC conductivity discussed in the previous section, the results of the determination of the static DP of ILs cannot always be reliable, as evidenced by the distribution of ε_s values published in the literature; nevertheless, it is safe to assume that the static DP of protic ILs is substantially higher than the aprotic ε_s (which at room temperature does not exceed, as a rule, 10-15 units);
- As in the case of protic VOCs, the significant static DP value of protic ILs (which can reach several tens of units) is due to the formation of a relatively stable network of hydrogen bonds, not only between ILs molecules, but also between their ions and molecules;
- The rather complex type of DR and OKE spectra for ILs is due to the various polarization processes that occur in a wide frequency range from megahertz to distant terahertz; these processes include the polarization of IL molecules (α-relaxation), their constituent ions, and the more complex structures formed in ILs and their solutions;
- Accurate identification of these polarization processes is hindered again by the interfering influence of the DC conductivity in the case of molecular (dipole) polarization and the low intensity of ion polarization peaks; as the temperature rises as in the case of VOCs, the absorption maxima of different polarizations shift towards higher frequencies;
- The proximity of the activation energies for dielectric relaxation, viscosity and DC-conductivity processes indicates the commonality of the polarization and translational

movement of particles (molecules and ions) in ILs; some general patterns of these processes are discussed in the next section, which also analyzes the distances between the molecules and the ions in ILs.

Electrical Conductivity (DC Conductivity) and Dielectric Properties of Ionic Liquids

It is very important for theory and practice to establish the relationship between the DC conductivity and dielectric characteristics of the ILs. The increase in DC conductivity of electrolyte solutions in polar solvents as the temperature increases is due to the decrease in the viscosity of the ions and the increase in their kinetic energy and is not directly associated with the dielectric properties of the solvent or solution. The theoretical description of the effect of solvent and temperature on DC conductivity of solutions is usually based on the use of the Walden rule, Equation (5).

For ILs, there is a linear dependence of log λ from log η, but the average slope is not equal to one, as expected for the ideal Walden rule [66], but is 0.92 ± 0.02 [100]. Therefore, the empirical Vogel-Fulcher-Tammann (VFT), Equation (4), [99] and also Arrhenius equation, Equation (2), [104] are generally used in the analysis of temperature dependencies of DC conductivity of ILs.

The temperature dependencies of DC conductivity in Arrhenius coordinates (log κ – $1/T$) are not linear [105], so it is believed that the VFT Equation (4) better describes the temperature dependence of the DC conductivity of ILs [104]. It should be noted, however, that the disadvantage in the Arrhenius equation (nonlinearity of dependences of log κ – $1/T$) is easily eliminated if the possibility of the temperature dependence existence of activation energy is taken into account [16]. Analysis of the measurement results of the specific EC of ILs carried out in this work in the temperature range of 20-210°C showed that the activation energy of DC conductivity decreases from ~ 30-40 (30°C) to ~ 15-18 (200°C) kJ/mol depending on the nature of ILs. It should be noted here that there is also a decrease in the activation energy of DC conductivity for aqueous electrolyte solution [106].

For aqueous solutions of inorganic salts, it has also been found that as the temperature rises, their specific conductivity increases in direct proportion to the ratio of absolute DP $\varepsilon_s\varepsilon_0$ to the DR time τ, i.e., proportional to the limiting HF EC κ_∞ [80], Equation (44):

$$\kappa = K\frac{\varepsilon_s\varepsilon_0}{\tau} = K\kappa_\infty. \tag{44}$$

The proportionality described by Equation (44) is also performed for solutions of inorganic salts in non-aqueous solvents [94]. It is of interest to verify the existence of the proportionality for ILs and their solutions described by Equation (44). To this end, on the basis of the values published in the work [65], which were obtained as a result of studies by dielectric spectroscopy (DR) and OKE spectroscopy, the values of the limiting HF EC of the EAN IL were calculated, and the κ – κ_∞ dependencies, Figure 18a. The DC conductivity values necessary to construct these dependencies were also taken from the work of Sonnleitner et al. [65].

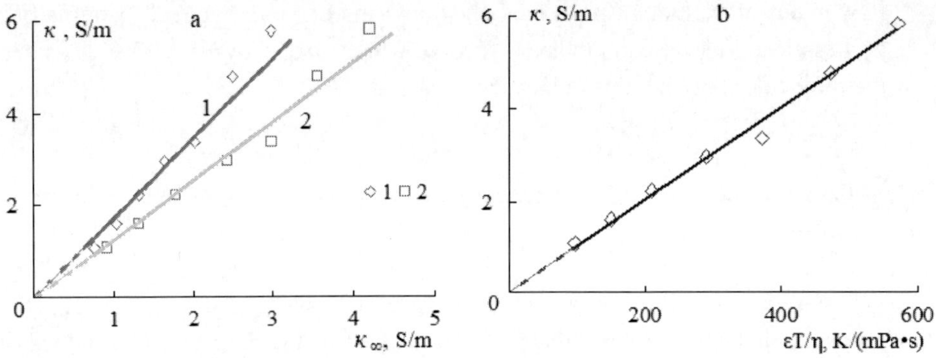

Figure 18. Dependence of DC conductivity of EAN on κ_∞ of an IL (a) calculated from the results of DR (1) and OKE (2); and on the $\varepsilon_s T/\eta$ ratio (b); 278.15 - 338.15 K temperature interval.

The τ values obtained in the work [65] by the DR and OKE methods differ significantly, and their difference increases with increasing temperature ($\tau_{DR/OKE}$=1.22 at 278.15 K and 1.50 at 338.15 K). According to the data presented in Figure 18a, in general, the proportionality described by Equation (44) is observed. With an increase in the temperature, DC conductivity increases in proportion to κ_∞. Slight deviations from linear dependence are observed at 328.15 and 338.15 K temperatures. Most likely, these deviations are caused by errors in determining the relaxation time at these temperatures.

According to the SED Equation (33), the dipole relaxation time τ is proportional to the η/T ratio. Therefore, it is possible to check the fairness of the Equation (47) for ILs by replacing the relaxation time τ with the ratio η/T. In this case, it is necessary to expect direct proportionality already in the $\kappa - \varepsilon_s T/\eta$ coordinates. This dependency is shown in Figure 18b. In these coordinates, the DC values are better placed on a straight line, which is probably due to the fact that the viscosity of ILs η is determined more precisely than the dipole relaxation time τ.

The existence of the proportionality described by Equation (44) suggests using the following Equation (45), instead of Walden rule, Equation (5) [60]:

$$\lambda\frac{\tau}{\varepsilon_s} = const \text{ or } \frac{\lambda}{\kappa_\infty} = const. \tag{45}$$

According to the expression (45), the ratio λ/κ_∞ should not depend on temperature, same for the inverse value κ_∞/λ, which has a concentration dimension (mol·m^{-3}). If there are no experimental data on the dielectric relaxation times τ in the expression (45), this value can be replaced by the ratio η/T. In this case, Equation (46) is obtained:

$$\frac{\lambda\eta}{\varepsilon_s T} = const. \tag{46}$$

Equation (46) differs from the Walden rule in that it takes into account the effect of not only the viscosity η on the molar conductivity λ, but also the static permittivity ε_s and the temperature *T*.

The proportionalities described by Equations (45) and (46) are also applicable for the cases of ethylammonium formate (EAF), butylammonium formate (BAF), and pentylammonium formate (PAF), as well as the pseudo-IL triethylamine + formic acid (molar ratio 1:2, TAF)), Figures 19 and 20. In these Figures, the $\kappa - \varepsilon_s T/\eta$ and $\kappa - \kappa_\infty$ dependences are shown, constructed on the basis of the DC conductivity κ, viscosity η values and also ε_s and τ (using which the value κ_∞ was calculated) provided in the work of [67].

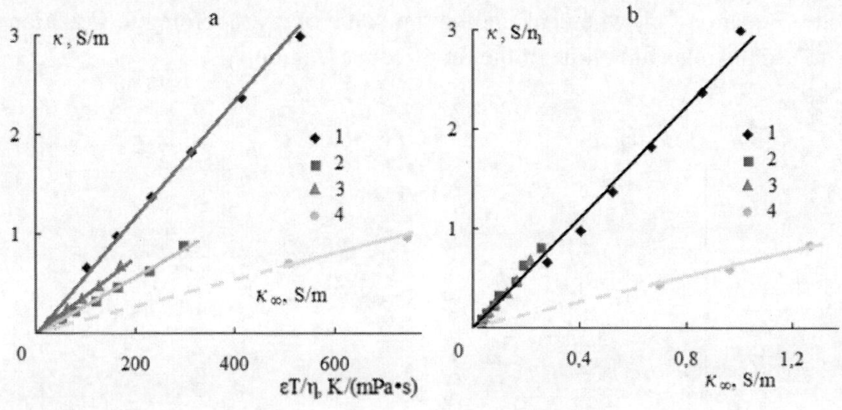

Figure 19. Dependence of DC conductivity of EAF (1) and BAF (2), PAF (3) and TAF on the $\varepsilon_s T/\eta$ ratio (a) and on κ_∞ of ILs (b).

For all ILs (1, 2, 3) and pseudo-IL (4), Figure 19a, the DC conductivity increases in direct proportion to the $\varepsilon_s T/\eta$ ratio (since the viscosity of pseudo-IL is significantly lower than the viscosity of ILs; only two points for 278.15 and 288.15 K are plotted for this mixture). The correlation coefficient of the considered $\kappa - \varepsilon_s T/\eta$ dependencies (Figure 19a) for all ILs exceeds 0.99. Thus, Equation (47) can be used to analyze the temperature dependence of DC conductivity of ILs:

$$\frac{\kappa\eta}{\varepsilon_s T} = const, \tag{47}$$

which differs from the Equation (46) in that in it the molar conductivity λ is replaced by the specific conductivity κ.

The $\kappa - \kappa_\infty$ dependence (Figure 19b) is more universal than the $\kappa - \varepsilon_s T/\eta$ dependence (Figure 19a). In the 278.15 - 338.15 K temperature range, this dependence is unified for EAF, BAF and PAF and is described by Equation (44), in which the coefficient K for EAF, BAF and PAF is equal to 2.8 ± 0.1; 3.0 ± 0.2 and 2.9 ± 0.1, respectively. Pseudo-IL triethylamine + formic acid (molar ratio 1:2) does not fit on this dependence, however, the proportionality described by Equation (34) is observed for this mixture. The DC conductivity of the mixture is 68% of κ_∞.

From the data presented in Figure 19b, it also follows that the DC conductivity of the ILs κ is on average three times higher than their limiting HF EC κ_∞ ($K \approx 3.0 \pm 0.2$). This fact requires special consideration and goes beyond the scope of this work. It should be noted here that according to Table 4 for MAF and BAF, $\kappa \approx \kappa_\infty$, in aqueous solutions of electrolytes, the DC conductivity of the salts constitutes some part of the κ_∞ of water and in a wide temperature range its temperature dependence is also described by Equation (44) [60].

To calculate the κ_∞ of ILs according to Equation (15), the relaxation times of the faster processes (τ_2 and τ_3) given in the work [67] can be used. The value κ_∞ received in this case, is tens and hundreds of times higher than the DC conductivity κ of the ILs. As the accuracy of determination of the τ_2 and τ_3 values is significantly lower, there is a wide distribution of the points on the dependences $\kappa - \kappa_{\infty,2}$, Figure 20. Therefore, for the analysis of the dependencies described by Equation (44) when calculating the value of κ_∞ according to Equation (15), it is proposed to use the relaxation time of the first (Debye) region τ_1.

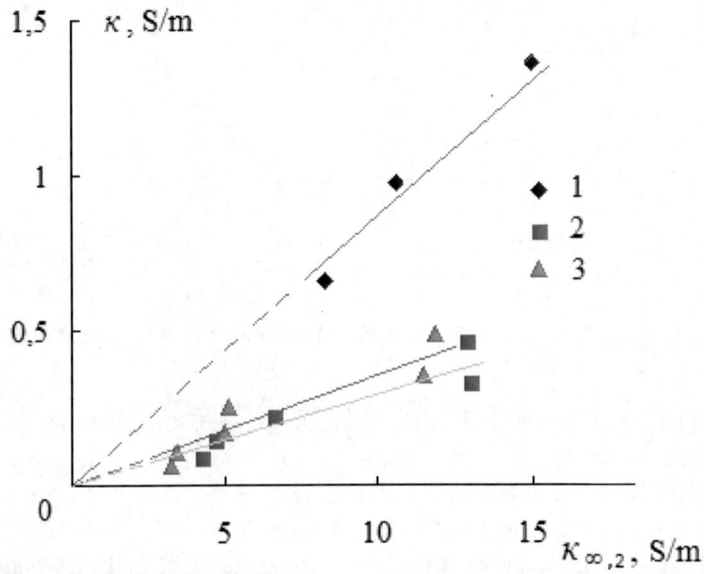

Figure 20. Dependence of DC conductivity of EAF (1), BAF (2) and PAF (3) on $\kappa_{\infty,2}$ of ILs.

In conclusion for the dependency analysis $\kappa - \kappa_\infty$, it can be noted that the proportionality described by Equation (44) is also observed for IL solutions in polar solvents [60], Figure 21.

In all of the solutions shown in Figure 21, the DC conductivity κ increases in a directly proportional way to the HF EC κ_∞ for all the IL solutions. It should be noted that the dependencies $\kappa - \kappa_\infty$ for both the 0.02 M solution of $[P_{6\,6\,6\,14}]Cl$ in acetonitrile (1) and dimethylsulfoxide (2), Figure 21b, fit into one line. Since the HF EC κ_∞ of acetonitrile is 3-3.5 times larger than the κ_∞ of dimethyl sulfoxide, the DC conductivity of the acetonitrile solutions is also 3-3.5 times larger than that of dimethyl sulfoxide solutions. A comparison of the $\kappa - \kappa_\infty$ dependencies for the solutions of ILs in VOCs and for aqueous solutions of inorganic salts carried out in the work [107] showed that despite the different nature of the temperature dependence of κ_∞ solvents (Figure 22a), the dependence of the DC conductivity κ on the limiting HF EC κ_∞ of a solvent turns out to be unified (Figure 22b).

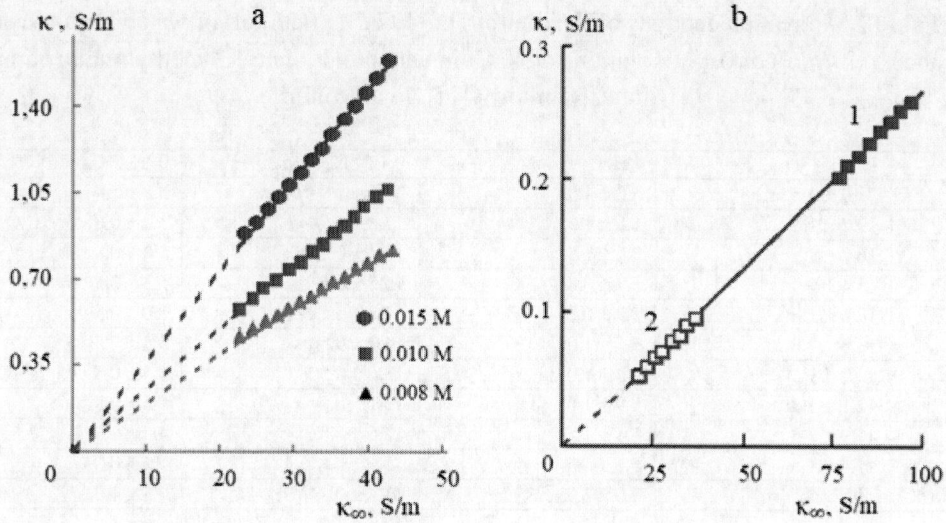

Figure 21. Dependencies $\kappa - \kappa_\infty$ for solutions of 1-butyl-3-methylimidazolium bistriflamide, [C_4mim][NTf_2], in dimethylformamide (a) and 0.02 M solutions of trihexyltetradecylphosphonium chloride, [P_{66614}]Cl, in acetonitrile (1) and dimethylsulfoxide (2) (b).

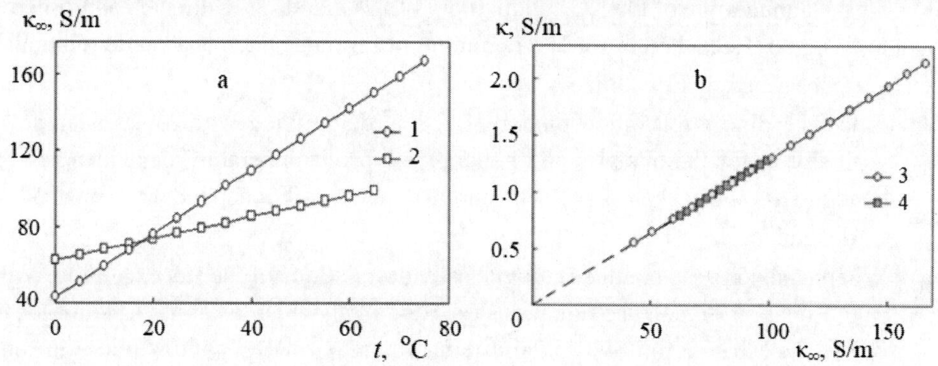

Figure 22. Temperature curves of the limiting HF EC of water (1) and acetonitrile (2) (a) and DC EC of aqueous 0.1 M NaCl (3) and 0.1 M 1-butyl-3-methylimidazolium triflate, [C_4mim][OTf], in acetonitrile (4) solutions *vs.* limiting HF EC of solvents (b) [107], reprinted with permission from [107].

On a single curve, similar to that shown in Figure 22b, the dependencies $\kappa - \kappa_\infty$ for 0.1 M solutions of 1-octyl-3-methylimidazolium triflate ([C_8mim][OTf]) in acetonitrile and $NaNO_3$ in water are demonstrated. The uniform dependence $\kappa - \kappa_\infty$ for IL solutions in acetonitrile and aqueous solutions of inorganic salts, Figure 22b, takes place in spite of the fact that in the temperature range of 5 – 60°C, the values of activation energies of the DC conductivity E_κ of these solutions significantly differ, Table 7.

In the 5-60°C temperature range, the E_κ of aqueous salt solutions are reduced from 21.6 to 10.7 kJ/mol, and the E_κ of IL solutions are reduced from only 5.8 to 5.4 kJ/mol. It should be noted here that the proportionality described by Equation (44) is performed for ILs and their solutions in solvents, as well as for aqueous solutions of inorganic salts, since for these systems the activation energy of the limiting HF EC in the temperature range under consideration is equal to the activation energy of the DC conductivity.

Table 7. Activation energies of the limiting HF EC $E\kappa_\infty$ (kJ/mol) of water, acetonitrile and E_κ (kJ/mol) of 0.1 M solutions of NaCl in water and 1-butyl-3-methylimidazolium triflate, [C4mim][OTf], in acetonitrile

t, °C	H_2O	CH_3CN	0.1 M NaCl	0.1 M [C4mim][OTf]
5	21.0 ± 1.5	5.8 ± 1.5	21.6 ± 1.0	5.8 ± 1.0
10	19.2 ± 1.5	5.2 ± 1.5	17.8 ± 0.8	5.8 ± 0.8
15	16.7 ± 1.5	5.0 ± 1.5	16.4 ± 0.8	5.8 ± 0.8
20	15.4 ± 1.0	5.7 ± 0.8	15.5 ± 0.5	5.3 ± 0.5
25	14.4 ± 1.0	5.9 ± 0.8	14.8 ± 0.5	5.6 ± 0.5
30	13.7 ± 1.0	5.1 ± 1.0	14.2 ± 0.5	5.5 ± 0.5
35	12.9 ± 1.5	5.9 ± 1.0	13.6 ± 0.5	5.5 ± 0.5
40	12.1 ± 1.5	5.9 ± 1.0	13.1 ± 0.8	5.5 ± 0.8
45	11.9 ± 1.5	5.1 ± 1.5	12.6 ± 0.8	5.5 ± 0.8
50	11.4 ± 1.5	5.1 ± 1.5	11.9 ± 0.8	5.5 ± 0.8
55	11.0 ± 1.5	5.0 ± 1.5	11.4 ± 0.8	5.5 ± 1.0
60	10.7 ± 2.0	5.1 ± 1.5	10.7 ± 1.0	5.4 ± 1.0

The following can be concluded from this section:

- As the temperature of the DC conductivity of ILs increases in direct proportion to their limiting HF EC, wherein for alkylammonium formates there is a single relationship $\kappa - \kappa_\infty$ described by Equation (44);
- Since the dipole relaxation time τ of ILs is determined with lower accuracy than viscosity η, for the analysis and generalization of temperature dependencies of DC conductivity κ of ILs, it is preferable to use the $\kappa - \varepsilon_s T/\eta$ dependencies instead of $\kappa - \kappa_\infty$ dependencies;
- When analyzing the conductivity-viscosity correlations of the ILs excepting Walden's rule, Equation (5) can use Equations (46) and (47), which take into account the effect of the dielectric permittivity ε_s and temperature T on DC conductivity, not only of viscosity;
- The proportionality between κ and κ_∞ described by Equation (44) is observed not only in pure ILs, but also in their solutions in polar solvents; these dependencies can be used to estimate and predict the DC conductivity values of ILs;
- The activation energy of the limiting HF EC of polar solvents coincides with the E_κ of IL solutions and inorganic salts in these solvents, which explains the proportionality between κ and κ_∞ described by Equation (44).

Association, Solvation and Distance between Molecules and Ions in Ionic Liquids

The processes of association and solvation in ionic liquids and their solutions are determined by the dielectric characteristics of the liquid environment. The association of ions in particular depends on the distance between the oppositely charged ions, which in turn decreases with increasing concentration. The solvation of ions is largely determined by the presence of free solvent molecules, which are capable of coordinating around the ions. The amount of such solvent molecules also decreases with increasing concentration. An analysis of these two important factors, which determinee association and solvation, is carried out in this section.

The distance between the ions decreases with increasing concentration, and at a certain concentration, a solvent molecule can no longer be located between the ions. If the energy of the Coulombic attraction exceeds the kinetic energy of the ions, contact ion pairs (CIP) are formed. In addition to CIP in ILs and their solutions, the formation of solvate-separated ion pairs (SIP) is possible. In particular, if the distance between the centers of the oppositely charged ions d in the liquid phase is equal to the sum of their radii ($R_{K^+} + R_{An^-}$) and the diameter of the solvent molecule d_S, then ion pairs separated by one solvent molecule (SIP) can be formed, Figure 23a.

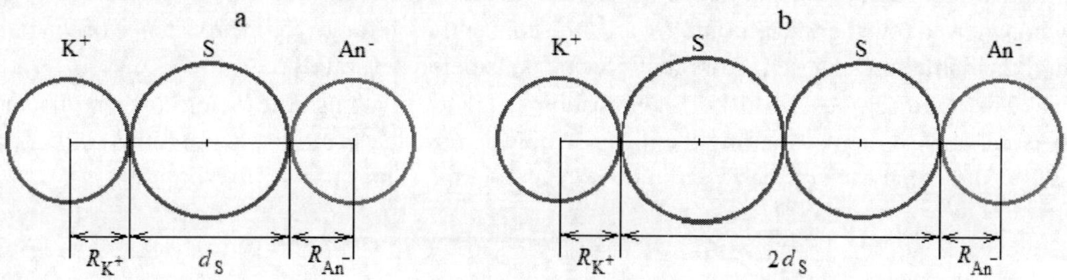

Figure 23. The distance between ions in the formation of solvate-separated ion pairs of one (a) and two (b) solvent molecules; R_{K^+} и R_{An^-} are respectively the radii of the cation and anion, d_S is the diameter of a solvent molecule.

In this case, one ion is solvated (in aqueous solutions of inorganic electrolytes, these are usually multicharge cations). Similarly, ion pairs in which both ions (cation and anion) are solvated, are formed if the distance between the centers of ions d is equal to the sum of the radii of the ions and two diameters of the solvent molecule, Figure 23b. The ions in these ionic pairs (2SIP) are separated by two solvent molecules. Contact ion pairs (CIP), as mentioned above, are formed in a solution if the distance between the ion centers d is smaller than the diameter of the solvent molecules. Thus, the following geometric boundaries of formation of the three types of ion pairs can be established.

For contact ion pairs:

$$d < R_{K^+} + d_S + R_{An^-}. \quad (48)$$

For ion pairs separated by one solvent molecule:

$$R_{K^+} + d_S + R_{An^-} \leq d < R_{K^+} + 2d_S + R_{An^-}. \quad (49)$$

For ion pairs separated by two solvent molecules:

$$d \geq R_{K^+} + 2d_S + R_{An^-}. \quad (50)$$

The second factor that affects the solvation and association process is the ratio between the number of ions and the number of solvent molecules capable of solvating them. When the number of solvent molecules in solution N is equal to the sum of the coordination numbers of the cation (N_{K+}) and the anion (N_{An-}), both ions are solvated in the solution and ion pairs separated by two solvent molecules (2SIP) can be formed, Figure 23b. The distance between the ion centers in this case is subject to the condition (49). Ion pairs separated by one solvent molecule (SIP) are formed if the solvent molecules are sufficient to solvate only one ion (cation or anion). Such ion pairs are formed if the condition described by inequality (49) is satisfied.

Finally, contact ion pairs (CIP) are formed if solvent (N) molecules in the solution are not even enough to solvate one of the ions ($N<N_{K+,An-}$). The formation of such ion pairs will occur if the distance between the ions in the solution is less than the diameter of the solvent molecule, i.e., the condition described by the inequality (48) is met. Each of the factors discussed above (the distance between ions and the number of solvent molecules per ion) separately do not allow an unambiguous conclusion about the nature of the association and solvation processes, but when viewed together, they point towards the concentration range for the existence of ion pairs and their structure. These factors will now be considered separately.

The approach described by Shcherbakov et al. [60] was used to determine the distance between ions. When evaluating the distance between particles (molecules and ions) in ILs, we will assume that each of the N particles is located in the center of its cube, Figure 24.

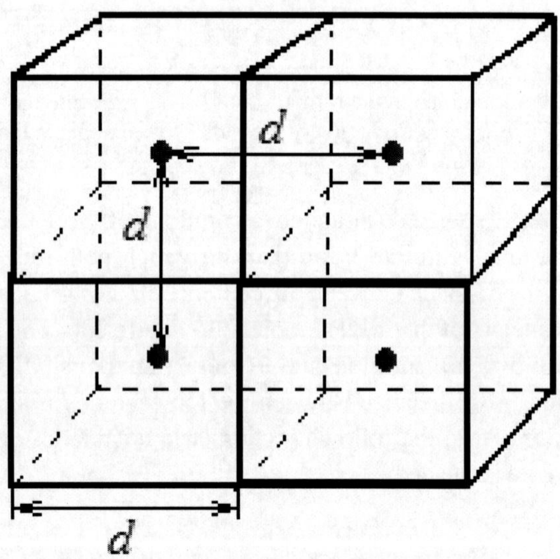

Figure 24. Arrangement of particles (molecules and ions) in the center of cubes.

The number of cubes is equal to the number of particles, and the volume of one cube V_1 is equal to the total volume V divided by the number of particles N:

$$V_1=V/N \tag{51}$$

The average distance between particles is easy to calculate if the total volume V and the number of particles N are known. For example, for one mole of ideal gas under normal conditions:

$$V=V_o=22.4\text{ L} = 22.4\cdot 10^{-3}\text{ m}^3 \tag{52}$$

The volume V_1 in which one molecule of the ideal gas is located is found by dividing the molar volume V_o by the Avogadro number N_A:

$$V_1 = V_o/N_A = 22.4\cdot10^{-3}/(6.02\cdot10^{23}) = 37.2\cdot10^{-27}\ \mathrm{m}^3 \tag{53}$$

The average distance between the two molecules will be equal to the length of the cube edge d (Figure 24):

$$d = \sqrt[3]{V_1} = 33.4\cdot10^{-10}\ \mathrm{m} = 33.4\ \mathring{A}. \tag{54}$$

To calculate the distance between the centers of molecules in an IL, its molar volume V_m and Avogadro's number are used. Since the degree of dissociation of the ILs is not known, two limiting cases are considered, when dissociation is absent (the number of particles is N_A) and when all ILs molecules are completely dissociated (the number of particles is $2N_A$). In the first case, the distance between the centers of the ILs d_1 molecules is:

$$d_1 = \sqrt[3]{\frac{V_m}{N_A}} = 1.184\sqrt[3]{V_m}\cdot10^{-10}\ \mathrm{m} = 1.184\sqrt[3]{V_m}\ \mathring{A}. \tag{55}$$

In the second case, the distance between the ion centers will be:

$$d_2 = \sqrt[3]{\frac{V_m}{2N_A}} = 0.940\sqrt[3]{V_m}\cdot10^{-10}\ \mathrm{m} = 0.940\sqrt[3]{V_m}\ \mathring{A}. \tag{56}$$

The molar volume in Equations (55) and (56) is expressed as cm^3.

Since the degree of dissociation in the ILs is unknown, the distance between the particles (molecules and ions) in the ILs will be in the range of:

$$1.184\sqrt[3]{V_m} \le d \le 0.940\sqrt[3]{V_m}\ \mathring{A} \tag{57}$$

For liquid water (V_m=18 см^3), using Equation (55), the following expression is obtained:

$$d = 1.184\sqrt[3]{18} = 3.10\ \mathring{A}. \tag{58}$$

The obtained value is slightly larger than the diameter of the water molecule (2.76 Å). However, the calculated value of the diameter of the water molecule according to Equation (55) is the same, if the dense packaging of its molecules is taken into account (in general, the ball occupies only a part of the volume of the cube). Instead of molar volume, it is necessary to put $0.7V_m$ into Equation (55).

As a result, Equation (55) for estimating the distance between the centers of two molecules in the liquid phase takes the form of:

$$d = \sqrt[3]{\frac{0.7V_{\mathrm{m}}}{N_{\mathrm{A}}}} = 1.0513\sqrt[3]{V_{\mathrm{m}}}\cdot 10^{-10}\ \mathrm{m} = 1.0513\sqrt[3]{V_{\mathrm{m}}}\ \mathrm{\mathring{A}}. \tag{59}$$

In the case of water, using Equation (59), results in $d = 1.0513\sqrt[3]{18} = 2.756\ \mathrm{\mathring{A}}$. This value coincides with the diameter of the water molecule (2.76 Å). Equation (59) may be used to estimate the size of the liquid non-electrolyte molecules, for example the VOCs. The factor 1.0513 in Equation (59) is approximately equal to half the sum of the factors in Equations (57) and (58). Therefore, Equation (59) can also be proposed to estimate the average distance between the particles (molecules and ions) in ILs. The distance between the ions in solution *d* can be estimated by knowing the molar concentration of the solute *c* (mol/L). 1 L of the solution is considered. In general, if *n* ions are formed during the dissociation of the one electrolyte molecule, their concentration in the solution will be *nc*. The volume per particle (per ion) in the solution will be:

$$V_1 = \frac{10^{-3}}{N_{\mathrm{A}}nc}\ \mathrm{m}^3. \tag{60}$$

The distance between the ions in the solution will be equal to the cubic root of V_1:

$$d = \sqrt[3]{\frac{10^{-3}}{N_{\mathrm{A}}nc}} = \sqrt[3]{\frac{10^{-3}}{6.02\cdot 10^{23}nc}} = \frac{11.84}{\sqrt[3]{nc}}\ \mathrm{\mathring{A}}. \tag{61}$$

It follows from expression (61) that for a 1 M non-electrolyte solution (the case where ILs molecules do not dissociate), the distance between the molecules is 11.84 Å. If all the molecules in the 1M IL solution dissociate completely, then the distance between the ions will be: $d=11.84/(2)^{1/3}=11.84/1{,}26=9.40$ Å. Equation (61) can be used to calculate the distance between ions in different types of electrolyte solutions. Since the dissociation of the ILs always produces two ions, the IL solutions Equation (62) takes the following form:

$$d = \sqrt[3]{\frac{10^{-3}}{2N_{\mathrm{A}}c}} = \frac{9.40}{\sqrt[3]{c}}\ \mathrm{\mathring{A}}. \tag{62}$$

From Equation (62), it is possible to find a concentration *c* (mol/L) in the solution of ILs between ions which are separated by a certain distance *d* (Å):

$$c = \frac{10^{-3}}{2N_{\mathrm{A}}d^3} = \frac{8,30\cdot 10^2}{d^3},\ \mathrm{mol/l}. \tag{63}$$

This expression can be used to estimate the concentrations of ILs at which various types of ion pairs (CIP, SIP and 2SIP) are formed in their solutions in VOCs. The *d* values for these ion pairs are determined by Expressions (48), (49) and (50). The equations given in this section are used for the analysis of the concentration dependence of specific conductivity for four IL solutions in acetonitrile: 1-butyl-3-methylimidazolium bis{(trifluoromethyl)sulfonyl}amide, $[C_4mim][NTf_2]$; 1-butyl-3-methylimidazolium trifluoromethanesulfonate triflate, $[C_4mim]$ [OTf]; 1-octyl-3-methylimidazolium triflate $[C_8mim][OTf]$ and 1-butyl-3-methylpyridinium bis{(trifluoromethyl)sulfonyl}amide, $[C_4m_\beta py][NTf_2]$.

These ILs were synthesized at the Queen's University Belfast (Northern Ireland). Their specific conductivity was measured over a wide range of concentrations in the temperature range of 20-65°C. With an increase in concentration, the specific EC κ passes through the maximum, while for each IL, the experimental values of the the reduced electrical conductivity of the κ/κ_{max} fit well on a single curve (κ_{max} is the maximum specific EC at a given temperature). Figure 25 shows the κ/κ_{max} – *c* relationship for $[C_4mpy][NTf_2]$ acetonitrile solutions as an example.

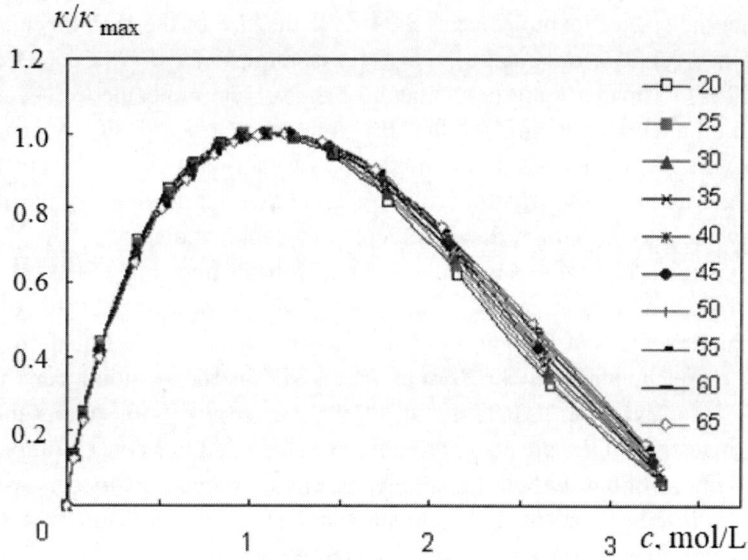

Figure 25. Dependence of EC κ/κ_{max} solutions of $[C_4mpy][NTf_2]$ in acetonitrile on the molar concentration *c* in the 20 – 65°C temperature range; temperature values (°C) are shown in figure.

One curve in the κ/κ_{max} – *c* coordinates also fits the values of the reduced EC for the other three studied ILs. If the given c/c_{max} concentration (c_{max} is the concentration corresponding to the maximum EC at this temperature) is used as an argument, the values for all four ILs are located on the same curve [60]. The considered patterns of change for specific ECs in the κ/κ_{max} – *c* and κ/κ_{max} – c/c_{max} coordinates were previously established for aqueous solutions of strong [58] and weak [59] electrolytes.

Using the equations obtained in this section, it is possible to find out the nature of the maximum in the concentration dependencies of the specific EC of the ILs under consideration, Figure 25. For the four ILs shown in Table 8 the molar mass (M), density (ρ) and molar volume (V_m) calculated from the molecular size Equation (59) ($d_{(ILs)}$), the concentration corresponding to the maximum specific EC ($c(\kappa_{max})$) and some other parameters, which are discussed below.

Table 8. Values of the molar mass (M), density (ρ) and molar volume (V_m) are calculated from the molecular size Equation (59) ($d_{(ILs)}$), the concentration corresponding to the maximum specific EC ($c(\kappa_{max})$), distance (d) corresponding to maximum specific EC, concentration corresponding to SIP formation (c_{SIP}) and the molar concentrations in pure ILs (c_{ILs})

IL	M, g/mol	ρ, g/cm^3	V_m, cm^3	$d_{(ILs)}$, Å	$c(\kappa_{max})$/mol/L	d, Å	c_{SIP}, mol/L	c_{ILs}, mol/L
[C_4mim][NTf_2]	419.4	1.4403	291.2	6.97	1.06	9.22	0.639	3.43
[C_4mim][OTf]	288.3	1.3083	220.4	6.35	1.54	8.15	0.762	4.54
[C_8mim][OTf]	344.4	1.1923	288.8	6.95	1.03	9.31	0.643	3.46
[C_4mpy][NTf_2]	430.4	1.4187	303.4	7.06	1.15	8.97	0.624	3.30

The distance d between the centers of the IL ions at the concentration $c(\kappa_{max})$ corresponding to the maximum specific EC can be estimated from Equation (62) (Table 8), assuming that all ILs molecules are dissociated. For [C_4mim][NTf_2], for example, d=9.40/(1.06)$^{1/3}$=9.22 Å. The values of d corresponding to the maximum specific EC of the solutions are also presented in Table 8. Based on the molar volume of acetonitrile (52.13 cm^3) according to Equation (42), it was found that the size of its molecule is 3.94 Å. If the size of the IL molecule is subtracted from the obtained value d (for [C_4mim][NTf_2], for example, 9.22-6.97=2.25 Å), then the value obtained (2.34 Å) is smaller than the diameter of the solvent molecule (3.94 Å). Thus, at the c_{max} concentration corresponding to the EC maximum on the EC dependence of the concentration, the inequality (48) is satisfied, i.e., contact ion pairs (CIP) are formed in the solution. In the calculations carried out, it is assumed that the diameter of the IL molecule is approximately equal to the sum of the diameters of its cation and anion.

Solvate-separated ion pairs containing a single solvent molecule (SIP) can be formed in solution under the condition described by inequality (49). Substituting the sum of the size of IL and acetonitrile molecules in Equation (63) instead of the value d in Equation (63), the cSIP concentration corresponding to the formation of SIP in the solution can be found. For [C_4mim][NTf_2], for example, d=6.97+3.94=10.91 Å; c_{SIP}=830/10.91^3=0.639 mol/L. The c_{SIP} values for all investigated ILs are also presented in Table 8. At c_{SIP} concentrations (Table 8) in acetonitrile solutions of the studied ILs, an increase in the specific EC occurs with an increase in concentration (Figure 25). Therefore, it can be concluded that SIP ion pairs separated by a single solvent molecule are not formed in the tested ILs solutions.

The degree of dissociation of molecules into ions in pure ILs is not known. In the above ILs, the maximum concentration of undissociated molecules c_{ILs} is 3.30 – 4.54 mol/L, Table 8. Since the concentration of the ILs decreases during dissociation and the number of ions doubles compared to the number of dissociated molecules, the concentration of ions and undissociated molecules becomes commensurable even with a small dissociation of ILs. For [C_4mim] [NTf_2], for example, at a degree of dissociation α=20%, the concentration of ions will be equal to 1.372 mol/L, and the concentration of undissociated molecules will decrease to 2.744 mol/L. There are only two undissociated molecules per ion at this degree of dissociation. With a high degree of dissociation of the ILs, the concentration of undissociated ILs molecules is significantly lower than the concentration of ions formed. The ions formed during dissociation can no longer be separated by undissociated molecules.

For all ILs, the concentration of ions is equal to the concentration of undissociated molecules at a dissociation degree of 25%. Therefore, in pure ILs, it seems that it does not make sense to talk about the solvation of the formed ions, as is customary in the case of aqueous and non-aqueous electrolyte solutions (in pure water the concentration of its molecules is 55.5 mol/L, therefore, in 1 M KCl solution, subject to complete dissociation of the electrolyte for each ion, there are an average of 27.7 water molecules). In pure ILs with their substantially higher molecular weight, ions cannot be surrounded by undissociated molecules. In these ILs, it is possible to form ion-molecular aggregates in which there is one or two undissociated IL molecules per IL ion. Believing that only the non-dissociated ILs are capable of conducting electric current, we conclude that the degree of dissociation in pure ILs is not large. At least, it cannot appear to exceed 50%, since with this degree of dissociation, two IL ions are formed from one IL molecule, and one non-dissociated molecule remains that can separate these ions and prevent the contact ion pair formation.

Summarizing the above, the following can be noted:

- The formation of ion pairs of different structures (CIP, SIP and 2SIP) in ILs solutions is possible under conditions, described by inequalities (48), (49) and (50);
- Equations (55) and (56) allow us to estimate the distance between molecules and ILs ions in solutions. Equation (63) allows us to determine the concentration *c* at which a certain value of distance occurs in IL solutions between ions;
- The presence of a maximum in the dependence of the specific EC on the IL concentration is associated with the formation of contact ion pairs in solution;
- The formation of solvate-separated ionic pairs (SIP and 2SIP) is unlikely in ILs solutions; at concentrations where one or two solvent molecules can be placed between ILs ions, the specific EC increases with the increase in ILs content;
- In pure ILs, the concentration of undissociated molecules is not high and is commensurate with the concentration of ILs ions; ions remain separated by one non-dissociated IL molecule at a dissociation degree that does not exceed 50%.

Dielectric Characteristics of Ionic Liquid Solutions in Polar Solvents

As noted above, accurate measurements of the dielectric characteristics of conductive solutions became possible when, in the middle of the last century, there was a transition from the radio frequency (meter wavelength range) into the microwave range (centimeter and millimeter wavelength range) [72, 108]. When calculating the dielectric characteristics (ε_s and τ) based on the values measured at individual frequencies ε'', it was necessary to introduce a correction for the DC conductivity κ, Equation (49).

Analysis of the relaxation processes when applying a microwave field to an electrolyte solution was carried out, as a rule, using pie diagrams (Figures 5 and 6). Depending on the nature of the dissolved electrolyte and its concentration in aqueous solutions of inorganic salts, pie diagrams were obtained indicating either the absence or presence of a relaxation time distribution [109]. Modern equipment currently makes it possible to obtain almost the entire dielectric spectrum of IL solutions, ranging from several MHz to tens and hundreds of THz. The analysis and interpretation of the dielectric spectrum of IL solutions uses the same approach that was previously developed for electrolyte solutions [108]. Subsequently, this approach was supplemented with new interpretations of individual regions of the spectrum obtained over a wide frequency range [101].

As shown in the work [101], the wideband dielectric spectrum of electrolyte solutions can detect even a weak ionic association that occurs in aqueous inorganic salt solutions. In these solutions, three types of ion pairs have been established - contact (CIP) and solvate-separated (SIP, 2SIP) (Figure 22). Analysis of the obtained spectra makes it possible to obtain the concentrations of all the present ion pairs and to estimate the constants of their association. One important circumstance should be paid attention to. If in ILs the contributions of additional relaxation processes (4, 5, 6, Figure 14) are offset to the higher frequency region than the main process (Debye region of relaxation of the IL molecules 3, Figure 14), then in electrolyte solutions and ILs solutions, additional regions of relaxation of the ion pairs are shifted to the region of lower frequencies, compared with the Debye region of relaxation of the polar solvent [23, 24, 101, 110-112]. One of the first publications that claims an additional low-frequency region of ion pair relaxation is the case of a 1M aqueous solution of beryllium sulfate [113]. The results of various studies on the dielectric relaxation processes in IL solutions will now be considered.

The ionic association of $[C_4mim]Cl$ solutions in acetonitrile and water was investigated not only by dielectric radio spectroscopy at a concentration of $c \leq 1.3$ M at T = 298.15 K, but also by measuring the molar conductivity Λ of diluted solutions ($c \leq 0.006$ M) in the $273.15 \lesssim T \leq 313.15$ K temperature range [24]. As a result of the conductometric studies, it was found that moderate ion pairing occurs in acetonitrile. In this case, when the temperature increases, the K_a association constant increases from 56.0 (273.15 K) to 68.2 (313.15 K). In water, $[C_4mim]Cl$ shows weak pairing (unlike in acetonitrile solutions). For $[C_4mim]Cl$ in water, K_a decreases from 7.2 (278.15 K) to 4.5 (313.15). Two modes in aqueous solution and three modes for solutions in acetonitrile were found in the dielectric spectrum of $[C_4mim]Cl$ solutions. Low-frequency modes (at ~ 1 GHz for water and at ~ 2 GHz for acetonitrile) were attributed to the rotational diffusion of the $[C_4mim]Cl$ contact ion pairs. For aqueous solutions, high-frequency relaxation (~ 20 GHz) is a complex mode due to volumetric water and, possibly, free cations. For solutions in acetonitrile, high-frequency relaxation (~ 45 GHz) is due solely to an undisturbed solvent. Relaxation in the intermediate frequency region (~11 GHz) was also found in acetonitrile solutions. Free cations and the acetonitrile molecules solvating them are responsible for this region [24]. The evaluation of the effective solvation numbers performed in this work showed that the cation $[C_4mim]^+$ is more strongly solvated in acetonitrile - effective solvation numbers of $[C_4mim]Cl$ in acetonitrile are about twice as large as in water.

According to the expression (34), the λ_0/κ_∞ ratio should not depend on the temperature, i.e., when the temperature increases, the molar conductivity of $[C_4mim]Cl$ solutions λ_0 in water and acetonitrile should increase directly in proportion to the limiting HF EC κ_∞ of the polar solvent. The $\lambda_0 - \kappa_\infty$ relationships for the $[C_4mim]Cl$ solutions in water and acetonitrile are presented in Figure 26.

As with non-aqueous inorganic salt solutions [94], the molar conductivity $[C_4mim]Cl$ increases with increasing temperature in proportion to the HF EC of water (1) and acetonitrile (2), Figure 26. Thus, the $\lambda_0 - \kappa_\infty$ relationships shown in Figure 26 confirm the validity of Equation (45), which can be used, as well as Walden's rule, in the analysis of the temperature dependence of the molar conductivity of IL solutions. Equation (45), as well as Equation (44) discussed above, confirm the close relationship of the ion migration process (κ and λ_0 values) with the dielectric characteristics of the solvent (ε_s and τ), the ratio of which determines the value of its HF EC, Equation (15).

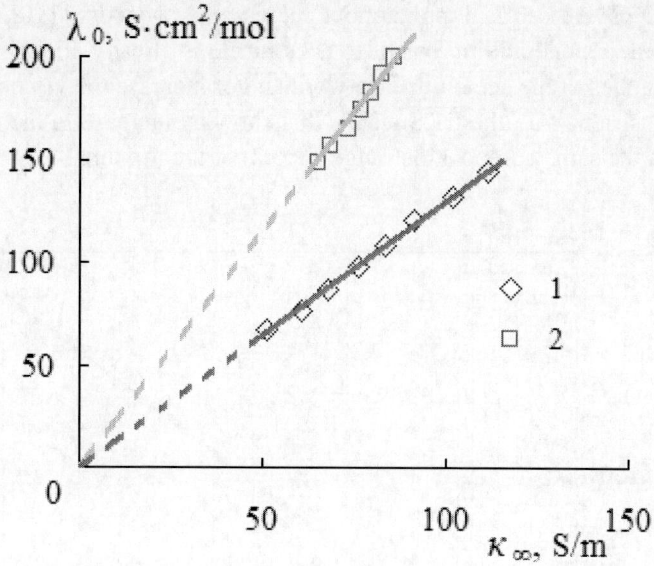

Figure 26. Dependence of molar conductivity of $[C_4mim]Cl$ solutions in water (1) and in acetonitrile (2) on the limiting HF EC of the polar solvent.

Solutions of $[C_6mim][NTf_2]$ in acetonitrile (AN) were investigated by dielectric spectroscopy (DRS) and ultrafast transient absorption techniques using the 120-apo-b-carotenoic-120-acid, 120 CA, molecular probe [110]. The combination of these methods, according to the authors, provides a more complete idea of the polarity and dynamics of the $[C_6mim][NTf_2]$ - acetonitrile mixtures. In the studied frequency range (0.2-89 GHz), the IL spectrum is modeled by the Cole-Cole low-frequency mode superposition (CC) with a center at ~ 0.6 GHz frequency and a small Debye (D) relaxation with a center at ~ 200 GHz frequency. For $[C_6mim][NTf_2]$, as for all imidazolium ILs, the low-frequency mode can be attributed to a jumping reorientation of the cation to large angles. The high-frequency mode is only a formal description of the significant contributions of librations and intermolecular oscillations that prevail in the far infrared region. The dielectric spectra of $[C_6mim][NTf_2]$ - (AN) mixtures are well approximated by the combination of CC + D modes.

In the work [110], it has been found that as the concentration of the IL increases, the static permittivity of the solutions greatly decreases (from ~ 35.9 for acetonitrile to 12.7 for pure $[C_6mim][NTf_2]$). The average dielectric relaxation time τ_{ev} *vice versa*, is greatly increased (from 3.3 ps for acetonitrile to 243 ps for pure $[C_6mim][NTf_2]$). A smooth extrapolation of the ε_s to the DP value of pure acetonitrile is not possible with a decrease in the IL content [110]. At the same time, an interesting pattern is observed on the dependence of the inverse HF EC κ_∞ on the concentration of $[C_6mim][NTf_2]$, Figure 27 (κ_∞ values were calculated by us based on the ε_s and τ_{ev} values of solutions in acetonitrile given in this work).

The concentration range of $1/\kappa_\infty$ can be divided into two regions: diluted - region 1 ($x<0.2$) and concentrated - region 2 ($x>0.2$) solutions. It should be pointed out that with a low IL content, discrete ion pairs affect the properties of the solution. In this area, the DR spectrum is dominated by the contribution of acetonitrile. At $x>0.2$, IL begins to have a significant effect on the DR spectrum. The dependence on the composition of the internal conversion time constant τ_1 [110] is similar in nature to that shown in Figure 27: in the interval of mole fractions

0≤x<0.2, τ_l≈40±3 ps. At x>0.2, a monotonous increase in τ_l occurs [110]. As the authors of this work point out, a small additive of IL to acetonitrile is already enough to slow down the reorientation dynamics of the acetonitrile network and/or increase the viscosity of the mixture. The pattern is consistent with the DRS results that show an increase in the mean longitudinal relaxation time in the same composition range (x<0.2) of the mixture.

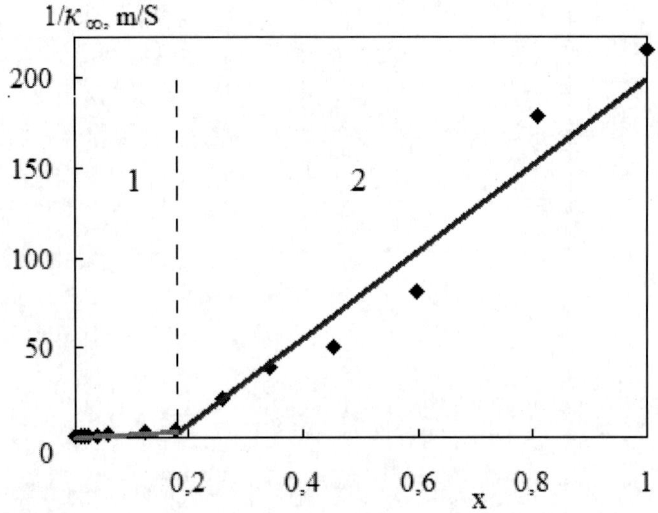

Figure 27. Inverse HF EC κ_∞ *versus* concentration of [C_6mim][NTf_2].

Wideband dielectric spectroscopy was used to examine the solutions of 1-alkyl-3-methylimidazolium (alkyl: ethyl, butyl, or hexyl) tetrafluoroborates in acetonitrile at 25°C [112]. The dielectric spectra of all studied mixtures as well as in the work [110] are approximated by the superposition of only two processes: the low-frequency Cole-Cole (CC) mode with the center at ~2 GHz and the higher-frequency Debye (D) relaxation with the center at ~50 GHz. For mixtures with molar fractions of IL x_{IL} < 0.2, an additional low frequency Debye process was used, i.e., the model D + CC + D was applied [112]. A detailed analysis of the CC and D models showed that they are composed of different relaxation processes. The slower CC process centered at ~2 GHz is due to the reorientation of the dipolar imidazolium cations. The reorientation of the "slow" acetonitrile molecules, which have lost their mobility as a result of solvation of these cations, is also superimposed on this process. The slow CC relaxation at the mole fraction of ILs x≤0.2 may also be caused by the relaxation of the contact ion pairs (CIPs). The fast Debye mode with a center at a frequency of ~ 50 GHz is due to the relaxation of the almost free acetonitrile molecules. The rapid intermolecular oscillations and librations of the ILs are also possible in this region by overlapping this relaxation region with a low-frequency wing of very fast intermolecular oscillations and librations of ILs [112].

Thus, as in the previous work [110], the entire concentration region of the IL test solutions in acetonitrile can be divided into two regions. At low concentrations (x<0.2), the ILs behave as conventional weakly associated electrolytes, and at higher (x>0.2) concentrations, they behave like expanded ("lubricated") ILs. The transition region (x≈0.2) is characterized by the redissociation of ion pairs and the establishment of an IL-like structure. In this concentration region (0.1<x<0.3), for solutions of all the studied ILs in acetonitrile, there are extreme values of τ_1 and τ_2 relaxation times for the various processes [112]: the τ_1 value reaches its minimum value and the τ_2 reaches its maximum value. This boundary region at the x≈0.2 concentration is also characteristic of the solutions of $[C_6mim][NTf_2]$ in acetonitrile [110], Figure 27.

The use of different combinations of the Debye (D) and Cole-Cole (CC) models, in particular the CC + D and D + CC + D, allowed to also estimate the association constants K_a of the ILs in acetonitrile [110]. The obtained K_a values are arranged in the following order:

$[C_2mim][BF_4] > [C_4mim][BF_4] > [C_6mim][BF_4]$

Although the obtained association constants exceed the K_a values calculated from the precision conductometric measurements, they confirmed that the ion association of $[C_nmim][BF_4]$ in acetonitrile is rather weak and it is broadly comparable with that of the simple $NaBPh_4$, $LiClO_4$, Bu_4NClO_4, Bu_4NBPh_4 and LiCl electrolytes in this solvent [112].

The dielectric relaxation and the dynamics of solvation of 1-butyl-3-methylimidazolium tetrafluoroborate mixtures with water have been investigated at 25.00 ± 0.05°C [23]. The DR spectra were recorded in the 200 MHz to 89 GHz frequency range with 17 different $[C_4mim][BF_4]$ - H_2O mixture compositions. The tested solutions were measured immediately after preparation to reduce the errors associated with $[BF_4]^-$-ion hydrolysis [23]. The dielectric spectra obtained in this work were modeled using a superposition of the four Debye-type processes with dielectric relaxation times ranging from 1 to 300 ps. The Dynamic Stokes shift measurements on the solute coumarin 153 (C153) were also measured to study the solvation processes. These measurements for the 7 formulations at 20.5°C were made by combining fluorescence upconversion (80 fs resolution) and time-correlated single photon counting data (20 ns range). As a result of the analysis of the obtained data, it was found that all the properties measured continuously changed depending on the composition of the mixture, if it was expressed not in molar, but in volume fractions. The solvation response functions predicted by a simple model of the dielectric continuum were similar to but ~7-fold faster than the spectral response functions measured with C153.

Despite the use of a cubic polynomial in the equation for the static permittivity ε_s of the studied mixture, when expressing the concentration $[C_4mim][BF_4]$ in volume fractions, the ε_s decreases approximately linearly with its growth, which cannot be said about the transport properties - the viscosity η and the electrical conductivity κ. With an increase in the IL concentration, the specific conductivity passes through the maximum that occurs at a volume fraction of the IL ~ 0.5. This pattern is characteristic of aqueous mixtures of ILs [57]. In general, the authors of this work noted that the dielectric characteristics of the mixtures $[C_4mim][BF_4]$ with water with an increase in the concentration of the IL (expressed in mole or volume fractions) changed monotonously and there was no evidence for abrupt structural changes taking place at particular compositions from any of the data examined [23].

The dielectric relaxation of the ethylammonium nitrate (EAN) with acetonitrile (AN) mixtures at 25 °C was investigated over the entire composition range by wideband dielectric spectroscopy over a frequency range of $0{,}2 \le \nu/\text{GHz} \le 89$ in the work [114]. The dielectric spectra obtained in this work were split into two relaxation processes, both of which proved to be composite modes. In diluted solutions, the existence of the higher-frequency Debye relaxation with a center at a frequency of ~ 60 GHz, which is due to the rotational diffusion of acetonitrile molecules, has been established. At higher concentrations of EAN, according to the authors of the work [114], ultrafast intermolecular oscillations and librations of EAN prevailed in the relaxation processes. For concentrated EAN solutions, low-frequency relaxation is mainly associated with the abrupt reorientation of the ethylammonium cation. Contact ion pairs (CIPs) predominate in this frequency range in the diluted EAN solutions. From the amplitudes of the relaxation processes in the work [114], effective solvation numbers and ion pair concentrations were determined. The effective cation solvation number (~7) was found by extrapolation to the zero-tending mole fraction of EAN. This value drops to ≈0.2 as the IL content in the solution increases. Calculated from the dielectric data, the EAN association constant (K_{0A}=970 L mol^{-1}) was found to be more than 30 - 50 times the electrolyte 1:1 association constant in AN. This fact indicates the formation of contact ion pairs (CIPs) due to the formation of strong hydrogen bonds between anions and cations despite the strong solvation of cations.

The dielectric relaxation of ethylammonium nitrate (EAN) and its mixtures with nitromethane was investigated in the study [111] in two single-phase concentration regions of the EAN $0<x_{EAN} \leq 0.013$ and $0.4 \leq x_{EAN} <1$ at 25°C. Even in the title of this work, the authors noted the possibility of a strong ion association induced by hydrogen bonding. All the spectra obtained in this work were well described by the superposition of two relaxation processes of a complex nature. With a low EAN content ($<x_{EAN} \leq 0.013$), the solutions behaved as a strongly associated electrolyte, showing moderate solvation of ions by nitromethane molecules. At high EAN contents away from a miscibility gap ($x_{EAN} \geq 0.4$), the observed dynamics represented the typical behavior of a lubricated ionic liquid smoothly achieving pure EAN properties. This work provides a clear evidence that hydrogen bonds between ions are the main driving force behind the strong association of EAN to form contact ion pairs in nitromethane, as occurs in acetonitrile [114].

The pure nitromethane in this work [111] was also investigated in the temperature range of 5 to 65°C to explore the possibilities of its use as a calibration standard in dielectric spectroscopy. The resulting spectra in the 0,05 - 89 GHz frequency range were well approximated by one Debye equation. This means the dynamics of this dipole liquid is determined by the rotational diffusion of nitromethane dipoles close to the slip boundary conditions with only weak dipole-dipole correlations. Unfortunately, difficulties in obtaining reproducible purity samples make it difficult to use nitromethane as a calibration standard for dielectric measurements [111]. The reason for the miscibility gap of nitromethane with EAN in the $0.013<x_{EAN}<0.4$ concentration range is, according to the authors of the work [111], a lower acceptor number compared to acetonitrile (2.7 for NM *versus* 14.1 for AN), which leads to weaker solvation and stronger segregation in NM compared to AN [111].

Hefter and Buchner in their work on the dielectric spectroscopy of electrolyte solutions [101] noted that this method makes it possible to better understand the nature of ion-ion and ion-solvent interactions and will be useful for finding a weak ion association due to its unique ability to detect solvent-separated particles to quantify ion pairs. It also provides important information about ionic solvation. Dielectric relaxation spectroscopy (DRS) has a unique ability to detect "slow" water produced by hydrophilic and hydrophobic hydration of solutes. In aqueous electrolyte solutions, the authors used this method to estimate "effective" ion hydration numbers based on the dynamics of the closest water molecules. Great opportunities opened up for DRS to study the behavior of ionic liquids and their solutions in polar solvents. Despite the sufficiently low dielectric permeability of ILs, as is not surprising, they do not contain a significant amount of ion pairs. However, they, as shown by ultra-wideband DRS, exhibit extremely complex behavior, especially at frequencies in the THz region [101].

The characteristic features of most ILs are their negligible pressure, wide electrochemical windows, high inherent conductivities, high thermal and electrochemical stability, tuneable physicochemical properties, hence, some ILs could be used as environmentally sound, green electrolytes in batteries. All these beneficial characteristics of ILs open up opportunities for their broad application in their pure state and as solutions. Prior to this application of ionic liquids, it is important to know their fundamental physical and chemical properties. The significance and importance of these results become even more obvious if the fact that energy storage cells based on ILs are exposed to large temperature differences is considered. The knowledge of dielectric properties, viscosity and behaviour at different temperatures is crucial for the various applications, planning and preparing of production and optimization processes. Further studies of ILs and their solutions by the DRS method over a wide range of frequencies and temperatures will provide new information on the structure and ion-molecular processes surrounding ILs. This information will allow progress in the established practical applications of ILs, as well as leading to new areas.

Based on the analysis of the results of dielectric studies of IL solutions, the following can be drawn:

- The dielectric spectra of IL solutions in polar solvents are well approximated by the Cole − Cole (CC) mode superposition and higher-frequency Debye relaxation (D); for mixtures with molar fractions of IL $x_{IL} < 0.2$; an additional low-frequency Debye process is also possible, that is, the model D + CC + D;
- Unlike pure ILs, in which additional relaxation processes (CC) occur in the higher frequency region compared to the main Debye (D) molecular relaxation region, in IL solutions in polar solvents (acetonitrile, water) additional dispersion regions (CC) occur at lower frequencies compared to the molecular Debye (D) dielectric relaxation;
- The entire concentration region of the aprotic IL solutions in acetonitrile can be divided into two regions: at low concentrations ($x<0.2$), ILs behave as ordinary weakly associated electrolytes, and at higher concentrations ($x>0.2$) like expanded ("lubricated") ILs; the transition region ($x \approx 0,2$) may be associated with the redissociation of ion pairs and the establishment of an IL-like structure;
- Unlike ILs solutions in acetonitrile, the dielectric characteristics of [C_4mim][BF_4] mixtures with water change monotonically as the concentration increases, indicating that there are no sharp structural changes in the tested mixture ([C_4mim][BF_4]-H_2O).

Conclusion

The electrical conductivities of ionic liquids (ILs) and their mixtures with molecular solvents have been analysed. The electrical conductivity of pure ILs decreases with increased alkyl chain length on the cation as their viscosities increase too, and conversely increases by enlarging the anion size. This electrical conductivity increase is due to hydrogen bonding and charge delocalization. The dependence on the solvent composition (IL: solvent ratio) and temperature is additionally discussed. Electrical conductivities of ionic liquids were considered in dilute and concentrated ionic liquid solutions in polar solvents. From the analysis of molar conductivities and association constants, it was deduced that ILs behave like strong electrolytes in dilute solutions.

The association of ILs depends on the solvation nature of its ions, the permittivity of the solvent, and temperature; the association constants K_a of ILs in polar solvents increase with temperature. The thermodynamic characteristics of ion association are defined *via* various equations of the molar electrical conductivity concentration dependence are currently used, which can give inconsistent values of the association constants K_a. To compare the K_a values of different ILs in polar solvents of different nature, it is advisable to use the data obtained using the same electrical conductivity equation. The maximum electrical conductivity κ_{max} at a given temperature and the corresponding concentration c_{max} can be used to generalize the concentration and electrical conductivity temperature dependences of concentrated solutions of ILs in polar solvents.

In terms of dielectric properties of ionic liquids, the electrical conductivity and dielectric constant of ILs are independent of the frequency of the electromagnetic field in the low frequency domain. Therefore, it is necessary to take this relationship into account and consider the phenomena of permittivity dispersion and the high-frequency conductivity dispersion of ILs.

The specific electrical conductivity (EC) κ and static dielectric permittivity (DP) ε_s of ILs define the relationship of the conductivity currents to the displacement currents in an alternating electromagnetic field. When these currents are equal in ILs, an electrical (ionic) relaxation occurs, the time of which θ is equal to the ratio of absolute DP ($\varepsilon_s\varepsilon_o$) to the specific EC κ.

As a result of ionic relaxation, there is an apparent increase in the electrical conductivity with an increase in the frequency of the electromagnetic field ω; similarly, there is an apparent increase in the dielectric constant of ILs with increasing frequency; in fact, the specific EC κ and DP ε_s of the ILs retain their constant value in the region of frequencies corresponding to ionic relaxation, which is independent of the frequency of the electromagnetic field.

The ionic relaxation time of the diluted electrolyte solution is substantially shorter than the dipole dielectric relaxation time of water; in ILs, the ionic relaxation time θ becomes commensurate with the dipole dielectric relaxation time τ of the polar liquid.

The DC conductivity of the ILs is closely related to their dielectric characteristics, in particular, to the ratio of the absolute DP ($\varepsilon_s\varepsilon_o$) to the time DR ($\tau$), which is equal to the limiting HF EC of the ILs, the time of ionic relaxation θ is close to the time of the dipolar relaxation τ in ILs.

In terms of dispersion of the dielectric permittivity of ionic liquids, when analyzing the dielectric relaxation spectra of conducting ILs, correction for the DC conductivity is necessary. At comparatively high DC conductivity values of ILs, the ionic losses ε''_{ion} shield the dipole loss ε''_d, preventing identification of the molecular processes occurring in the ILs. When analyzing the dielectric relaxation spectra and determining the parameters of the relaxation time distribution, it is necessary to take into account the calculation errors of the dipole component of the dielectric loss coefficient ε''_d, which increases sharply with a decrease in frequency. The numerical value of the dielectric relaxation time τ can be estimated from the frequency corresponding to the maximum ε''_d.

When considering relaxation processes in ILs, in addition to dipole relaxation processes, it is also necessary to consider the relaxation of HF EC. Since ILs are characterized by high viscosities η, and therefore large dielectric relaxation times, the limiting HF EC of ILs is significantly lower than that of VOCs.

The static DP of protic ILs is substantially higher than the aprotic ε_s (which at room temperature does not exceed, as a rule, 10-15 units) and is due to the formation of a relatively stable network of hydrogen bonds, not only between ILs molecules, but also between their ions and molecules;

The rather complex type of DR and OKE spectra for ILs is due to the various polarization processes that occur in a wide frequency range from megahertz to distant terahertz; these processes include the polarization of IL molecules (α-relaxation), their constituent ions, and the more complex structures formed in ILs and their solutions.

Accurate identification of these polarization processes is hindered again by the interfering influence of the DC conductivity in the case of molecular (dipole) polarization and the low intensity of ion polarization peaks; as the temperature rises, the absorption maxima of different polarizations shift towards higher frequencies.

The proximity of the activation energies for dielectric relaxation, viscosity and DC-conductivity processes indicates the commonality of the polarization and translational movement of particles (molecules and ions) in ILs.

The temperature of the DC conductivity (electrical conductivity) of ILs increases in direct proportion to their limiting HF EC.

Since the dipole relaxation time τ of ILs is determined with lower accuracy than viscosity η, for the analysis and generalization of temperature dependencies of DC conductivity κ of ILs, it is preferable to use the $\kappa - \varepsilon_s T/\eta$ dependencies instead of $\kappa - \kappa_\infty$ dependencies.

When analyzing the conductivity-viscosity correlations of the ILs excepting Walden's rule, equations which take into account the effect of the dielectric permittivity ε_s and temperature T on DC conductivity, not only of viscosity should be considered.

The proportionality between κ and κ_∞ is observed in pure ILs and also in their solutions in polar solvents and can be used to estimate and predict the DC conductivity values of ILs. The activation energy of the limiting HF EC of polar solvents coincides with the E_κ of IL solutions and inorganic salts in these solvents.

The presence of a maximum in the dependence of the specific EC on the IL concentration is associated with the formation of contact ion pairs in solution. The formation of solvate-separated ionic pairs (SIP and 2SIP) is unlikely in ILs solutions; at concentrations where one or two solvent molecules can be placed between ILs ions, the specific EC increases with the increase in ILs content.

In pure ILs, the concentration of undissociated molecules is not high and is commensurate with the concentration of ILs ions; ions remain separated by one non-dissociated IL molecule at a dissociation degree that does not exceed 50%.

The dielectric spectra of IL solutions in polar solvents are well approximated by the Cole−Cole (CC) mode superposition and higher-frequency Debye relaxation (D); for mixtures with molar fractions of IL $x_{IL} < 0.2$; an additional low-frequency Debye process is also possible, that is, the model D + CC + D.

Unlike pure ILs, in which additional relaxation processes (CC) occur in the higher frequency region compared to the main Debye (D) molecular relaxation region, in IL solutions in polar solvents (acetonitrile, water) additional dispersion regions (CC) occur at lower frequencies compared to the molecular Debye (D) dielectric relaxation;.

The entire concentration region of the aprotic IL solutions in acetonitrile can be divided into two regions: at low concentrations (x<0.2), ILs behave as ordinary weakly associated electrolytes, and at higher concentrations (x>0.2) like expanded ("lubricated") ILs; the transition region (x≈0,2) may be associated with the redissociation of ion pairs and the establishment of an IL-like structure.

References

[1] Plechkova, N. V.; Seddon, K. R. Applications of ionic liquids in the chemical industry. *Chem. Soc. Rev.* 2008, 37, 123-150.

[2] Jin, H.; O'Hare, B.; Dong, J.; Arzhantsev, S.; Baker, G. A.; Wishart, J. F.; Benesi, A. J.; Maroncelli, M. Physical properties of ionic liquids consisting of the 1-butyl-3-methylimidazolium cation with various anions and the bis(trifluoromethylsulfonyl)imide anion with various cations. *J. Phys. Chem. B.* 2008, 112, 81-92.

[3] Galinski, M.; Lewandowski, A.; Stepniak, I. Ionic liquids as electrolytes. *Electrochim. Acta.* 2006, 51, 5567-5580.

[4] Schütter, C.; Neale, A. R.; Wilde, P.; Goodrich, P.; Hardacre, C.; Passerini, S.; Jacquemin, J.; Balducci, A. The use of binary mixtures of 1-butyl-l-methylpyrrolidinium bis {(trifluoromethyl)sulfonyl}imide and aliphatic nitrile solvents as electrolyte for supercapacitors. *Electrochim. Acta.* 2016, 220, 146-155.

[5] Marcus, Y., *Ionic Liquid Properties*, 2016, Springer International Publishing.

[6] Stoppa, A.; Zech, O.; Kunz, W.; Buchner, R. The Conductivity of Imidazolium-Based Ionic Liquids from (-35 to 195)°C. A. Variation of Cation's Alkyl Chain. *J. Chem. Eng. Data.* 2010, 55, 1768-1773.

[7] Hapiot, P.; Lagrost, C. Electrochemical reactivity in room-temperature ionic liquids. *Chem. Rev.* 2008, 108, 2238-2264.

[8] Wu, K. J.; Luo, H. T.; Yang, L. J. Structure-Based Model for Prediction of Electrical Conductivity of Pure Ionic Liquids. *AIChE J.* 2016, 62, 3751-3762.

[9] Leys, J.; Rajesh, R. N.; Menon, P. C.; Glorieux, C.; Longuemart, S.; Nockemann, P.; Pellens, M.; Binnemans, K. Influence of the anion on the electrical conductivity and glass formation of 1-butyl-3-methylimidazolium ionic liquids. *J. Chem. Phys.* 2010, 133.

[10] Avent, A. G.; Chaloner, P. A.; Day, M. P.; Seddon, K. R.; Welton, T. Evidence for Hydrogen-Bonding in Solutions of 1-Ethyl-3-Methylimidazolium Halides, and its Implications For Room-Temperature Halogenoaluminate(III) Ionic Liquids. *J. Chem. Soc. Dalton Trans.* 1994, 3405-3413.

[11] Elaiwi, A.; Hitchcock, P. B.; Seddon, K. R.; Srinivasan, N.; Tan, Y. M.; Welton, T.; Zora, J. A. Hydrogen-bonding in imidazolium salts and its implications for ambient-temperature halogenoaluminate(III) ionic liquids. *J. Chem. Soc. Dalton Trans.* 1995, 3467-3472.

[12] Iolitec. *1-Butyl-3-methylimidazolium thiocyanate, >98%.* 2022. https://iolitec.de/en/products/ionic_liquids/catalogue/imidazolium-based/il-0063-hp (accessed.

[13] Yoshida, Y.; Baba, O.; Saito, G. Ionic liquids based on dicyanamide anion: Influence of structural variations in cationic structures on ionic conductivity. *J. Phys. Chem. B.* 2007, 111, 4742-4749.

[14] Nishida, T.; Tashiro, Y.; Yamamoto, M. Physical and electrochemical properties of 1-alkyl-3-methylimidazolium tetrafluoroborate for electrolyte. *J. Fluor. Chem.* 2003, 120, 135-141.

[15] Sangoro, J.; Iacob, C.; Serghei, A.; Naumov, S.; Galvosas, P.; Karger, J.; Wespe, C.; Bordusa, F.; Stoppa, A.; Hunger, J.; Buchner, R.; Kremer, F. Electrical conductivity and translational diffusion in the 1-butyl-3-methylimidazolium tetrafluoroborate ionic liquid. *J. Chem. Phys.* 2008, 128.

[16] Artemkina, Y. M.; Shcherbakov, V. V.; Plechkova, N. V.; Seddon, K. R. Electrical Conductivity of Some Ionic Liquids. *Chem. Ind. Today.* 2010, 6, 13-17.

[17] Ueno, K.; Tokuda, H.; Watanabe, M. Ionicity in ionic liquids: correlation with ionic structure and physicochemical properties. *Phys. Chem. Chem. Phys.* 2010, 12, 1649-1658.

[18] Yuan, W. L.; Yang, X.; He, L.; Xue, Y.; Qin, S.; Tao, G. H. Viscosity, Conductivity, and Electrochemical Property of Dicyanamide Ionic Liquids. *Front. Chem.* 2018, 6.

[19] Nancarrow, P.; Al-Othman, A.; Mital, D. K.; Dopking, S. Comprehensive analysis and correlation of ionic liquid conductivity data for energy applications. *Energy.* 2021, 220.

[20] Marcus, Y.; Hefter, G. Ion pairing. *Chem. Rev.* 2006, 106, 4585-4621.

[21] Angell, C. A.; Byrne, N.; Belieres, J. P. Parallel developments in aprotic and protic ionic liquids: Physical chemistry and applications. *Accounts Chem. Res.* 2007, 40, 1228-1236.

[22] Stassen, H. K.; Ludwig, R.; Wulf, A.; Dupont, J. Imidazolium Salt Ion Pairs in Solution. *Chem. Eur. J.* 2015, 21, 8324-8335.

[23] Zhang, X. X.; Liang, M.; Hunger, J.; Buchner, R.; Maroncelli, M. Dielectric Relaxation and Solvation Dynamics in a Prototypical Ionic Liquid plus Dipolar Protic Liquid Mixture: 1-Butyl-3-Methylimidazolium Tetrafluoroborate plus Water. *J. Phys. Chem. B.* 2013, 117, 15356-15368.

[24] Bešter-Rogač, M.; Stoppa, A.; Hunger, J.; Hefter, G.; Buchner, R. Association of ionic liquids in solution: a combined dielectric and conductivity study of bmim Cl in water and in acetonitrile. *Phys. Chem. Chem. Phys.* 2011, 13, 17588-17598.

[25] Boruń, A. Conductance and ionic association of selected imidazolium ionic liquids in various solvents: A review. *J. Mol. Liq.* 2019, 276, 214-224.

[26] Robinson, R. A.; Stokes, R. H., *Electrolyte Solutions*, 2002, Dover Publications.

[27] Fuoss, R. M.; Kraus, C. A. Properties of electrolytic solutions. IV. The conductance minimum and the formation of triple ions due to the action of coulomb forces. *J. Am. Chem. Soc.* 1933, 55, 2387-2399.

[28] Fuoss, R. M.; Shedlovsky, T. Extrapolation of Conductance Data for Weak Electrolytes. *J. Am. Chem. Soc.* 1949, 71, 1496-1498.

[29] Covington, A. K.; Pethybridge, A. D. Electrolyte solutions. *Annu. Rep. Prog. Chem., Sect. A. Inorg. and Phys. Chem.* 1977, 74, 5-21.

[30] Fuoss, R. M. Review of the theory of electrolytic conductance. *J. Solution Chem.* 1978, 7, 771-782.

[31] Barthel, J. Electrolytes in non-aqueous solvents. *Pure Appl. Chem.* 1979, 51, 2093-2124.

[32] Vorobyov, A. F.; Shcherbakov, V. V.; Ksenofontova, N. A. The nature of electrical conductivity and the association of ions in electrolyte solutions. Thermodynamic properties of solutions. *Proceedings Moscow Mendeleev Chem. Eng. Inst.* 1980, 111, 21-34.

[33] Safonova, L. P., "Conductometry." In *Theorectical and Experimental Methods of Solution Chemistry*, 2011, Problems of solution chemistry; pp 464-524.

[34] Barthel, J.; Krienke, H.; Kunz, W., *Physical Chemistry of Electrolyte Solutions: Modern Aspects* 1998, Dr. Dietrich Steinkopff Verlag GmbH & Co. KG Springer-Verlag.

[35] Fuoss, R. M.; Xsia, K. L. Association of 1-1 salts in water. *PNAS USA.* 1967, 57, 1550-1557.

[36] Justice, J. C. Interpretation for distance parameter of Fuoss-Onsager conductance equation in case of ionic association. *Electrochim. Acta.* 1971, 16, 701-712.

[37] Fuoss, R. M. Conductance-concentration function for paired ion model. *J. Phys. Chem.* 1978, 82, 2427-2440.

[38] Quint, J.; Viallard, A. Electrical conductance of electrolyte mixtures of any type. *J. Solution Chem.* 1978, 7, 533-548.

[39] Pethybridge, A. D.; Taba, S. S. Precise conductimetric studies on aqueous-solutions of 2-2 electrolytes. 2. Analysis of data for $MgSO_4$ in terms of new equations from Fuoss and from Lee and Wheaton. *J. Chem. Soc. Faraday Trans.* 1980, 76, 368-376.

[40] Lee, W. H.; Wheaton, R. J. Conductance of symmetrical, unsymmetrical and mixed electrolytes. 2. Hydrodynamic terms and complete conductance equation. *J. Chem. Soc. Faraday Trans.* 1978, 74, 1456-1482.

[41] Lee, W. H.; Wheaton, R. J. Conductance of symmetrical, unsymmetrical and mixed electrolytes. 3. Examination of new model and analysis of data for symmetrical electrolytes. *J. Chem. Soc. Faraday Trans.* 1979, 75, 1128-1145.

[42] Boruń, A.; Bald, A. Ionic association and conductance of ionic liquids in dichloromethane at temperatures from 278.15 to 303.15 K. *Ionics.* 2016, 22, 859-867.

[43] Boruń, A.; Bald, A. Conductometric Studies of 1-Ethyl-3-methylimidazolium Tetrafluoroborate and 1-Butyl-3-methylimidazolium Tetrafluoroborate in 1-Propanol at Temperatures from (283.15 to 318.15) K. *Int. J. Electrochem. Sci.* 2014, 9, 2790-2804.

[44] Voroshylova, I. V.; Smaga, S. R.; Lukinova, E. V.; Chaban, V. V.; Kalugin, O. N. Conductivity and association of imidazolium and pyridinium based ionic liquids in methanol. *J. Mol. Liq.* 2015, 203, 7-15.

[45] Bešter-Rogač, M.; Hunger, J.; Stoppa, A.; Buchner, R. Molar Conductivities and Association Constants of 1-Butyl-3-methylimidazolium Chloride and 1-Butyl-3-methylimidazolium Tetrafluoroborate in Methanol and DMSO. *J. Chem. Eng. Data.* 2010, 55, 1799-1803.

[46] Bešter-Rogač, M.; Stoppa, A.; Buchner, R. Ion Association of Imidazolium Ionic Liquids in Acetonitrile. *J. Phys. Chem. B.* 2014, 118, 1426-1435.

[47] Kalugin, O. N.; Voroshylova, I. V.; Riabchunova, A. V.; Lukinova, E. V.; Chaban, V. V. Conductometric study of binary systems based on ionic liquids and acetonitrile in a wide concentration range. *Electrochim. Acta.* 2013, 105, 188-199.

[48] Boruń, A.; Bald, A. Conductometric Studies of 1-Ethyl-3-methylimidazolium Tetrafluoroborate and 1-Butyl-3-methylimidazolium Tetrafluoroborate in N,N-Dimethylformamide at Temperatures from (283.15 to 318.15) K. *J. Chem. Eng. Data.* 2012, 57, 475-481.

[49] Boruń, A.; Fernandez, C.; Bald, A. Conductance Studies of Aqueous Ionic Liquids Solutions [emim][BF_4] and [bmim][BF_4] at Temperatures from (283.15 to 318.15) K. *Int. J. Electrochem. Sci.* 2015, 10, 2120-2129.

[50] Lee, W. H.; Wheaton, R. J. Conductance of symmetrical, unsymmetrical and mixed electrolytes. 1. Relaxation terms. *J. Chem. Soc. Faraday Trans.* 1978, 74, 743-766.

[51] Wang, H. Y.; Wang, J. J.; Zhang, S. L.; Pei, Y. C.; Zhuo, K. L. Ionic Association of the Ionic Liquids [C_4mim][BF_4], [C_4mim][PF_6], and [$C_n$mim]Br in Molecular Solvents. *Chem. Phys. Chem.* 2009, 10, 2516-2523.

[52] Boruń, A.; Bald, A. Triple Ion Formation in Solutions of [emim][BF_4] and [bmim][BF_4] in Dichloromethane at Various Temperatures. A New Method of Analysis of Conductivity Data. *Int. J. Electrochem. Sci.* 2016, 11, 7714-7725.

[53] Boruń, A. Conductometric studies of [emim][BF_4] and [bmim][BF_4] in propan-2-ol. Association of ionic liquids in alcohols. *J. Mol. Liq.* 2017, 240, 717-722.

[54] Shcherbakov, V. V.; Artemkina, Y. M.; Plechkova, N. V.; Seddon, K. R. The electrical conductivity of solutions of some ionic liquids in acetonitrile. *Chem. Ind. Today.* 2011, 12, 36-41.

[55] Casteel, J. F.; Amis, E. S. Specific conductance of concentrated solutions of magnesium salts in water-ethanol system. *J. Chem. Eng. Data.* 1972, 17, 55-59.

[56] Stoppa, A.; Hunger, J.; Buchner, R. Conductivities of Binary Mixtures of Ionic Liquids with Polar Solvents. *J. Chem. Eng. Data.* 2009, 54, 472-479.

[57] Woodward, C. E.; Harris, K. R. A lattice-hole theory for conductivity in ionic liquid mixtures: application to ionic liquid plus water mixtures. *Phys. Chem. Chem. Phys.* 2010, 12, 1172-1176.

[58] Shcherbakov, V. V. Regularity of Conductivity of Concentrated Aqueous Solutions of Strong Electrolytes. *Russ. J. Electrochem.* 2009, 45, 1292-1295.

[59] Artemkina, Y. M.; Shcherbakov, V. V. Electrical conductivity of associated electrolyte-water systems. *Russ. J. Inorg. Chem.* 2010, 55, 1487-1489.

[60] Shcherbakov, V. V.; Artemkina, Y. M.; Akimova, I. A.; Artemkina, I. M. Dielectric Characteristics, Electrical Conductivity and Solvation of Ions in Electrolyte Solutions. *Materials.* 2021, 14.

[61] Turton, D. A.; Sonnleitner, T.; Ortner, A.; Walther, M.; Hefter, G.; Seddon, K. R.; Stana, S.; Plechkova, N. V.; Buchner, R.; Wynne, K. Structure and dynamics in protic ionic liquids: A combined optical Kerr-effect and dielectric relaxation spectroscopy study. *Faraday Discuss.* 2012, 154, 145-153.

[62] Weingärtner, H. The static dielectric permittivity of ionic liquids. *J. Mol. Liq.* 2014, 192, 185-190.

[63] Weingärtner, H., "Dielectric Properties of Ionic Liquids: Achievements So Far and Challenges Remaining." In *Ionic Liquids Further UnCOILed: Critical Expert Overviews*, 1st ed.; Plechkova, N. V., Seddon, K. R. Eds.; 2014, John Wiley & Sons Inc.; pp 235-258.

[64] Krüger, M.; Huang, M. M.; Bründermann, E.; Weingärtner, H.; Havenith, M. Combined THz and Microwave Dielectric Spectroscopy of Intermolecular Interactions in Homologous Protic Ionic Liquids. *IEEE Trans. Terahertz Sci. Technol.* 2011, 1, 313-320.

[65] Sonnleitner, T.; Turton, D. A.; Hefter, G.; Ortner, A.; Waselikowski, S.; Walther, M.; Wynne, K.; Buchner, R. Ultra-Broadband Dielectric and Optical Kerr-Effect Study of the Ionic Liquids Ethyl and Propylammonium Nitrate. *J. Phys. Chem. B.* 2015, 119, 8826-8841.

[66] Wojnarowska, Z.; Paluch, M. Recent progress on dielectric properties of protic ionic liquids. *J. Phys. Condens. Matter.* 2015, 27.

[67] Nazet, A.; Buchner, R. Dielectric response and transport properties of alkylammonium formate ionic liquids. *J. Chem. Phys.* 2018, 148.

[68] Akhadov, Y. Y., *Dielectric Properties of Binary Solutions: A Data Handbook*, 1980, Elsevier; Pergamon Press.

[69] Akhadov, Y. Y., *The Dielectric Parameters of Pure Liquids: A Handbook*, 1999, MAI, Moscow.

[70] Janković, B.; Manić, N.; Buchner, R.; Płowaś-Korus, I.; Pereiro, A. B.; Amado-González, E. Dielectric properties and kinetic analysis of nonisothermal decomposition of ionic liquids derived from organic acid. *Thermochim. Acta.* 2019, 672, 43-52.

[71] Petrowsky, M.; Frech, R. Concentration dependence of ionic transport in dilute organic electrolyte solutions. *J. Phys. Chem. B.* 2008, 112, 8285-8290.

[72] Hasted, J. B.; Ritson, D. M.; Collie, C. H. Dielectric Properties of Aqueous Ionic Solutions. Part I. *J. Chem. Phys.* 1948, 16, 1-11.

[73] Ritson, D. M.; Hasted, J. B. Dielectric Properties of Aqueous Ionic Solutions. *J. Chem. Phys.* 1948, 16, 11-21.

[74] Shcherbakov, V. V. Analysis of the Electrochemical Cell Impedance Allowing for the Solution Capacitance. *Russ. J. Electrochem. C/C Electrokhimiia.* 1998, 34, 111-114.

[75] Jaeger, J., *Methods of Measurement in Electrochemistry [Russian translation]*, 1977, Mir.

[76] Falkenhagen, H.; Hertz, H. G.; Eberling, W., *Theorie der Elektrolyte* [*Electrolyte theory*], 1971, Hirzel Verlag.

[77] Anderson, J. E. The Debye-Falkenhagen effect - experimental fact or fiction. *J. Non-Cryst. Solids.* 1994, 172, 1190-1194.

[78] Kaatze, U. Complex permittivity of water as a function of frequency and temperature. *J. Chem. Eng. Data.* 1989, 34, 371-374.

[79] Lobo, V. M. M.; Quaresma, J. L., *Handbook of Electrolyte Solutions*, 1989, Elsevier.

[80] Shcherbakov, V. V.; Ermakov, V. I.; Artemkina, Y. M. Dielectric Characteristics of Water and Electric Conductivity of Aqueous Electrolytes. *Russ. J. Electrochem.* 2017, 53, 1301-1306.

[81] Von Hippel, A. R., *Dielectrics and Waves*, 1995, Artech House.

[82] Böttcher, C. J. F.; Bordewijk, P., *Dielectrics in Time-Dependent Fields*, 1978, Elsevier.

[83] Debye, P., *Polar Molecules*, 1929, Chemical Catalog. Co.

[84] Dote, J.; Kivelson, D.; Schwartz, R. N. A molecular quasi-hydrodynamic free-space model for molecular rotational relaxation in liquids. *J. Phys. Chem.* 1981, 85, 2169-2180.

[85] Cole, K. S.; Cole, R. H. Dispersion and Adsorption in Dielectrics I. Alternating Current Characteristics. *J. Chem. Phys.* 1941, 9, 341-351.

[86] Davidson, D. W.; Cole, R. H. Dielectric Relaxation in Glycerine. *J. Chem. Phys.* 1950, 18, 1417.

[87] Hall, D. G.; Cole, R. H. Dielectric polarization of sulfuric acid solutions. *J. Phys. Chem.* 1981, 85, 1065-1069.

[88] Debye, P.; Falkenhagen, H. The dispersion of conductivity and dielectric constant of strong electrolytes. *Phys. Z.* 1928, 29, 121-132.

[89] Shcherbakov, V. V.; Ermakov, V. I. Complex and limiting high-frequency conductances of concentrated electrolyte solutions. *Russ. J. Phys. Chem.* 1977, 51, 1784-1787.

[90] Shcherbakov, V. V.; Ermakov, V. I. High-frequency conductivity of electrolyte solutions and dielectrics. *Russ. J. Electrochem.* 1977, 13, 1091-1092.

[91] Shcherbakov, V. V. Dispersion of high-frequency conductivity of polar solvents. *Russ. J. Electrochem. C/C Electrokhimiia.* 1994, 30, 1367-1373.

[92] Shcherbakov, V. V. Frequency dispersion of electroconductivity of electrolyte solutions. *Russ. J. Electrochem.* 1996, 32, 580-583.

[93] Shcherbakov, V.; Artemkina, Yu.; Ermakov, V. *Electrolyte Solutions, Electric Conductivity of Solutions and Dielectric Properties of Polar Solvents*, 2012, Saarbrucken, Palmarium Academic Publishing.

[94] Shcherbakov, V. V.; Artemkina, Y. M. Limiting equivalent electrical conductivity of inorganic salt solutions and dielectric properties of a polar solvent. *Russ. J. Inorg. Chem.* 2013, 58, 968-971.

[95] Rybinska-Fryca, A.; Sosnowska, A.; Puzyn, T. Prediction of dielectric constant of ionic liquids. *J. Mol. Liq.* 2018, 260, 57-64.

[96] Wakai, C.; Oleinikova, A.; Ott, M.; Weingärtner, H. How polar are ionic liquids? Determination of the static dielectric constant of an imidazolium-based ionic liquid by microwave dielectric spectroscopy. *J. Phys. Chem. B.* 2005, 109, 17028-17030.

[97] Schröder, C.; Sonnleitner, T.; Buchner, R.; Steinhauser, O. The influence of polarizability on the dielectric spectrum of the ionic liquid 1-ethyl-3-methylimidazolium triflate. *Phys. Chem. Chem. Phys.* 2011, 13, 12240-12248.

[98] Hunger, J.; Stoppa, A.; Buchner, R.; Hefter, G. Dipole Correlations in the Ionic Liquid 1-N-Ethyl-3-N-methylimidazolium Ethylsulfate and Its Binary Mixtures with Dichloromethane. *J. Phys. Chem. B.* 2009, 113, 9527-9537.

[99] Sonnleitner, T.; Turton, D. A.; Waselikowski, S.; Hunger, J.; Stoppa, A.; Walther, M.; Wynne, K.; Buchner, R. Dynamics of RTILs: A comparative dielectric and OKE study. *J. Mol. Liq.* 2014, 192, 19-25.

[100] Schreiner, C.; Zugmann, S.; Hartl, R.; Gores, H. J. Fractional Walden Rule for Ionic Liquids: Examples from Recent Measurements and a Critique of the So-Called Ideal KCl Line for the Walden Plot. *J. Chem. Eng. Data.* 2010, 55, 1784-1788.

[101] Hefter, G.; Buchner, R. Dielectric relaxation spectroscopy: an old-but-new technique for the investigation of electrolyte solutions. *Pure Appl. Chem.* 2020, 92, 1595-1609.

[102] Hunger, J.; Stoppa, A.; Thoman, A.; Walther, M.; Buchner, R. Broadband dielectric response of dichloromethane. *Chem. Phys. Lett.* 2009, 471, 85-91.

[103] Schrödle, S.; Annat, G.; MacFarlane, D. R.; Forsyth, M.; Buchner, R.; Hefter, G. Broadband dielectric response of the ionic liquid N-methyl-N-ethylpyrrolidinium dicyanamide. *Chem. Commun.* 2006, 1748-1750.

[104] Xu, H.; Zhang, D. D. Viscosities and Conductivities of [BMIM]$Zn(Ac)_xCl_y$ ($x = 0, 1, 2, 3$; $y = 3, 2, 1, 0$) Ionic Liquids at Different Temperatures. *J. Chem.* 2017, 2017.

[105] Grishina, E. P.; Kudryakova, N. O.; Ramenskaya, L. M.; Fadeeva, Y. A. The Temperature Effect on the Transport Properties of 1-Alkyl-3-methylimidazolium Bis(trifluoromethylsulfonyl)imide Ionic Liquids. *Russian Journal of Physical Chemistry A.* 2018, 92, 724-729.

[106] Artemkina, Y. M.; Shcherbakov, V. V.; Akimova, I. A. The Temperature Dependence of the Electrical Conductivity Activation Energy of the of Aqueous Electrolyte Solutions. *Mater. Sci. Forum.* 2021, 1031, 228-233.

[107] Shcherbakov, V. V.; Artemkina, Y. M.; Pleshkova, N. V.; Seddon, K. R. Ultimate High-Frequency Conductivity of Solvent and Electroconductivity of Electrolyte Solutions. *Russ. J. Electrochem.* 2009, 45, 922-924.

[108] Hasted, J. B., *Aqueous Dielectrics*, 1973, Chapman and Hall.

[109] Ermakov, V. I.; Levin, V. V.; Hubetzov, S. B.; Shcherbakov, V. V. To the question of the nature of the dielectric permittivity of solutions of electrolytes. *Russ. J. Phys. Chem.* 1975, 49, 1749-1752

[110] Lohse, P. W.; Bartels, N.; Stoppa, A.; Buchner, R.; Lenzer, T.; Oum, K. Dielectric relaxation and ultrafast transient absorption spectroscopy of $[C_6mim]^+[Tf_2N]^-$/acetonitrile mixtures. *Phys. Chem. Chem. Phys.* 2012, 14, 3596-3603.

[111] Nazet, A.; Weiβ, L.; Buchner, R. Dielectric relaxation of nitromethane and its mixtures with ethylammonium nitrate: Evidence for strong ion association induced by hydrogen bonding. *J. Mol. Liq.* 2017, 228, 81-90.

[112] Stoppa, A.; Hunger, J.; Hefter, G.; Buchner, R. Structure and Dynamics of 1-*N*-Alkyl-3-*N*-Methylimidazolium Tetrafluoroborate plus Acetonitrile Mixtures. *J. Phys. Chem. B.* 2012, 116, 7509-7521.

[113] Pottel, R. Die komplexe Dielektrizitätskonstante wäβriger Lösungen einiger 2-2-wertiger Elektrolyte im Frequenzbereich 0,1 bis 38 GHz) [*The complex dielectric constant of aqueous solutions of some divalent electrolytes in the frequency range 0.1 to 38 GHz)*]. *Ber. Bunsenges. Phys. Chem.* 1965, 69, 363-378.

[114] Sonnleitner, T.; Nikitina, V.; Nazet, A.; Buchner, R. Do H-bonds explain strong ion aggregation in ethylammonium nitrate plus acetonitrile mixtures? *Phys. Chem. Chem. Phys.* 2013, 15, 18445-18452.

Chapter 4

The Molecular Design of Ionic Liquids with Properties Tailored for CO_2 Capture

Rima Biswas*
Process Simulation Research Group, School of Chemical Engineering,
Vellore Institute of Technology, Vellore, Tamil Nadu, India

Abstract

The increasing concentration of greenhouse gases, especially carbon dioxide, is the main cause of global warming. Carbon capture, utilization, and sequestration (CCUS) strategy appears to be effective in reducing levels of carbon dioxide in the atmosphere. Despite great advances in this area, commercial CO_2 capture techniques based on absorption phenomena still have significant limitations. High capital expenses for CO_2 capture, low absorption and desorption rates, usage of corrosive solvents, and solvent losses because of evaporation are amongst the fundamental obstructions.

Ionic liquids (ILs) are potential candidates for CO_2 capture due to their unique properties, including negligible vapor pressures, high thermal stabilities, and other interesting properties. In this chapter, we discuss the physicochemical properties of CO_2 and ILs, the capture of CO_2 using ILs, characterization of CO_2-IL systems, CO_2 solubility and selectivity in ILs, and the mechanism of CO_2 capture with ILs at the molecular level. The relevant thermodynamic properties, mass transfer correlations, and the different factors which influence effective CO_2 absorption are also elaborated.

Keywords: ionic liquids, physicochemical properties, thermodynamic properties, CO_2 capture, absorption, CO_2-IL interaction mechanism

Abbreviations

CCS	Carbon capture and storage
MDEA	Methyl diethanolamine
MEA	Monoethanolamine
AMP	2-amino-2-methylpropanol
LLE	Liquid-liquid equilibria

* Corresponding Author's Email: tima.biswas@vit.ac.in.

In: Properties and Applications of Ionic Liquids
Editors: Noureddine Abidi and Julia L. Shamshina
ISBN: 979-8-88697-635-9

VLE	Vapor-liquid equilibria
$[C_6H_4F_9mim][Tf_2N]$	1-Methyl-3-(nonafluorohexyl)imidazolium Bis[(trifluoromethyl)sulfonyl]imide
$[C_2mim][TfO]$	1-Ethyl-3-methylimidazolium trifluoromethane-sulfonate
$[C_2mim][EtSO_4]$	1-Ethyl-3-methylimidazolium ethyl sulfate
$[C_2mim][NTf_2]$	1- Ethyl-3-mehylimidazolium bis(trifluoromethylsulfonyl)Imide
$[C_2mim][BF_4]$	1-Ethyl-3-methylimidazolium tetrafluoroborate
$[C_2mim][TFA]$	1-Ethyl-3-methylimidazolium trifluoroacetate
$[C_2mim][OAc]$	1-Ethyl-3-methylimidazolium acetate
$[C_4mim][NO_3]$	1-N-Butyl-3-methylimidazolium nitrate
$[C_4mim][PF_6]$	1-Butyl-3-methyl-imidazolium hexafluorophosphate
$[C_4mim][BF_4]$	1-Butyl-3-methyl-imidazolium tetrafluoroborate
$[C_4mim]$ [DCA]	1-Butyl-3-methyl-imidazolium dicyanamide
$[C_4mim]$ [SCN]	1-Butyl-3-methylimidazolium thiocyanate
$[C_4mim][TFA]$	1-Butyl-3-methylimidazolium trifluoroacetate
$[C_4mim][OAc]$	1-Butyl-3-methylimidazolium acetate
$[C_4mim][TFES]$	1-Butyl-3-methylimidazolium 1,1,2,2-tetrafluoroethanesulfonate
$[C_4mim][PRO]$	1-Butyl-3-methylimidazolium propionate
$[C_4mim][ISB]$	1-Butyl-3-methylimidazolium isobutyrate
$[C_4mim][TMA]$	1-Butyl-3-methylimidazolium trimethylacetate
$[C_4mim][LEV]$	1-Butyl-3-methylimidazolium levulinate
$[C_4mim][SUC]$	1-Butyl-3-methylimidazolium succinamate
$[C_4mim][IAAc]$	1-Butyl-3-methylimidazolium iminoacetic acid acetate
$[C_4mim][[TfO]$	1-Butyl-3-methylimidazolium trifluoromethane-sulfonate
$[C_4mim][IDA]$	Bis(1-butyl-3-methylimidazolium)iminodiacetate
$[C_6mim][[PF_6]$	1-Hexyl-3-methylimidazolium hexafluorophosphate
$[C_6mim][BF_4]$	1-Hexyl-3-methylimidazolium tetrafluoroborate
$[C_6mim][Tf_2N]$	1-Hexyl-3-methylimidazolium bis(trifluoromethylsulfonyl)imide
$[C_6mim][TfO]$	1-Hexyl-3-methylimidazolium trifluoromethane-sulfonate
$[C_6mim][FAP]$	1-Hexyl-3-methylimidazolium tris(pentafluoroethyl)trifluoro phosphate
$[C_6mim][Tf_2N]$	1-Hexyl-3-methylimidazolium bis(trifluoromethylsulfonyl)imide
$[C_6mim][FEP]$	1-Hexyl-3-methylimidazolium tris(pentafluoroethyl)trifluorophosphate
$[C_8mim][PF_6]$	1- Octyl-3-methylimidazolium hexafluorophosphate
$[C_8mim][BF_4]$	1- Octyl-3-methylimidazolium tetrafluoroborate
$[C_8mim][TfO]$	1-Octyl-3-methylimidazolium trifluoromethane-sulfonate
$[C_8mim][BF_4]$	1- Octyl-3-methylimidazolium tetrafluoroborate
$[C_1mim][MP]$	1,3-Dimethylimidazolium methylphosphonate
BHEAA	Bis(2-hydroxyethyl)-ammonium acetate
BHEAL	Bis(2-hydroxyethyl)-ammonium lactate
HHEMEA	2-Hydroxy-N-(2-hydroxyethyl)-N-methylethanaminium acetate
HHEMEL	2-Hydroxy-N-(2-hydroxyethyl)-N-methylethanaminium lactate
HEF	2-Hydroxyethyl ammonium formate
HEA	2-Hydroxyethyl ammonium acetate
HEAF	2-(2-Hydroxyethoxy)-ammonium formate

HEAA	2-(2-Hydroxyethoxy)-ammonium acetate
HEAL	2-(2-Hydroxyethoxy)-ammonium lactate
HEL	2-Hydroxyethyl ammonium lactate
THEAA	Tri-(2-hydroxyethyl)-ammonium acetate
THEAL	Tri-(2-hydroxyethyl)-ammonium lactate
[THTDP]Cl	Trihexyltetradecylphosphonium chloride
[THTDP][NTf_2]	Trihexyltetradecylphosphonium bis(trifluoromethylsulfonyl)imide
[MeBuPy][BF_4]	N-Butyl-4-methylpyridinium tetrafluoroborate
[MeBuPy][DCA]	N-Butyl-3-Methylpyridinium dicyanamide
[MeBuPy][SCN]	N-Butyl-4-Methylpyridinium thiocyanate
[MeBuPyrr][SCN]	1-Butyl-1-Methylpyrrolidinium thiocyanate
[MeBuPyrr][DCA]	1-Butyl-1-methylpyrrolidinium dicyanamide
[MeBuPyrr][TFA]	1-Butyl-1-methylpyrrolidinium trifluoroacetate
[TBP][FOR]	Tetrabutylphosphonium formate
[BmPyrr][FEP]	1-Butyl-1-methylpyrrolidinium tri(pentafluoroethyl)trifluorophosphate
[N-BuPy][BF_4]	N-Butylpyridinium tetrafluoroborate
[pNH_2Bim][PF_6]	1-(3-third amino)-3-butyl imidazole hexafluorophosphate
[C_4mbpy]Br	1-butyl-3-methylpyridinium bromide

Introduction

Global climate change has gotten a lot of attention in the last few years [1]. Carbon dioxide emissions into the atmosphere are to blame for climate change and global warming [2, 3]. Carbon capture and storage (CCS) is one of the most important endeavors for mitigating the effects of global warming. The development of CO_2 capture and sequestration technology is emphasized in this solution. CCS is a concept that focuses on reducing CO_2 emissions into the atmosphere through industrial operations including ammonia production, natural gas processing, and cement production, to mention a few. CCS includes CO_2 capture, separation, transportation, and CO_2 storage. The most critical and cost-effective phase is CO_2 capture and separation. Physical and chemical absorption, selective adsorption by solid adsorbents, hydration separation, and membrane separation are all methods that can be used to capture CO_2 from a gas mixture [4].

Pre-combustion, post-combustion, and oxyfuel-combustion procedures are the three primary kinds of carbon capture techniques. The fuel is burned with air and oxygen in the post-combustion and oxy-combustion procedures, respectively. The heat generated by these technologies is utilized to power a generator. The composition of the flue-gas is the fundamental distinction between the post-combustion and oxy-combustion processes. The flue-gas in the latter process contains highly concentrated CO_2, which appears to be suited for underground storage. However, in order to extract the carbon from the flue-gas, further process units are required in the post-combustion approach [5]. The fuel is gasified to create syngas during pre-combustion [6]. The water-gas-shift reactors are fed by syngas, which contains CO and H_2. CO and H_2 are converted into additional H_2 and CO_2 in the presence of steam. The energy is then generated by combusting CO_2 gathered from the outflow stream.

In comparison to the other two approaches, post-combustion is a more straightforward retrofit option for existing power plants [5]. Pre-combustion and oxy-combustion technologies can be used in new power plants [7]. The following are some of the strategies for extracting CO_2 from the three conditioning processes that are currently being researched. Absorption happens within the bulk of the material by physical or chemical interaction. CO_2 reacts with chemical absorbents, producing covalent bonds between the molecules [8]. Adsorption, as opposed to absorption, occurs at the material's surface. This interaction might take place physically (Van der Waals) or chemically.

Solid materials with large surface areas, such as activated carbons, zeolites, ion-exchange resins, and silica gel are common adsorbers. These can be used to capture CO_2 by the separation process [9]. Membrane separation technology is based on the physical or chemical interaction of certain gases with the membrane material. The rate of gases passing through the membrane can be controlled by modifying the membrane [10]. There are many different types of gas separation membranes available, including zeolites, porous inorganic membranes, and polymeric membranes, some of which are employed on an industrial scale and might be used in the CO_2 capture process [11, 12]. Cryogenic separation is a cooling and condensation process. This has the advantage of allowing direct liquid CO_2 production, which benefits transportation possibilities [13].

Various solvents have been used in the chemical absorption of CO_2 capture techniques [14]. Amines are commonly employed as absorbents for carbon capture activities due to their high reactivity with CO_2 molecules, high absorption capacity, and excellent thermal stability [15]. Methyl diethanolamine (MDEA), Monoethanolamine (MEA), and 2-amino-2-methylpropanol (AMP) are the most frequently used absorbents for CO_2 capture systems among all amines [16]. Amines have a number of disadvantages, including high equipment corrosion rates, high capital costs, and amine degradation by O_2, NO_2, and SO_2 present in flue gases, which results in high energy consumption and high absorbent make-up rate during high-temperature absorbent regeneration processes [17]. Several investigations have been carried out to replace amines with solvents that have lower volatility, lower corrosive properties, greater thermal stability, lower degradation rate, and cheaper regeneration cost [16]. Currently, certain ILs have been presented as a viable amine substitute solvent in carbon capture procedures [18]. The physical properties of ILs are mentioned in Table 1.

Table 1. The physical properties of ILs suitable for amine replacement in CO_2 capture processes [19]

Property	General characteristics
Thermal stability	high
Viscosity	< 100 cP, workable
Dielectric constant	< 30
Specific conductivity	< 10 mS/cm, good
Vapor pressure	negligible
Freezing temperature	< 100 °C
Liquidous temperature	> 200 °C
Polarity	moderate
Solvency	strong
Catalytical character	excellent (for organic reactions)

Generally, the conventional ILs have captured the carbon by physical absorption process. Although changing cations or anions in ILs can improve CO_2 solubility, their CO_2 capture capability is still limited when compared to current commercial carbon capture systems such as amine-based solvents [20]. To use appropriate ILs in various applications, a thorough understanding of the molecular structure and accurate determination of thermodynamic and physical properties such as density, viscosity, solubility, selectivity, liquid-liquid equilibria (LLE), and vapor-liquid equilibria (VLE) are required. Understanding the phase behavior of ILs-gas systems is necessary to use ILs in gas separation procedures.

Thermodynamic modeling with proper equations of state is a well-developed tool for obtaining phase equilibrium conditions. Many research has done on the solubility of gas in imidozolium-based ILs with $[BF_4]^-$ cations and $[PF_6]^-$ anions [21]. Gas solubility in ILs can be improved by increasing the operating pressure and expanding the interface area of gas/ILs systems. Maginn et al. [18] investigated the solubility of nine distinct gases in $[C_4mim][PF_6]$ IL and discovered that water and CO_2 interact strongly. CO_2 has a higher solubility in ILs as compared to the solubility of ethane, ethylene, CO, methane, nitrogen, and hydrogen in ILs.

Yokozeki et al. [22] correlated the solubility of CO_2 in eighteen different ILs to fluids properties and operating conditions by using the Redlich-Kwong equation of state. Zhao et al. [23] investigated four low-viscous diamino protic ILs with fluorine-substituted phenolic anions for CO_2 capture. CO_2 solubilities in the protic ILs were determined using an isochoric saturation method at temperatures ranging from 303 K to 333 K and pressure upto 1.5 bar. The applications of the novel anion-functionalized protic ILs in the CO_2 absorption have been investigated by Zhu et al. [24] The substituents in imidazole-based anion have a substantial impact on the CO_2 capture.

Much research has been done into developing suitable ILs for capturing CO_2 from flue gases and natural gas. There are also few studies on the physiochemical factors and physical properties of IL systems. As a result, the goal of this review is to study the several ILs for CO_2 absorption and the physical properties and thermodynamic behaviors of CO_2 and ILs.

Ionic Liquids for CO_2 Capture

ILs are materials that are comprised of organic/inorganic anions and large organic cations and have melting temperatures below 100°C [25]. ILs have been extensively investigated in basic and industrial applications throughout the last decade. Figure 1 shows an example of some of the common cations and anions in ILs. Many of ILs have several distinctive and diverse properties such as low vapor pressure, high thermal and chemical stability, tuneable nature, big electrochemical window, and excellent solvent capabilities for a variety of polar and nonpolar molecules. Because of these properties, research into producing and applying ILs has spanned several industries over the last decade, including solar cells, electrolytes, electropolishing and electroplating, lubricants, and biomass processing [26-31]. The huge number of ILs that can be synthesized in the lab and acquired commercially has made this viable [32] Basic ILs are provided by companies such as Proionic, Sigma-Aldrich, Solvionic, Merck, Iolitec, and, Sachem which can also assist in the creation and development of ILs for specialized purposes [30, 33]. As a result, these compounds have generated an interesting new medium for upcoming technological applications that are already available commercially.

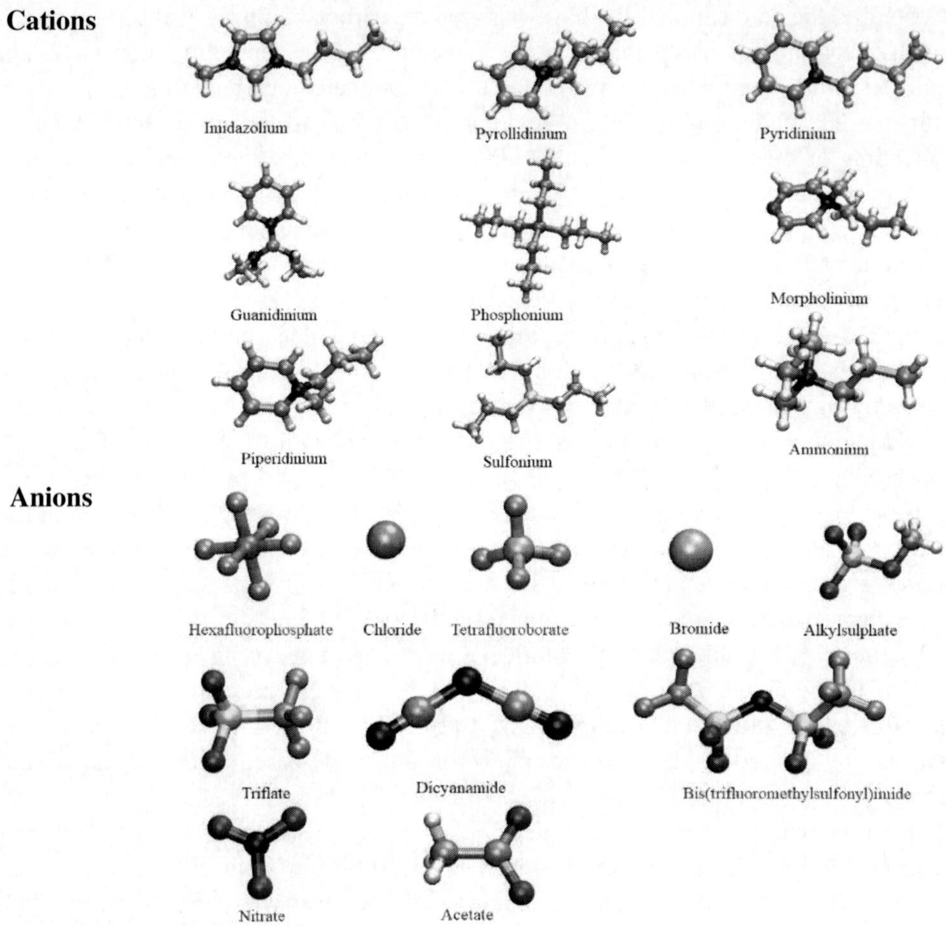

Figure 1. Structures of cations and anions of most common ILs.

The Effect of Anion and Cation

Assessment of the basic building components, that is, cation/anion combinations must be examined in order to design an effective procedure for trapping CO_2 in ILs. CO_2 capture has been shown to increase when ILs with CO_2-philic groups on the anion, such as carbonyls or fluorines, are synthesized. The findings are summarised in Table 2.

Table 2. Effects of anions in different ILs [37]

Classifications	Anions	Solubility of CO_2 in IL
Fluorinated anions	Hexafluorophosphate [PF_6]	Relatively high
	Tetrafluoroborate [BF_4]$^-$	
	Bis(trifluoromethylsulfonyl) imide [Tf_2N]$^-$	
	Trifuoromethanesulfonate [TfO]	
	Tris(trifluoromethylsulfonyl) methide [Methide]	
Nonfluorinated anions	Nitrate [NO_3]$^-$	Low
	Dicyanamide [DCA]$^-$	

The origin of high solubility has been found to be substantially dependent on the choice of anion in recent investigations. Aki et al. [34] used seven ILs to investigate the effect of the anion. The cation 1-butyl-3-methylimidazolium $[C_4mim]^+$ was present in all of them. They also looked into the effects of the cations on CO_2 solubility, finding that increasing the alkyl chain on the cation resulted in a modest increase in solubility, which became more noticeable at higher pressures. Muldoon et al. [35] found that the CO_2 solubility increases with increasing fluorinated alkyl chains in the imidazolium-based cations. They compared $[C_6mim][Tf_2N]$ to $[C_6H_4F_9mim][Tf_2N]$ and discovered that presence of fluorinated alkyl chains resulted in increased CO_2 solubility.

In general, research on CO_2 solubility in ILs has concentrated on imidazolium-based compounds. CO_2 solubilities in two ILs, such as $[THTDP][Tf_2N]$ and [THTDP]Cl, were reported by Carvalho et al. [36]. They discovered that the solubility measures were significantly higher than those of contemporary imidazolium-based ILs. Although imidazolium is the stable and commercially available cation, it is clear that other bases can be used to produce further upgrades and possibilities.

Imidazolium-Based IL for CO_2 Capture

Many studies have shown that CO_2 has a high solubility in imidazolium-based ILs. Cadena et al. investigated the mechanisms for the high affinity between imidazolium-based ILs and CO_2. They used both pure and pairs of imidazolium-based ILs including $[C_4mim][BF_4]$ and $[C_4mim][BF_4]$; $[C_2\ C_1mim][Tf_2N]$) and $[C_2mim][Tf_2N]$; and $[C_4C_1mim][PF_6]$) and $[C_4][PF_6]$. Cadena et al. showed that CO_2 solubility in ILs occurs due to the creation of a strong network of anions and cations, with CO_2 filling the fluid's interstices. The $[PF_6]^-$ anion was found to have the strongest interactions. This research found that CO_2 organizes strongly around the $[PF_6]^-$ anion in a "tangent like" configuration that maximizes beneficial interactions, but that modest variations in cation structure had little effect on the dominant CO_2-anion interactions. The CO_2 solubility for imidazolium based ILs is shown in Table 3.

Table 3. CO_2 solubility for imidazolium-based ILs

IL	T (K)	P (bar)	Solubility (mol CO_2/mol IL)	References
$[C_2mim][TfO]$	303.85	149	0.626	[41]
	303.85	15	0.261	[41]
$[C_2mim][EtSO_4]$	333	94.61	0.457	[42]
$[C_2mim][NTf_2]$	303	10	0.225	[37]
	333	10	0.144	[37]
$[C_2mim][BF_4]$	298.15	8.75	0.106	[37]
$[C_2mim][NTf_2]$	298.15	9.03	0.209	[37]
	298.1	19.99	0.390	[22]
$[C_2mim][TFA]$	298.1	19.99	0.282	[22]
$[C_2mim][OAc]$	298.1	19.99	0.428	[22]
$[C_4mim][NO_3]$	323	92.62	0.530	[42]
$[C_4mim][PF_6]$	313	96.67	0.729	[42]
	303	10	0.166	[37]
	333	10	0.101	[37]
	298.15	6.66	0.122	[43]

Table 3. (Continued)

IL	T (K)	P (bar)	Solubility (mol CO_2/mol IL)	References
[C_4mim][BF_4]	303 333	10 10	0.146 0.089	[37] [37]
[C_4mim] [DCA]	303 333	10 10	0.143 0.099	[37] [37]
[C_4mim] [SCN]	303 333	10 10	0.097 0.066	[37] [37]
[C_4mim][TFA]	298.1	19.99	0.301	[22]
[C_4mim][OAc]	298.1	19.99	0.455	[22]
[C_4mim][TFES]	298.0	19.9	0.285	[22]
[C_4mim][PRO]	298.2	19.9	0.390	[22]
[C_4mim][ISB]	298.2	20	0.403	[22]
[C_4mim][TMA]	298.1	19.9	0.431	[22]
[C_4mim][LEV]	298.1	19.9	0.460	[22]
[C_4mim][SUC]	298.1	19.9	0.232	[22]
[C_4mim][IAAc]	298.1	19.9	0.191	[22]
[C_4mim][[TfO]	303.85 303.85	160 11.5	0.672 0.273	[41] [41]
[C_4mim][IDA]	298.1	19.9	0.395	[22]
[C_6mim][[PF_6]	298.15	9.27	0.167	[43]
[C_6mim][BF_4]	298.15	8.99	0.163	[43]
[C_6mim][Tf_2N]	298.15	8.59	0.236	[37]
[C_6mim][TfO]	303.85 303.85	180 12.5	0.717 0.288	[41] [41]
[C_6mim][FAP]	298.1	19.99	0.493	[22]
[C_6mim][Tf_2N]	298.1	19.74	0.433	[22]
[C_6mim][FEP]	283.5	17.99	0.517	[44]
[C_8mim][PF_6]	313	92.67	0.755	[42]
[C_8mim][BF_4]	313	92.90	0.708	[42]
[C_8mim][TfO]	303.85 303.85	180 15.8	0.741 0.344	[41] [41]
[C_8mim][BF_4]	303 333	10 10	0.187 0.121	[37] [37]
[C_1mim][MP]	313.35 313.45	95 34	0.475 0.162	[45] [45]

In general, the solubility of CO_2 in all ILs increases with rising CO_2 pressure and decreases with increasing temperature, according to the findings. The researchers also discovered that CO_2 solubility is highly influenced by the anion used. One of these RTILs, [C_4mim][OAc], has been discovered to have a chemical interaction with CO_2 [38-40]. Yokozeki et al. [22] performed CO_2 solubility tests for 18 RTILs in 2008, with eight of them demonstrating chemical absorption processes. They discovered that all [C_2mim][OAc], [C_4mim][OAc], [C_4mim][ISB], [C_4mim][PRO], [C_4mim][LEV], and [C_4mim][TMA] RTILs that show high chemical absorption with CO_2 include the anion [X-COO]$^-$, where X is $CH_2CH_2C{=}OCH_3$ for strong chemical absortion and CH_2NHCH_2COOH for physical absorption. Generally, the traditional RTILs with acidic or basic functions have a significant impact on CO_2 absorption.

Ammonium-Based ILs on CO_2 Capture

Despite the fact that some of the ILs discussed here have a great potential for absorbing CO_2 with adequate solubility and selectivity, they still have limits in terms of operability and cannot be compared to the traditional solvents used in industry. The negligible vapor pressure of ILs can be considered a benefit for the gas absorption process in terms of environmental issues. However, CO_2 absorption rate is a function of solvent viscosity, which can be limited by such solvents' high viscosity. As a result, scientists have attempted to create solvents that combine the beneficial features of ILs with the high CO_2 absorption performance of amines.

Synthesizing ILs is one strategy for achieving this goal while also increasing absorption rate and CO_2 loading. ILs with appended amino groups appear to be promising for CO_2 capture. Even at low CO_2 partial pressures, amine-functionalized ILs may capture up to 33 mol% of CO_2 through chemical complexation, which is much higher than ILs normally utilized for CO_2 absorption [46] ILs are a viable choice for CO_2 absorption because of their high CO_2 solubility performance paired with their unique properties [47]. ILs with an attached amine group, such as those shown in Table 4, have been extensively studied. Bates et al. [46] created amine-functionalized IL [pNH$_2$Bim][PF$_6$] and discovered that it chemically reacts with CO_2. The CO_2 combines with the amine on the IL, which then reacts with another amine to generate a double salt of ammonium carbamate. For every two ILs, one CO_2 is obtained using this method. Traditional aqueous amines also exhibit this 1:2 capture process at the molecular level. Theoretically, when amino groups are exclusively attached to the anion, a 1:1 ratio can be achieved, allowing for a more efficient process.

Table 4. CO_2 solubility for ammonium-based ILs

IL	T (K)	P (bar)	Solubility (mol CO_2/mol IL)	References
Bis(2-hydroxyethyl)-ammonium acetate (BHEAA)	298.15	15.15	0.1076	[51]
	298.15	5.48	0.0391	[51]
Bis(2-hydroxyethyl)-ammonium lactate (BHEAL)	298.15	15.12	0.0835	[51]
	298.15	3.46	0.0192	[51]
2-Hydroxy-N-(2-hydroxyethyl)-N-methylethanaminium acetate (HHEMEA)	298.15	15.42	0.0761	[51]
	298.15	6.15	0.0300	[51]
2-Hydroxy-N-(2-hydroxyethyl)-N-methylethanaminium lactate (HHEMEL)	298.15	15.2	0.077	[51]
	298.15	3.4	0.017	[51]
2-Hydroxyethyl ammonium formate (HEF)	303	78.9	0.308	[52]
	303	4.4	0.034	[52]
2-Hydroxyethyl ammonium acetate (HEA)	303	90.1	0.400	[52]
	303	8.9	0.068	[52]
2-(2-Hydroxyethoxy)-ammonium formate (HEAF)	303	72.8	0.190	[52]
	303	6.6	0.030	[52]
2-(2-Hydroxyethoxy)-ammonium acetate (HEAA)	303	65.7	0.486	[52]
	303	7.6	0.088	[52]
2-(2-Hydroxyethoxy)-ammonium lactate (HEAL)	303	73.2	0.264	[52]
	303	12.4	0.070	[52]
2-Hydroxyethyl ammonium lactate (HEL)	303	82	0.242	[52]
	303	7.8	0.041	[52]
Tri-(2-hydroxyethyl)-ammonium acetate (THEAA)	303	82.5	0.256	[52]
	303	10.3	0.053	[52]
Tri-(2-hydroxyethyl)-ammonium lactate (THEAL)	303	70.9	0.461	[52]
	303	9.6	0.100	[52]

Task-specific ionic liquids (TSILs) have the ability to absorb CO_2 both chemically and physically [48-50]. Chemical absorption, similar to that of aqueous amines, occurs at low pressures (usually less than 2 bars). Physical absorption dominates the capture mechanism when the majority of the chemical bonding has occurred; this is especially true at high pressures, whereas aqueous amines reach their absorption limitations at low pressure. This demonstrates how the absorption properties of ILs with amine functionalities combine the desirable properties of physical and chemical solvents.

RTILs and ammonium-based IL are resilient and have a high absorption capacity in general; nevertheless, the intense synthesis required, high viscosity, and the fact that the IL functions as both the capture material and the dispersion limit their use. This topic was first examined by Camper et al. [47]. They created a $[C_nmim][Tf_2N]$ RTIL solution containing 16% v/v MEA and discovered that it is capable of capturing one mole of CO_2 per two moles of MEA at low CO_2 partial pressures in a quick and reversible manner. The RTIL solution precipitates a MEA-carbamate, which aids in driving the capture process [47]. The effect of temperature, pressure, and various anion and cation combinations on ILs is shown in Tables 4 and 5.

Table 5. CO_2 solubility for phosphonium, pyridinium and pyrrolidinium ILs

IL	T(K)	P(bar)	Solubility (mol CO_2/mol IL)	References
[THTDP]Cl	302.55	149.95	0.800	[36]
	313.27	5.17	0.200	[36]
[THTDP][NTf_2]	296.58	721.85	0.879	[36]
	293.2	6.12	0.308	[36]
[MeBuPy][BF_4]	303	10	0.144	[37]
	333	10	0.096	[37]
[MeBuPy][DCA]	303	10	0.143	[37]
	333	10	0.068	[37]
[MeBuPy][SCN]	303	10	0.096	[37]
	333	10	0.063	[37]
[MeBuPyrr[SCN]	303	10	0.097	[37]
	333	10	0.060	[37]
[MeBuPyrr][DCA]	303	10	0.120	[37]
	333	10	0.061	[37]
[MeBuPyrr][TFA]	303	10	0.167	[37]
	333	10	0.103	[37]
[TBP][FOR]	298.1	19.9	0.348	[22]
[BmPyrr][FEP]	283.5	18.00	0.498	[44]
[N-BuPy][BF_4]	323	92.35	0.581	[42]

Biodegradability and Thermal Stability of ILs

The biodegradability of ILs is one of the most important characteristics of their environmental impact. According to the recommendations of the Organization for Economic Cooperation and Development, rapidly biodegradable compounds undergo rapid and complete mineralization. Weak or extensive molecular cleavage is viewed as primary or ultimate biodegradability potential, respectively [53]. Biodegradability is considerably increased when chemical groups such as hydroxyl, carboxylic acid groups, amides, esters, aldehyde, or linear alkyl chains are present. This can be used to purposefully develop biodegradable ILs [54]. Adding an ester group to the side chain of many pyridinium [55] or imidazolium ILs can improve the main biodegradability of the cation [56]. Even when unsaturated groups (alkyl) or a sulfonate group were introduced, alkyl chains with four carbon atoms were shown to be poorly biodegradable [57].

Even though the intermediate chemical product (3-methylpyridine) could not be definitively identified, Pham et al. [58] reported a degradation process for [C_4mbpy]Br by hydroxylation of the side chain. Nonetheless, pyridinium ILs have a higher biodegradability potential than imidazolium ILs [59]. The research has revealed that a variety of possible degradation pathways exist, each of which is reliant on the microbial population and its metabolic capabilities. Furthermore, some phosphonium ILs have been found to be highly resistant to microbial attack. In this situation, trihexylphosphine and tricyclohexylphosphine-derived cations are examples. Because they had distinct ester side chains, these ILs mixed with various anions [60]. The cholinium cation and cholinium alkanoates, on the other hand, have been found as easily and highly biodegradable compounds [61].

Because there is a growing trend in studies focused on this topic, particularly the imidazolium family and their high persistence and ecotoxicological risk, there is a growing tendency in studies focusing on this topic [62]. Ionic interactions influence their sorption and desorption in soil, and their mobility influences lipophilicity, side chain length, and Coulombic interactions [63]. Nonetheless, filamentous fungi appear to play a crucial part in biodegradation, owing to the ability of many environmental fungal strains (which are widely found in soil) to resist high ILs concentrations [61]. In conclusion, the majority of the research found that the cation and anion have fundamentally different destiny mechanisms and follow different degradation paths [53].

Effective Parameters for CO_2 Absorption

The Effect of Anion and Cation Overall Mass Transfer Coefficients

The relationship between overall mass transfer coefficients and individual mass transfer coefficients was characterized by Strigle [64] as follows (Eq. (1)):

$$\frac{1}{K_G} = \frac{1}{k_G} + \frac{H}{k_L} \tag{1}$$

Here k_G and K_G are the individual and overall gas-phase mass transfer coefficients. k_L represents individual liquid-phase mass transfer coefficients. H indicates Henry's law constant. Overall mass transfer coefficients can be represented as (Eq. (2)) [65]:

$$K_G = \left(\frac{N_A}{[P(y_A - y_A^*)]}\right) \tag{2}$$

Here y_A is the mole fraction of A in the gas bulk, and y_A* represents the mole fraction of A in the gas phase in equilibrium. P denotes the pressure of the system. K_Ga_v is more useful for describing absorption rates in absorption equipment like the column. As a result, the mass transfer coefficient is more prominent in the absorption column unit volume rather than the interfacial area unit, as seen below, in (Eq. (3)):

$$K_G a_v = \left(\frac{N_A a_v}{[P(y_A - y_A^*)]}\right) \quad (3)$$

The term $N_A a_V$ is derived from absorption tests in columns in which the concentration profile of the absorbed component A in the gas phase is measured along with the column height, allowing the $K_G a_V$ values to be computed. $K_G a_V$, can be defined as seen below, in (Eq. (4)) [66]:

$$K_G a_v = \left(\frac{G_I}{[P(y_{A,G} - y_A^*)]}\right)\left(\frac{dY_{A,G}}{dZ}\right) \quad (4)$$

At pressure P, the inert gas flow rate G_I is an experimental operating condition. Along the column, the concentration of CO_2 in the gas stage ($y_{A,G}$) can be measured. The concentration of solute A in equilibrium with the bulk concentration C_{AL} is denoted by y_A*. The term $\left(\frac{dY_{A,G}}{dZ}\right)$ represents the mole ratio of solute concentration gradient [66]. For the first time, Zhang et al. [67] explored CO_2 absorption with ILs in a rotating packed bed (RPB) contactor. To describe the mechanism of gas-liquid mass transfer, they presented a model based on penetration theory.

According to their model, the increase of gas-liquid interface in RPB allows for more mass transfer. The absorption of CO_2 in NH_2-functionalized ILs is higher than in regular [C_4mim][BF_4], according to Galán Sánchez et al. [68]. The fundamental reason is the reaction between the NH_2 group and CO_2, which results in a 1.2–2.3 enhancement factor (depends on the concentration of IL). They determined the kinetic constant ($k_{1.1}$) of the reaction between functionalized IL and CO_2. The rates of kinetic constants range from 32.4 to 87.6 l/mol.s. These constants are affected by the amount of IL present and the temperature. Galán Sánchez et al. [69] reported a mechanism for the reaction between CO_2 and NH_2-ILs based on the concept of a single-step reaction between CO_2 and primary amine groups that creates carbamate and quaternary ammonium species with 1 : 2 ratio of CO_2 : NH_2-IL [69].

Henry's Constant

Henry's constant is required to calculate the mass transfer coefficient in IL systems [70]. Henry's constant is a critical parameter that cannot be directly measured. Based on Henry's constant, a great number of correlations have been presented in the literature to explain the solubility of CO_2 [68]. Henry's law constants for potential ILs are presented in Table 6.

The gas solubility of RTILs reduces as temperature and pressure drop. The gas solubility can be represented in terms of a Henry coefficient under ideal conditions. Methane absorption into ILs is significantly lower than CO_2 absorption. Zhang et al. [70] discovered that when ILs are added to the system, Henry's constant lowers. They indicated that the reason is primarily due to increased viscosity as a result of introducing more ILs to the system.

Table 6. Henry's law constant for potential ILs [71]

IL	T(K)	CO_2 (bar)	CH_4 (bar)
[C_4mim][PF_6]	298	51.2±1.5	-
	303	55.8±1.4	935±90
	318	75.4±1.2	-
	333	95.2±1.5	1470±165
IL	T(K)	CO_2 (bar)	CH_4 (bar)
[C_4mim][BF_4]	298	52.7±1.0	-
	303	59.4±1.6	1057±145
	316	72.4±1.2	-
	333	101.6±2.8	1670±258
	344	125.4±4.2	-
[C_2mim][Tf_2N]	303	38.0±1.6	1020±190
	333	63.3±1.5	1680±45
[C_4mim][DCA]	303	60.3±1.2	-
	333	94.4±3.5	-
	343	111.4±4.8	-
[BuMePyrr][TFA]	303	53.2±1.1	620±50
	333	84.8±1.4	-

Physical Properties of ILs

Because of their wide range of applications, ILs have gotten a lot of attention. These salts have a low melting point and contain a variety of cations and anions. As a result of the diversity of these ions, ILs can be molecularly designed to be used in a variety of applications [72-76]. The physiochemical parameters of ILs, such as viscosity, diffusivity, density, and surface tension must be measured in order to establish their suitability for diverse applications. Numerical simulations of these qualities that are reliable and exact can aid in the application of ILs at high pressures and temperatures. Viscosity as an effective feature of ILs has been studied in the literature using Eyring's total speed theory as well as molecular dynamics simulations [77, 78]. Only a few cations and anions have been studied for viscosity under high pressure. Harris et al. [79, 80] used a falling body viscometer to estimate [C_4mim] [BF_4], [C_4mim] [PF_6], and [C_8mim] [PF_6] at 200 MPa, using Litovitz and VFT equations. Kanakubo et al. [81] used the same ions to determine molar conductivity and diffusion coefficient, as well as viscosity, using the Stokes-Einstein equation.

Different groups have measured and correlated the diffusivity of gases like CO_2 and hydrofluorocarbons. Morgan et al. [82] compared CO_2 diffusivity and solubility in phosphonium- and imidazolium-based ILs. Diffusivity measurements are carried out with the use of a diffusion cell. The primary operation of this cell is to diffuse gas via a membrane into a chamber containing a sample of the IL. Due to phase behavior and high solubility of CO_2 in ILs, the intermolecular forces between anions and cations decreases in presence of CO_2. The performance of an IL-based solution as a solvent is influenced by surface tension. In some circumstances, the actual influence of surface tension can be negated by the impact of reaction rate.

According to Tsai et al. [83], lowering the surface tension from roughly 72 to 35 mN/m impacted the usable mass transfer area at the very least. Because of the decreased surface stress, it was assumed that this was due to a prohibition associated with capillary processes. Under low surface tension conditions, packing with a larger breadth has a considerably better improvement within the usable mass transfer region than packing with a smaller expanse.

Zhang et al. [67] found that the mass transfer coefficient in the RPB contactor is much higher than in the packed bed column. They also studied the effect of temperature, rotating speed, and liquid and gas flow rate on the mass transfer. Their penetration theory-based model accurately represented the mechanism of gas-liquid mass transfer and was found to be in good agreement with experimental results. Zhang et al. [84] also studied the effect of the addition of water on CO_2 bubble behavior in IL systems. Their findings suggest that adding a modest amount of water to a system can increase mass transfer by lowering the viscosity of the IL. Their findings suggest that ILs in a water-mixing solution perform better in CO_2 absorption.

Conclusion

In terms of environmental, economic, and technical concerns, CO_2 collection processes face a number of obstacles. Novel CO_2 removal methods, such as absorption with ILs have been developed and studied using thermodynamic and transport phenomena. Solubilities of CO_2 in different imidazolium-based ILs were reviewed in this work. The CO_2 capture method employing ILs is briefly presented in this review work, which also highlights the absorption mechanisms. This study focused on CO_2 capture through chemical absorption by ILs and their mixtures from the aspect of their physiochemical characteristics and mass transfer correlation. ILs are called the new alternative solvents due to their tunable physicochemical character, low heat capacity, thermal stability, negligible vapor pressure high and CO_2 solubility. Imidazolium-based ILs (such as and $[C_2mim][Tf_2N]$, $[C_4mim][PF_6]$, $[C_4mim][BF_4]$) were identified as a well-known structure and proposed as a good mixing with other traditional solvents in this review. Important characteristics such as anions and cations type, ions length (size), and operating conditions all have an impact on CO_2 solubility and selectivity.

The availability, affordability, compatibility, and purity of ILs as a green solvent for CO_2 capture are the primary difficulties. These issues are now being encountered on a laboratory scale, and solutions must be found before they can be scaled up to industry. At this time, the benefits and drawbacks of ILs and amines appear to be evenly balanced. High CO_2 solubility, cheap cost, the minimal energy input for regeneration, environmental friendliness, and long-term reusability are the major criteria for optimum CO_2 capture methods. Amphetamines now have the advantages of high CO_2 solubility and inexpensive cost. However, the possibility exists because of the large variety of ILs that can be generated and they can be synthesized in different ways. Recently amines have the advantage of low cost and having a high CO_2 solubility. Furthermore, the cost is expected to reduce as the research and commercialization of ILs in other industries increases.

References

[1] Brosch, T. Affect and emotions as drivers of climate change perception and action: A review. *Current Opinion in Behavioral Sciences* 2021, 42, 15-21. https://doi.org/10.1016/j.cobeha.2021.02.001

[2] Zhang, K.; Ruben, G. B.; Li, X.; Li, Z.; Yu, Z.; Xia, J.; Dong, Z. A comprehensive assessment framework for quantifying climatic and anthropogenic contributions to streamflow changes: A case

study in a typical semi-arid North China basin. *Environmental Modelling & Software* 2020, 128, 104704. https://doi.org/10.1016/j.envsoft.2020.104704.

[3] Zhu, B.; Ma, S.; Xie, R.; Chevallier, J.; Wei, Y.-M. Hilbert Spectra and Empirical Mode Decomposition: A Multiscale Event Analysis Method to Detect the Impact of Economic Crises on the European Carbon Market. *Computational Economics* 2018, 52, 105-121. https://doi.org/10.1007/s10614-017-9664-x.

[4] Shafeeyan, M. S.; Daud, W. M. A. W.; Shamiri, A. A review of mathematical modeling of fixed-bed columns for carbon dioxide adsorption. *Chemical Engineering Research and Design* 2014, 92, 961-988. https://doi.org/10.1016/j.cherd.2013.08.018.

[5] Chen, Y.; Mutelet, F.; Jaubert, J.-N. l. Modeling the solubility of carbon dioxide in imidazolium-based ionic liquids with the PC-SAFT equation of state. *The Journal of Physical Chemistry B* 2012, 116, 14375-14388. https://doi.org/10.1021/jp309944t.

[6] Bailera, M.; Lisbona, P.; Romeo, L. M.; Espatolero, S. Power to Gas projects review: Lab, pilot and demo plants for storing renewable energy and CO2. *Renewable and Sustainable Energy Reviews* 2017, 69, 292-312. https://doi.org/10.1016/j.rser.2016.11.130.

[7] Sreedhar, I.; Vaidhiswaran, R.; Kamani, B. M.; Venugopal, A. Process and engineering trends in membrane based carbon capture. *Renewable and Sustainable Energy Reviews* 2017, 68, 659-684. https://doi.org/10.1016/j.rser.2016.10.025.

[8] Gui, X.; Tang, Z.; Fei, W. CO2 capture with physical solvent dimethyl carbonate at high pressures. *Journal of Chemical & Engineering Data* 2010, 55, 3736-3741. https://doi.org/10.1021/je1002708.

[9] Chaffee, A. L.; Knowles, G. P.; Liang, Z.; Zhang, J.; Xiao, P.; Webley, P. A. CO2 capture by adsorption: materials and process development. *International Journal of Greenhouse Gas Control* 2007, 1, 11-18. https://doi.org/10.1016/S1750-5836(07)00031-X.

[10] Kentish, S. E.; Scholes, C. A.; Stevens, G. W. Carbon dioxide separation through polymeric membrane systems for flue gas applications. *Recent Patents on Chemical Engineering* 2008, 1, 52-66.

[11] Zhang, Y.; Sunarso, J.; Liu, S.; Wang, R. Current status and development of membranes for CO2/CH4 separation: A review. *International Journal of Greenhouse Gas Control* 2013, 12, 84-107. https://doi.org/10.1016/j.ijggc.2012.10.009.

[12] Kamble, A. R.; Patel, C. M.; Murthy, Z. V.P. A review on the recent advances in mixed matrix membranes for gas separation processes. *Renewable and Sustainable Energy Reviews* 2021, 145, 111062. https://doi.org/10.1016/j.rser.2021.111062.

[13] Burt, S.; Baxter, A.; Baxter, L. In *Cryogenic CO2 capture to control climate change emissions*, Proceedings of the 34th International Technical Conference on Clean Coal & Fuel Systems, 2009.

[14] Rochelle, G. T. Amine scrubbing for CO2 capture. *Science* 2009, 325, 1652-1654. doi: 10.1126/science.1176731.

[15] Riemer, P. W.; Ormerod, W. G. International perspectives and the results of carbon dioxide capture disposal and utilisation studies. *Energy Conversion and Management* 1995, 36, 813-818. https://doi.org/10.1016/0196-8904(95)00128-Z.

[16] Kothandaraman, A., *Carbon Dioxide Capture by Chemical Absorption: A Solvent Comparison Study*. Citeseer: 2010; Vol. 72.

[17] Yu, C.-H.; Huang, C.-H.; Tan, C.-S. A review of CO2 capture by absorption and adsorption. *Aerosol and Air Quality Research* 2012, 12, 745-769. https://doi.org/10.4209/aaqr.2012.05.0132.

[18] Anthony, J. L.; Maginn, E. J.; Brennecke, J. F. Solubilities and thermodynamic properties of gases in the ionic liquid 1-n-butyl-3-methylimidazolium hexafluorophosphate. *The Journal of Physical Chemistry B* 2002, 106, 7315-7320. https://doi.org/10.1021/jp020631a.

[19] Johnson, K. E. What's an ionic liquid? *The Electrochemical Society Interface* 2007, 16, 38.

[20] Shannon, M. S.; Bara, J. E. Properties of alkylimidazoles as solvents for CO2 capture and comparisons to imidazolium-based ionic liquids. *Industrial & Engineering Chemistry Research* 2011, 50, 8665-8677. https://doi.org/10.1021/ie200259h.

[21] Ramdin, M.; de Loos, T. W.; Vlugt, T. J. State-of-the-art of CO2 capture with ionic liquids. *Industrial & Engineering Chemistry Research* 2012, 51, 8149-8177. https://doi.org/10.1021/ie3003705.

[22] Yokozeki, A.; Shiflett, M. B.; Junk, C. P.; Grieco, L. M.; Foo, T. Physical and chemical absorptions of carbon dioxide in room-temperature ionic liquids. *The Journal of Physical Chemistry B* 2008, 112, 16654-16663. https://doi.org/10.1021/jp805784u.

[23] Zhao, T.; Zhang, X.; Tu, Z.; Wu, Y.; Hu, X. Low-viscous diamino protic ionic liquids with fluorine-substituted phenolic anions for improving CO2 reversible capture. *Journal of Molecular Liquids* 2018, 268, 617-624. https://doi.org/10.1016/j.molliq.2018.07.096.

[24] Zhu, X.; Song, M.; Xu, Y. DBU-based protic ionic liquids for CO2 capture. *ACS Sustainable Chemistry & Engineering* 2017, 5, 8192-8198. https://doi.org/10.1021/acssuschemeng.7b01839.

[25] Mehraein, I.; Riahi, S. The QSPR models to predict the solubility of CO2 in ionic liquids based on least-squares support vector machines and genetic algorithm-multi linear regression. *Journal of Molecular Liquids* 2017, 225, 521-530. https://doi.org/10.1016/j.molliq.2016.10.133.

[26] Seddon, K. R. A taste of the future. *Nature Materials* 2003, 2, 363-365. https://doi.org/10.1038/nmat90 7.

[27] Tan, S. S. Y.; MacFarlane, D. R. Ionic liquids in biomass processing. *Ionic Liquids*, 2009, 311-339.

[28] Abbott, A. P.; McKenzie, K. J.; Ryder, K. S., Electropolishing and electroplating of metals using ionic liquids based on choline chloride. *ACS Publications* 2007, 10.1021/bk-2007-0975.ch013.

[29] Wasewar, K. L., Ionic Liquids: The Smart Materials in Process Industry. In *Handbook of Smart Materials, Technologies, and Devices: Applications of Industry 4.0*, Springer: 2021; pp 1-28.

[30] Greer, A. J.; Jacquemin, J.; Hardacre, C. Industrial applications of ionic liquids. *Molecules* 2020, 25, 5207. https://doi.org/10.3390/molecules25215207.

[31] Nasirpour, N.; Mohammadpourfard, M.; Heris, S. Z. Ionic liquids: Promising compounds for sustainable chemical processes and applications. *Chemical Engineering Research and Design* 2020, 160, 264-300. https://doi.org/10.1016/j.cherd.2020.06.006.

[32] Stark, A.; Ott, D.; Kralisch, D.; Kreisel, G.; Ondruschka, B. Ionic liquids and green chemistry: a lab experiment. *Journal of Chemical Education* 2010, 87, 196-201. https://doi.org/10.1021/ed8000396.

[33] Shiflett, M. B., *Commercial Applications of Ionic Liquids*. Springer: 2020; Vol. 605.

[34] Aki, S. N.; Mellein, B. R.; Saurer, E. M.; Brennecke, J. F. High-pressure phase behavior of carbon dioxide with imidazolium-based ionic liquids. *The Journal of Physical Chemistry B* 2004, 108, 20355-20365. https://doi.org/10.1021/jp046895+.

[35] Muldoon, M. J.; Aki, S. N.; Anderson, J. L.; Dixon, J. K.; Brennecke, J. F. Improving carbon dioxide solubility in ionic liquids. *The Journal of Physical Chemistry B* 2007, 111, 9001-9009. https://doi.org/10.1021/jp071897q.

[36] Carvalho, P. J.; Alvarez, V. H.; Marrucho, I. M.; Aznar, M.; Coutinho, J. A. High carbon dioxide solubilities in trihexyltetradecylphosphonium-based ionic liquids. *The Journal of Supercritical Fluids* 2010, 52, 258-265. https://doi.org/10.1016/j.supflu.2010.02.002Get.

[37] Torralba-Calleja, E.; Skinner, J.; Gutiérrez-Tauste, D. CO2 capture in ionic liquids: a review of solubilities and experimental methods. *Journal of Chemistry* 2013, 2013. https://doi.org/10.1155/2013/473584.

[38] Szala-Bilnik, J.; Crabtree, E.; Abedini, A.; Bara, J. E.; Turner, C. H. Solubility and diffusivity of CO2 in ionic polyimides with [C (CN) 3] x [oAc] 1− x anion composition. *Computational Materials Science* 2020, 174, 109468. https://doi.org/10.1016/j.commatsci.2019.109468.

[39] Durak, O.; Zeeshan, M.; Keskin, S.; Uzun, A. [BMIM][OAc] coating layer makes activated carbon almost completely selective for CO2. *Chemical Engineering Journal* 2022, 437, 135436. https://doi.org/10.1016/j.cej.2022.135436.

[40] Gurau, G.; Rodríguez, H.; Kelley, S. P.; Janiczek, P.; Kalb, R. S.; Rogers, R. D. Demonstration of chemisorption of carbon dioxide in 1, 3-dialkylimidazolium acetate ionic liquids. *Angewandte Chemie* 2011, 123, 12230-12232. https://doi.org/10.1002/ange.201105198.

[41] Shin, E.-K.; Lee, B.-C. High-pressure phase behavior of carbon dioxide with ionic liquids: 1-alkyl-3-methylimidazolium trifluoromethanesulfonate. *Journal of Chemical & Engineering Data* 2008, 53, 2728-2734. https://doi.org/10.1021/je8000443.

[42] Blanchard, L. A.; Gu, Z.; Brennecke, J. F. High-pressure phase behavior of ionic liquid/CO2 systems. *The Journal of Physical Chemistry B* 2001, 105, 2437-2444. https://doi.org/10.1021/jp003309d.

[43] Kim, Y.; Choi, W.; Jang, J.; Yoo, K.-P.; Lee, C. Solubility measurement and prediction of carbon dioxide in ionic liquids. *Fluid Phase Equilibria* 2005, 228, 439-445. https://doi.org/10.1016/j.fluid.2004.09.006.

[44] Zhang, X.; Liu, Z.; Wang, W. Screening of ionic liquids to capture CO2 by COSMO-RS and experiments. *AIChE Journal* 2008, 54, 2717-2728. https://doi.org/10.1002/aic.11573.

[45] Revelli, A.-L.; Mutelet, F.; Jaubert, J.-N. High carbon dioxide solubilities in imidazolium-based ionic liquids and in poly (ethylene glycol) dimethyl ether. *The Journal of Physical Chemistry B* 2010, 114, 12908-12913. https://doi.org/10.1021/jp1057989.

[46] Bates, E. D.; Mayton, R. D.; Ntai, I.; Davis, J. H. CO2 capture by a task-specific ionic liquid. *Journal of the American Chemical Society* 2002, 124, 926-927. https://doi.org/10.1021/ja017593d.

[47] C amper, D.; Bara, J. E.; Gin, D. L.; Noble, R. D. Room-temperature ionic liquid− amine solutions: tunable solvents for efficient and reversible capture of CO2. *Industrial & Engineering Chemistry Research* 2008, 47, 8496-8498. https://doi.org/10.1021/ie801002m.

[48] Qu, Y.; Zhao, Y.; Li, D.; Sun, J. Task-specific ionic liquids for carbon dioxide absorption and conversion into value-added products. *Current Opinion in Green and Sustainable Chemistry* 2022, 100599. https://doi.org/10.1016/j.cogsc.2022.100599.

[49] Shahrom, M. S. R.; Wilfred, C. D.; Taha, A. K. Z. CO2 capture by task specific ionic liquids (TSILs) and polymerized ionic liquids (PILs and AAPILs). *Journal of Molecular Liquids* 2016, 219, 306-312. https://doi.org/10.1016/j.molliq.2016.02.046.

[50] Yang, Z.-Z.; Zhao, Y.-N.; He, L.-N. CO2 chemistry: task-specific ionic liquids for CO2 capture/activation and subsequent conversion. *RSC advances* 2011, 1, 545-567. 10.1039/C1RA00307K.

[51] Kurnia, K.; Harris, F.; Wilfred, C.; Mutalib, M. A.; Murugesan, T. Thermodynamic properties of CO2 absorption in hydroxyl ammonium ionic liquids at pressures of (100–1600) kPa. *The Journal of Chemical Thermodynamics* 2009, 41, 1069-1073. https://doi.org/10.1016/j.jct.2009.04.003.

[52] Yuan, X.; Zhang, S.; Liu, J.; Lu, X. Solubilities of CO2 in hydroxyl ammonium ionic liquids at elevated pressures. *Fluid Phase Equilibria* 2007, 257, 195-200. https://doi.org/10.1016/j.fluid.2007.01.031.

[53] Petkovic, M.; Seddon, K. R.; Rebelo, L. P. N.; Pereira, C. S. Ionic liquids: a pathway to environmental acceptability. *Chemical Society Reviews* 2011, 40, 1383-1403. https://doi.org/10.1039/C004968A

[54] Boethling, R.; Sommer, E.; DiFiore, D. Designing small molecules for biodegradability. *Chemical reviews* 2007, 107, 2207-2227. https://doi.org/10.1021/cr050952t.

[55] Harjani, J. R.; Singer, R. D.; Garcia, M. T.; Scammells, P. J. Biodegradable pyridinium ionic liquids: design, synthesis and evaluation. *Green Chemistry* 2009, 11, 83-90. 10.1039/B811814K.

[56] Gathergood, N.; Garcia, M. T.; Scammells, P. J. Biodegradable ionic liquids: Part I. Concept, preliminary targets and evaluation. *Green Chemistry* 2004, 6, 166-175. https://doi.org/10.1039/B315270G.

[57] Stolte, S.; Abdulkarim, S.; Arning, J.; Blomeyer-Nienstedt, A.-K.; Bottin-Weber, U.; Matzke, M.; Ranke, J.; Jastorff, B.; Thöming, J. Primary biodegradation of ionic liquid cations, identification of degradation products of 1-methyl-3-octylimidazolium chloride and electrochemical wastewater treatment of poorly biodegradable compounds. *Green Chemistry* 2008, 10, 214-224. https://doi.org/10.1039/B713095C.

[58] Pham, T. P. T.; Cho, C.-W.; Jeon, C.-O.; Chung, Y.-J.; Lee, M.-W.; Yun, Y.-S. Identification of metabolites involved in the biodegradation of the ionic liquid 1-butyl-3-methylpyridinium bromide by activated sludge microorganisms. *Environmental Science & Technology* 2009, 43, 516-521. https://doi.org/10.1021/es703004h.

[59] Grabińska-Sota, E.; Kalka, J. An assessment of the toxicity of pyridinium chlorides and their biodegradation intermediates. *Environment International* 2003, 28, 687-690. https://doi.org/10.1016/S0160-4120(02)00112-5.

[60] Atefi, F.; Garcia, M. T.; Singer, R. D.; Scammells, P. J. Phosphonium ionic liquids: design, synthesis and evaluation of biodegradability. *Green Chemistry* 2009, 11, 1595-1604. https://doi.org/10.1039/B913057H.

[61] Petkovic, M.; Ferguson, J. L.; Gunaratne, H. N.; Ferreira, R.; Leitao, M. C.; Seddon, K. R.; Rebelo, L. P. N.; Pereira, C. S. Novel biocompatible cholinium-based ionic liquids—toxicity and biodegradability. *Green Chemistry* 2010, 12, 643-649. https://doi.org/10.1039/B922247B.

[62] Studzińska, S.; Kowalkowski, T.; Buszewski, B. Study of ionic liquid cations transport in soil. *Journal of Hazardous Materials* 2009, 168, 1542-1547. https://doi.org/10.1016/j.jhazmat.2009.03.029.

[63] Mrozik, W.; Jungnickel, C.; Ciborowski, T.; Pitner, W. R.; Kumirska, J.; Kaczyński, Z.; Stepnowski, P. Predicting mobility of alkylimidazolium ionic liquids in soils. *Journal of Soils and Sediments* 2009, 9, 237-245. https://doi.org/10.1007/s11368-009-0057-1.

[64] Strigle, R. F. *Random Packings and Packed Towers: Design and Applications*. 1987.

[65] Geankoplis, C. J. *Transport Processes and Separation Process Principles:(Includes Unit Operations)*. Prentice Hall Press

[66] Demontigny, D.; Tontiwachwuthikul, P.; Chakma, A. Parametric studies of carbon dioxide absorption into highly concentrated monoethanolamine solutions. *The Canadian Journal of Chemical Engineering* 2001, 79, 137-142. https://doi.org/10.1002/cjce.5450790121.

[67] Zhang, L.-L.; Wang, J.-X.; Xiang, Y.; Zeng, X.-F.; Chen, J.-F. Absorption of carbon dioxide with ionic liquid in a rotating packed bed contactor: mass transfer study. *Industrial & Engineering Chemistry Research* 2011, 50, 6957-6964. https://doi.org/10.1021/ie1025979.

[68] Galán Sánchez, L. *Functionalized ionic liquids: absorption solvents for carbon dioxide and olefin separation*. 2008. 10.6100/ir639177.

[69] Sánchez, L. G.; Meindersma, G.; De Haan, A. Kinetics of absorption of CO2 in amino-functionalized ionic liquids. *Chemical Engineering Journal* 2011, 166, 1104-1115. https://doi.org/10.1016/j.cej.2010.12.016.

[70] Zhang, X.; Bao, D.; Huang, Y.; Dong, H.; Zhang, X.; Zhang, S. Gas–liquid mass-transfer properties in CO2 absorption system with ionic liquids. *AIChE Journal* 2014, 60, 2929-2939. https://doi.org/10.1002/aic.14507.

[71] Babamohammadi, S.; Shamiri, A.; Aroua, M. K. A review of CO2 capture by absorption in ionic liquid-based solvents. *Reviews in Chemical Engineering* 2015, 31, 383-412. https://doi.org/10.1515/revce-2014-0032.

[72] Kaur, G.; Kumar, H.; Singla, M. Diverse applications of ionic liquids: A comprehensive review. *Journal of Molecular Liquids* 2022, 118556. https://doi.org/10.1016/j.molliq.2022.118556.

[73] Raman, A. P. S.; Singh, M. B.; Jain, P.; Chaudhary, P.; Bahadur, I.; Lal, K.; Kumar, V.; Singh, P. An Update on Synthesis, Properties, Applications and Toxicity of the ILs. *Journal of Molecular Liquids* 2022, 119989. https://doi.org/10.1016/j.molliq.2022.119989.

[74] Li, Z.; Mangolini, F. Recent Advances in Nanotribology of Ionic Liquids. *Experimental Mechanics* 2021, 61, 1093-1107. https://doi.org/10.1007/s11340-021-00732-7.

[75] Correia, D. M.; Fernandes, L. C.; Fernandes, M. M.; Hermenegildo, B.; Meira, R. M.; Ribeiro, C.; Ribeiro, S.; Reguera, J.; Lanceros-Méndez, S. Ionic liquid-based materials for biomedical applications. *Nanomaterials* 2021, 11, 2401. https://doi.org/10.3390/nano11092401.

[76] Wang, Z.; He, S.; Nguyen, V.; Riley, K. E. Ionic liquids as "Green solvent and/or electrolyte" for energy interface. *Engineered Science* 2020.

[77] Glasstone, S.; Laidler, K. J.; Eyring, H. *The Theory of Rate Processes; The Kinetics of Chemical Reactions, Viscosity, Diffusion and Electrochemical Phenomena*; McGraw-Hill Book Company: 1941.

[78] Kelkar, M. S.; Maginn, E. J. Effect of temperature and water content on the shear viscosity of the ionic liquid 1-ethyl-3-methylimidazolium bis (trifluoromethanesulfonyl) imide as studied by atomistic simulations. *The Journal of Physical Chemistry B* 2007, 111, 4867-4876. https://doi.org/10.1021/jp0686893.

[79] Harris, K. R.; Woolf, L. A.; Kanakubo, M. Temperature and pressure dependence of the viscosity of the ionic liquid 1-butyl-3-methylimidazolium hexafluorophosphate. *Journal of Chemical & Engineering Data* 2005, 50, 1777-1782. https://doi.org/10.1021/je050147b.

[80] Harris, K. R.; Kanakubo, M.; Woolf, L. A. Temperature and pressure dependence of the viscosity of the ionic liquids 1-hexyl-3-methylimidazolium hexafluorophosphate and 1-butyl-3-methylimidazolium bis (trifluoromethylsulfonyl) imide. *Journal of Chemical & Engineering Data* 2007, 52, 1080-1085. https://doi.org/10.1021/je700032n.

[81] Kanakubo, M.; Harris, K. R.; Tsuchihashi, N.; Ibuki, K.; Ueno, M. Effect of pressure on transport properties of the ionic liquid 1-butyl-3-methylimidazolium hexafluorophosphate. *The Journal of Physical Chemistry B* 2007, 111, 2062-2069. https://doi.org/10.1021/jp067328k.

[82] Morgan, D.; Ferguson, L.; Scovazzo, P. Diffusivities of gases in room-temperature ionic liquids: data and correlations obtained using a lag-time technique. *Industrial & Engineering Chemistry Research* 2005, 44, 4815-4823. https://doi.org/10.1021/ie048825v.
[83] Tsai, R. E.; Schultheiss, P.; Kettner, A.; Lewis, J. C.; Seibert, A. F.; Eldridge, R. B.; Rochelle, G. T. Influence of surface tension on effective packing area. *Industrial & Engineering Chemistry Research* 2008, 47, 1253-1260. https://doi.org/10.1021/ie070780l.
[84] Zhang, X.; Dong, H.; Bao, D.; Huang, Y.; Zhang, X.; Zhang, S. Effect of small amount of water on CO2 bubble behavior in ionic liquid systems. *Industrial & Engineering Chemistry Research* 2014, 53, 428-439. https://doi.org/10.1021/ie4020827.

Chapter 5

The Properties and Applications of Surface-Active Ionic Liquids

Javad Saien*
and Mona Kharazi†
Department of Applied Chemistry, Bu-Ali Sina University, Hamedan, Iran

Abstract

Efforts to improve the interfacial activity of surfactants and the demand for low interfacial tensions (IFT)/critical micelle concentrations (CMCs) for operations under harsh conditions, have persuaded scientists to search for effective and resistive materials. For this aim, surface-active ionic liquids (SAILs) with amphiphilic nature and unique properties have received much interest in recent decades, due to their high interfacial activity, recyclability, environmental friendliness, and stability under high salinity and temperature conditions. Although the information on properties and use of SAILs is limited, a comprehensive compilation is required to understand state-of-the-art technological advances in SAILs area.

Accordingly, this chapter first describes some of the basic concepts and characteristics of SAILs and then explains how the IFT and CMC change in the presence of SAILs. Then, a comprehensive overview on the effects of alkyl chain length, head group, and anion present in the chemical structure of SAILs is presented. A comparison of the performance of different structural SAILs is also provided. The following section is devoted to the review of influence of operating conditions such as temperature, presence of electrolytes, pH, and nature of phases, on SAILs performance. Thereafter, a brief discussion on relevant aspects of green chemistry is provided. Developments in the use of SAILs in different processes such as emulsion stabilization, suspension of inherent insoluble materials, as well as solid-phase extraction are also included.

The current chapter attempts to provide an outlook of SAILs applications and will serve as an inspiration for further extensive research on SAILs.

Keywords: surface-active ionic liquids, interfacial tension, micelle, stabilization, dispersion, extraction

* Corresponding Author's Email: saien@basu.ac.ir.
† Co-corresponding Author's Email: kharazy91@yahoo.com.

In: Properties and Applications of Ionic Liquids
Editors: Noureddine Abidi and Julia L. Shamshina
ISBN: 979-8-88697-635-9

Abbreviations

CMC	Critical micelle concentration
IFT	Interfacial tension
IL	Ionic liquid
N_{agg}	Micelle aggregation number
O/W	Oil-in-water
PAHs	Polycyclic aromatic hydrocarbons
PIT	Phase inversion temperature
SAIL	Surface-active ionic liquid
SFT	Surface tension
SPE	Solid phase extraction
Surfactant	Surface-active agent
W/O	Water-in-oil

Cation Part of SAILs

$[BT]^+$	Butyrolactam
$[C_nmim]^+$	1-Alkyl-3-methylimidazolium
$[C_2mim]^+$	1-Ethyl-3-methylimidazolium
$[C_4mim]^+$	1-Butyl-3-methylimidazolium
$[C_5mim]^+$	1-Pentyl-3-methylimidazolium
$[C_6mim]^+$	1-Hexyl-3-methylimidazolium
$[C_8mim]^+$	1-Octyl-3-methylimidazolium
$[C_9mim]^+$	1-Nonyl-3-methylimidazolium
$[C_{10}mim]^+$	1-Decyl-3-methylimidazolium
$[C_{12}mim]^+$	1-Dodecyl-3-methylimidazolium126
$[C_{14}mim]^+$	1-Tetradecyl-3-methylimidazolium
$[C_{16}mim]^+$	1-Hexadecyl-3-methylimidazolium
$[C_{18}mim]^+$	1-Octadecyl-3-methylimidazolium
$[C_6C_4im]^+$	1-Hexyl-3-butylimidazolium
$[C_8C_4im]^+$	1-Octyl-3-butylimidazolium
$[C_8C_8im]^+$	1-Octyl-3-octylimidazolium
$[C_{10}C_4im]^+$	1-Decyl-3-butylimidazolium
$[C_{12}C_4im]^+$	1-Dodecyl-3-butylimidazolium
$[C_{12}C_{12}im]^+$	1,3-Didodecylimidazolium
$[C_{16}C_4im]^+$	1-Hexadecyl-3-butylimidazolium
$[C_nim\text{-}C_m\text{-}imC_n]$	1,1'-(Alkane-1,4-diyl)bis(3-alkyl-1H-imidazol-3-ium)
$[C_nim\text{-}C_4\text{-}imC_n]$	1,1'-(Butane-1,4-diyl)bis(3-alkyl-1H-imidazol-3-ium)
$[C_4im\text{-}C_2\text{-}imC_4]^+$	1,1'-(Ethane -1,4-diyl)bis(3-butyl-1H-imidazol-3-ium)
$[C_4im\text{-}C_4\text{-}imC_4]^+$	1,1'-(Butane-1,4-diyl)bis(3-butyl-1H-imidazol-3-ium)
$[C_4im\text{-}C_6\text{-}imC_4]^+$	1,1'-(Hexane -1,4-diyl)bis(3-butyl-1H-imidazol-3-ium)
$[C_4im\text{-}C_{10}\text{-}imC_4]^+$	1,1'-(Decane-1,4-diyl)bis(3-butyl-1H-imidazol-3-ium)
$[C_8im\text{-}C_4\text{-}imC_8]^+$	1,1'-(Butane-1,4-diyl)bis(3-octyl-1H-imidazol-3-ium)

$[C_{10}im\text{-}C_2\text{-}imC_{10}]^+$	1,1'-(Ethane -1,4-diyl)bis(3-decyl-1H-imidazol-3-ium)
$[C_{10}im\text{-}C_4\text{-}imC_{10}]^+$	1,1'-(Butane-1,4-diyl)bis(3-decyl-1H-imidazol-3-ium)
$[C_{12}im\text{-}C_4\text{-}imC_{12}]^+$	1,1'-(Butane-1,4-diyl)bis(3-dodecyl-1H-imidazol-3-ium)
$[C_{12}im\text{-}C_{10}\text{-}imC_{12}]^+$	1,1'-(Decane-1,4-diyl)bis(3-dodecyl-1H-imidazol-3-ium)
$[Mpy]^+$	1-Methylpyridinium
$[C_8py]^+$	1-Octylpyridinium
$[C_{16}py]^+$	1-Hexadecylpyridinium
$[C_{18}py]^+$	1-Octadecylpyridinium
$[C_nmpy]^+$	1-Alkyl-3-methylpyridinium
$[C_8mpy]^+$	1- Octyl-3-methylpyridinium
$[C_4mpyr]^+$	1-Butyl-1-methylpyrrolidinium
$[C_{10}mpyr]^+$	1-Decyl-1-methylpyrrolidinium
$[C_{12}mpyr]^+$	1-Dodecyl-1-methylpyrrolidinium
$[C_{14}mpyr]^+$	1-Tetradecyl-1-methylpyrrolidinium
$[C_{16}mpyr]^+$	1-Hexadecyl-1-methylpyrrolidinium
$[C_{18}mpyr]^+$	1-Octadecyl-1-methylpyrrolidinium
$[CP]^+$	Caprolactam
$[PA]^+$	Propylammonium
$[P4441]^+$	Tributy-lmethylphosphonium
$[P666,14]^+$	Trihexyl-tetradecylphosphonium
$[3\text{-}HPA]^+$	3-Hydroxypropylammonium

Anion Part of SAILs

$[Ac]^-$	Acetate
$[AlCl_4]^-$	Tetrachloroaluminate
$[AS]^-$	Alkyl sulfate
$[BF_4]^-$	Tetrafluoroborate
$[CF_3COO]^-$	Trifluoroacetate
$[CF_3SO_3]^-$	Trifluoromethanesulfonate
$[CH_3COO]^-$	Acetate
$[C_6H_{13}COO]^-$	Hexanoate
$[DS]^-$	Dodecylsulfonate
$[HCOO]^-$	Formate
$[NTf_2]^-$	Bis(trifluoromethylsulfonyl) imide
$[OAc]^-$	Acetate
$[OTf]^-$	Trifluoromethanesulfonate
$[PF_6]^-$	Hexafluorophosphate
$[SCN]^-$	Thiocyanate
$[TA]^-$	Trifluoroacetate
$[Tos]^-$	P-toluenesulfonate
SAILs	Surface-active ionic liquids
BAF	Butylammonium formate
$[C_{16}IPS]$	3-(1-hexadecyl-3-imidazolio) propanesulfonate

[C_{16}IPS-Nsa]	3-(1-hexadecyl-3-imidazolio) propanesulfonate β-naphthalene sulfonate

Conventional Surfactants

CTAB	Cetyltrimethylammonium bromide
C_nTAB	Alkyltrimethylammonium bromides
C_nTAC	Alkyltrimethylammonium chloride
DTMAC	Dodecyltrimethylammonium chloride
SAS	Sodium alkyl sulfates

Introduction

Amphiphilic ionic liquids (ILs) with different hydrophobic alkyl chains and hydrophilic head groups in their chemical structure are termed as surface-active ionic liquids, abbreviated "SAILs". Investigations on the performance of SAILs in various fields of emulsification, foaming, colloid formation, chemistry at the interface, and oil and petroleum/water interfacial behavior have confirmed that they can significantly reduce the interfacial tension (IFT) and exhibit low value of critical micelle concentration (CMC). Accordingly, compared to analog structured conventional surfactants, SAILs exhibit higher interfacial activity.

SAILs can be designed to exhibit remarkable physical and chemical properties like ionic conductivity, recyclability, non-flammability, low vapor pressure (negligible at room temperature), wide liquid state range, high solvation capability, high polarity, good catalytic properties, and high stability under harsh conditions. Further, due to environmentally friendly nature, most of them are considered as "green materials".

By a deliberate choice of appropriate cation/anion for the structure of SAILs, their physical and chemical properties such as hydrophobicity, viscosity, density, and stability can be easily altered for task-specific purposes. In this regard, different types of SAILs, including Anionic, Catanionic, Multicationic, and Zwitterionic have been introduced and their superior interfacial activities have been scrutinized. These perspectives have elaborated SAIL products in various fields like analytical chemistry, biotechnology, emulsion stabilization, and separations. Nonetheless, operational conditions affect the characteristics of SAILs and their interfacial activities.

In this chapter a brief explanation of the basic concepts and characteristics of SAILs is first presented. Thereafter, the performance of SAILs in IFT reduction and micelle formation are provided by considering the influence of alkyl chain length, head groups, and counterions, as well as the influence of different operational conditions: temperature, presence of electrolytes, pH, and the nature of phases. The following section is focused on the green chemistry aspects of SAILs in terms of their toxicity and biodegradability. Subsequently, a concise discussion about the most important and favorable applications of SAILs is given. The most relevant fields are emulsion stabilization, stabilizing suspensions of some inherently insoluble materials, and solid-phase extraction. Whenever it has been possible, the characteristics and the outstanding performance of the new types of SAILs have been considered in order to elucidate the benefits of these materials. Finally, in the last section, the crucial challenges and possible research directions and conclusions are presented.

Basic Concepts

Interfacial Tension

Interfacial tension (IFT) is the work required to create a unit of new surface between two immiscible fluids. This property is considered as surface tension (SFT) when one of the phases is a gas or air. IFT is a measure of molecular cohesive forces at the interface of immiscible liquids, stemming from unbalanced forces that molecules bear from adjacent molecules. IFT plays a vital role in designing and operating with immiscible fluids as well as the hydrodynamic and contact of the phases. Therefore, the study of the IFT provides comprehensive information on the interactions and involved forces, which is helpful in understanding different aspects of extraction, emulsification, stabilization, foaming agents, and petroleum processes [1]. It is important to find a way to reduce the IFT for better operations in many industries. In this regard, a methodology is adsorption of surfactants at the interface to diminish the resultant force between adjacent molecules.

Surfactants and Their Properties

Surface-active agents, briefly surfactants, are compounds with both hydrophilic (lipophobic or water liking) and hydrophobic (lipophilic or oil liking) portions, that enable them to have "amphiphilic" nature. The presence of ionic or polar hydrophilic group aids water solubility while the presence of organic hydrophobic group promotes oil solubility.

When surfactants are dissolved in a binary organic-aqueous system, they are adsorbed in a way that their hydrophilic group is located in the aqueous phase while the hydrophobic group is located in the oil phase. A typical orientation of the adsorbed surfactant molecules at the interface is shown in Figure 1. Such orientation results in interruption of molecular interactions at the interface and decrease of the IFT.

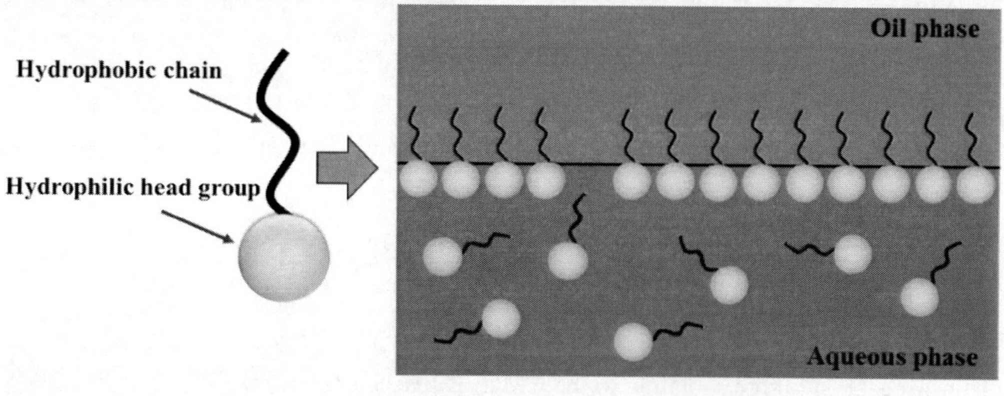

Figure 1. Schematic of a surfactant molecule and orientation of molecules at the interface.

The solubility is another relevant parameter. It depends on the structure of surfactants. Higher polarity and large number of polar groups in the hydrophilic portion favors solubility in the aqueous phase, whereas long alkyl chains and branching chains facilitate the solubility in the oil phase.

At the beginning of the adsorption process, the rate of adsorption is higher than desorption, and the IFT decreases constantly. By measuring the instant IFTs during the process, a "dynamic IFT" could be obtained which represents the rate of surfactant transfer from bulk to the interface and provides information about the adsorption kinetic. Given enough time, the surfactants entirely accumulate at the interface. At this time, surfactant adsorption and desorption rates will be the same and the IFT will be in equilibrium, known as "equilibrium IFT".

Micelle and Critical Micelle Concentration

Above a certain concentration, surfactant monomers are gathered in a way that their hydrophilic portions extend to the aqueous environment while the hydrophobic portions are centered, known as "micelle formation". This configuration minimizes the free energy of a surfactant solution. The concentration at which micelles begin to form is known as the critical micelle concentration, abbreviated CMC. At a CMC, there is an equilibrium between surfactant monomers and micelles.

There is a similar phenomenon for oil-soluble surfactants, known as "reverse micelle". In this case, to minimize the free energy, the hydrophilic portions are assembled in the micelle cores and the hydrophobic portions expand away into the oil environment. Typical shapes of a micelle and a reverse micelle are presented in Figure 2.

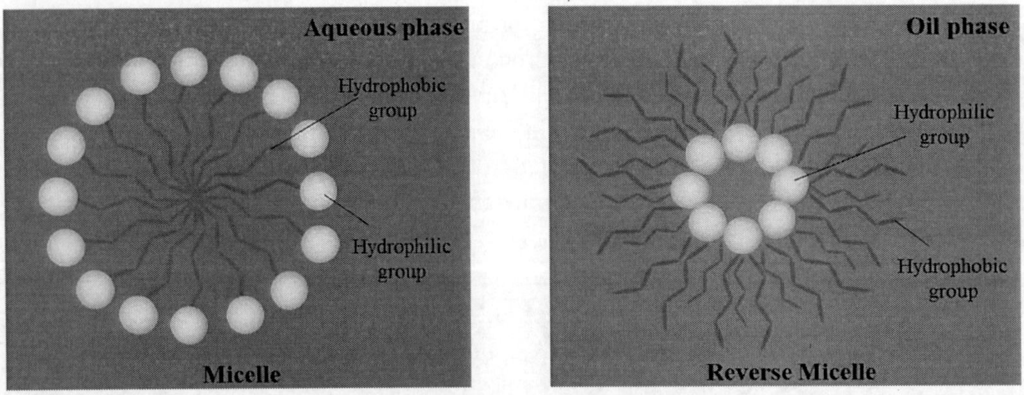

Figure 2. Schematic of a micelle and a reverse micelle formation. Reprinted from [2] with permission from Springer.

Surfactants and Interfacial Tension

The adsorption of surfactants at the interface provides a significant reduction in the IFT until a distinct breakpoint (corresponding to the CMC) is reached. Above the CMC, the addition of surfactants creates micelles and the total number of adsorbed molecules at the interface does not change, thus, the IFT remains almost constant. For a better illustration, the pattern of IFT change versus surfactant concentration is schematically presented in Figure 3. It is noteworthy that the presence of the minimum point before CMC (as depicted in Figure 3), implies adsorption of surface-active impurities at the interface at concentrations below the CMC, after which impurities dissolve in the micelles [3].

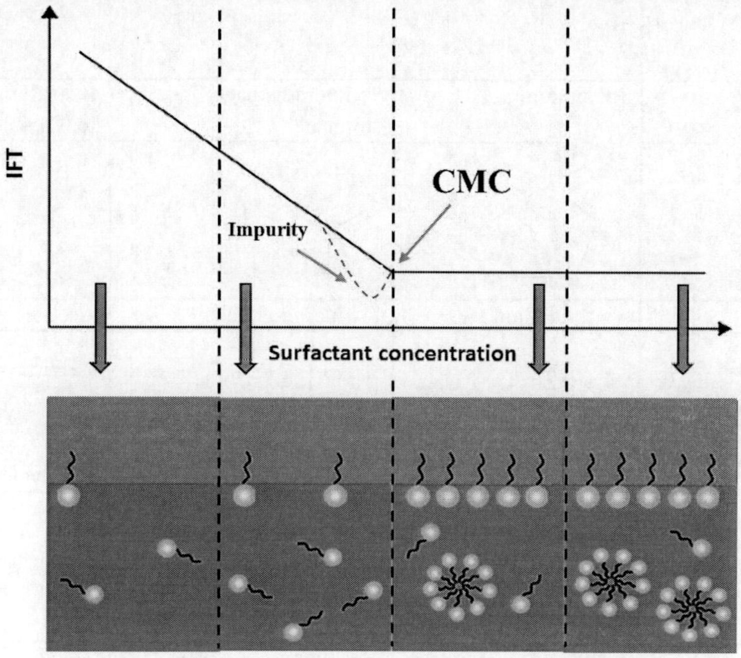

Figure 3. The variation of IFT versus surfactant concentration corresponding to surfactant adsorption at the interface and micelle formation.

Structure and Classification of SAILs

Chemical Structure

Ionic liquids (ILs) consist of organic cations and inorganic/organic anions that can be designed to exhibit prominent properties such as high chemical and thermal stability, low vapor pressure, recyclability, and environmentally friendly nature. The steric asymmetry of involved ions causes mismatching and forming unsteady crystal lattices. Therefore, their melting point is by definition below 100°C and many ILs remain liquid at room temperature (Room Temperature Ionic Liquids, RTILs).

In the structure of ILs, organic cations such as imidazolium, pyridinium, pyrazolium, pyrrolidinium, sulfonium, phosphonium, ammonium, guanidinium, piperidinium, and cholinium are coupled with various inorganic or organic anions, ranging from simple inorganic ions (like halides) to more complex organic species such as trifluoromethanesulfonate, $[OTf]^-$, acetate, $[CH_3COO]^-$, bis(trifluoromethanesulfonyl)imide, $[NTf_2]^-$, trifluoroacetate, $[TA]^-$, tetrachloroaluminate, $[AlCl_4]^-$, thiocyanate, $[SCN]^-$, hexafluorophosphate, $[PF_6]^-$, tetrafluoroborate, $[BF_4]^-$. Some conventional cations and anions in ILs are presented in Table 1.

Table 1. Some typical cations and anions in the structure of ILs

Cation		Anion	
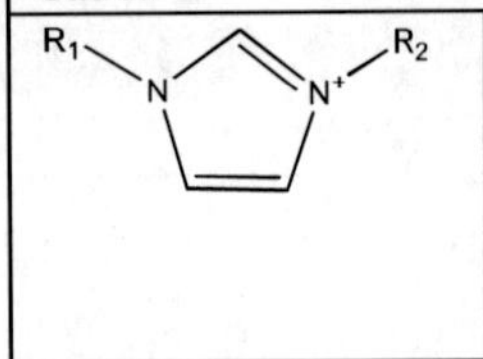	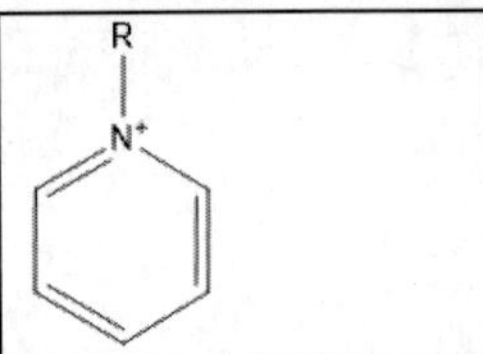	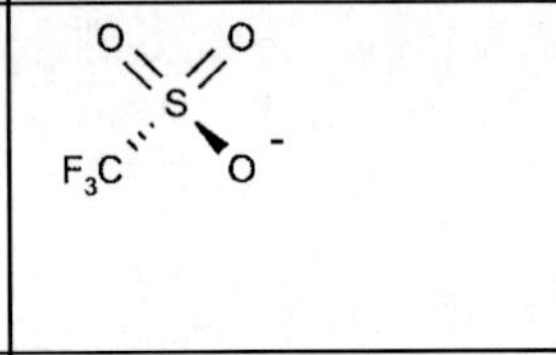	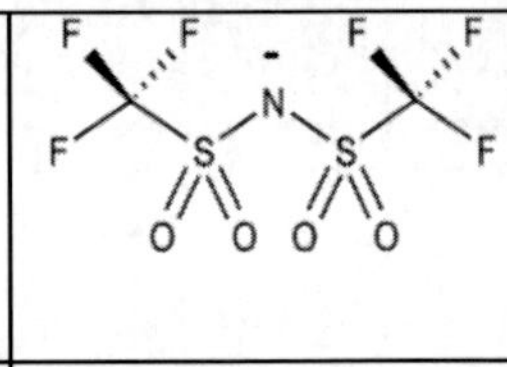
Imidazolium	Pyridinium	Trifluoromethane-sulfonate	Bis(trifluoromethanesul fonyl) imide
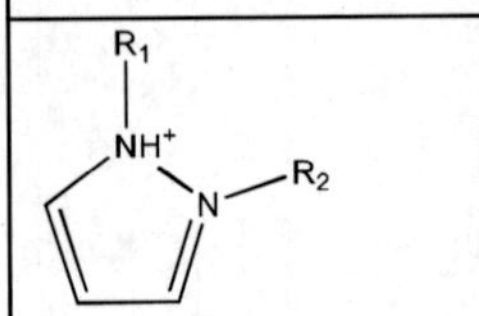	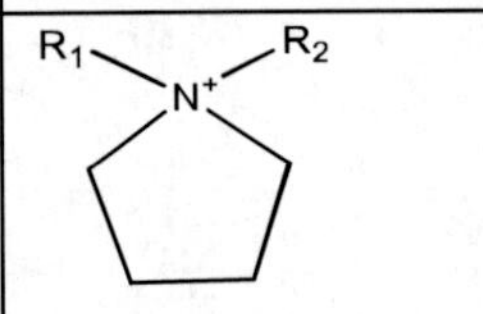	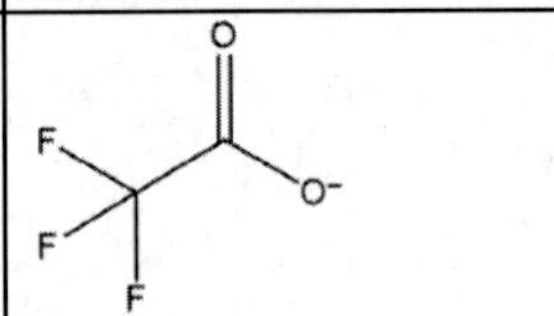	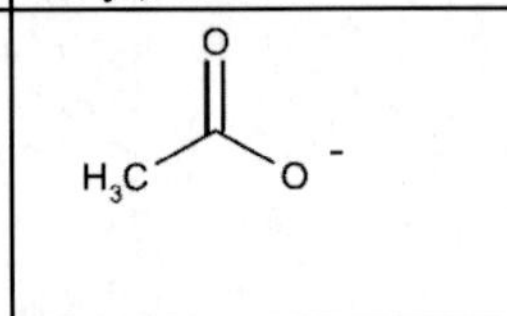
Pyrazolium	Pyrrolidinium	Trifluoroacetate	Acetate
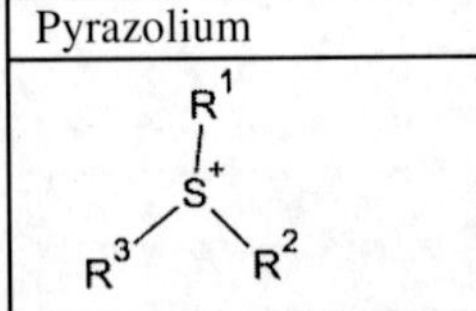	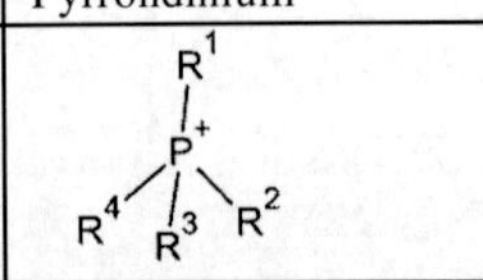	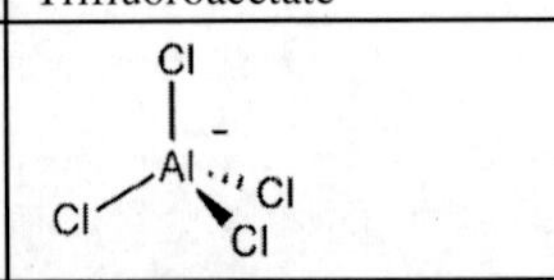	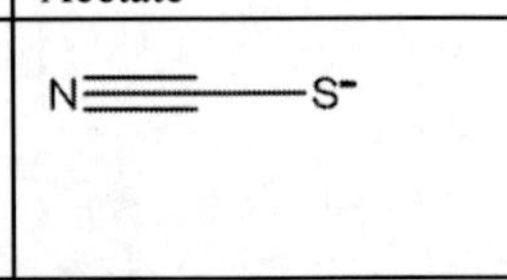
Sulfonium	Phosphonium	Tetrachloroaluminate	Thiocyanate
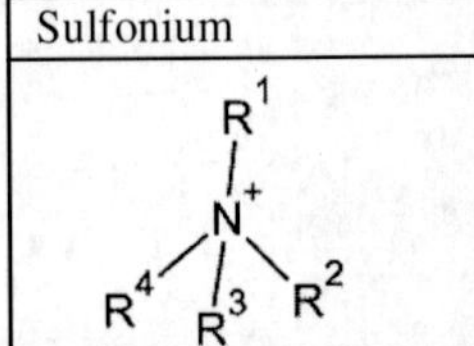	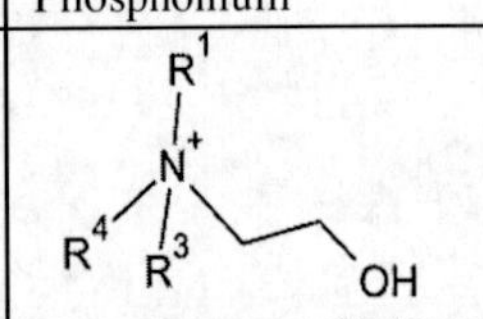	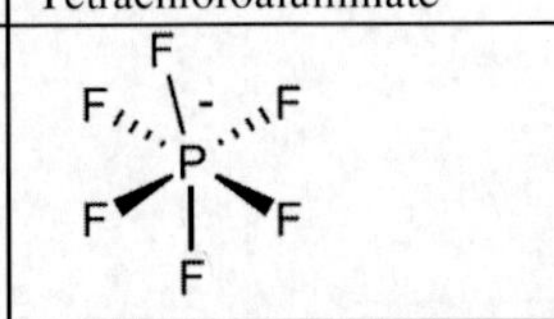	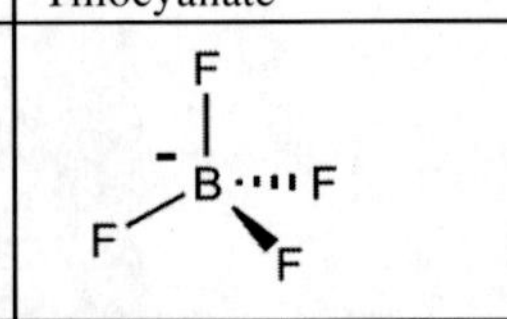
Ammonium	Cholinium	Hexafluorophosphate	Tetrafluoroborate

Surface-active Ionic Liquids (SAILs) exhibit amphiphilic properties. Similar to conventional surfactants, SAILs consist of hydrophilic head group with appended hydrophobic chain(s), and counterions in their molecular structure. SAILs exhibit two distinct behaviors of adsorption and orientation at the oil-water interfaces causing IFT reduction and creating aggregates in the bulk of the solutions. From a practical point of view, these are useful in different processes like creating microemulsions, stabilizing suspensions, dispersing some inherent insoluble materials in solutions, and extracting specific compounds based on aggregates. For better comparison, the molecule of a typical SAIL is shown in Figure 4. Relevantly, a structure with larger head group exhibits higher hydrophobicity and weaker hydration tendency which assist its migration toward the interface.

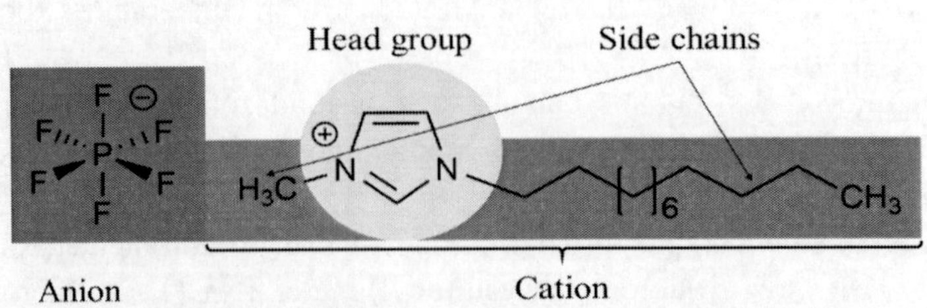

Figure 4. Molecular structures of an imidazolium SAIL, $[C_{12}mim][PF_6]$.

IFT Reduction and Aggregation Formation

Upon adsorption at the interface, the hydrophilic portion of SAILs resides in the polar phase and the hydrophobic one is in the non-polar phase. This orientation reduces the intermolecular forces and consequently reduces IFT. The higher the concentration of SAILs is, the more orientation variations will take place until the surface becomes saturated; higher concentrations may not give lower IFTs. At this point, which corresponds to the CMC, SAIL molecules start to form aggregates (as a micelle or a reverse micelle) in the bulk [4]. Then, by raising the concentration, the number of aggregates increases with nearly no interface effect. Different trends of IFT variation with SAIL concentration (linear or limber) have been observed depending on their structure, nature of involved phases, and the rate of adsorption.

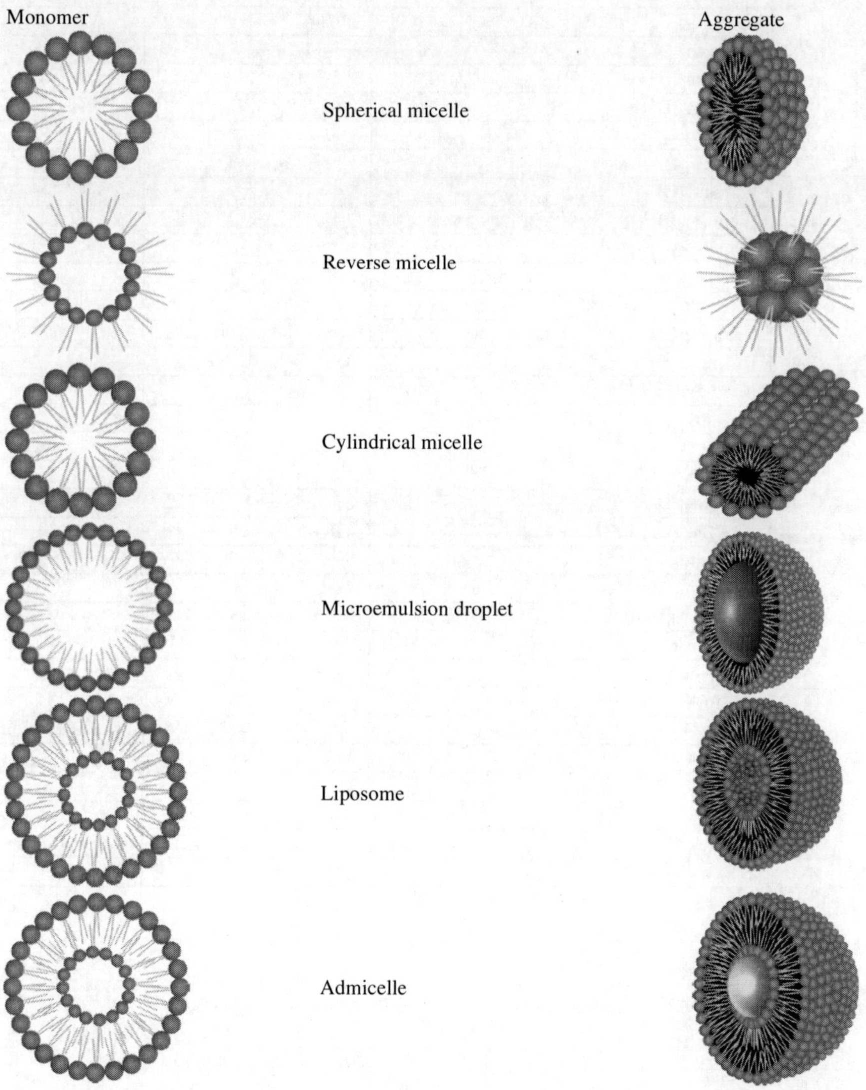

Figure 5. Schematic of different SAIL monomers and aggregates.

Table 2. The CMC and N_{agg} values for different types of SAILs at 298 K

Type	SAIL	CMC (mM)	N_{agg}
Cationic	$[C_{6}mim]$[1]$[Cl]$	900	
	$[C_{8}mim][Cl]$	90-234	
	$[C_{10}mim][Cl]$	10-59.9	40-48
	$[C_{12}mim][Cl]$	13.17-20	58-67
	$[C_{14}mim][Cl]$	2.98-3.68	79-89
	$[C_{16}mim][Cl]$	0.86-1.3	97-118
	$[C_{18}mim][Cl]$	0.4-0.45	99-118
	$[C_{4}mim][Br]$	800-900	
	$[C_{6}mim][Br]$	400-797	
	$[C_{8}mim][Br]$	121-150	21-27
	$[C_{9}mim][Br]$	74	45
	$[C_{10}mim][Br]$	29.3-41	27-47
	$[C_{12}mim][Br]$	4.3-11.2	37-44
	$[C_{14}mim][Br]$	2.5-2.9	48-61
	$[C_{16}mim][Br]$	0.55-0.84	64-76
	$[C_{8}mim][I]$	100-150	
	$[C_{4}mim][BF_{4}]$	729-820	10-27
	$[C_{8}mim][BF_{4}]$	28	
	$[C_{12}mim][BF_{4}]$	7.6-9.2	
	$[C_{16}mim][BF_{4}]$	1.37[a]	
	$[C_{8}py]$[2]$[Cl]$	274	
	$[C_{16}py][Cl]$	1.07	94
	$[C_{8}mpy][Cl]$	505	
	$[C_{10}mpyr]$[3]$[Br]$	55.5	
	$[C_{12}mpyr][Br]$	13.6	
	$[C_{14}mpyr][Br]$	3.3	
	$[C_{16}mpyr][Br]$	0.83	
	$[C_{18}mpyr][Br]$	0.25	
	$[C_{6}C_{4}im]$[4]$[Br]$	271	
	$[C_{8}C_{4}im][Br]$	41	
	$[C_{10}C_{4}im][Br]$	6.3	
	$[C_{12}C_{4}im][Br]$	2.4	
	$[C_{16}C_{4}im][Br]$	0.1	
	$[C_{8}C_{8}im][Br]$	5.6	
	$[C_{12}C_{12}im][Br]$	0.1	
Gemini	$[C_{10}im\text{-}C_{2}\text{-}imC_{10}]$[5]$[Br]_{2}$	149	
	$[C_{8}im\text{-}C_{4}\text{-}imC_{8}][Br]_{2}$	323	
	$[C_{10}im\text{-}C_{4}\text{-}imC_{10}][Br]_{2}$	117	
	$[C_{12}im\text{-}C_{4}\text{-}imC_{12}][Br]_{2}$	72	
	$[C_{4}im\text{-}C_{10}\text{-}imC_{4}][Br]_{2}$	2	
	$[C_{12}im\text{-}C_{10}\text{-}imC_{12}][Br]_{2}$	0.6	
Zwitterionic	$[C_{14}(C_{3}SO_{3})im]$	0.057-0.1	65

[a] Measured at 313 K.

[1] $[C_nmim]^+$: 1-Alkyl-3-methylimidazolium
[2] $[C_npy]^+$: 1-Alkylpyridinium
[3] $[C_nmpyr]^+$: 1-Alkyl-1-methylpyrrolidinium
[4] $[C_nC_4im]^+$: 1-Alkyl-3-butylimidazolium
[5] $[C_nim\text{-}C_m\text{-}imC_n]$: 1,1'-(Alkane-1,4-diyl)bis(3-alkyl-1H-imidazol-3-ium)

Though the shape and properties of the SAIL micelles are similar to the conventional surfactants, their bigger head group provides a special bulky structure. It leads to lesser micelle aggregation number (N_{agg}), which determines the number of monomers (SAIL particles) involved in the formation of an aggregate unit. In this manner, SAILs can create micelles at lower concentrations than conventional surfactants do. Accordingly, the corresponding CMC values, determined by the surface tension and the conductivity measurements, as well as the N_{agg} values, determined by various techniques such as fluorescence quenching, molecular geometry and electrical conductivity are summarized in Table 2.

Although a spherical shape is the most common geometry of the SAIL micelles; bilayer sheet, disk-shaped, ellipsoidal, and cylindrical shapes are also possible. For oil-soluble SAILs, a reverse micelle is formed in non-polar solutions. Meanwhile, SAILs can generate other aggregation forms around the particles such as microemulsion droplets, liposomes, and admicelles. For better visualization, the shapes of different SAIL aggregates as well as the monomer of these aggregates are schematically depicted in Figure 5.

Classification of SAILs

SAILs are known as "designer molecules" with respect to altering/modifying their cations and/or anions. In this regard, various types of SAILs have been utilized with specific properties. The following classification is appropriate based on the nature of the surface-active moiety and the number of head groups in the SAIL structure.

- *"Cationic"* type where the surface-active moieties are located on the cation. These are the most typical SAILs.
- *"Anionic"* type, composed of anions that possess long alkyl chains (branched or linear), coupled with a rather large heterocyclic cation. As an important advantage, Anionic SAILs have lower toxicity compared to other SAILs.
- *"Catanionic"* type, in which both the cation and the anion contain long alkyl chains, with surface-active character. This type of SAILs can decrease the IFT of the systems to ultra-low values (around 10^{-3} mN/m). The asymmetry in the cation and anion, as well as presence of linear or branched alkyl chains, are the factors affecting these SAILs' activity.
- *"Multicationic"* type, with more than one hydrophilic head group, connected by a spacer; these groups can be the same or different. Each hydrophilic head group is appended with hydrophobic alkyl chains. Such structure provides good hydrophobicity, desired micellization, and is of low toxicity. Here, both the length of alkyl chains and length and structure of the spacer play an important role in the hydrophobicity and molecular orientation at the interface.
- *"Zwitterionic"* type, where at least two functional groups are present: one having a positive charge and the other having a negative charge linked together by a bridge. With this structure, the electrostatic repulsion of charged groups is minimized giving ordered and intense molecular orientations at the interface. These properties can totally reduce the IFTs to ultra-low values by using trace amounts of SAILs.

It is worth noting that nanostructure Multicationic SAILs such as 1,1'-(butane-1,4-diyl)bis(3-alkyl-1H-imidazol-3-ium), [$C_n$im-C_4-imC_n] n = 4, 8, and 12, have been recently designed and synthesized at low costs. The products have been successfully tested to reduce the IFT and micelle formation in oil-water system. Investigation on this product is in its infancy and more research is needed in this area.

Effects of Molecular Structure on the Performance of SAILs

There is a strong relationship between the chemical structure of SAILs and their performance. Parts of molecular structure including alkyl chain, head group, and counterions contribute in the SAIL characteristics. For example, miscibility and thermal stability of SAILs could be improved by altering the structure of anions, whereas, their hydrophobicity, interfacial activity, aggregate formation, viscosity, and density can be altered with head groups and/or the lengths of alkyl chains. In this regard, a summary of the SAIL characteristics and properties corresponding to different parts of the molecular structure is listed in Table 3. Discussions on the effect of alkyl chain length, head group, and anions on the functionality of SAILs are presented subsequently.

Table 3. SAIL properties corresponding to different cations and anions [5]

Chemical formula		Abbreviation	MW	mp point	Density	Visco-sity
Cation	Anion		(g/mol)	(°C)	(g/cm³)[a]	(cP)[a]
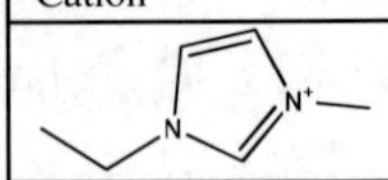	$[BF_4]^-$	$[C_2mim][BF_4]$	197.8	6	1.25	66
	$[PF_6]^-$	$[C_2mim][PF_6]$	256.1	58 - 62	1.37	450
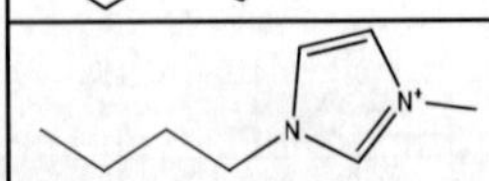	$[BF_4]^-$	$[C_4mim][BF_4]$	225.8	– 82	1.21	233
	$[PF_6]^-$	$[C_4mim][PF_6]$	284.2	10	1.37	400
	$[Br]^-$	$[C_4mim][Br]$	218.9	60	1.13	solid
	$[Cl]^-$	$[C_4mim][Cl]$	146.5	89	1.12	solid
	$[CF_3SO_3]^-$	$[C_4mim][CF_3SO_3]$	260.0	16	1.29	90
	$[NTf_2]^-$	$[C_4mim][NTf_2]$	487.9	– 4	1.42	52
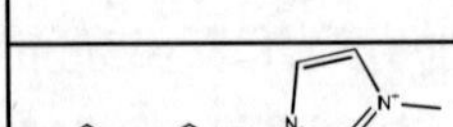	$[BF_4]^-$	$[C_5mim][BF_4]$	240.1	– 88	1.23	321
	$[BF_4]^-$	$[C_6mim][BF_4]$	254.1	– 82	1.07	211
	$[PF_6]^-$	$[C_6mim][PF_6]$	312.0	– 61	1.30	800
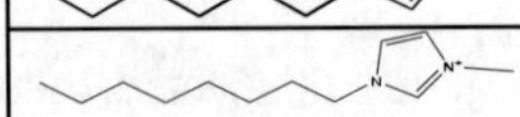	$[BF_4]^-$	$[C_8mim][BF_4]$	281.8	– 79	1.11	440
	$[Cl]^-$	$[C_8mim][Cl]$	230.5	0	1.00	16
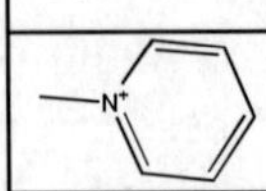	$[NTf_2]^-$	$[Mpy][NTf_2]$	416.0	0	1.44	39
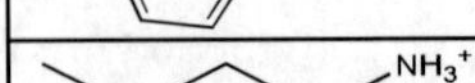	$[HCOO]^-$	BAF	91.0	– 10	0.99	12
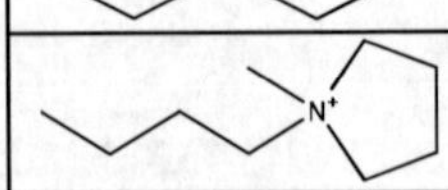	$[NTf_2]^-$	$[C_4mpyr][NTf_2]$	422.0	– 50	1.40	71

[a] Measured at 25°C.

Alkyl Chain

The alkyl chain length and the number of branches are important factors affecting hydrophobicity, solubility, interfacial activity, and micelle formation of SAILs. Larger number of methylene bridges (–CH_2–) increase the hydrophobicity and interfacial activity of SAILs, which favor their adsorption and reduce the IFT. The longer hydrophobic chain also makes a greater van der Waals attraction between the SAIL molecules and non-polar phase molecules, leading to more stable adsorption at the interface. Similar behavior is observed if the second alkyl chain is present in the structure of SAILs. Comparing the interfacial activity of different imidazolium SAILs of [$C_n$mim][Cl] in the *n*-butyl acetate-water system indicates maximum IFT reductions by using SAILs with n = 6, 7, and 8 which provides 20, 26, and 33% reductions, respectively. Hence, these reductions reach 67, 70 and 74% by using SAILs of n = 12, 14 and 16, respectively [6, 7]. The trend of IFT reduction with alkyl chain length for [$C_n$mim][Cl] is illustrated in Figure 6.

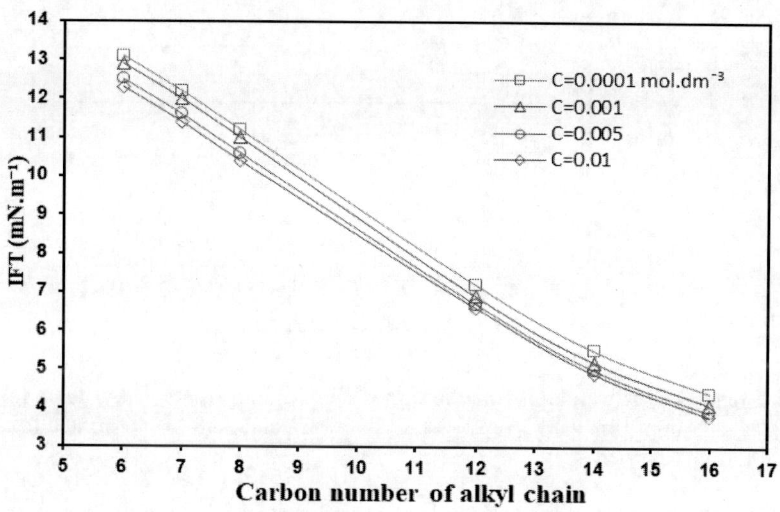

Figure 6. IFT variation as a function of carbon number of alkyl chain for [$C_n$mim][Cl] SAIL. Reprinted from [9] with permission from Elsevier.

Indeed, the required SAIL concentration for declining IFT to a certain value decreases with increase of the alkyl chain length. From economic point of view, this is attractive for industries since little amount of surfactant is required. Studies show that the minimum interface area per adsorbed molecule (A_{min}) is reduced sharply with alkyl chain length, indicating compact monolayer arrangements at the interface [2].

For Multicationic SAILs, the IFT is influenced not only by the alkyl chain length but also by the type and length of the spacer in the molecular structure. It has been demonstrated that longer alkyl chains drastically reduce the IFT; but the length of the spacer has a more complex effect. For instance, the IFT values in the presence of [$C_n$im-C_m-imC_n][Br]$_2$ as a Gemini SAIL (a branch of Multicationic SAILs), with m (number of alkyl chain carbon atoms) = 4, 8 and 12 and n (spacer) = 2, 4 and 6, in the crude oil-water system, have higher effectiveness with the longer chains [8]. Noteworthy, the C_4 spacer SAIL shows better performance than the two others at high concentrations owing to the fact that the spacer length not only influences the hydrophobicity of SAILs but also alters the number of adsorbed molecules. In this regard, short spacer SAILs exhibit a particular orientation since the alkyl chains remain perpendicular to the interface and the number of adsorbed SAILs increases. However, long spacer SAILs lay extended along the interface and thus, the number of adsorbed molecules will be limited. This implies that the spacer length has a complicated influence, and a longer spacer does not necessarily give higher effectiveness.

Considering aggregation, the longer alkyl chains provide lower CMCs under compact arrangements, due to higher hydrophobicity and lower solubility in aqueous solutions. In this regard, a sharp decrease of CMC with increase of the number of carbons in homologous series of 1-alkyl-3-methyl-imidazolium halide, [$C_n$mim][X] ([X] = [Cl] and [Br]) has been reported (Figure 7).

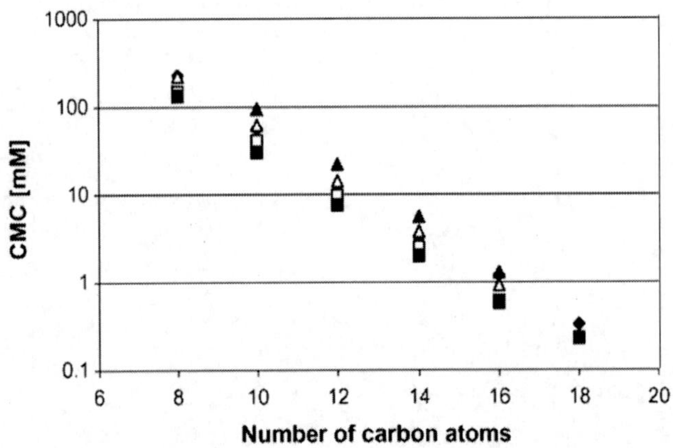

Figure 7. The trend of CMC variation with the carbon atoms number in the alkyl chain for [$C_n$mim][Cl]. (◆), [$C_n$mim][Br] (□), C_nTAC (▲), C_nTAB (△), and SAS (■). Reprinted from [3] with permission from Elsevier.

In addition, the length of the second alkyl substituent in SAILs plays similar positive role. For instance, [$C_{12}C_{12}$im][Br], gives a much lower CMC than [C_{12}mim][Br] [3]. Investigations also show that for [$C_n$mim][X] and [$C_n$mpy][X] SAILs ([X] = [Cl], [Br], and [I]), micelle aggregates are not evident for n = 4. Transitory aggregates with no detectable structure are relevant for n = 6 and ordinarily oblate ellipsoidal micelles are formed for SAILs with longer chain lengths. Meanwhile, the concentrations higher than CMC do not lead to considerable change in the shape and the size of the aggregates [10]. This is in contrast to the aggregates formed by the SAILs with short and medium alkyl chains (n = 4 and 6), where higher concentrations can lead to large and elongated micelle structures [11]. Further, by increasing the number of alkyl branches, the micelle aggregation number (N_{agg}) and the minimum interface area per adsorbed molecule (A_{min}) decline.

Head Group

The head group structure can affect the hydrophilicity of the SAILs. In general, bulky heterocyclic head groups with the π-electron rings give higher IFT reductions than those with simple groups. This is due to poor hydration as well as to a lesser electrostatic repulsion force by these head groups, a facile migration and greater adsorption at the interface. In this regard, the most widely studied SAILs for IFT reduction are heterocylic nitrogen-contained cations such as imidazolium, pyridinium, pyrazolium, thiazolium, pyrrolium, and oxazolium. However, guanidinium, sulfonium, pyrrolidinium, phosphonium, and ammonium head groups have also been introduced in recent years. Figure 8 shows examples of nitrogen-containing heterocyclic ring head groups.

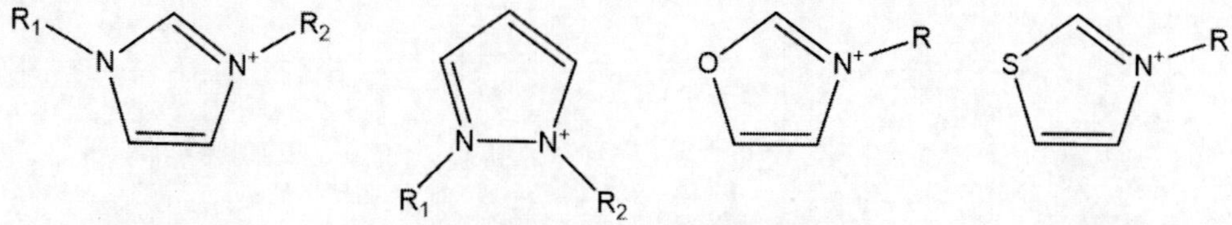

Figure 8. Examples of nitrogen-containing heterocyclic ring head groups in SAILs. Reprinted from [2] with permission from Springer.

Among the imidazolium, thiazolium, piperidinium, and pyrrolidinium SAILs, imidazolium head group are effective in reducing the IFT compared to piperidinium and pyrrolidinium [10]. The reason is that the π-electron system in the ring, reduces repulsions between head groups and facilitates SAILs adsorption at the interface. Meanwhile, thiazolium head group leads to IFT variations and micelle formations close to those of imidazolium, indicating that replacement of the N-atom with S- in the head groups, despite improving aromaticity, does not significantly alter the activity of SAILs. In addition, good biodegradability has been demonstrated for imidazolium and pyridinium head groups, highlighting them as environmentally friendly surfactants [12].

Concerning CMC, the head group has a definite influence on the micelle stability and aggregation number. SAILs with larger head groups provide lesser CMCs and aggregation numbers, due to lower hydration and faster saturation interface. Accordingly, lower CMCs have been reported for the imidazolium head group compared to piperidinium and pyrrolidinium [10]. Meanwhile, as can be seen in Table 2, the Multicationic and Zwitterionic SAILs exhibit low CMCs and aggregation numbers compared to monocationic SAILs, because of their bulky structure.

Anion

The kind of anions has an influence on the miscibility of SAILs in polar/non-polar solvents and their interfacial activity albeit with a less extent compared to alkyl chain and the head group. By changing anion, task-specific SAILs can be prepared for specific applications. Some common inorganic and organic anions in SAILs are tetrafluoroborate $[BF_4]^-$, hexafluorophosphate $[PF_6]^-$, tetrachloroaluminate $[AlCl_4]^-$, bis(trifluoromethylsulfonyl)imide $[NTf_2]^-$, p-toluenesulfonate $[Tos]^-$, trifluoromethanesulfonate $[OTf]^-$, trifluoroacetate $[TA]^-$, acetate $[Ac]^-$, trifluoroacetate $[CF_3COO]^-$, and halides. Figure 9 shows typical $[C_4mim]^+$ - imidazolium SAILs with different anions.

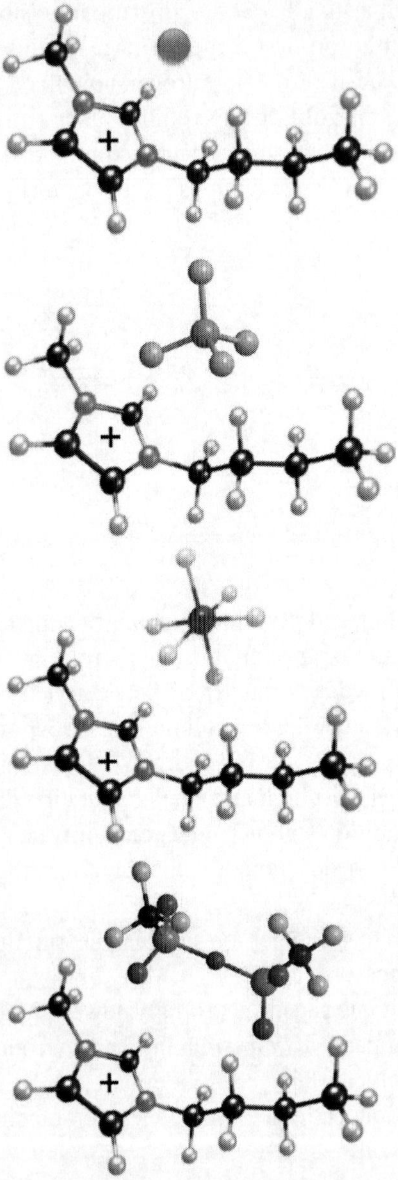

$[C_4mim][Cl]$

$[C_4mim][BF_4]$

$[C_4mim][PF_6]$

$[C_4mim][NTf_2]$

Figure 9. The chemical structure of $[C_4mim]^+$ cation imidazolium SAILs with different anions.

Molecular dynamic simulation confirms that bulkiness of anions with greater polarizability has a strong effect on the IFT variation. This is due to easier penetration of cations into the electron cloud of anions, leading to electrostatic repulsion between head groups to assist the adsorption. Big anions are less prone to effective hydration, giving easier adsorption at the interface and adapting ultimate low IFT values. These effects eventually contribute to aggregation at very low concentrations. In this regard, the interfacial activity of $[C_{16}mim][X]$, with X = $[I]^-$, $[Br]^-$ and $[Cl]^-$, increases in the sequence: $[I]^- > [Br]^- > [Cl]^-$ consistent with stronger polarizability and weaker hydration [13]. For better judgment, the relatively close and far spaces between SAIL molecules at the interface with $[Br]^-$ and $[Cl]^-$ anions, respectively, are illustrated schematically in Figure 10.

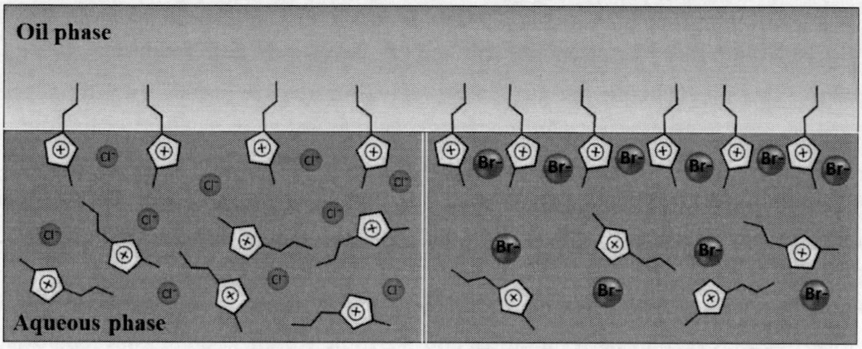

Figure 10. Schematic configuration of typical SAILs with different anions at the interface.

Concerning the CMC, the value is lower for SAILs with bigger anions, attributed to high polarizability, which leads to more adsorption on the aggregates and less electrostatic repulsion between the head groups. For instance, by using $[C_{12}mim]^+$ SAILs, anions have been effective in CMC reduction as $[OAc]^- > [Br]^- > [Cl]^-$ [14].

Another significant effect of anions on CMC is relevant to the solubility of SAILs. The bulky anions can limit the solubility of SAILs prior to aggregation. For example, a study on the self-aggregation behavior of imidazolium SAILs, $[C_nmim]$, $2 \leq n \leq 14$, with different anions of $[Cl]^-$, $[PF_6]^-$ and $[NTf_2]^-$ show that micelle formation is detected only for SAILs with $n > 8$ and with $[Cl]^-$ anion, whereas no micelle has been detected for $[PF_6]^-$ and $[NTf_2]^-$, which can be attributed to low solubility of these SAILs in aqueous solutions [15].

Considering alkyl chain in the anions, longer chain anions show better performance in reducing IFT due to greater hydrophobicity and poor hydration, which results in higher adsorption at the interface. A study performed with various cation SAILs of butyrolactam $[BT]^+$, caprolactam $[CP]^+$, propylammonium $[PA]^+$, and 3-hydroxypropylammonium $[3\text{-}HPA]^+$ indicated that different anions are effective in IFT reduction ($[C_6H_{13}COO]^- > [CH_3COO]^- > [HCOO]^-$) [16] (Table 4).

Table 4. The range of IFT values (mN/m) for crude oil-water system in the presence of different SAILs at 298.15 K

SAIL	Dosage (mg/dm³)						
	0	50	100	500	1000	5000	10000
[BT][HCOO]	34.26	24.12	23.96	23.18	22.24	19.51	16.89
[BT][CH_3COO]	34.26	21.75	21.13	20.18	19.23	16.42	14.03
[BT][$C_6H_{13}COO$]	34.26	20.31	19.66	18.54	17.13	15.07	12.95
[CP][HCOO]	34.26	20.36	19.84	18.49	17.04	15.38	13.26
[CP][CH_3COO]	34.26	19.24	18.85	17.56	15.68	13.94	12.45
[CP][$C_6H_{13}COO$]	34.26	17.68	16.95	15.64	14.34	12.77	11.39
[PA][HCOO]	34.26	24.85	23.81	21.77	20.38	18.57	17.21
[PA][CH_3COO]	34.26	21.56	21.03	19.84	18.94	16.84	15.11
[3-HPA][HCOO]	34.26	20.74	19.64	18.69	17.58	15.79	14.33
[3-HPA][CH_3COO]	34.26	20.51	19.86	18.69	17.43	15.73	14.12

Comparison of the Performance of Different Structure SAILs

A minor change in the structure of SAILs brings a strong influence on the IFT reduction and CMC formation. A longer alkyl chain, a bulkier head group with π-electron ring, and a higher polarizing anion strengthen the interfacial activity of SAILs. This is due to the weaker hydration and less electrostatic repulsion between the head groups. For SAILs with the same hydrophilic head group, the alkyl chain length can play a more important role than the anions. Meaningfully, the Catanionic, Multicationic, and Zwitterionic SAILs exhibit a higher interfacial activity than monocationic analogs owing to their stronger amphiphilic nature.

For Catanionic and Multicationic types, multiple alkyl chains in both the anion and cation groups as well as ordered and packed arrangement at the interface improve their effectiveness. These characteristics make them promising materials for reducing IFT and CMC, which make them suitable for applications in drug delivery, nanoparticle synthesis, and environmental remediations [2]. Nevertheless, branching and asymmetric alkyl chains are basic factors that affect their performance. A comparison between a series of Gemini imidazolium SAILs with monocationic analogs indicate that Gemini SAILs strongly reduce the IFT even under low concentrations [17].

In Zwitterionic SAILs, the presence of cationic and anionic groups in the same molecule minimizes the electrostatic repulsion of the charged head groups, resulting in a denser molecular arrangement at the interface and further reducing the IFT than with monocationic analogs. Accordingly, this kind of SAILs can reduce the IFT to ultra-low values. Moreover, Anionic SAILs can sometimes reduce the IFT stronger than monocationic analogs; however, their activity depends on various parameters such as nature of involved phases and structure of the cation and the anion. Table 5 summarizes the performance of different types of SAILs in various phase systems.

Table 5. Performance of different SAILs for different chemical systems*

Type	SAIL	Phase system	IFT range (mN/m)	CMC (mol/dm^3, unless stated otherwise)	Ref.
Cationic	$[C_8Py][Cl]$	Crude oil-water	27.6 - 19.1	1000 (mg/dm^3)	[18]
	$[C_{18}Py][Cl]$		27.6 - 1.8	500 (mg/dm^3)	
	$[C_{12}mim][Cl]$		27.6 - 6.3	500 (mg/dm^3)	
	$[C_{18}mim][Cl]$		27.6 - 1.2	100 (mg/dm^3)	
Anionic	$[C_4mim][DS]$	Water-benzene	20 - 0.9		[19]
		Water-decane	20 - 2.5		
		Water-tetradecane	20 – 2.5		
	[P4441][DS]	Crude oil-water	48 - 0.1	9×10^{-4}	[20]
Zwitterionic	$C_{16}IPS$	Crude oil-water	20 - 1		[21]
	$C_{16}IPS$-Nsa		20 - 0.1		
Cationic	$[C_{16}mim][Br]$		20 - 3.7		
	$[C_{16}mpyr][Br]$		20 - 5.8		
Gemini	$[C_4im\text{-}C_2\text{-}imC_4][Br]_2$	Crude oil-water	31.8 - 13.3	0.820	[9]
	$[C_4im\text{-}C_4\text{-}imC_4][Br]_2$		31.8 - 8.8	0.999	
	$[C_4im\text{-}C_6\text{-}imC_4][Br]_2$		31.8 - 11.5	0.905	
	$[C_8im\text{-}C_4\text{-}imC_8][Cl]_2$	Water-model oil	18 - 10	$>1\times10^{-2}$	[22]
	$[C_{10}im\text{-}C_4\text{-}imC_{10}][Cl]_2$		15 - 5.9	4.6×10^{-3}	
	$[C_{12}im\text{-}C_4\text{-}imC_{12}][Cl]_2$		13 - 4.78	1.2×10^{3}	
	$[C_8im\text{-}C_4\text{-}imC_8][Br]_2$	Crude oil-water	31.8 - 0.7	0.2	[23]
	$[C_{12}im\text{-}C_4\text{-}imC_{12}][Br]_2$	Crude oil-water	31.8 - 1.2×10^{-2}	1.88×10^{-2}	[24]
Catanionic	$[P666,14][(i\text{-}C_8)_2PO_2]$	Water-n-dodecane	52.2 - 5.09		[25]

* Temperature is 303.2 K for Refs [21, 22], 313.2 K for [19], 353.2 K for [18], and 298.2 K for others.

Effects of Operating Conditions on the Performance of SAILs

Considering the effects of the operating conditions on the performance of SAILs, the important effects of temperature, pH, presence of electrolytes, and the nature of involved phases are discussed in this section. In each case, the changes in IFT and micelle formation are considered with respect to the influence of SAILs under different operating conditions.

Effect of Temperature

The temperature influences both the IFT reduction and micelle formation and can be employed for process intensification. In most cases, the temperature reduces the IFT; however, in some cases, IFT may rise with the temperature. The breakdown of iceberg structure of the ordered water molecules around the hydrophobic portion of SAILs occurs with elevated temperature, assisting the mobility toward the interface. In addition, the kinetic agitation of molecules is improved with the temperature, weakening the intermolecular forces and consequently, reducing the IFT. In this regard, studies of different types of SAILs such as imidazolium, proline, aromatics and heterocycles in various chemical systems demonstrated higher IFT reductions with temperature. The variation of different imidazolium-based SAILs with temperature is shown in Figure 11.

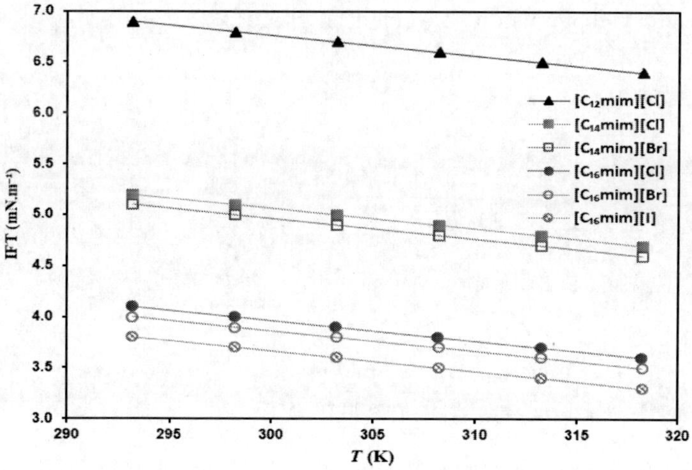

Figure 11. Variation of IFT with temperature for different imidazolium SAILs. Reprinted from [9] with permission from Elsevier.

In contrary to these findings, some aromatic heterocyclic head groups such as pyridinium and imidazolium with more than 8 carbon atoms in their alkyl chain lead to an increase of IFT with temperature [19]. Changes in the distribution of SAILs between phases as well as "phase inversion" at a specific temperature (PIT) are responsible for this behavior [26]. At temperatures below the PIT, the adsorption of SAILs is intensified with temperature, resulting in lower IFTs. Upon reaching PIT, the adsorbed SAILs reach a maximum amount with minimum IFT value and beyond this temperature, adsorbed SAIL molecules diffuse into the oil phase. Thus, the number of SAIL molecules at the interface decreases and the IFT increases.

The temperature also exerts complex effects on the CMC. In most of cases, micelles are formed at low concentrations as the temperature increases. Nevertheless, in some other cases, the CMC first decreases to reach a minimum value (at T_{min}) and then rises, i.e., forming a U-shape variation [3]. This variation in the CMC can be explained by two competitive phenomena: dehydration of hydrophilic portions and interaction of hydrophobic portions. As the temperature rises, dehydration of hydrophilic head groups reduces the solubility of SAILs and facilitates the formation of micelles at low concentrations. On the other hand, heat elevates the kinetic mobility of the hydrophobic portion of SAILs, which prevents accumulation of molecules to form micelle, and CMC increases. Consequently, the effect of the first phenomenon is dominant at low temperatures, and by passing T_{min}, the second phenomenon will be reached. In general, T_{min} value depends on the hydrophilicity of head groups and hydrophobicity of the alkyl chains as well as the nature of counterions.

Effect of pH

Another essential parameter that influences the performance of SAILs is the pH. Almost all investigations indicate that the increase in the pH is accompanied by extra decrease in IFT. This is due to the tendency of hydroxyl ions (OH^-) to adsorb at the interface, which reduces the electrostatic repulsion between SAIL molecules and creating opportunities for SAIL adsorption. Figure 12 indicates the effect of hydroxyl ions on the adsorption of SAILs by reducing the electrostatic repulsion between head groups.

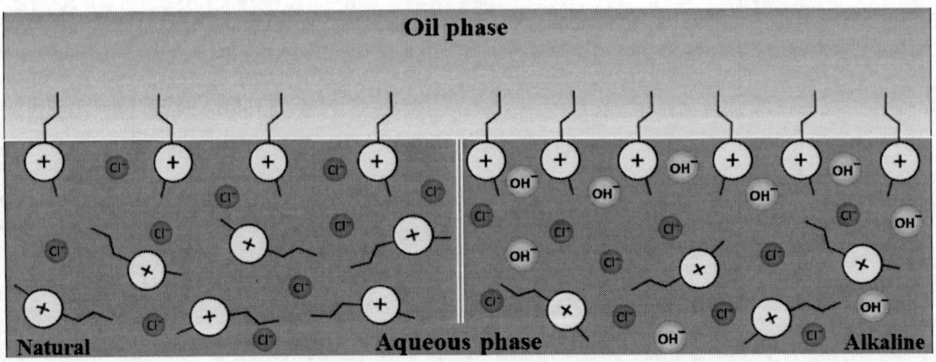

Figure 12. Schematic representation of the simultaneous adsorption of hydroxyl ions and SAILs at the interface.

As the pH increases, micelles are formed at lower concentrations. The hydroxyl ions facilitate the aggregation of SAILs by neutralizing the positive charges of micelles. In this regard, an investigation on a branch of imidazolium SAILs, [$C_n$mim][Cl], on the toluene-water system has been performed within pH of (5.5 - 8.5) indicating significantly higher reductions in both IFT and CMC to about 13 and 30%, respectively [27]. Also 52.9 and 67% reductions in IFT and CMC with a Gemini SAIL [C_8im-C_4-imC_8][Br]$_2$ in the crude oil-water system within the pH range of (2.5 - 9.5) have been achieved [23]. Based on these findings, it has been confirmed that the pH can provide a positive effect by assisting adsorption of SAILs.

Effect of Electrolytes

SAILs remain stable in the presence of electrolytes and there is a positive synergistic effect between SAILs and electrolytes for the IFT reduction and micelle formation. This is attributed to the well-known "salting out" effect and neutralization of the charge of SAIL ions at the interface.

In aqueous solutions, the cation and anion of electrolytes and those of SAILs compete for water capture, but smaller and harder inorganic ions, with a higher surface charge density provide stronger hydration [28]. This is due to hydrogen bonding with water and the formation of a cage-like structure around the ions as schematically shown in Figure 13 for Na^+ cation. Thus, the number of water molecules around the large SAIL ions will be decreased, as well as their solubility in bulk, favoring the adsorption of SAILs at the interface. As can be seen in Figure 14, the sequence of ion strength in hydration varies according to the Hofmeister series. For instance, sodium halides have greater effect than calcium halides on diminishing the solubility of SAILs.

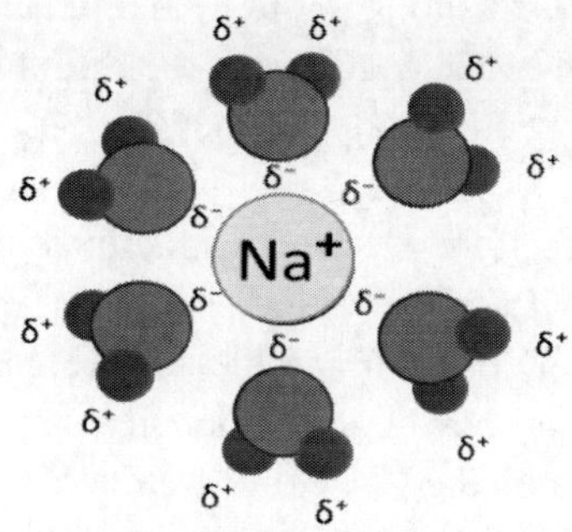

Figure 13. Schematic of the cage-like structures formation surrounding the Na^+ cation in water media.

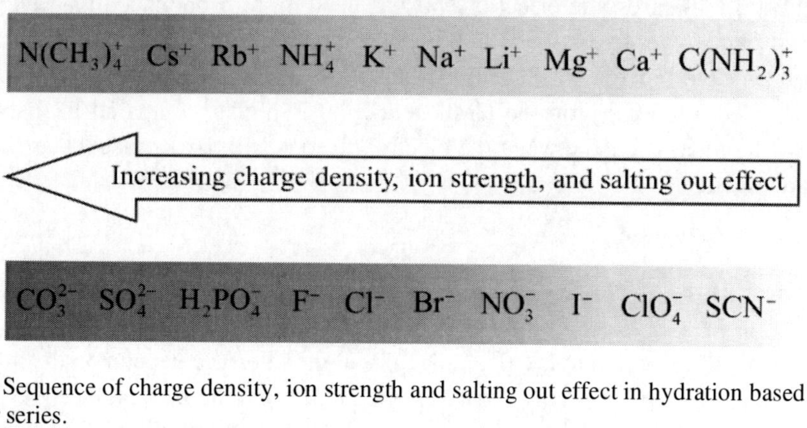

Figure 14. Sequence of charge density, ion strength and salting out effect in hydration based on the Hofmeister series.

Another remarkable effect is that the interaction of SAIL ions with the oppositely charged ions of electrolytes results in rather tight packing at the interface. In the contrary case of interaction of SAIL ions with no added electrolytes, electrostatic repulsion between SAILs' head groups is observed (schematically illustrated in Figure 15).

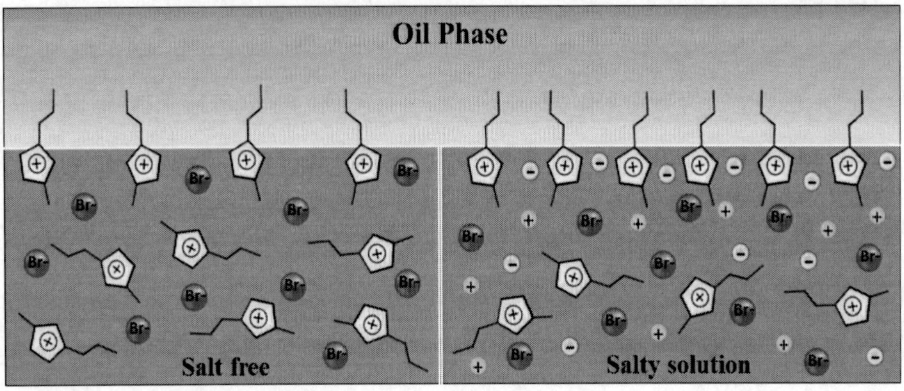

Figure 15. The change in the electrostatic repulsion between head groups of a typical SAIL.

These phenomena significantly reduce the IFT in the simultaneous presence of SAILs and salts. It has been shown that the IFT of crude oil-water is significantly reduced with addition of K_2SO_4, $MgSO_4$, KCl, and NaCl salts to [C_8Py][Cl] and [C_{18}Py][Cl] SAILs [29]. Interestingly, the positive synergistic effect of NaCl, $MgCl_2$, and their mixture with the [C_8im-C_4-imC_8][Br]$_2$ Gemini imidazolium SAIL has been reported with about 60% higher IFT reductions compared to salt-free case [23].

Through the salting out effect, the electrolytes facilitate micelle formation by dehydrating the hydrophilic portion of SAILs. In addition, salt ions reduce the electrostatic repulsion between the charged portions, by neutralizing the surface charge of the micelles, resulting in lower CMCs. Low hydrated ions show high efficiencies in neutralizing the system charges. In this regard, the effect of halides on the CMC of different SAILs has been reported to be in the order of $[Cl]^- < [Br]^- < [I]^-$ [30]. Thus, the addition of KCl, KBr, NaCl, and NaBr to aqueous solutions of [C_{16}mim][Br], [C_{16}mim][C]l, and [C_{12}mim][Cl] resulted in 89% higher CMC reduction [27]. In all cases, the effect of Na^+ has been more pronounced than K^+ due to low charge density and lower hydration which is consistent with the Hofmeister series. The same effect has been detected for Na_2SO_4 and NaCl in solutions containing [C_{10}mim][Cl] [15]. Overall, the effect of salinity on the CMC reduction is higher compared to that of the pH. Noteworthy, the interactions between the electrolyte ions and SAIL micelles are currently of great research interest.

Effect of the Nature of Phases

The adsorption and orientation of SAIL molecules at the interface depend on the interaction of SAILs and the molecules of involved phases. Generally, interactions include the polar/polar, non-polar/non-polar, and non-polar/polar contacts. The nature of phases, polarity, and molecular structure of SAILs can alter the solubility and interfacial activity. An example for this matter is [C_4mim][DS]: for aliphatic systems of water-decane and water-tetradecane, IFT is reduced from 12 to about 2.5 mN/m; whereas, for the aromatic system of water-benzene, IFT is reduced from 12 to about 0.9 mN/m. The latter case is attributed to the π-π interaction between [C_4mim]$^+$ and the benzene ring, which stabilizes the SAIL molecules at the interface.

These observations are consistent with the most negative calculated interaction energies between the $[C_4mim]^+$ and benzene molecules.

Due to various parameters involved in the process, the behavior of SAILs in contact with different phases is complicated. For instance, the longer alkyl chain length of ions are, the stronger the miscibility of SAILs in a non-polar phase would be. Conversely, the ionic nature of the head groups favors the dissolution of SAILs in polar phases. In addition, probable hydrogen bonds between water molecules and electronegative atoms of the SAIL anions and/or the hydrogen atoms in the SAIL cations (such as imidazolium and pyridinium) facilitate the solubility of SAILs in aqueous phase.

SAILs and Green Chemistry

In recent decades, the principles of green chemistry have been rationalized for design of processes and application of chemicals. In this case, SAILs have been recognized as green materials because of their low vapor pressure and recyclability. However, their toxicity and biodegradability levels are the most important factors that need to be considered. Attempts have been made to evaluate the toxicity of SAILs for human health and in different environmental scenarios [31]. Generally, the balance of hydrophobicity and hydrophilicity of the molecules, the length of the alkyl chain, the charge density of the head group, and the type of the counterions, all contribute to the toxicity and biodegradability of SAILs.

The toxicity of surfactants, particularly SAILs, is attributed to their adsorption into biological membranes, driven by nonspecific hydrophobic interactions, which disrupt cellular membranes and cause acute or chronic effects on sensitive organisms. The rate of diffusion of SAILs into cell walls greatly advances with the length of the hydrophobic chain, so that the longer alkyl chain is, the easier for the compound to enter the cell of organisms, causing membrane damage and cell death. Therefore, to reduce the toxicity of SAILs, special attention should be paid to employing suitable alkyl chain length. Meanwhile, the newly introduced amino acid-based SAILs with natural biological moieties in their structure, can be considered as one of the green alternatives for a wide range of commercial opportunities [31]. Furthermore, pyridinium SAILs show low toxicity and exhibit an antimicrobial activity that depends on the hydrophobicity and hydrophilicity, molecular size, and the charge density of their head group [31].

Along with toxicity, biodegradation is also essential in the green chemistry of SAILs [12]. The structure has definite effect on the biodegradability of SAILs. In Figure 16 (a), different categories of SAILs are illustrated according to their biodegradability [31]. In this regard, SAILs with long and/or functionalized alkyl chains (such as alcohol or carboxylic acid) show better biodegradation rates. Based on mineralization capability, imidazolium and pyridinium SAILs indicate good biodegradability, which makes them good surfactant candidates [12]. In addition, the two potential breakdown sites of amide hydrolysis and ester hydrolysis provide mineralizable transformation products, which are represented typically for pyridinium ethyl ester derivatives in Figure 16 (b) [31].

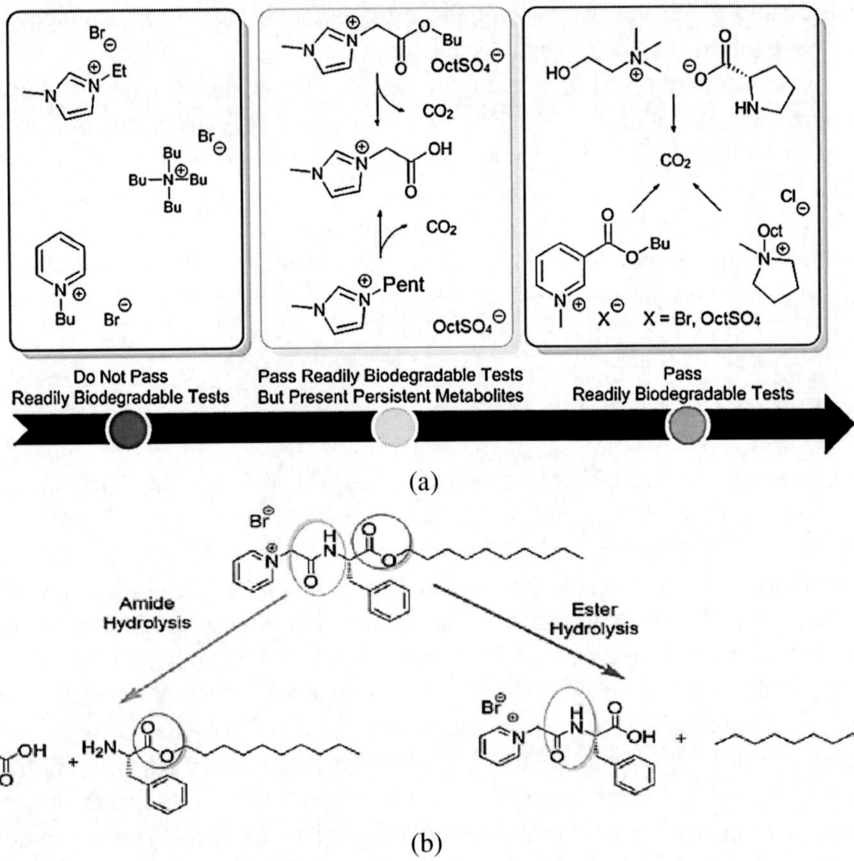

Figure 16. Structural features in biodegradability of SAILs (a), and potential sites for the breakdown of a pyridinium SAIL (b). Reprinted from [31] with permission from Royal Society of Chemistry (RSC).

Results from a comprehensive study on both toxicity and biodegradability of a series of SAILs with different cationic head groups of pyridinium, imidazolium, and cholinium and the alkyl chain lengths (from n = 2 to n =16 carbon atoms), have revealed that medium-chain pyridinium and imidazolium SAILs (n = 6 to 8) are the greenest ones with low toxicity and high biodegradability [31]. Lower toxicity of the Multicationic SAILs for organisms has been confirmed, which is due to their stretched conformation, limiting the incorporation with cell membranes. Short alkyl chain Gemini SAILs with polyethylene glycol spacer showed a low toxicity [32] while the high thermal stability of this kind of SAILs reduce their biodegradability. Hence, complementary research on the molecular design is required to improve their degradation capability.

Despite the mentioned favorable outcomes, a number of results indicate that some SAILs with halogen anions such as $[BF_4]^-$, $[PF_6]^-$, $[CF_3SO_3]^-$, and $[AlCl_4]^-$ produce environmental hazards and some others such as long-chain morpholinium cations are toxic [33, 34]. Therefore, studies in this area are required and the green chemistry aspects of SAILs should be considered by investigating other parameters such as sustainability index, recyclability, and prevention of waste/by-products before commercial scale utilization. Thus, despite variety of SAILs are still being explored in different fields and lots of efforts being made to improve their environmental aspects, a real breakthrough will not be possible until more attention is paid to the design and production of more degradable and less toxic SAILs.

Trends in Application of SAILs

Due to the amphiphilic nature and the versatility in their preparation, SAILs can enhance the performance of different processes such as stabilization and separations (extraction). This section provides an overview of trends and developments in different fields. Noteworthy, despite many investigations on the properties of SAILs, commercialization efforts are still in their infancy.

Emulsion Stabilization

An emulsion is a dispersion of one phase in another phase with which it is not miscible in the form of fine droplets. Microemulsions are practically interesting for different uses. Danielsson and Lindman [35] defined a microemulsion as "a system of water, oil, and amphiphilic compound, which has a single optically isotropic and thermodynamically stable liquid solution". The droplet size of microemulsions is smaller than that of conventional emulsions but bigger than typical micelles. To create droplets and form water in oil (W/O) or oil in water (O/W) emulsions, the interface must be divided into new small parts with larger total interface areas. Therefore, to have stable microemulsions, the IFT of the system must be low. In this regard, SAILs, with amphiphilic nature capable of forming protective layers, can be used to create stable microemulsions which are electrostatically and/or sterically stabilized. Some significant applications of microemulsions include reaction catalysis, nanoparticle synthesis, and drug delivery.

To form O/W emulsions, the polar portion of SAILs remain in the aqueous phase and the non-polar extend into the oil phase, while for a W/O emulsion, such arrangement is reversed (Figure 17). The emulsifiers are used for preventing coalescence of droplets. In this case, the surface of droplets gets covered with SAILs, and droplets become electrically charged. The electric charge on droplets (either positive or negative) separate them electrostatically and keep them stable. The bulky and the large SAIL head groups could sterically inhibit the coalescence of droplets.

The molecular structure of SAILs has an influence on the morphology and droplet size of microemulsions. In term of the alkyl chain length, the interface disruption happens more easily with the long chain SAILs and thus facilitate formation of small droplets. Similarly, longer alkyl chains in counterions contribute to microemulsion formation due to stronger amphiphilic nature. Further, the head group of SAILs plays a major role in determining whether microemulsions will be formed in a polar or non-polar phase. In addition, by altering the polarity of the counterions, the effectiveness of SAILs in the formation of microemulsions changes. In this regard, results show that imidazolium SAILs have an influence in the order of $[C_4mim][NTf_2] > [C_4mim][PF_6] > [C_4mim][BF_4]$ [36].

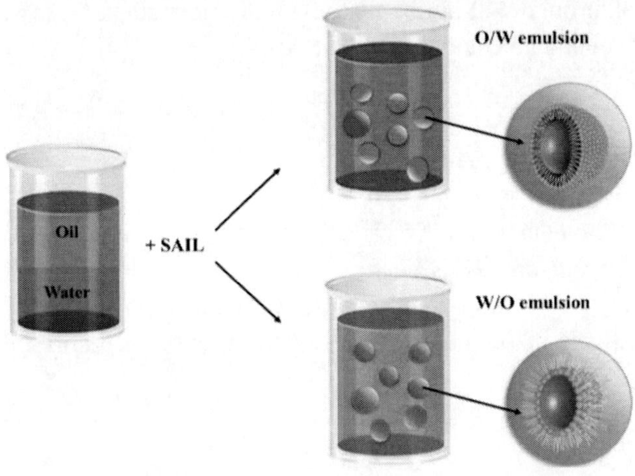

Figure 17. Formation of oil-in-water and water-in-oil emulsions by using SAILs.

Suspension Stabilization

While stabilizing suspensions, SAILs adsorb on the surface and form stable particles in the solutions. Hemimicelles or admicelles, formed by adsorbing SAILs, can disperse solids in liquids by forming protective layers on the surface of particles. Hydrophobic chains, as protective groups in SAILs, increase the stability of suspensions by decreasing the van der Waals attractive forces. A hemimicelle is a monolayer of a surfactant at the particle surface; whereas, an admicelle is a surrounding bilayer (Figure 18). It is important to note that there is no chemical bond between the solid surfaces and the SAIL groups in these structures and such arrangements are formed at concentrations below the CMCs.

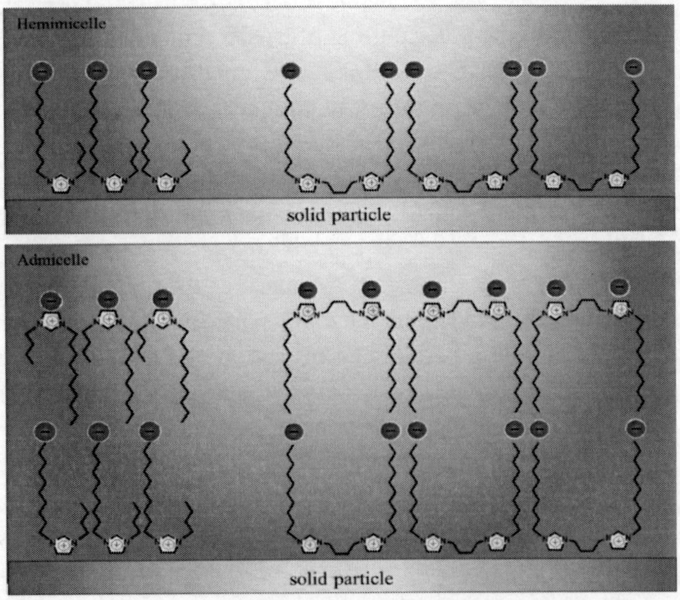

Figure 18. Schematic of hemimicelle and admicelle layers of Cationic and Gemini SAILs on a solid particle.

Recently SAILs have been utilized for dispersing insoluble materials like lipophilic substances, metals, and metal oxides without using additional stabilizers. In this respect, stabilization of metals has been an interesting topic in nanomaterial engineering for particle synthesis and microstructure formation. SAILs have also been used to produce 3D semiconductor microparticles as well as very fine metal nanoparticles such as gold [37]. The morphology of different structure metal nanoparticles, prepared using different SAILs, is shown in Figure 19.

Figure 19. Morphology of Au nanodendrites (a), Bi_2S_3 flower collections (b), and Au polyhedral particles (c), prepared by using different SAILs. Reprinted from [38, 39] with permission from American Chemical Society (ACS).

The chemical structures of SAILs greatly influence the stability and morphology of the particles in dispersions. Considering the tailorable SAIL structures, it has been confirmed that a change in the cationic or anionic portion of SAILs significantly alters the properties of particles. Accordingly, long alkyl chains assist the stability of dispersion by making a protective hydrophobic layer around the particles. The presence of long chains also results in ordered structures with fewer adsorbed SAILs. Meanwhile, the tendency for micelle formation is increased by diminishing the solubility in the polar bulk phase. Thus, the amount of SAILs on the particles decreases. Conversely, short alkyl chain SAILs are not big enough to remain at the surface of particles and they cannot create sufficient repulsion between particles. Given these results, to achieve the best performance, selecting an appropriate alkyl chain length is important.

SAILs with ionized head groups, such as phosphonium and ammonium show strong interaction with metal particles. Different head groups can affect the thickness of the formed layers. As mentioned earlier, Catanionic and Anionic SAILs that possess amphiphilic anions can electrostatically stabilize particles by creating anionic layer around positively charged metals and metal oxides. Nonetheless, the cationic portion of these SAILs are less effective in stabilizing the particles.

In general, by manipulating SAILs by selecting the proper alkyl chains as well as cations and anions, different types of solid particles can be stabilized in various solutions. Imidazolium SAILs with $[BF_4]^-$, $[PF_6]^-$, and $[NTf_2]^-$ anions are often used for these aims. Indeed, more research is needed to understand how the structure of SAILs affects the particle size, morphology, and the stability of dispersions.

Solid Phase Extraction

In the solid phase extraction (SPE) process, the separation is assisted by linking SAILs on the surface of materials and creating suspensions in solutions. The particles can then be centrifuged or separated by another liquid as an eluent. The SPE is utilized in bio-extraction and analyzing methods via concentrating and purifying samples to isolate the analytes of interest from matrices such as blood, water, and soil. Additionally, because of the creation of immiscible metal particles, SAILs can separate metal ions from the aqueous media selectively even for very similar metals.

In a SPE, SAILs form hemimicelles and admicelles around the hydrophilic solid particles, as schematically presented in Figure 20. Thus, the aliphatic alkyl chain of SAILs makes the particles hydrophobic and turn into cloudy solutions. Therefore, the solid particles can form a sediment by centrifugation. The particles could also be dispersed in an organic solvent. After separating loaded particles, they can be eluted by washing with a non-polar solvent, which disrupts the junction of SAILs and particles, leading to the perfect recovery of particles. If the hydrophobic particles exist in an organic phase, the alkyl chain of SAILs adsorb on the surface of the particles and their polar head group extend into the organic phase. This arrangement produces hydrophilic particles, which can be extracted from the organic phase.

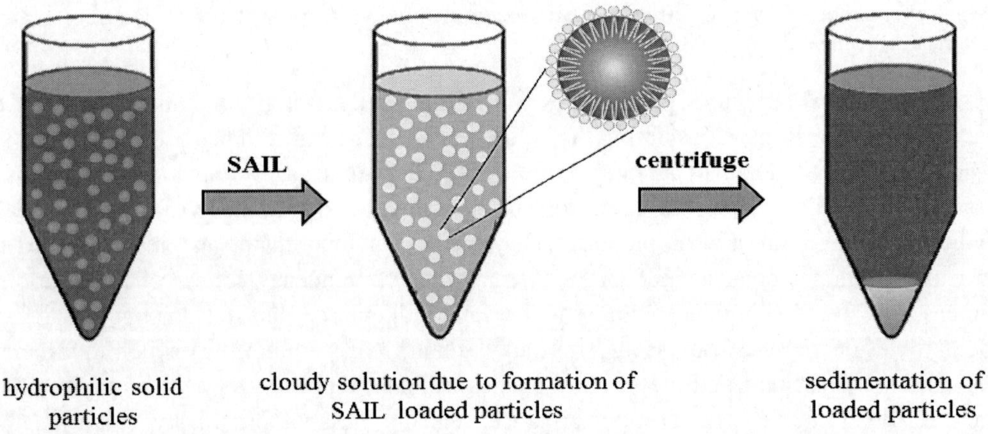

Figure 20. Schematic of solid phase extraction by using SAILs.

Desirable SPE conditions such as short extraction time, high efficiency and production of pure products is attainable by using appropriate SAILs. For instance, the use of [C_{16}mim][Br] in SPE of three polycyclic aromatic hydrocarbons (PAHs) from soil samples and comparing with CTAB indicates a maximum separation of PAHs with CTAB of only half of that with the SAIL [40]. It can be attributed to the delocalized charge and large size of the imidazolium head group, providing strong interaction with multiple aromatic rings.

Suitable alkyl chain and head group, as well as counterion, make the SAILs as suitable SPE extracting agents. Results demonstrate that alkyl chain length drastically affects the extraction efficiency due to the change in the solubility and the size of aggregates. In addition, the extraction of particles is favored with alkyl chain length. Reasonably, low amounts of a long alkyl chain SAIL are required for this aim.

The type of head group, on the other hand, has an influence on the solubility in aqueous or organic solutions, shape and size of the aggregates, and aggregation stability. Up to date, SAILs with different head groups of piperidinium, pyrrolidinium, and pyridinium have been successfully used for SPE; however, the imidazolium type with $[BF_4]^-$ and $[Br]^-$ counterions have been the most commonly used.

Finally, the choice of the counterion needs to be considered. In this regard, the extraction of trace amounts of Pt(IV) and Pu(IV) from acidic solutions using long chain phosphonium SAILs with different anions of $[P_{666,14}][Cl]$, $[P_{666,14}][BF_4]$, and $[P_{666,14}][PF_6]$, have been examined. Results demonstrate that $[P_{666,14}][Cl]$ is a better choice for extracting metals due to its higher hydrophilicity [41].

Conclusion

In this chapter, the properties and applications of surface-active ionic liquids were evaluated. Amphiphilic SAILs have the capability to reduce the IFT and the CMC in different systems, establishing new opportunities for different applications. It has been revealed that the structure of SAILs and the operational parameters have important effects.

Considering the structure of SAILs, selecting long alkyl chains leads to effective adsorption and consequently low IFT achievements. The bulkiness and π-electron ring head groups provide higher interfacial activity. In addition, heterocyclic head groups have significant effects on reducing IFTs and that the large counterions with significant polarizability enhance the performance of SAILs. For these reasons, the CMC of SAILs are diminished and micelles are formed more easily with long alkyl chains, bulky heterocyle head groups, and large counterions. Catanionic, Multicationic, and Zwitterionic types of SAILs show higher interfacial activity than single analogous surfactants, making them good candidates for different applications.

Results of investigations on different operating conditions suggest that application of temperatures higher than room temperature enhance molecular agitation and mobility of adsorbing SAILs. For some SAILs, however, this trend is reversed. In addition, the adsorption tendency and effectiveness raise with the pH, which is due to decreasing electrostatic repulsion between the SAIL head groups with hydroxyl ions. In the presence of electrolytes, the low hydration of SAILs bring about significantly higher IFT reductions due to salting-out effect as well as the reduction of electrostatic repulsion between the head groups. Regarding CMC, the temperature has a complicated effect, but salinity and pH exert positive effects on the micelle formation at low concentrations.

Concerning the toxicity and biodegradability of SAILs, it was reported that SAILs can be considered as good alternatives to conventional toxic surfactants. In this regard, pyridinium and imidazolium SAILs with medium alkyl chain as well as Multicationic SAILs show lower toxicity. Further, by changing the length and functional groups in the alkyl chains, the toxicity and biodegradability of SAILs can be modified.

The interesting properties of SAILs have found desired applications in stabilizing and extracting materials. In emulsion stabilization, addition of SAILs by reducing the IFT and creating electrostatic and steric protective layers, create stable microemulsions and stabilize immiscible mixtures. In the case of stabilizing suspensions, SAILs can disperse insoluble substances into solutions by adsorbing on the surface of the particles and forming hemimicelles or admicelles. Meanwhile, the separation of particles in solutions is highly feasible by linking SAILs on the surface of particles.

The design of new SAILs as stabilizing and separating agents has to be done to achieve ultra-low IFT values, to produce multiple functional emulsifiers, and to create stable micellar structures. Branched or multi-branched alkyl chains in the cationic and anionic portions and altering the head groups could be manipulated for improving the characteristics and the application of SAILs. Furthermore, the investigation on the feasibility study for large scales is important in the future. In general, it has been rationalized that SAILs have the potential to solve a variety of industrial issues.

References

[1] Giustiniani, A., Drenckhan, W., Poulard, C. Interfacial tension of reactive, liquid interfaces and its consequences. *Advances Colloid Interface Sci.* 2017, *247*, 185–197.

[2] Kharazi, M., Saien, J., Asadabadi, S. Review on amphiphilic ionic liquids as new surfactants: From Fundamentals to applications. *Topics Curr. Chem.* 2022, *380*, 1–44.

[3] Łuczak, J., Hupka, J., Thöming, J., Jungnickel, C. Self-organization of imidazolium ionic liquids in aqueous solution. *Colloids Surfaces A*2008, *329*, 125–133.

[4] Asadabadi, S., Saien, J., Khakizadeh, V. Interface adsorption and micelle formation of ionic liquid 1-hexyl-3-methylimidazolium chloride in the toluene+ water system. *J. Chem. Thermodyn.* 2013, *62*, 92–97.

[5] Han, D., Row, K. H. Recent applications of ionic liquids in separation technology. *Molecules* 2010, *15*, 2405–2426.

[6] Saien, J., Kharazi, M., Asadabadi, S. Adsorption behavior of short alkyl chain imidazolium ionic liquids at n-butyl acetate + water interface: experiments and modeling. *Iran. J. Chem. Eng.* 2015, *12*, 59–74.

[7] Saien, J., Kharazi, M., Asadabadi, S. Adsorption behavior of long alkyl chain imidazolium ionic liquids at the n-butyl acetate + water interface. *J. Mol. Liq.* 2015, *212*, 58–62.

[8] Kharazi, M., Saien, J., Yarie, M., Zolfigol, M. A. Different spacer homologs of gemini imidazolium ionic liquid surfactants at the interface of crude oil-water. *J. Mol. Liq.* 2019, *296*, 111748.

[9] Saien, J., Kharazi, M. A comparative study on the interface behavior of different counter anion long chain imidazolium ionic liquids. *J. Mol. Liq.* 2016, *220*, 136–41.

[10] Sastry, N. V., Vaghela, N. M., Aswal, V. K. Effect of alkyl chain length and head group on surface-active and aggregation behavior of ionic liquids in water. *Fluid Phase Equi.* 2012, *327*, 22–29.

[11] Goodchild, I., Collier, L., Millar, S. L., Prokeš, I., Lord, J. C., Butts, C. P., Bowers, J., Webster, J. R., Heenan, R. K. Structural studies of the phase, aggregation and surface behaviour of 1-alkyl-3-methylimidazolium halide + water mixtures. *Journal Colloid Interface Sci.*, 2007, *307*, 455–468.

[12] Docherty, K. M., Dixon, J. K., Kulpa Jr, C. F. Biodegradability of imidazolium and pyridinium ionic liquids by an activated sludge microbial community. *Biodegradation* 2007, *18*, 481–493.

[13] Saien, J., Asadabadi, S. Alkyl chain length, counter anion and temperature effects on the interfacial activity of imidazolium ionic liquids: Comparison with structurally related surfactants. *Fluid Phase Equil.* 2015, *386*, 134–139.

[14] Rodríguez-Palmeiro, I., Rodríguez-Escontrela, I., Rodríguez, O., Arce, A., Soto, A. Characterization and interfacial properties of the surfactant ionic liquid 1-dodecyl-3-methyl imidazolium acetate for enhanced oil recovery. *RSC Adv.* 2015, *5*, 37392–37398.

[15] Blesic, M., Marques, M. H., Plechkova, N. V., Seddon, K. R., Rebelo, L. P. N., Lopes, A. Self-aggregation of ionic liquids: micelle formation in aqueous solution. *Green Chem.* 2007, *9*, 481–490.

[16] Sakthivel, S., Chhotaray, P. K., Velusamy, S., Gardas, R. L., Sangwai, J. S. Synergistic effect of lactam, ammonium and hydroxyl ammonium based ionic liquids with and without NaCl on the surface phenomena of crude oil/water system. *Fluid Phase Equil.* 2015, *398*, 80–97.

[17] Kharazi, M., Saien, J., Yarie, M., Zolfigol, M. A. Promoting activity of gemini ionic liquids surfactant at the interface of crude oil-water. *Petroleum Res.* 2021, *31*, 20–22.

[18] Manshad, A. K., Rezaei, M., Moradi, S., Nowrouzi, I., Mohammadi, A. H. Wettability alteration and interfacial tension (IFT) reduction in enhanced oil recovery (EOR) process by ionic liquid flooding. *J J. Mol. Liq.* 2017, *248*, 153–162.

[19] Zhou, H., Zhu, Y., Peng, T., Song, Y., An, J., Leng, X., Yi, Z., Sun, Y., Jia, H. Systematic study of the effects of novel halogen-free anionic surface-active ionic liquid on interfacial tension of water/model oil system. *J. Mol. Liq.* 2016, *223*, 516–520.

[20] Rodríguez-Escontrela, I., Rodríguez-Palmeiro, I., Rodríguez, O., Arce, A., Soto, A. Characterization and phase behavior of the surfactant ionic liquid tributylmethylphosphonium dodecylsulfate for enhanced oil recovery. *Fluid Phase Equil.* 2016, *417*, 87–95.

[21] Jia, H., Lian, P., Liang, Y., Zhu, Y., Huang, P., Wu, H., Leng, X., Zhou, H. Systematic investigation of the effects of zwitterionic surface-active ionic liquids on the interfacial tension of a water/crude oil system and their application to enhance crude oil recovery. *Energy Fuels* 2018, *32*, 154–160.

[22] Zhou, H., Liang, Y., Huang, P., Liang, T., Wu, H., Lian, P., Leng, X., Jia, C., Zhu, Y., Jia, H. Systematic investigation of ionic liquid-type gemini surfactants and their abnormal salt effects on the interfacial tension of a water/model oil system. *J. Mol. Liq.* 2018, *249*, 33–39.

[23] Saien, J., Kharazi, M., Yarie, M., Zolfigol, M. A. A Systematic investigation of a surfactant type nano gemini ionic liquid and simultaneous abnormal salts effects on the crude oil/water interfacial tension. *Ind. Eng. Chem. Res.* 2019, *58*, 3583–94.

[24] Kharazi, M., Saien, J., Yarie, M., Zolfigol, M. A. The superior effects of a long chain gemini ionic liquid on the interfacial tension, emulsification and oil displacement of crude oil-water. *J. Petroleum Sci. Eng.* 2020, *195*, 107543.

[25] Rodríguez-Escontrela, I., Rodríguez-Palmeiro, I., Rodríguez, O., Arce, A., Soto, A. Phase behavior of the surfactant ionic liquid trihexyltetradecylphosphonium bis(2,4,4-trimethylpentyl)phosphinate with water and dodecane. *Colloids Surfaces A* 2015, *480*, 50–59.

[26] Saien, J., Kharazi, M., Pino, V., Pacheco-Fernández, I. Trends offered by ionic liquid-based surfactants: applications in stabilization, separation processes, and petroleum industry. *Separation Purification Rev.* 2022, 2052094.

[27] Asadabadi, S., Saien, J. Effects of pH and salinity on adsorption of different imidazolium ionic liquids at the interface of oil–water. *Colloids Surfaces A* 2016, *489*, 36–45.

[28] Saien, J., Asadabadi, S. Salting out effects on adsorption and micellization of three imidazolium-based ionic liquids at liquid–liquid interface. *Colloids Surfaces A* 2014,*444*,138–143.

[29] Najimi, S., Nowrouzi, I., Manshad, A. K., Farsangi, M. H., Hezave, A. Z., Ali, J. A., Keshavarz, A., Mohammadi, A. H. Investigating the effect of [C_8Py]Cl and [C_{18}Py]Cl ionic liquids on the water/oil interfacial tension by considering Taguchi method. *J. Petroleum Explor. Prod. Tech.* 2019, *9*, 2933–2941.

[30] Varade, D., Joshi, T., Aswal, V. K., Goyal, P. S., Hassan, P. A., Bahadur, P. Effect of salt on the micelles of cetyl pyridinium chloride. *Colloids Surfaces A* 2005, *259*, 95–101.

[31] Kapitanov, I. V., Jordan, A., Karpichev, Y., Spulak, M., Perez, L., Kellett, A., Kümmerer, K., Gathergood, N. Synthesis, self-assembly, bacterial and fungal toxicity, and preliminary biodegradation studies of a series of L-phenylalanine-derived surface-active ionic liquids. *Green Chem.* 2019, *21*, 1777–1794.

[32] Steudte, S., Bemowsky, S., Mahrova, M., Bottin-Weber, U., Tojo-Suarez, E., Stepnowski, P., Stolte, S. Toxicity and biodegradability of dicationic ionic liquids. *RSC Adv.* 2014, *4*, 5198–5205.

[33] Bera, A., Belhaj, H. Ionic liquids as alternatives of surfactants in enhanced oil recovery—a state-of-the-art review. *J. Mol. Liq.* 2016, *224*, 177–188.

[34] Ławniczak, Ł., Materna, K., Framski, G., Szulc, A., Syguda, A. Comparative study on the biodegradability of morpholinium herbicidal ionic liquids. *Biodegradation* 2015, *26*, 327–340.

[35] Danielsson, I., Lindman, B. The definition of microemulsion, *Colloids Surfaces* 1981, *3*, 391–392

[36] Rao, V. G., Mandal, S., Ghosh, S., Banerjee, C., Sarkar, N. Ionic liquid-in-oil microemulsions composed of double chain surface-active ionic liquid as a surfactant: temperature dependent solvent

and rotational relaxation dynamics of coumarin-153 in [Py][TF2N]/[C4mim][AOT]/benzene microemulsions. *J. Phys. Chemistry, B* 2012, *116*, 8210–8221.

[37] Łuczak, J., Paszkiewicz, M., Krukowska, A., Malankowska, A., Zaleska-Medynska, A. Ionic liquids for nano-and microstructures preparation. Part 1: Properties and multifunctional role. *Adv. Colloid Interface Sci.* 2016, *230*, 13–28.

[38] Jiang, J., Yu, S. H., Yao, W. T., Ge, H., Zhang, G. Z. Morphogenesis and crystallization of Bi_2S_3 nanostructures by an ionic liquid-assisted templating route: synthesis, formation mechanism, and properties. *Chem. Materials* 2005, *17*, 6094–6100.

[39] Qin, Y., Song, Y., Sun, N., Zhao, N., Li, M., Qi, L. Ionic liquid-assisted growth of single-crystalline dendritic gold nanostructures with a three-fold symmetry. *Chem. Materials*2008, *20*, 3965–3972.

[40] Zhang, Q., Yang, F., Tang, F., Zeng, K., Wu, K., Cai, Q., Yao, S. Ionic liquid-coated Fe_3O_4 magnetic nanoparticles as an adsorbent of mixed hemimicelles solid-phase extraction for preconcentration of polycyclic aromatic hydrocarbons in environmental samples. *Analyst* 2010, *135*, 2426–2433.

[41] Myasoedova, G. V., Molochnikova, N. P., Mokhodoeva, O. B., Myasoedov, B. F. Application of ionic liquids for solid-phase extraction of trace elements. *Analytical Sci.* 2008, *24*, 1351–1353.

Chapter 6

Surface Active Ionic Liquids in the Petroleum Industry

**Javad Saien*
and Mona Kharazi†**
Department of Applied Chemistry,
Bu-Ali Sina University, Hamedan, Iran

Abstract

Crucial challenges in petroleum industry have persuaded scientists to improve existing strategies on enhanced oil recovery (EOR) by utilizing more effective materials than currently available. Relevantly, amphiphilic surface-active ionic liquids (SAILs) with suitable properties like proper interfacial activity, recyclability, environmental friendliness, and stability under high temperature and salinity conditions are known as appropriate materials. Although much attention has been paid to various applications of SAILs, their applications in the petroleum industry require a comprehensive review.

In this frame, the present chapter focuses on the performance of SAILs in petroleum industry. For this purpose, the characteristics of SAILs are first discussed. Different applications of SAILs in surfactant flooding, wettability alteration, and oil displacement are then surveyed. Thereafter, discussions about the effects of the type of crude oil, the interaction of SAILs with reservoir rocks, and comparison with conventional surfactants are provided. Whenever it is relevant, the activity of SAILs is explained in terms of chemical structure and operating conditions such as temperature and salinity. The characteristics and outstanding performance of the new types of SAILs are also regarded to highlight the benefits of these materials.

Finally, the challenges and possible future directions for the use of SAILs and conclusions are gathered in the last section. Generally, this chapter provides in-depth information about the application of SAILs in petroleum industry.

Keywords: surface active ionic liquids, petroleum industry, enhanced oil recovery, surfactant flooding, wettability alteration, oil displacement

* Corresponding Author's Email: saien@basu.ac.ir.
† Corresponding Author's Email: kharazy91@yahoo.com.

In: Properties and Applications of Ionic Liquids
Editors: Noureddine Abidi and Julia L. Shamshina
ISBN: 979-8-88697-635-9

Abbreviations

API	American Petroleum Institute
CMC	Critical micelle concentration
EOR	Enhanced oil recovery
IFT	Interfacial tension
IL	Ionic liquid
LPG	Liquefied petroleum gas
OOIP	original oil in place
PIT	Phase inversion temperature
RTIL	Room temperature ionic liquid
SAIL	Surface-active ionic liquid
SARA	Saturates, aromatics, resins and asphaltenes
Surfactant	Surface-active agent
θ	Contact angle
	Cation part of SAILs
[BT]	Butyrolactam
[$C_n$mim]	1-Alkyl-3-methylimidazolium
[C_4mim]	1-Butyl-3-methylimidazolium
[C_6mim]	1-Hexyl-3-methylimidazolium
[C_8mim]	1-Octyl-3-methylimidazolium
[C_{10}mim]	1-Decyl-3-methylimidazolium
[C_{12}mim]	1-Dodecyl-3-methylimidazolium
[C_{16}mim]	1-Hexadecyl-3-methylimidazolium
[C_{18}mim]	1-Octadecyl-3-methylimidazolium
[$C_n$im-C_4-imC_n]	1,1'-(Butane-1,4-diyl)bis(3-alkyl-1H-imidazol-3-ium)
[C_8im-C_4-imC_8]	1,1'-(Butane-1,4-diyl)bis(3-octyl-1H-imidazol-3-ium)
[C_{12}im-C_4-imC_{12}]	1,1'-(Butane-1,4-diyl)bis(3-dodecyl-1H-imidazol-3-ium)
[CP]	Caprolactam
[C_{16}mpyr]	1-Hexadecyl-1-methylpyrrolidinium
[C_8py]	1-Octylpyridinium
[C_{12}py]	1-Dodecylpyridinium
[C_{18}py]	1-Octadecylpyridinium
[Et_2NH_2]	Diethylammonium
[Et_3NH]	Triethylammonium
[PA]	Propylammonium
[P144,4]	Tributyl-methylphosphonium
[P666,14]	Trihexyl-tetradecylphosphonium
[3-HPA]	3-Hydroxypropylammonium
	Anion part of SAILs
[Ac]	Acetate
[BF_4]	Tetrafluoroborate
[CF_3COO]	Trifluoroacetate
[CH_3COO]	Acetate
[$C_6H_{13}COO$]	Hexanoate

[DS]	Dodecylsulfonate
[HCOO]	Formate
[HSO_4]	Hydrogen sulfate
[H_2PO_4]	Dihydrogen phosphate
[$(i\text{-}C_8)_2PO_2$]	Bis(2,4,4-trimethylpentyl) phosphinate
[NTf_2]	Bis(trifluoromethylsulfonyl)imide
[OTf]	Trifluoromethanesulfonate,
[PF_6]	Hexafluorophosphate
	SAILs
Ammoeng 102	Tetra-alkyl ammonium sulfate
[C_{16}IPS]	3-(1-hexadecyl-3-imidazolio) propanesulfonate
[C_{16}IPS-Nsa]	3-(1-hexadecyl-3-imidazolio) propanesulfonateβ-naphthalene sulfonate
	Conventional surfactants
AN-120	Dodecyl benzene sulfonic acid + Nonylphenolethoxylate
CTAB	Cetyltrimethylammonium bromide
DTMAC	Dodecyltrimethylammonium chloride
NX-610	Dodecyl dimethyl amine betaine
NX-1510	Dodecyl dimethyl amine betaine + Octadecyl amine ethoxylate
NX-2760	Dodecyl dimethyl amine betaine + Octadecyl amine ethoxylated betaine
SDBS	Sodium dodecylbenzenesulfonate
SDS	Sodium dodecyl sulfate
Tergitol 15-S-5	Ethoxylated C11~14 secondary alcohol
Tergitol 15-S-7	Ethoxylated C11~14 secondary alcohol
Tergitol 15-S-9	Ethoxylated C11~14 secondary alcohol
Triton X-100	Polyoxyethylene octyl phenyl ether

Introduction

In addition to many challenges facing the petroleum industry such as declining crude oil reserves and the low possibility of finding new oil fields, only one-thirds of the original oil in place (OOIP) can be recovered with conventional methods. Thus, inevitably, the petroleum industry has to look for enhanced oil recovery (EOR) techniques like surfactant flooding that focuses on the interfacial tension (IFT) reduction, wettability alteration and oil displacement. However, the low resistance and weak performance of conventional surfactants at high temperatures and salinity conditions, as well as their tendency to adsorb on the surface of reservoir rocks, have persuaded scientists to search for more effective and economical materials in EOR.

Surface active ionic liquids (SAILs) can be considered as an important alternative in EOR. Most of the SAILs show strong interfacial activity when compared to conventional surfactants with the same molecular structure. SAILs can reduce IFT by adsorption at the oil-water interface and can alter the wettability of reservoir rocks by changing the surface properties. SAILs can provide the oil displacement, bringing about elevated crude oil mobility in the reservoirs and thus achieving a higher oil production. Relevantly, the unique properties of

SAILs such as stability at high temperatures and salinity, and environmentally friendly nature make them suitable candidates for surfactant flooding process. In addition, high molecular weight SAILs can increase the viscosity of the injected fluids and provide sweeping efficiency in EOR. Furthermore, SAIL molecules are adjustable by choosing the appropriate cation/anion in their structure, and task-specific SAILs can be synthesized for utilization in EOR. In this regard, Anionic, Catanionic, Multicationic, and Zwitterionic types of SAILs have been examined and their superior properties have been scrutinized.

Although SAILs demonstrate desired outcomes in the petroleum industry, screening their performance under real conditions is an essential step before using in the large-scale EOR operations. Therefore, the entire injected fluid properties and reservoir conditions such as the nature of crude oil and the interaction of SAILs with the surface of reservoir rocks must be regarded for assessing use of SAILs in EOR.

Despite many attempts, a comprehensive report has not been dedicated to this area yet. Therefore, a comprehensive compilation is required to understand the state of the art to date. This chapter initially provides a brief explanation about the SAIL characteristics and their applicability in petroleum industry. Thereafter, the performance of SAILs in various aspects of EOR such as surfactant flooding, wettability alteration, and oil displacement from solid surfaces under different conditions is surveyed. Relevantly, effects of the type of crude oil and interaction of SAILs with the surface of reservoir rocks are considered. Whenever appropriate, the effects of the chemical structure of SAILs and the operating conditions of temperature and salinity are discussed. The characteristics and outstanding performance of the new types of SAILs are also regarded to highlight the benefits of these materials. To select the most suitable surfactants and display higher interfacial activity, a brief comparison between SAILs and conventional surfactants is subsequently provided. Finally, the crucial challenges as well as possible directions and the conclusions are summarized in the last section. In general, the purpose of this chapter is to provide in-depth information on the use of SAILs in the petroleum industry and to pave the ways for producing better kinds of SAILs that may fulfill the requirement of the EOR processes.

Characteristics of Surface Active Ionic Liquids

Structure and Properties

Ionic liquids (ILs) are composed of organic cations and inorganic/organic counter anions. Steric asymmetry of the ions causes mismatch and unstable crystal lattices that reduces the melting point. ILs have melting points below 100°C, and room temperature ionic liquids (RTILs) specifically refer to those which are liquid at room temperature or even below. Outstanding properties of many ILs such as high chemical and thermal stability even under harsh conditions, low vapor pressure, recyclability, non-flammability, and environmentally friendly manner suggest high number of applications for these materials.

IL properties can be altered by changing/modifying different parts of their molecular structure giving them capability of "designer materials." In this regard, many organic cations can be coupled with various inorganic or organic anions. For better illustration, structures of typical IL cations and anions are shown in Figure 1.

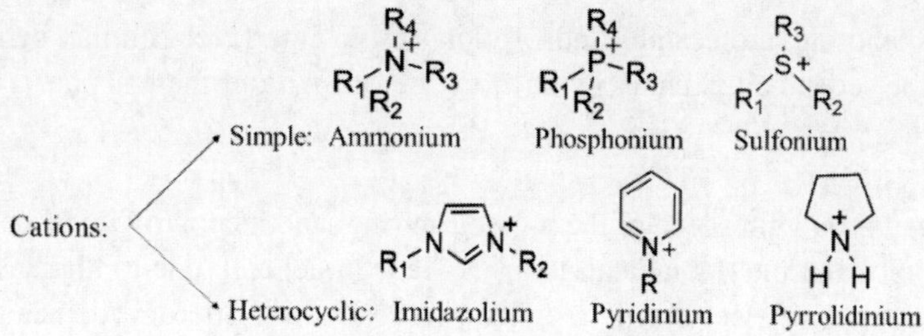

Anions: BF_4^-, PF_6^-, X^- (X = Cl, Br, I), OH^-, CH_3COO^-, $PhSO_3^-$

Figure 1. A number of usual IL cations and counter anions [1].

Surface-active ionic liquids, briefly "SAILs," are a class of ILs that have both the hydrophilic and the hydrophobic groups in their structure, resulting in "amphiphilic" character of the compounds. The ionic or polar hydrophilic group tend to be in the aqueous phase and organic hydrophobic group in the oil phase. Thus, SAILs are adsorbed at the interface of oil-water, causing decrease in the intermolecular forces between the phases and consequently reducing the interfacial tension (IFT), which refers to the molecular cohesive forces at the interface of liquids [2].

IFT depends on the concentration of SAILs, their structure, the nature of involved phases, and the rate of adsorption. Presence of bulky heteroatomic ring in SAILs provides higher hydrophobicity and weaker hydration and leads to easier migration toward the interface and higher interfacial activity compared to conventional surfactants. Adsorption at the interface continues until the surface becomes saturated and SAILs start to form micelles in the bulk. This concentration is referred to as critical micelle concentration (CMC). After reaching CMC, higher number of aggregates with nearly no interface effect will be generated and the IFT remains almost constant.

On the other hand, adsorption of SAILs on the surface of rocks changes the properties and the surface wettability can be changed depending on the orientation of SAILs at the interface. Also, oil displacement might occur due to the presence of SAILs. These are discussed in detail during the following sections.

Impact of the Structure of SAILs

Generally, SAILs have three major parts: head group, alkyl chain, and anion in their molecule. By altering these parts, the characteristics of the SAILs could be modified for specific applications. The major SAIL properties of hydrophobicity, interfacial activity, miscibility and thermal stability could be improved. In this regard, an overview on SAIL characteristics with respect to different molecular parts is explained in following relevant cases.

Alkyl Chain

The alkyl chain length and the number of involved branches are important factors for SAILs. Long alkyl chains provide higher adsorption, hydrophobicity, and interfacial activity, however, solubility in aqueous phase decreases with alkyl chain increase. A longer hydrophobic chain also causes greater van der Waals attraction between the SAIL molecules and the crude oil phase, which leads to more stable adsorption at the interface. Similar behavior has been observed for the second alkyl chain in the structure of SAILs if there is one.

Head Group

The size of head group, presence of the π-electron ring, and its polarity have definite effects on the hydrophilicity and interfacial activity of SAILs. In general, due to higher hydrophobicity, poor hydration and low electrostatic repulsion forces, bulky heterocyclic head groups with π-electron rings result in higher interfacial activity compared to those with simple groups like ammonium. In this regard, the most widely studied SAILs are imidazolium, pyridinium, pyrrolidinium, pyrazolium, thiazolium, and guanidinium.

Counter Anion

The type of counter anion affects the miscibility of SAILs in polar/non-polar solvents and their interfacial activity to a lesser extent compared to the alkyl chain and the head group. Studies indicate that the bulkier anions with greater polarizability increase the interfacial activity of SAILs. This is due to the fact that big anions are less prone to hydration and adsorb at the interface easier, which reduces the electrostatic repulsion between head groups. Some usual inorganic and organic anions are hexafluorophosphate $[PF_6]^-$, tetrafluoroborate $[BF_4]^-$, bis(trifluoromethylsulfonyl)imide $[NTf_2]^-$, trifluoromethanesulfonate, $[OTf]^-$, acetate, $[Ac]^-$, and halides.

Classification

Various types of SAILs with specific properties and different structures have been utilized in the petroleum industry. Generally, SAILs can be classified based on the nature of the surface-active moiety and the number of head groups (Figure 2).

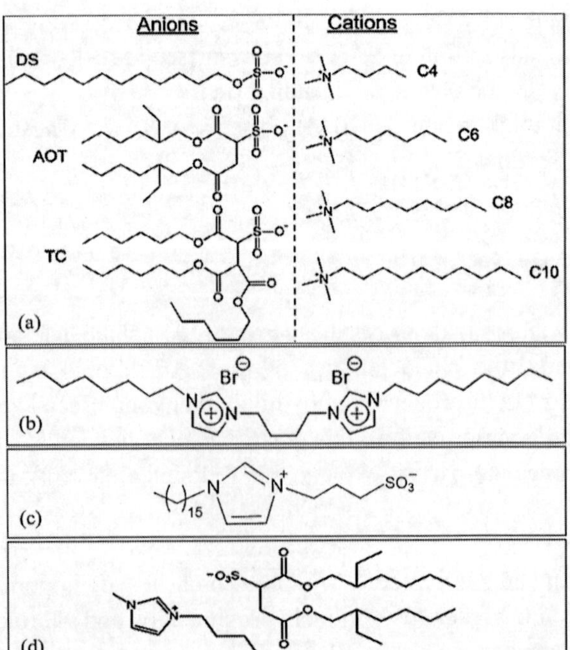

Figure 2. Catanionic (a), Multicationic (b), Zwitterionic (c) and Anionic (d) types of SAILs [1].

Accordingly, new types of SAILs are classified into Cationic, Catanionic, Multicationic, Zwitterionic, and Anionic, which are schematically shown in Figure 2. These types of SAILs exhibit better interfacial activity compared to conventional cationic SAILs; however, investigation of these types of SAILs is in its infancy and needs comprehensive studies.

Enhanced Oil Recovery and SAILs

Principles

Crude oil is known as the natural petroleum source consisting of carbon, nitrogen, oxygen, sulfur elements and traces of metals. Crude oil is naturally formed in the earth porous rock deposits, "oil reservoirs." Such complex mixture of crude oil is separated into four components of Saturates (S), Aromatics (A), Resins (R), and Asphaltenes (A) abbreviated as SARA. Different crude oils are basically classified based on their composition, i.e., percentage of SARA, and American Petroleum Institute (API) degree.

Although energy consumption is growing worldwide and it is estimated to double by 2040 [3], declining crude oil reserves, low probability of exploring new fields and meeting crude oil demands have posed serious challenges to the petroleum industry. Use of primary and secondary recovery techniques can extract only one-thirds of the original OOIP. To solve this problem, most of the mature oil fields in the world need EOR to maintain their production level. EOR is the process of injecting gases, chemicals, and/or applying thermal energy in order to recover excess crude oil. Classification of different EOR methods is presented in Figure 3.

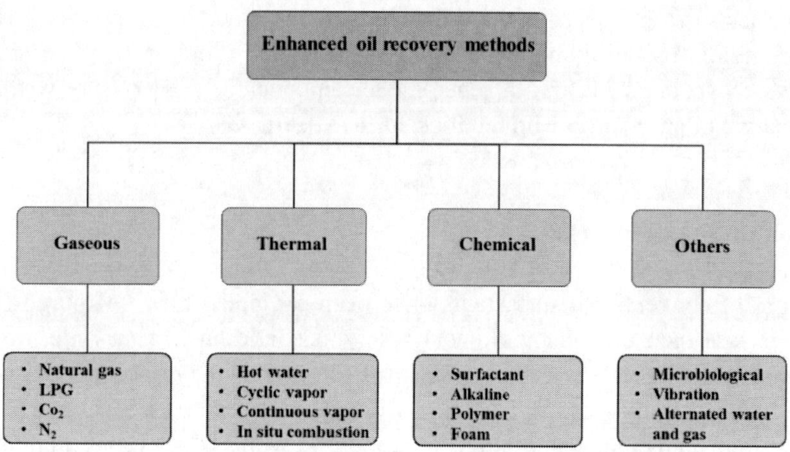

Figure 3. Classification of different EOR methods.

Among those, chemical EORs are effective and low-cost solutions. In chemical EORs strategies, selected chemicals like surfactants, alkalis, polymers, and organic solvents are utilized. Surfactant flooding is a major chemical EOR technique that deals with reducing IFT to enhance oil mobilization through porous rocks and directs the residual oil flows toward the production well. Figure 4 illustrates a schematic of surfactant flooding EOR in reservoirs.

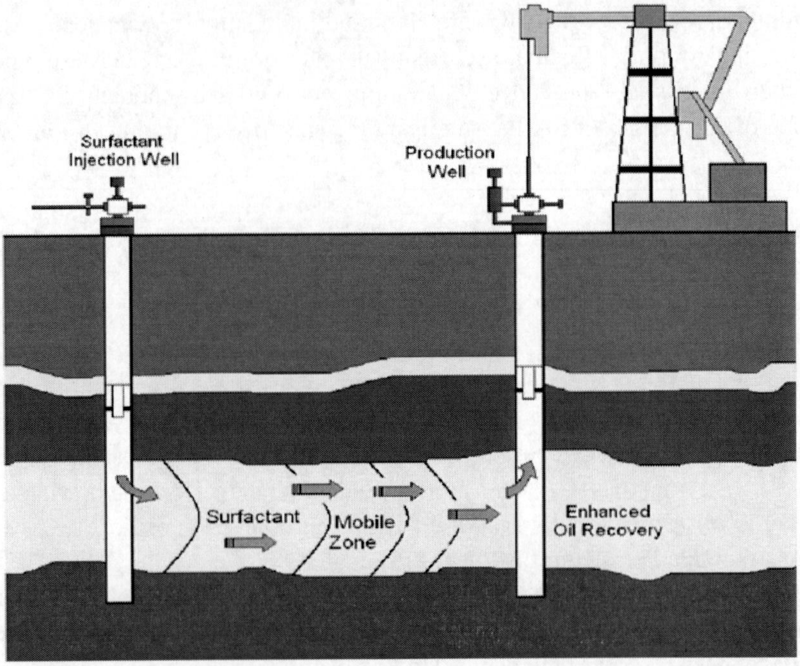

Figure 4. A schematic of the surfactant flooding EOR in reservoirs [4].

The wettability alteration and the oil displacement are two other important phenomena using which the oil recovery during EOR might be improved. By adsorption on the rock surface, surfactants can alter the wettability of the reservoir rocks from oil-wet to water-wet or displace oil from the rock surface. This smoothens the crude oil mobilization in reservoirs and enhances the oil production. Measuring the contact angle helps determining wettability alteration from oil-wet to water-wet or *vice versa*. In general, one can conclude higher water-wettability when the contact angle of an oil drop with the rock surface decreases.

Application of SAILs in EOR

Various types of conventional surfactants have been examined for EOR purposes, however, most of them lose their effectiveness under reservoir conditions. Meanwhile, from technical and economical points of view, a low surfactant adsorption on the reservoir rocks is considered to be an advantage. In this regard, SAILs are regarded as efficient materials. The stability of SAILs under high temperature and salinity conditions as well as their strong amphiphilic nature make them good candidates for EOR. Accordingly, SAILs cause IFT reduction during surfactant flooding by adsorbing and orienting at the oil-water interface and facile the wettability alteration and the oil displacement by changing the properties of the rock surface. The role of SAILs in EOR process is schematically illustrated in Figure 5.

Additionally, high molecular weight SAILs facile the mobility of crude oils by elevating the viscosity of the injected fluid, providing high sweeping efficiency. SAILs are also suitable for thermal oil recovery due to high thermal stability, and are able to improve oil production especially for heavy crude oils. In the same manner, aromaticity of some SAILs can help increase oil production. As other advantages, SAILs are able to decrease the water consumption during oil extraction, assist separation of sand and clay from crude oil, and reduce the corrosion.

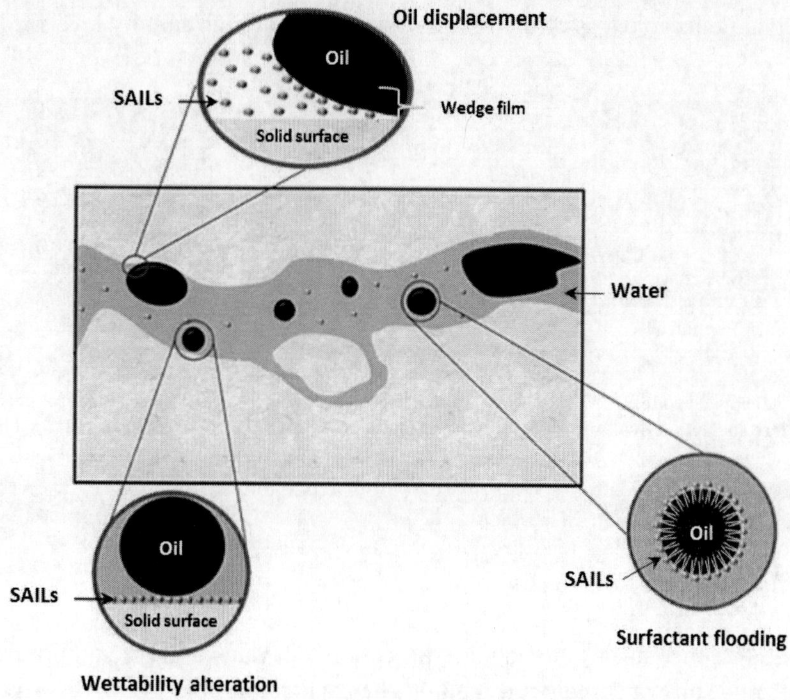

Figure 5. Schematic of different SAIL applications in EOR operations [5].

Despite the high potential of SAILs in EOR process, the entire injected fluid system and reservoir conditions such as the nature of crude oil, type of rocks, and solubility of SAILs in brine are important parameters to be considered in operations. Another important factor is the interaction of SAILs with the surface of the reservoir rocks. Therefore, scrutiny of SAILs is an essential step before using them in EOR operations or predicting the behavior of SAILs. These are deliberated in detail in the following sections of this chapter.

Surfactant Flooding

A high level IFT between crude oil and water is the essential reason for the low mobilizing of residual oil into reservoirs. IFT represents the extent of molecular cohesive forces at the interface of immiscible liquids stemming from unbalanced forces between molecules at the interface. IFT has a major role in sketching and operating of immiscible fluids and affects the fluids' hydrodynamics upon contact between phases. Therefore, it is important to find a way to reduce the IFT in reservoirs. In the surfactant flooding method, SAIL molecules are adsorbed at the crude oil-water interface. By reducing IFT to an ultralow value, the capillary number (which represents the relative effects of viscous drag forces versus cohesive interfacial forces for crude oil-water system in the reservoirs) raises [1]. As a result, the forces which trap oil drops in the pores are weakened and crude oil production from reservoirs increase. The mechanism of oil recovery based on IFT reduction is schematically presented in the Figure 6. By utilizing imidazolium SAILs, for instance, it has been possible to recover about 90% of the extra-heavy crude oil from Utah and Canadian tar sands, and SAILs have been recycled up to 5 times without noticeable loss of their performance even at high salinity levels [6].

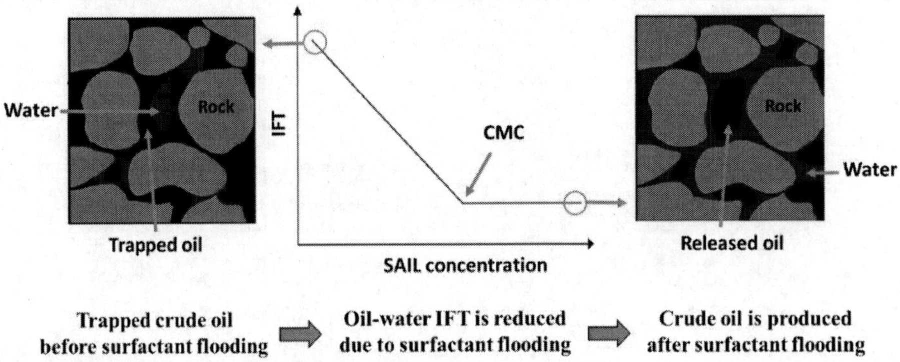

Figure 6. Mechanism of crude oil recovery based on IFT reduction.

Effects of the Structure of SAILs

Owing to the fact that surfactant flooding process is relevant to the adsorption capability of SAILs at the crude oil-water interface, a minor change in the structure of SAILs like alkyl chain length, identity of a head group and counter anion affect the system's IFT and the process efficiency.

In this regard, SAILs with longer hydrophobic alkyl chains are capable to reduce IFT to a larger extent. This can be ascribed to the higher adsorption at the interface as well as the stronger van der Waals attraction between the alkyl chain of SAILs and the hydrophobic moiety of the crude oil. Also, branched alkyl chains significantly reduce the IFT between crude oil and water.

Importantly, low amounts of the long chain SAILs are required to determine IFT value, making performance evaluation of SAILs economical and environmentally beneficial [7]. Thus, comparing the performance of [C_8mim][Cl], [C_{12}mim][Cl], [C_8py][Cl], and [C_{12}py][Cl], indicates higher efficiency of SAILs possessing longer alkyl chains [8, 9]. Figure 7 shows the IFT variation as a function of alkyl chain length in [$C_n$im-C_4-imC_n][Br]$_2$ at different concentrations.

By examining the effect of head groups, it has been found that SAILs with bulky heterocyclic head groups and delocalized charge rings such as imidazolium or thiazolium display a superior activity in reducing IFT. This phenomenon is attributed to the higher hydrophobicity of SAILs with heterocyclic head groups as well as the strong interaction between the aromatic compounds of the crude oil and the aromatic head group of the SAILs. Appropriately, using [C_4mim][H_2PO_4] and [C_6mim][HSO_4], with imidazolium head groups, the IFT has been reduced up to 37.8 and 42.4%, respectively; however, using [Et_2NH_2][H_2PO_4] and [Et_3NH][HSO_4] with the quaternary ammonium head group the reductions have been just 32.6 and 37.7%, respectively [11, 12].

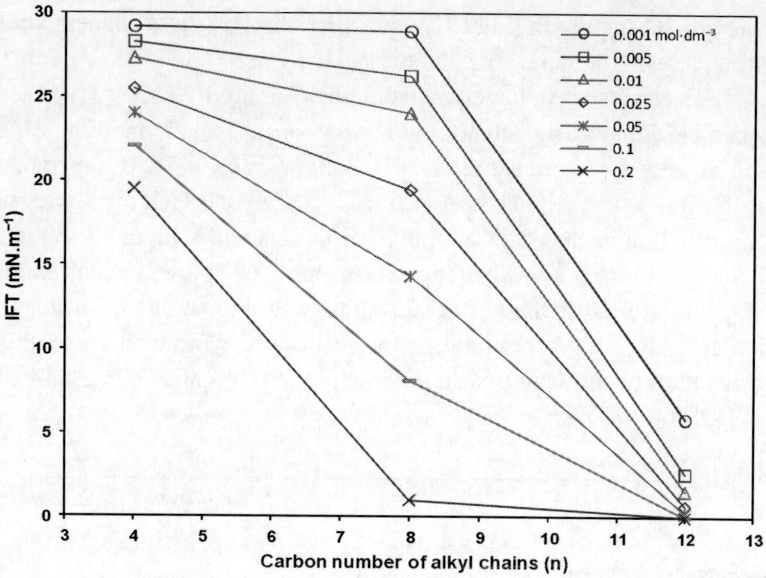

Figure 7. IFT variation as a function of alkyl chain length for $[C_n im\text{-}C_4\text{-}imC_n][Br]_2$ [10].

In the case of counter anion, higher interfacial activity is obtained with increase of polarizability. This is due to the greater interaction of polarizable counter anions with the SAIL-cations, which decreases repulsion between the cationic head groups at the interface and provides further IFT reductions [13]. For instance, by using $[C_4mim][X]$ ILs, the efficiency has been appeared in the sequence of $[H_2PO_4]^-$ > $[PF_6]^-$ > $[BF_4]^-$ > $[HSO_4]^-$ [11]. The length of alkyl chain on the counter anions also affects IFT reduction and hence process efficiency As a case, for the applied cations of $[CP]^+$, $[BT]^+$, $[PA]^+$, and $[3\text{-}HPA]^+$, the IFT reduction has been effective in the order of $[C_6H_{13}COO]^-$ > $[CH_3COO]^-$ > $[HCOO]^-$ [14]. Such improved performance of SAILs with longer alkyl chain on anions is attributed to the interaction of alkyl chains and the crude oil, assisting the adsorption of SAILs at the interface.

The newly introduced types of SAILs show higher efficiency compared to the usual Cationic SAILs. Interestingly, ultra-low IFT values can be achieved with Catanionic, Zwitterionic, and Multicationic SAILs. Noteworthy point is that using co-surfactants for achieving ultra-low IFTs is more or less usual, but no co-surfactant has been used for these types of SAILs [1]. These outcomes are attributed to the stronger amphiphilic nature of the aforementioned SAILs that is a consequence of longer alkyl chains, reduction of electrostatic repulsion between charged moieties, and close molecular orientation at the interface.

For Catanionic types, branched alkyl chains in both the anion and cation groups improve their capability. In this regard, use of $[P_{1444}][DS]$ has leaded the crude oil-water IFT to reach about 0.1 $mN.m^{-1}$, which is not comparable with about 10 $mN.m^{-1}$, relevant to $[C_{12}py][Cl]$, $[C_8mim][Cl]$, and $[C_{12}mim][Cl]$ with the same concentrations [15].

In Zwitterionic SAILs, there is a covalent bond between the cationic and anionic groups in each molecule. This neutralizes the electrostatic repulsion between charged molecules and creates intense molecular arrangement at the interface. A comparison of the interfacial activity of $[C_{16}IPS]$ and $[C_{16}IPS\text{-}Nsa]$ Zwitterionic SAILs with $[C_{16}mpyr][Br]$ and $[C_{16}mim][Br]$ Cationic SAILs (of the same alkyl chains length), has revealed that $[C_{16}IPS]$ and $[C_{16}IPS\text{-}Nsa]$ SAILs were able to reduce the IFT from 20 to 1 and 0.1 $mN.m^{-1}$, respectively; whereas for the same chain length [C_{16}mpyr][Br] and [C_{16}mim][Br] SAILs these values were 5.8 and 3.7 mN.m^{-1}, respectively, at the same concentrations [16].

Ultra-low IFTs can also be attained by using the Gemini SAILs, as a type of Multicationic SAILs. This can be achieved by extending the alkyl chain length, and high efficiency can be achieved even at very low concentration of Gemini SAILs [17]. Comparison between the efficiency of [C_{12}im-C_4-imC_{12}][Br]$_2$ and [C_8im-C_4-imC_8][Br]$_2$ Gemini SAILs with [C_{12}mim][Cl] and [C_8mim][Cl] mono analog Cationic SAILs is shown in Figure 8. As it could be seen, Gemini SAILs were able to provide better performance even at low concentrations and the long alkyl chain [C_{12}im-C_4-imC_{12}][Br]$_2$ has been able to reduce the crude oil-water IFT from 32 to about 10^{-3} mN.m^{-1} [10, 18]. Accordingly, aforementioned types of SAILs can be considered as the new generation of surfactants with distinct properties, capable in establishing ultra-low IFT values in EOR operations.

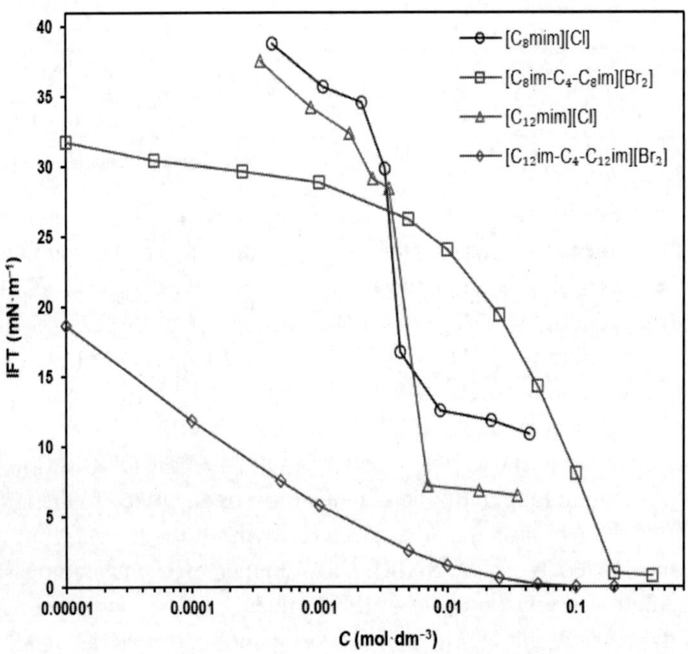

Figure 8. IFT variation versus bulk concentration of Gemini SAILs in comparison to mono analogs for crude oil-water system [10].

Effects of Temperature and Salinity

Unlike the conventional surfactants, SAILs remain stable at high temperatures and in most cases, temperature assists the IFT reduction. This effect has been attributed to the facile migration of SAILs toward the interface by breaking down the ordered iceberg structure of water molecules around the hydrophobic parts of the SAILs. Also, by elevating the agitation kinetic of molecules, the intermolecular forces at the interface are weakened leading to IFT reduction. Thus, reasonably, dominant high temperatures in reservoirs enhance the efficiency of surfactant flooding with SAILs. In this regard, different imidazolium, proline, aromatics and heterocyclic types of SAILs have been examined in the crude oil-water systems, and higher IFT reductions have been demonstrated as temperature rises.

On the other hand, in some cases, particularly in the presence of aromatic heterocyclic head groups (such as pyridinium and imidazolium) and with long alkyl chains (more than 8 carbon atoms), an ascending trend of variation is observed for IFT versus temperature [1]. This can be interpreted with respect to the "phase inversion" (PIT) at a specific temperature. At temperatures below PIT, adsorption of SAILs at the interface is favored as temperature increases, giving lower IFT values. Upon reaching PIT, the maximum adsorption of SAILs happens, providing a minimum IFT value, and by exceeding temperature from PIT, the adsorbed SAIL molecules diffuse into the oil phase. Thus, the number of SAILs at the interface decreases and the IFT increases. For better representation, the variation of the crude oil-water IFT with temperature is listed in Table 1 for different SAILs. As can be seen in this table, for the tetraalkyl ammonium sulfate Ammoeng 102 as an non- aromatic SAIL and [C_4mim][Cl] as a short alkyl chain imidazolium SAIL, IFT values decreased to a larger extent with the temperature increase. However, for long chain imidazolium and pyridinium SAILs, increase in temperature results in a slight increase in the IFT values [19]. It is important to note that all of the mentioned SAILs have been stable at high temperatures even to ~ 90°C.

Table 1. The IFT variation of crude oil-water system with temperature in the presence of different SAILs

Temperature (°C)	IFT ($mN.m^{-1}$)					
	[C_4mim][Cl]	[C_8mim][Cl]	[C_{12}mim][Cl]	[C_8py][Cl]	[C_{12}py][Cl]	Ammoeng 102
15	22.6					
20		1.3	0.5	12.1	0.9	4.8
25	19.1					
30		1.7	0.9	13.2	1.5	
35	18.6					
40		2.1	1.35	15.6	2.1	4.2
45	17.9					
50		2.4	2.1	16.1	2.5	
60		2.9	2.7	16.45	2.8	3.5
80						2.9
90						2.3

Another important parameter is the media salinity. SAILs not only remain stable under high salinity conditions, but also benefit from synergistic effect of salts resulting in IFT reduction. This is practically valuable in petroleum industry since saline sea waters are usually used for flooding in reservoirs. The reason for this phenomenon can be attributed to the well-known "salting out effect" and neutralizing the charge of SAIL ions at the interface. In aqueous solutions, the cation and anion of electrolytes and those of SAILs compete for water capturing, but smaller and harder electrolyte ions with a higher surface charge density provide strong hydration. It can be attributed to the hydrogen bonding in the water and formation of a cage-like structure surrounding the ions. Therefore, by adding salts, the number of reachable water molecules around the big SAIL ions decreases and consequently their solubility in the bulk diminishes, favoring adsorption at the interface [20]. In addition, the electrostatic repulsion between SAIL ions is compensated with the opposite charge of salt ions. This, in turn, leads to easier SAIL adsorption and significantly more IFT reductions. The effect of salt ions on the IFT of oil-water system is schematically presented in Figure 9.

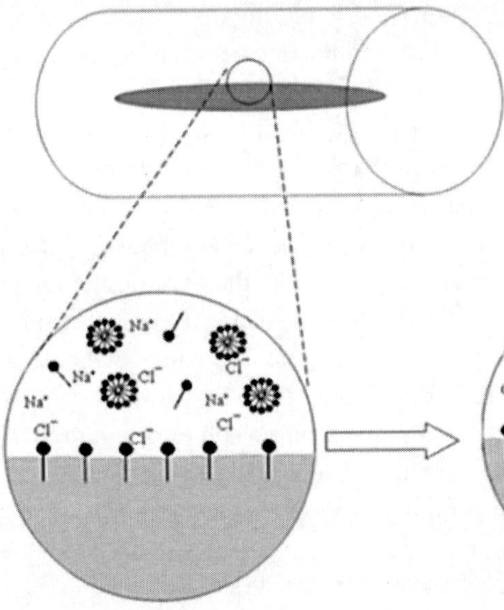

Figure 9. Schematic of the effect of salt ions on the IFT reduction of oil-water system [8].

In this framework, the results of research on the pyridinium and imidazolium SAILs showed that these SAILS can reduce the IFT of crude oil-natural brine more than crude oil-distilled water [8]. Specifically, SAILs have demonstrated significantly higher IFT reduction when NaCl salt was present at high concentrations. For instance, as can be seen in Table 2, IFT reduction with 1000 mg.dm^{-3} of NaCl, was about 44% higher than the salt-free condition [14]. Similar results have been obtained for the Gemini imidazolium SAIL of $[C_8im\text{-}C_4\text{-}imC_8][Br]_2$ in the presence of NaCl, $MgCl_2$, and their mixture within the concentration range of 0.1 to 1.5 mol.dm^{-3}. The IFT of crude oil-water system has experienced 23, 51, and 60% more reductions by using this SAIL when, respectively, monovalent and divalent salts as well as their mixtures been added [21].

Table 2. IFT values with different NaCl concentrations for Crude oil-water system in the presence of 1000 mg.dm^{-3} of SAILs at 298.15 K

NaCl Concentration	IFT (mN.m^{-1})		
Na Cl concentration (mg.dm^{-3})	$[BT][C_6H_{13}COO]$	$[CP][C_6H_{13}COO]$	$[3\text{-}HPA][CF_3COO]$
0	17.13	14.34	16.46
50	16.54	14.21	16.21
100	15.82	14.05	15.86
500	15.23	13.71	15.49
1000	14.67	13.42	15.32
5000	13.85	12.68	15.58
10000	13.43	11.45	13.54
50000	12.13	9.85	12.49
100000	9.85	8.57	10.81
200000	9.53	8.11	9.24
Maximum IFT reduction (%)	44.37	43.44	43.86

Wettability Alteration

Wettability relates to the interaction of the surface of rocks and present liquids in the reservoirs (crude oil or brine). This is known as one of the key parameters controlling the remaining OOIP. The wettability alteration in EOR, which corresponds to the converting the reservoir rock from oil-wet to water-wet significantly enhances the oil extraction.

Wettability can be assessed from the contact angle between the solid surface and the fluid as well as the three involved IFTs in the phase boundaries. The contact angles (θ) for different wettability conditions of the solid-water-oil system is schematically shown in Figure 10.

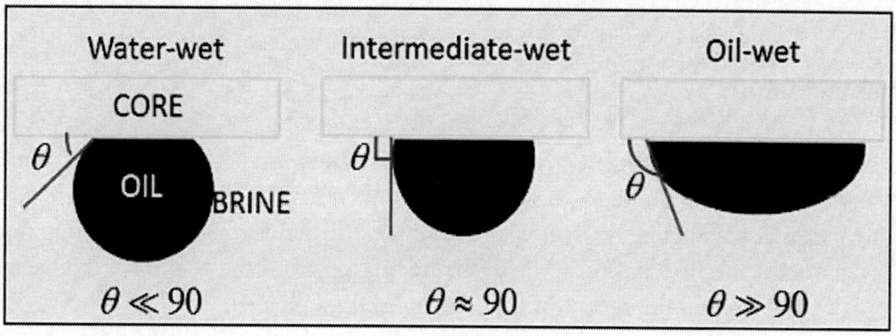

Figure 10. Schematic of different contact angles for different wettability conditions of the solid-water-oil system [22].

Solids with a contact angle of less than 90° are categorized as hydrophilic surface (water-wet), the contact angle of 90° indicates a moderate situation (intermediate-wet), and a contact angle of more than 90° represents a hydrophobic surface (oil-wet). It should be noted that for the oil drops formed in the aqueous phase and on contact with the solid surface, the external contact angle with respect to the aqueous phase is considered.

In the wettability alteration, adsorption of SAILs on the surface of reservoirs converts the oil-wet rocks to partial or completely water-wet. The proposed mechanism for this phenomenon is that upon adsorption of SAILs on the hydrophobic surface of the reservoir rocks, the alkyl chains are directed towards the rock surface while the polar parts are directed towards the solution. With this configuration, an oil-wet surface changes to water-wet. The charge density, structure and concentration of SAILs, the surface charge of the rocks and the nature of crude oil affect this phenomenon. For a better explanation, the effect of wettability alteration on the crude oil mobility and in contact with rock surface is schematically illustrated in Figure 11.

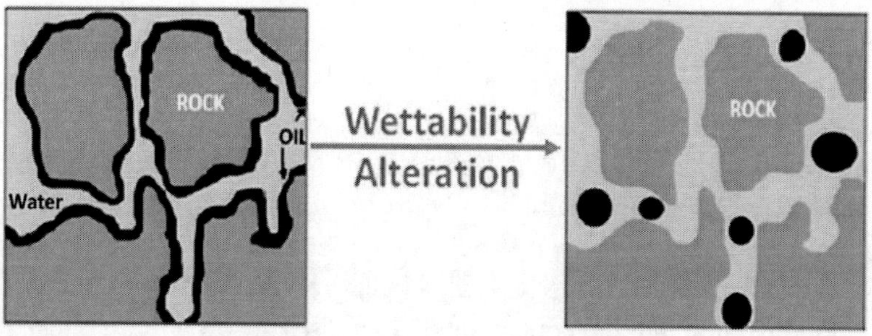

Figure 11. Schematics of wettability alteration for the mobility of the crude oil in reservoirs [23].

Relevantly, a change in the structure of SAILs significantly affects the wettability alteration. In this regard, the alkyl chain length seems to play an important role. Accordingly, studies on the effect of different imidazolium-based SAILs of [$C_n$mim][X] n = 4, 6 and 8, X = $[Cl]^-$, $[Br]^-$ and $[HSO_4]^-$, on the wettability of the quartz surface in contact with a surrogate oil (containing pentane, hexane, heptane, octane, decane, and dodecane) have revealed that SAILs with longer hydrocarbon chain change better the wettability from oil-wet to water-wet [24]. The effect of the SAIL head groups on wettability alteration of the carbonate surface has also been studied by measuring the contact angle of the oil drops in the presence of two long alkyl chain SAILs [C_{18}Py][Cl] and [C_{18}mim][Cl]. Results show that imidazolium-based SAILs can change wettability to water-wet better than the pyridinium-based ones [25].

Oil Displacement

Oil displacement from the surface of reservoir rocks, in the presence of SAILs, is an interesting phenomenon that is similar to the action of nanoparticles in the media; hence, it is not attainable with conventional surfactants. By decreasing the adhesion between the crude oil and the surface of reservoir rocks, the residual oil to be easily separated from the rocks and higher crude oil mobility is established in the reservoirs. The proposed mechanism for this observation is that upon adsorption of SAILs on both the surface of crude oil drops and the surface of rocks, a wedge film is formed in the three-phase boundaries of oil/aqueous/solid, which ultimately separates the crude oil from the solid surface and leads to oil displacement. For better illustration, the corresponding mechanism as well as the shape of the crude oil drops before and after oil displacement are presented in Figure 12.

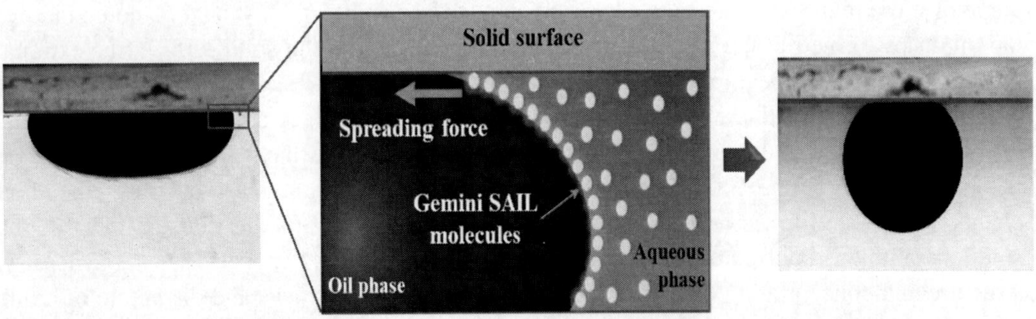

Figure 12. Schematic of the wedge-film formation at the three-phase boundaries that leads to oil displacement [10].

Formation of a SAIL layer on the oil drops and the rock surface and subsequently oil displacement can occur with long chain SAILs. In this regard, use of different alkyl chain length and effective concentration of imidazolium SAILs, [$C_n$mim][Cl], n = 2, 4, 6, 8, 10 and 12, and within the concentration range of 1000 - 8000 $mg{\cdot}dm^{-3}$, has provided oil displacements by using [C_{12}mim][Cl], [C_{10}mim][Cl] and high concentrations of [C_8mim][Cl]; however, this has not been detected for SAILs with shorted alkyl chain lengths (n = 2, 4, and 6) [26].

The oil displacement by using a long chain imidazolium Gemini SAIL [C_{12}im-C_4-imC_{12}]$[Br]_2$ demonstrated that the contact angle increases at low SAIL concentrations, but oil displacement occurred at high concentrations (more than $1.0{\times}10^{-3}$ $mol{\cdot}dm^{-3}$) [10]. Figure 13 shows the expansion of the oil drops on the surface of a solid in the presence of a long chain Gemini SAIL. High level contact angles with long chain SAILs, particularly at low concentrations, can be ascribed to the reduction in the IFT and disturbing the interface bonding, which results in spreading oil drops on the rock surface. Further increase in concentration leads to the diffusion of SAIL particles into the three-phase region of oil/water/solid phases (see Figure 12) and results in oil displacement.

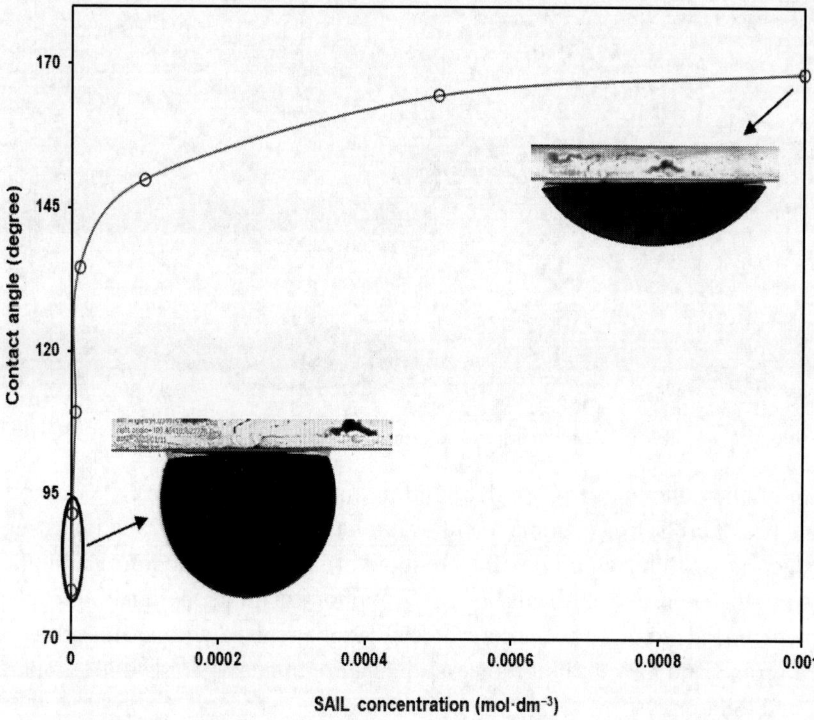

Figure 13. The expansion of oil drops on the solid surface in the presence of $[C_{12}im\text{-}C_4\text{-}imC_{12}][Br_2]$ long chain Gemini SAIL [10].

Impact of Reservoir Characteristics

Crude Oil

The nature of crude oil has a profound impact on EOR. The SARA composition percentage, API degree, viscosity, acidity, presence of the salts, sulfur or other impurities are the most important factors affecting the efficiency of SAILs in EOR. In this regard, the closer the nature of SAILs to crude oil, the greater their performance will be. For example, SAILs with basic nature are better for low acidic crude oils, and SAILs with aromatic rings in their structure are more effective for crude oils with high percentage of aromatics. Figure 14 shows a simplified diagram for selecting suitable SAILs based on the crude oils' characteristics.

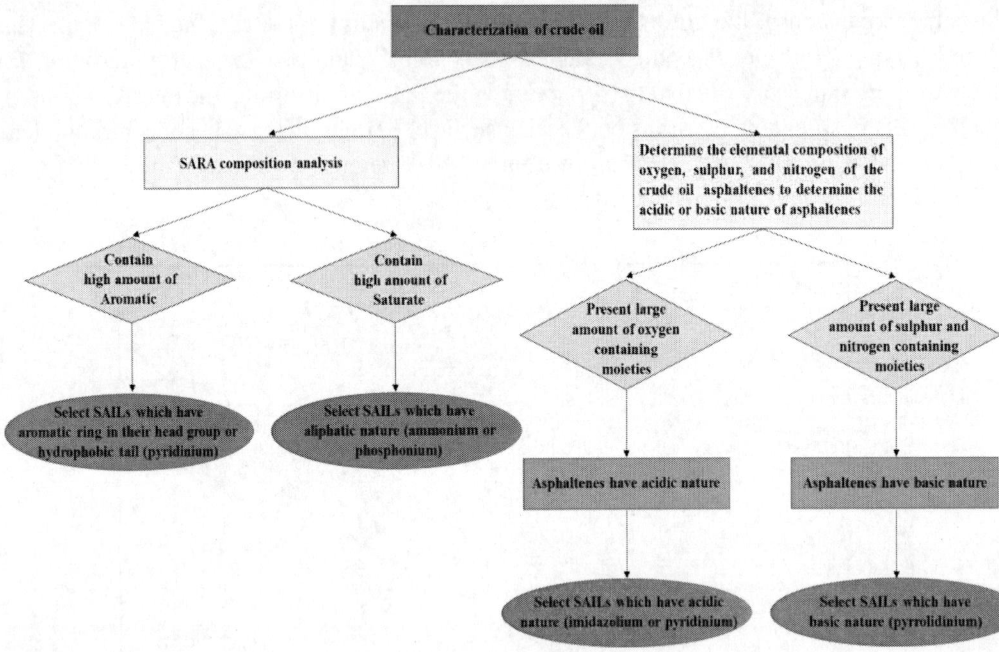

Figure 14. Selecting suitable SAILs based on the crude oil characteristics [27].

Relevantly, the potential of [C_8py][Cl] and [C_8mim][Cl] SAILs in IFT reduction of light, medium, and heavy crude oils in contact with water has been examined and it has been revealed that IFT reductions with heavy crude oil were more effective than medium and light ones [28]. This has been attributed to the affinity of SAIL monomers to asphaltene molecules in heavy crude oils, which leads to more reductions in IFT. Asphaltenes are polar molecules with acidic and alkaline groups and they exhibit rather acidic nature in heavy crude oils compared to those in medium and light ones.

Thus, imidazolium and pyridinium based SAILs of acidic nature, strongly interact with heavy crude oil asphaltenes, resulting in further reduction in the system IFT. In a rather fundamental study, by examining the effect of different systems of water and aromatic (benzene), alicyclic (cyclohexane), and aliphatic (hexane), all with the same number of carbon atoms, the profound effect of [C_4mim][DS], as an Anionic SAIL, has been elucidated. The higher IFT reduction with the aromatic hydrocarbon can be related to the π-π interaction between the aromatic ring of the benzene and the imidazolium SAIL [29]. Thus, SAILs with an aromatic ring in their head group interact effectively with the aromatic compounds in the crude oils and give significant IFT reduction.

Rock Surface

The interaction of SAILs with the reservoir rocks is another significant factor in evaluating the performance of SAILs. If SAILs were greatly adsorbed on rocks, significant amount of them would be lost during the surfactant flooding making the whole process economically unfeasible. The proposed mechanism for adsorption of SAILs on rocks is the electrostatic attraction between the charged rocks and SAIL molecules. Thus, screening SAILs for their retention in the reservoirs and the nature of rocks are essential for evaluating EOR processes.

In this regard, SAILs have been selected for different reservoirs. In general, Cationic SAILs are preferred for use in carbonate reservoirs due to the same surface charge of rocks and the SAILs. Also, Anionic SAILs work better in sandstone reservoirs due to the same surface charges. Noteworthy, Zwitterionic SAILs can be used in both cases.

For better assessment, lab scale core flooding experiments are usually performed with actual reservoir cores and with matching temperature and salinity conditions of the reservoirs. Small amounts of SAILs have been adsorbed on reservoir rocks in core flooding experiments, confirming a high oil recovery. Unfortunately, only a few studies have been devoted to this subject. Appropriately, a study on the adsorption of [C_{16}mim][Br] on the three alternating sand pack columns, confirmed the strong SAIL adsorption dependency on its concentration [30]. Another study on adsorption of imidazolium-based SAILs on carbonate rocks indicated low adsorptions (< 0.1 mg.m^{-2}) compared to those with pyridinium and pyrrolidinium-based ones [27].

Concerning salinity effect, the behavior of Ammoeng 102, confirmed that the higher media salinity led to higher adsorption of this Anionic SAIL on the sandstone rock [31]. Therefore, adsorption can sometimes be controlled by monitoring the salinity of the solutions. Alternatively, pH parameter can be monitored to find the best solution for the minimum adsorption of SAILs on the rocks.

Given the above points, characterizing the crude oils and rocks is a significant step in selecting the proper SAILs for EOR processes. This can be accomplished via screening, simulation, and modelling of reservoirs.

Comparing SAILs with Conventional Surfactants

In almost all cases, SAILs are more effective in reducing IFT of crude oil-water compared to conventional surfactants having the same alkyl chain lengths and counter anions. This is due to the bulkier head group or heteroatom ring structure, which give more hydrophobicity and weaker hydration tendency, assisting the migration of SAILs toward the interface. From a practical point of view, these are useful in EOR operations because of requiring low amounts of SAILs. Also, efficiency of SAILs can be improved by selecting suitable alkyl chains or counter anions and with large head groups, preferably nitrogen-containing heterocyclic rings with delocalized charges. For the aim of comparison, effects of eleven SAILs from three families (lactam, ammonium and hydroxylammonium) indicate that they can reduce the IFT better than the Tergitol 15-S-5, Tergitol 15-S-7, and Tergitol 15-S-9 non-ionic conventional surfactants [14, 32]. Relevantly, the maximum reduction in the crude oil-water IFT with [C_{16}mim][Br] and [C_{12}mim][Cl] have been reported to be 72%; however, this was about 68% for CTAB and DTMAC surfactants with similar SAIL structures but different head groups [33].

Noteworthy, unlike conventional surfactants that lose their efficiency under harsh temperature and salinity conditions, SAILs remain stable and can significantly reduce the IFT of crude oil and saline waters at high dominant temperatures in the reservoirs. Thus, as pointed in the above "surfactant flooding" section, SAILs are adaptable in using seawater for injecting into reservoirs, which is economically feasible.

The new types of SAILs exhibit more distinct capabilities for this purpose. As an example, the Anionic SAIL Ammoeng 102 has revealed a decreasing trend in IFT with increasing temperature from 20 to 90 °C and better performance compared to the commercial surfactant of Triton X-100 under similar conditions [19].

It is important to note that some conventional surfactants have been able to reduce the IFT to low values of 10^{-3} (sometimes to 10^{-4}) mN.m^{-1} irrespective of the involved phase systems, whereas using Cationic SAILs, orders of 10^{-2} mN.m^{-1} have been attainable [34]. Indeed, by tailoring specific SAIL structures (Zwitterionic, Catanionic, Multicationic, or Anionic), one can reach ultra-low IFT values. As an evidence, a study on the Catanionic SAIL of [$P_{666,14}$][(*i*-C_8)$_2PO_2$] showed high reductions in IFT compared to conventional surfactants for a surrogate oil and water chemical system [35]. Also, the Anionic SAIL of [C_4mim][DS] has effectively reduced the IFT of benzene-water system to 0.9 mN.m^{-1}, whereas, these values were, about 5 and 8.5 mN.m^{-1} respectively using SDBS and SDS conventional surfactants [29]. Adding to these, the long chain Multicationic SAILs have been able to reduce the IFT of crude oil-water system to ultra-low values [10].

Regarding the wettability alteration, SAILs can effectively change the wettability from oil-wet to water-wet. In this regard, comparison of the performance of imidazolium SAIL [C_{12}mim][Cl] with some conventional surfactants (AN-120, NX-610, NX-1510, NX-2760, and TR-880), by means of contact angle measurement, confirm that SAILs are superior in changing the wettability from oil-wet to water-wet [36]. Adding to these, oil displacement by SAILs, as was pointed above, is an interesting phenomenon that is similar to the action of nanoparticles and conventional surfactants may not provide this advantage.

Another remarkable point is that SAILs exhibit low adsorption on the charged reservoir rocks, which is in contrast to the conventional surfactants that tend to significantly adsorb on rock surface. In this regard, adsorption of long chain pyridinium, imidazolium, and pyrrolidinium SAILs has been compared with CTAB on the carbonate rocks and in the presence of monovalent and divalent salts (15 wt%). Results confirm that all the used SAILs are adsorbed much less than CTAB [27].

For better judgment, a comprehensive comparison on the potential applications of [C_{16}mim][Br] with the conventional analog CTAB surfactant for EOR operations is summarized in Table 3. Outcomes clearly indicate the suitability of the SAIL under similar reservoir conditions, and that the SAIL solution is much more effective.

Table 3. Performance of [C_{16}mim][Br] and CTAB in EOR operations under the same reservoir conditions [1]

Description	CTAB	[C_{16}mim][Br]
• IFT of crude oil-water (concentration 2000 mg.dm^{-3}, temperature 25°C, without additional salts)	8.4 mN.m^{-1}	5.7 mN.m^{-1}
• IFT of crude oil-brine (concentration 2000 mg.dm^{-3}, temperature 25°C, without additional salts)	7.7 mN.m^{-1}	3.8 mN.m^{-1}
• Additional oil recovery by surfactant flooding of sand pack column (concentration 3000 mg.dm^{-3}, temperature 25°C, salinity 3 wt%)	32.2%	71.3%
• Adsorption on carbonate rock (concentration 3500 mg.dm^{-3}, temperature 40°C, salinity 15 wt%)	25.14 mg.g^{-1} of rock	9.96 mg.g^{-1} of rock

Finally, it has to be mentioned that SAILs can be safely stored for long times without any decomposition and can be recycled several times. In a typical study, imidazolium SAILs have been recycled five times without remarkable loss in their efficiency even at high salinity levels [6]. Based on these results, SAILs are efficient materials for EOR processes.

Conclusion

The present chapter provides an overview on the properties of SAILs and highlights their effectiveness in the petroleum industry. The results clearly demonstrate that SAILs can be considered as good candidates in the EOR due to their high performance in IFT reduction, in wettability alteration and oil displacement.

In view of the SAIL structure, longer alkyl chain lengths in the cation or anion, and larger heterocyclic head groups with π-electron rings, as well as bulky counter anion with high polarizability, provide high interfacial activity. Besides, Catanionic, Multicationic, and Zwitterionic types usually exhibit higher interface activity compared to single analogous Cationic SAILs.

Operating conditions have to be considered too. SAILs remain stable under high temperature and salinity conditions. In most cases, temperature improves the IFT reduction; however, for some SAILs, this trend is reversed. In the presence of salts, the low hydration of SAILs, because of salting-out effect and reduction of electrostatic repulsion between the head groups, bring about high IFT reductions.

In the field of wettability alteration, the adsorption of SAILs on the surface of reservoirs changes the oil-wet rocks to water-wet. The charge density, structure, and concentration of SAILs affect this procedure and SAILs with longer hydrocarbon chain and bulkier size with higher concentrations become more efficient. Besides, displacement of the crude oil from the surface of rocks that provides higher crude oil mobility is a notable feature of the SAILs that is similar to the action of nanoparticles.

The nature of crude oils has a profound impact on the performance of SAILs. Thus, the appropriate SAILs must be selected based on the characteristics of the considered crude oil. Another important factor is the adsorption of SAILs on the surface of reservoir rocks prevents the economic feasibility of the EOR process. SAILs with different charges have different interactions with the rock surfaces. Accordingly, Cationic SAILs are preferred for carbonate reservoirs, while Anionic SAILs are suitable for sandstone reservoirs, and Nonionic or Zwitterionic SAILs can be used in both cases.

Comparison of SAILs with conventional surfactants indicates that most of the SAILs exhibit better interfacial activity compared to conventional surfactants with similar structures, due to presence of large cations or heteroatomic rings. Unlike the majority of conventional surfactants, which lose their effectiveness at high temperature and salinity conditions, SAILs retain their effectiveness and can be stored safely. Indeed, Zwitterionic, Catanionic, and Multicationic SAILs are more efficient. Comparison of the performance of SAILs with conventional surfactants confirm that long chain SAILs are superior for wettability alteration and oil displacement. Meanwhile, SAILs exhibit low adsorption on the reservoir rocks.

The content of this chapter highlights the SAILs benefits and shows that SAILs are good candidates for utilizing in petroleum industry. Certainly, before large scale usage, information must be provided, to ensure cost-efficiency. Consequently, it has been rationalized that SAILs can solve a variety of issues in the petroleum industry.

References

[1] Kharazi, M.; Saien, J.; Asadabadi, S. Review on amphiphilic ionic liquids as new surfactants: From Fundamentals to applications. *Top. Curr. Chem.* 2022, *380*, 1–44.

[2] Asadabadi, S.; Saien, J.; Khakizadeh, V. Interface adsorption and micelle formation of ionic liquid 1-hexyl-3-methylimidazolium chloride in the toluene+ water system. *J. Chem. Thermodyn.* 2013, *62*, 92–97.

[3] Kapustin, N. O.; Grushevenko, D. A. Global prospects of unconventional oil in the turbulent market: a long term outlook to 2040. *Oil Gas Sci. Technol.* 2018, *73*, 67–80.

[4] Cheraghian, G.; Hemmati, M.; Masihi, M.; Bazgir, S. An experimental investigation of the enhanced oil recovery and improved performance of drilling fluids using titanium dioxide and fumed silica nanoparticles. *J. Nanostructure Chem.* 2013, *3*, 1–9.

[5] Eltoum, H.; Yang, Y. L.; Hou, J. R. The effect of nanoparticles on reservoir wettability alteration: a critical review. *Petroleum Sci.* 2020, 1–18.

[6] Painter, P.; Williams, P.; Lupinsky, A. Recovery of bitumen from Utah tar sands using ionic liquids. *Energy Fuels* 2010, *24*, 5081–5088.

[7] Saien, J.; Kharazi, M.; Asadabadi, S. Adsorption behavior of short alkyl chain imidazolium ionic liquidsat n-butyl acetate+ water interface: experiments and modeling. *Iran. J. Chem. Eng.* 2015, *12*, 59–74.

[8] Zeinolabedini Hezave, A.; Dorostkar, S.; Ayatollahi, S.; Nabipour, M.; Hemmateenejad, B.Effect of different families (imidazolium and pyridinium) of ionic liquids-based surfactants on interfacial tension of water/crude oil system. *Fluid Phase Equilibr.* 2013, *360*, 139–145.

[9] Saien, J.; Kharazi, M.; Asadabadi, S. Adsorption behavior of long alkyl chain imidazolium ionic liquids at the n-butyl acetate+water interface. *J. Mol. Liq.* 2015, *212*, 58–62.

[10] Kharazi, M.; Saien, J.; Yarie, M.; Zolfigol, M. A. The superior effects of a long chain gemini ionic liquid on the interfacial tension, emulsification and oil displacement of crude oil-water. *J. Petroleum Sci. Eng.* 2020, *195*, 107543.

[11] Sakthivel, S.; Velusamy, S.; Gardas, R. L.; Sangwai, J. S. Use of aromatic ionic liquids in the reduction of surface phenomena of crude oil–water system and their synergism with brine. *Ind. Eng. Chem. Res.* 2015, *54*, 968–978.

[12] Sakthivel, S.; Velusamy, S.; Gardas, R. L;. Sangwai, J. S. Adsorption of aliphatic ionic liquids at low waxy crude oil–water interfaces and the effect of brine. *Colloid Surface A* 2015, *468*, 62–75.

[13] Saien, J.; Kharazi, M. A comparative study on the interface behavior of different counter anion long chain imidazolium ionic liquids. *J. Mol. Liq.* 2016, *220*, 136–41.

[14] Sakthivel, S.; Chhotaray, P. K.; Velusamy, S.; Gardas, R. L.; Sangwai, J. S. Synergistic effect of lactam, ammonium and hydroxyl ammonium based ionic liquids with and without NaCl on the surface phenomena of crude oil/water system. *Fluid Phase Equilibr.* 2015, *398*, 80–97.

[15] Rodríguez-Escontrela, I.; Rodríguez-Palmeiro, I.; Rodríguez, O.; Arce, A.; Soto, A. Characterization and phase behavior of the surfactant ionic liquid tributylmethylphosphonium dodecylsulfate for enhanced oil recovery. *Fluid Phase Equilibr.* 2016, *417*, 87–95.

[16] Jia, H.; Lian, P.; Liang, Y.; Zhu, Y.; Huang, P.; Wu, H.; Leng, X. Zhou, H. (2018). Systematic investigation of the effects of zwitterionic surface-active ionic liquids on the interfacial tension of a water/crude oil system and their application to enhance crude oil recovery. *Energy Fuels* 2018, *32*, 154–160.

[17] Kharazi, M., Saien, J., Yarie, M. Zolfigol, M. A. Different spacer homologs of gemini imidazolium ionic liquid surfactants at the interface of crude oil-water. *J. Mol. Liq.* 2019, *296*, 111748.

[18] Kharazi, M.; Saien, J.; Yarie, M.; Zolfigol, M. A. Promoting activity of gemini ionic liquids surfactant at the interface of crude oil-water. *Petroleum Res.* 2021, *31*, 20–22.

[19] Hanamertani, A. S.; Pilus, R. M.; Irawan, S. A Review on the application of ionic liquids for enhanced oil recovery. In *ICIPEG 2016*, edited by Springer, 2017, 133–147. Singapore.

[20] Saien, J.; Asadabadi, S. Salting out effects on adsorption and micellization of three imidazolium-based ionic liquids at liquid–liquid interface. *Colloid Surface A* 2014, *444*, 138–143.

[21] Saien, J.;Kharazi, M.; Yarie, M.; Zolfigol, M. A. A Systematic investigation of a surfactant type nano gemini ionic liquid and simultaneous abnormal salts effects on the crude oil/water interfacial tension. *Ind. Eng. Chem. Res.* 2019, *58*, 3583–94.

[22] Dashtaki, S. R. M.; Ali, J. A.; Manshad, A. K.; Nowrouzi, I.; Mohammadi, A. H.; Keshavarz, A. Experimental investigation of the effect of Vitagnus plant extract on enhanced oil recovery process using interfacial tension (IFT) reduction and wettability alteration mechanisms. *Journal Petrol. Explor.Prod. Tech.* 2020, *10*, 2895–2905.

[23] Shamsi Jazeyi, H.; Miller, C. A.; Wong, M. S.; Tour, J. M.; Verduzco, R. Polymer-coated nanoparticles for enhanced oil recovery. *J. Appl. Polym. Sci.* 2014, *131*, 1–13.

[24] Velusamy, S.; Sakthivel, S.; Sangwai, J. S. Effect of imidazolium-based ionic liquids on the interfacial tension of the alkane–water system and its influence on the wettability alteration of quartz under saline conditions through contact angle measurements. *Ind. Eng. Chem. Res.* 2017, *56*, 13521–13534.

[25] Manshad, A. K.; Rezaei, M.; Moradi, S.; Nowrouzi, I.; Mohammadi, A. H. Wettability alteration and interfacial tension (IFT) reduction in enhanced oil recovery (EOR) process by ionic liquid flooding. *J. Mol. Liq.* 2017, *248*, 153–162.

[26] Yahya, M.; Sangapalaarachchi, D. T.; Lau, E. Effects of carbon chain length of imidazolium-based ionic liquid in the interactions between heavy crude oil and sand particles for enhanced oil recovery. *J. Mol. Liq.* 2019, *274*, 285–292.

[27] Nandwani, S. K.; Malek, N. I.; Chakraborty, M.; Gupta, S. Insight into the application of surface-active ionic liquids in surfactant based enhanced oil recovery processes–a guide leading to research advances. *Energy Fuels* 2020, *34*, 6544–6557.

[28] Lashkarbolooki, M.; Ayatollahi, S. Investigation of ionic liquids based on pyridinium and imidazolium as interfacial tension reducer of crude oil− water and their synergism with $MgCl_2$. *J. Petrol. Sci. Eng.* 2018, 414–421.

[29] Zhou, H.; Zhu, Y.; Peng, T.; Song, Y.; An, J.; Leng, X.; Yi, Z.; Sun, Y.; Jia, H. Systematic study of the effects of novel halogen-free anionic surface active ionic liquid on interfacial tension of water/model oil system. *J. Mol. Liq.* 2016, *223*, 516–520.

[30] Nandwani, S. K.; Chakraborty, M.; Gupta, S. Chemical flooding with ionic liquid and nonionic surfactant mixture in artificially prepared carbonate cores: a diffusion controlled CFD simulation. *J. Petrol. Sci. Eng.* 2019, *173*, 835–843.

[31] Dahbag, M. B.; AlQuraishi, A.; Benzagouta, M. (Efficiency of ionic liquids for chemical enhanced oil recovery. *J. Petrol. Explor. Prod. Tech.* 2015, *5*, 353–361.

[32] Bera, A.; Ojha, K.; Mandal, A.; Kumar, T. Interfacial tension and phase behavior of surfactant-brine–oil system. *Colloid Surface A* 2011, *383*, 114–119.

[33] Saien, J.; Asadabadi, S. Alkyl chain length, counter anion and temperature effects on the interfacial activity of imidazolium ionic liquids: Comparison with structurally related surfactants. *Fluid Phase Equilibr.* 2015, *386*, 134–139.

[34] Bera, A.; Belhaj, H. Ionic liquids as alternatives of surfactants in enhanced oil recovery—a state-of-the-art review. *J. Mol. Liq.* 2016, *224*, 177–188.

[35] Rodríguez-Escontrela, I.; Rodríguez-Palmeiro, I.; Rodríguez, O.; Arce, A.; Soto, A. Phase behavior of the surfactant ionic liquid trihexyltetradecylphosphonium bis(2,4,4-trimethylpentyl)phosphinate with water and dodecane. *Colloid Surface A* 2015, *480*, 50–59.

[36] Nabipour, M.; Ayatollahi, S.; Keshavarz, P. Application of different novel and newly designed commercial ionic liquids and surfactants for more oil recovery from an Iranian oil field. *J. Mol. Liq.* 2017, *230*, 579–588.

Chapter 7

Nucleic Acids Stabilization with Ionic Liquids and Deep Eutectic Solvents

Jéssica S. Almeida
Emanuel V. Capela
Ana P. M. Tavares*
and Mara G. Freire†
CICECO – Aveiro Institute of Materials, Department of Chemistry,
University of Aveiro, Aveiro, Portugal

Abstract

The use of alternative solvents such as ionic liquids (ILs) and deep eutectic solvents (DES) in chemistry, chemical engineering, biochemistry, biotechnology, and biomedical applications has been increasing with remarkable improvements. Particularly, they present high relevance for biomolecules processing since their solubility, structure stability and biological activity can be significantly enhanced.

In this context, this chapter presents an overview on the use of ILs and DES as alternative solvents for the stabilization of nucleic acids given their relevance in research and biopharmaceuticals development. The most relevant IL- and DES-based formulations reported for nucleic acids are herein described, and the main molecular mechanisms and interactions underneath the increased stability promoted by these formulations are unveiled, while their key advantages for biotechnology and biomedical fields are highlighted.

In this chapter, readers from academies, research institutes and industries will gain useful information and deep knowledge on the designing of novel solvents for the processing of nucleic acids.

Keywords: alternative solvents, ionic liquids, deep eutectic solvents, stabilization, nucleic acids

---

* Corresponding Author's Email: aptavares@ua.pt.
† Corresponding Author's Email: maragfreire@ua.pt.

In: Properties and Applications of Ionic Liquids
Editors: Noureddine Abidi and Julia L. Shamshina
ISBN: 979-8-88697-635-9

Abbreviations

AA-ILs	Amino-acid-based Ionic Liquids
ABS	Aqueous Biphasic Systems
AgNPs	Silver Nanoparticles
ASOs	Antisense Oligonucleotides
aw	Water Activity
CAC	Critical Aggregation Concentration
CD	Circular Dichroism
COSMO-RS	Conductor-like Screening Model for Realistic Solvents
ct-DNA	Calf Thymus DNA
DESs	Deep Eutectic Solvents
DETMG	2,2-diethyl-1,1,3,3-tetramethylguanidinium acetate
DLS	Dynamic Lights Scattering
DMSO	Dimethyl Sulfoxide
dsDNA	Double Stranded DNA
EtG	Ethylene Glycol
FBS	Fetal Bovine Serum
FNAs	Framework Nucleic Acids
FTIR	Fourier-Transform Infrared Spectroscopy
GB-ILs	Good's Buffers Ionic Liquids
Gly	Glycerol
Gua-IL	Guanidinium tris(pentafluoroethyl)trifluorophosphate Ionic Liquid
HBA	Hydrogen Bond Acceptor
HBD	Hydrogen Bond Donor
HeLa	Cervical Cancer Cell line
HPLC	High Performance Liquid Chromatography
IBR	Integrated Biomarker Response
ILs	Ionic Liquids
ITC	Isothermal Titration Calorimetry
Kb	Binding Constant
LC50	Median Lethal Concentration
MD	Molecular Dynamic
MDA	Malondialdehyde
miRNA	Micro RNA
ncRNA	Noncoding RNA
NMe_3	Trimethylamine
NMR	Nuclear Magnetic Resonance
OH	Hydroxyl
PBS	Phosphate Buffer Solution
PCR	Polymerase Chain Reactions
PD	Propanediol
pGFP	Green Fluorescent protein plasmid
PI	Propidium Iodine

PPG	Polypropylene Glycol
RMSD	Root Mean Square Deviation
ROS	Reactive Oxygen Species
RT	Room Temperature
sDNA	DNA sodium salt
SEM	Scanning Electron Microscopy
siRNAs	Small Interfering RNAs
SOD	Superoxide Dismutase
sRNA	Small RNA
ssDNA	Single Stranded DNA
STM	Scanning Tunnelling Microscopy
TBABr	Tetrabutyl Bromide
Tm	Melting Temperature
TMA	Tetramethyl Ammonium
TMG	1,1,3,3- tetramethylguanidinium acetate
UVRR	Ultraviolet Resonance Raman
UV-Vis	Ultraviolet-Visible
VOCs	Volatile Organic Compounds
[Amim]Cl	1-allyl-3-methylimidazolium chloride
[Arg]	Arginine
[Asp]	Aspartic Acid
[BF_4]	Tetrafluoroborate
[C_2mim][Ac]	1-ethyl-3-methyl imidazolium acetate
[C_2mim]Cl	1-ethyl-3-methylimidazolium chloride
[C_2mim][Lac]	1-ethyl-3-methylimidazolium lactate
[C_2NH_2mim][BF_4]	1-aminoethyl-3-methylimidazolium tetrafluoroborate
[C_4mim][Ac]	1-butyl-3-methyl imidazolium acetate
[C_4mim]Br	1-butyl-3-methylimidazolium bromide
[C_4mim]Cl	1-butyl-3-methylimidazolium chloride
[C_4mim][Lac]	1-butyl-3-methyl imidazolium lactate
[C_4mim][$MeSO_4$]	1-butyl-3-methylimidazolium methyl sulfate
[C_4mim][NO_3]	1-butyl-3-methyl imidazolium nitrate
[C_4mim][PF_6]	1-butyl-3-methyl imidazolium hexafluorophosphate
[C_4mim][TfO]	1-butyl-3-methylimidazolium triflate
[C_6mim][BF_4]	1-hexyl-3-methylimidazolium tetrafluoroborate
[C_6mim]Br	1-hexyl-3-methylimidazolium bromide
[C_6mim][NO_3]	1-hexyl-3-methylimidazolium nitrate
[C_{10}mim]Cl	1-decyl-3-methylimidazolium chloride
[C_{12}mim]Cl	1-dodecyl-3-methylimidazolium chloride
[Ch][Ac]	Cholinium Acetate
[Ch]Br	Cholinium Bromide
[Ch]Cl	Cholinium Chloride
[Ch][DHP]	Cholinium Dihydrogen Phosphate
[Ch][For]	Cholinium Formate
[Ch][Gly]	Cholinium Glycine
[Ch][HEPES]	Cholinium 2-[4-(2-hydroxyethyl)piperazin-1-yl]ethanesulfonate

[Ch][IBA]	Cholinium Indole-3-Butyrate
[Ch][IAA]	Cholinium Indole-3-Acetate
[Ch][Lac]	Cholinium Lactate
[Ch][Lev]	Cholinium Levulinate
[Ch][MES]	Cholinium 2-(N-morpholino)ethanesulfonate
[Ch][NO_3]	Cholinium Nitrate
[Ch][Oct]	Cholinium Octanoate
[Ch][Pyr]	Cholinium Pyruvate
[Ch][TES]	Cholinium 2-[(2-hydroxy-1,1-bis(hydroxymethyl)ethyl)amino]ethanesulfonate
[Ch][Tricine]	Cholinium N-[tris(hydroxymethyl)methyl]glycinate
[$C_n$bim]Br	1-n-alkyl-3-butylimidazolium bromide
[$C_n$mim]Br	1-n-alkyl-3-methylimidazolium bromide
[Co(OIM)$_4$ $^{2+}$][Asp^-]$_2$	Tetra(1-octylimidazole)cobaltate- (II) di(L-aspartate)
[Co(OIM)$_4$ $^{2+}$][Glu^-]$_2$	Tetra(1-octylimidazole)cobaltate(II) di(L-glutamate)
[Glu]	Glutamic Acid
[HOmim][BF_4]	1-hydroxyethyl-3-methylimidazolium tetrafluoroborate
[Lys]	Lysine
[mim]Cl	1-methylimidazolium chloride
[mim][TfO]	1-methylimidazolium triflate
[MOC_2mim][BF_4]	1-methoxyethyl-3-methylimidazolium tetrafluoroborate
[N_{4444}]Br	Tetrabutylammonium Bromide
[P_{4444}]Br	Tetrabutylphosphonium Bromide
[PF_6]	Hexafluorophosphate

Introduction

In recent years, various solvents have been used for numerous processes in academia and industries [1]. Nevertheless, because of new environmental regulations, the challenges of using non-harmful and novel solvents have prompted a sustainable development of innovative products to address environmental issues [1–4]. In this regard, ionic liquids (ILs) have emerged as a novel class of alternative solvents. ILs, if properly designed, are able to reduce both economic costs and environmental pollution, when compared to the most traditional solvents, such as volatile organic compounds (VOCs) [1, 5–7].

ILs consist exclusively or almost exclusively of ions and they are typically formed by organic cations combined with organic or inorganic anions (Figure 1) that remain liquid at temperatures lower than 100°C [6]. As a new generation of non-aqueous solvents, multiple ILs have many favorable properties such as near-zero vapor pressure, a wide liquid range, low flammability, high ionic and thermal conductivity, high dissolution capability towards many substrates, high thermal and chemical stability, and a wide electrochemical potential window [6, 8, 9]. They are known as designer solvents due to the different combinations of anions and cations that can be made, hence they can be specifically designed for different purposes and applications [6, 8, 9]. Over the past two decades, ILs have gained enormous attention as reaction and stabilization media for biomolecules. However, their technical applicability and 'greenness' are sometimes limited [10, 11].

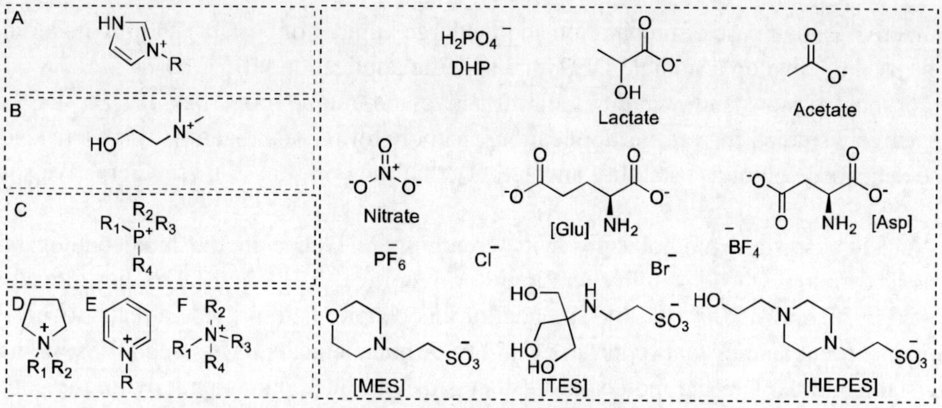

Figure 1. Representation of ILs cations (left) and anions (right) more commonly reported in the literature and in this chapter. Imidazolium (A), cholinium (B) and phosphonium (C) cations represent the classes of ILs more discussed herein. Pyrrolidinium (D), pyridinium (E) and tetraalkylammonium (F) cations are also broadly used in the synthesis of ILs.

Deep eutectic solvents (DESs) are also part of the novel class of alternative solvents and can be as attractive as ILs. In addition to sharing many of the desirable properties of ILs mentioned above, if properly designed, DESs may also present lower costs and enhanced biodegradability, biocompatibility and sustainability [10, 12–14]. DESs are eutectics, homogeneous mixtures of at least two compounds (Figure 2) that have a lower melting point than each of its individual compounds, usually prepared by mixing a hydrogen bond acceptor (HBA) and a hydrogen bond donor (HBD) with simultaneous heating and stirring until a liquid is formed [15, 16].

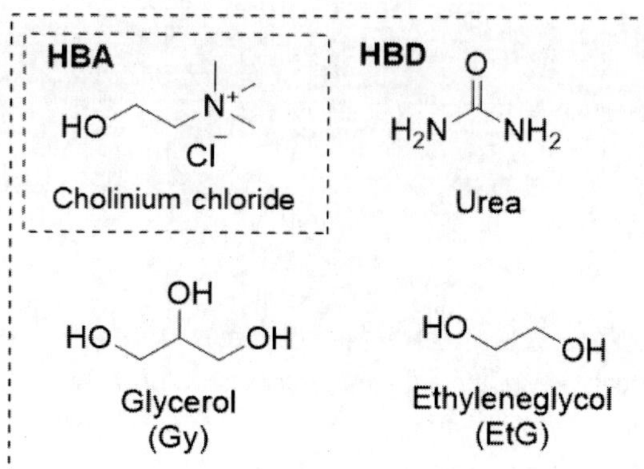

Figure 2. Representation of DESs main components reported in this chapter. As hydrogen bond acceptor (HBA) there's cholinium chloride ([Ch]Cl), and as hydrogen bond donors (HBDs) there are urea, glycerol (Gly) and ethylene glycol (EtG).

The DES remains a liquid at ambient temperature due to the formation of intermolecular hydrogen bonds, leading to a decreased freezing point [17]. Moreover, DESs play an important role in nature since they support the fact that, if properly prepared, most DESs can be nontoxic, nonvolatile, nonflammable and biodegradable liquids [18]. Another reason that makes DESs so attractive is their ease of preparation and the large number of possible starting materials that can be used to develop a suitable DES for a specific application [10].

For these reasons and owing to their attractive and unique properties, ILs and DESs have been actively studied for various applications, namely for the stabilization of nucleic acids due to the enhanced solubility, stability and activity that they present in these alternative solvents [19].

Nucleic acids are biomolecules with emerging relevance in the biotechnological and biomedical fields, whose stability and long-term storage in ILs and DESs has already been reported in several studies [8, 20]. The goal of this chapter is to provide a recent overview of important recent studies that report the stability of nucleic acids in ILs and DESs, in order to understand and explain the molecular mechanisms and interactions underlying the enhanced solubility and stability of this type of biomolecules in these novel and alternative solvents.

Information about the ILs and DESs-based formulations that best stabilize such molecules and their several types of structures, and which are the preferable conditions for this stabilization to occur is also provided. Furthermore, this chapter highlights key advantages of these biomolecules in ILs and DESs-based formulations in the fields of biotechnology and biomedicine. The sustainability, biodegradability, biocompatibility, ecotoxicity and cytotoxicity concerns are also addressed and discussed along this work, since they are very relevant features for the green credentials of such formulations. A schematic representation of the unique characteristics and main properties of ILs and DESs as alternative green solvents for the stabilization of nucleic acids is presented in Figure 3.

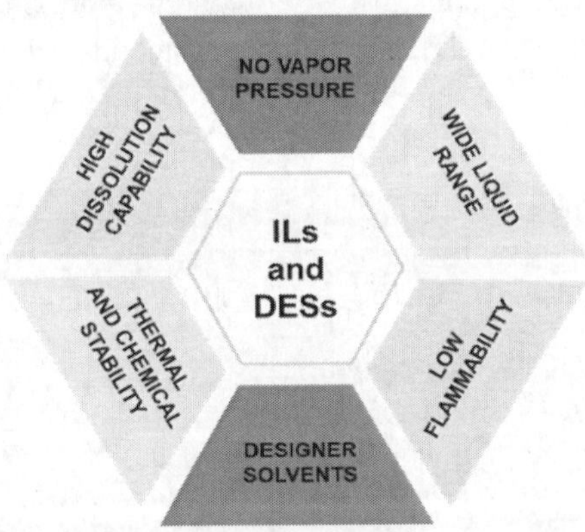

Figure 3. Schematic representation of the unique characteristics of ILs and DESs as alternative green solvents for the stabilization of nucleic acids.

Stability of Nucleic Acids in Alternative Solvents

Nucleic Acids

Genetic information and the blueprint of life are stored in the form of nucleic acids [21], macromolecules that contain a negatively charged sugar-phosphate backbone, and their bases possess aromatic rings, linked by hydrogen bonds [22]. The primary sequence of DNA, read from the canonical double helix, provides the code for RNA and protein synthesis. However, these already-information-rich molecules have several structural motifs found in biological systems that play critical roles in transcription, translation, nuclear organization and gene expression regulation [21, 22]. Due to their sequence selectivity and conformational polymorphism, nucleic acids emerge as powerful biological tools for biotechnological and therapeutic applications [21, 22].

DNA has several structural motifs, such as triplexes, rare DNA structures, formed when a third strand binds with sequence specificity to A*T and G*C Hoogsteen base pairs (the asterisk indicates the Hoogsteen base pair) in the major groove of a Watson–Crick base-paired DNA duplex [23]. Some proteins have evolved to recognize only one type of base pair, and they use intermolecular interactions to shift the equilibrium between Watson–Crick and Hoogsteen geometries [23]. Guanine (G)-rich sequences of DNA and RNA can associate into G-quartets, stabilized by Hoogsteen base pairing. Stacks of two or more of these G-quartets lead to a higher-order nucleic acid structural motif called a G-quadruplex [21]. These structures are highly abundant in regions of the human genome, like telomeres and oncogene promotor regions, suggesting a role in biology and therapeutics [21].

G-quadruplexes are also predominant in various aptamer structures, suggesting that they play vital stabilizing roles, emerging in biosensor and diagnostic applications with significant potential in various disciplines [21]. Similar to G-quadruplexes, C-rich regions form an intercalated quadruple-helical tetramolecular structure with loop regions called i-motifs, that can be found at promoter regions, telomeres, and centromeres [21]. Creative approaches are being pursued to incorporate i-motifs in cancer therapeutics. However, these i-motifs pursue a much greater role in the design of nanotechnological systems for analytical and biomedical purposes, relying on the structural transition of i-motif sequences as a result of pH changes [21].

RNA is well known for its crucial role in protein translation, but it also plays equally important roles in many other biological processes [21]. The commonly called "noncoding RNA" (ncRNA), protects and maintains the genome, processes RNA transcripts, is used as a defense mechanism against foreign DNA, exhibits enzymatic functions (i.e., ribozymes), and acts as structural element that organizes nuclear chromatin for gene expression [21]. Antisense oligonucleotides (ASOs), small interfering RNAs (siRNAs) and micro RNA (miRNA) are ncRNAs that can be used in the inhibition of gene expression and alter splicing to produce functional proteins, allowing the facile construction of potent multifunctional drugs and therapies that could substitute the drug cocktails in the future, and provide treatment for previously untreatable diseases, ranging from cancer to pandemic influenza or even to Alzheimer's disease [24, 25]. Beyond its therapeutic applications of targeting gene expression, RNA has also garnered great traction in the biotechnology space as a controllable nanomaterial for probes and tools [21].

Nucleic acids with short-term and long-term stability and biological activity are required for these applications [22]. Conventionally, nucleic acids are stored under refrigeration, and although they are reasonably stable in aqueous media, non-physiological pH, temperature, ionic strength and contaminating nucleases denature and disrupt the DNA helix [22]. Short-term storage is usually accomplished using high purity water or sterile phosphate buffered saline (PBS) solutions that need to be at temperatures below 4ºC [20], otherwise small volumes of water vaporize under open air conditions or at high temperature [22]. Long-term storage requires lyophilization and temperatures below -80ºC, or highly purified 100% formamide, which limits the subsequent applications of these nucleic acids and raise some toxicity and biocompatibility concerns [20]. Moreover, the energy costs associated with sample freezing makes the conventional storage strategies quite expensive both environmentally and economically. Most of the reactions involving nucleic acids as hybrid catalysts are usually carried out in aqueous solutions. However, a large number of organic substrates do not dissolve in aqueous medium, leading to long reaction times and low catalytic efficiencies [19].

Although it is often necessary to use partially or fully nonaqueous environments, DNA is found to lose structural integrity in organic solvents such as dimethyl sulfoxide (DMSO), formamide or methanol [19]. Therefore, it is extremely relevant to develop simple and sustainable strategies and protocols to maintain or increase nucleic acids stability and biological activity, ideally using mild temperatures and non-volatile, more benign and environmentally preferable solvents [19]. As presented in Figure 4, if suitably prepared, ILs and DESs have been emerged as relevant alternative solvents for nucleic acids, due to their attractive properties and interactions that they can potentially solve the drawbacks of the conventional media and traditional methods to stabilize and storage nucleic acids.

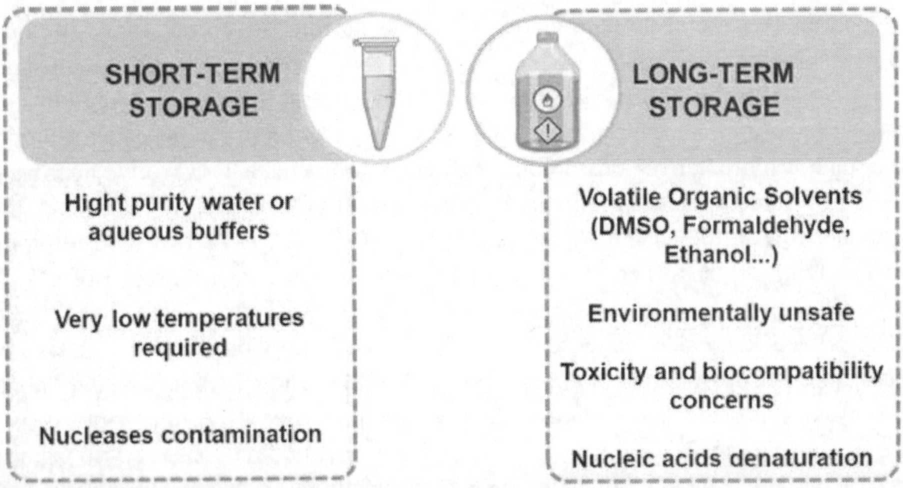

Figure 4. Schematic representation of the most common drawbacks of conventional methods used for the stabilization and storage of nucleic acids.

Hydrated ILs

Imidazolium-Based ILs

Several studies reporting imidazolium-based ILs combined with different anions that stabilize and interact with nucleic acids are addressed in this section [26–29]. Imidazolium-based ILs were pioneer ILs, abundantly reported in the literature and one of the leading classes of ILs to be investigated as stabilizers of biomolecules [1, 30]. For instance, Wang et al. [26] examined the binding of calf thymus DNA (ct-DNA) to 1-n-alkyl-3-methylimidazolium bromide family ILs ([$C_n$mim]Br (n=4, 6, 8, 10, and 12) in aqueous solutions using pyrene as the fluorescence probe, and reported that the binding Gibbs energy decreases linearly with the increase of the alkyl chain length of cations and the binding constants increase with the alkyl chain length. They further suggested that the main contributor to the DNA-IL interaction are the electrostatic interactions, although the hydrophobic interactions also increase with the alkyl chain length [26].

In a similar context, Jumbri et al. [27] reported the nature and mode of binding of 1-n-alkyl-3-butylimidazolium bromide ILs ([Cnbim]Br (n = 2, 4, 6) with ct-DNA [27]. The experiments were performed using a fixed 0.5 M IL solution at 25ºC [27]. The study predicted by Conductor-like Screening Model for Realistic Solvents (COSMO-RS), indicated that the smaller sized cation molecules with shorter hydrocarbon chain lengths could enter the DNA grooves to create a more hydrophobic microenvironment between the base pairs [27]. In addition, the docking studies revealed that the cation of ILs preferably binded to the minor groove of DNA [27]. Moreover, the change in alkyl chain lengths clearly influenced the binding energy between the ILs and ct-DNA.

The value of binding energy decreased as the length of the alkyl chain increased [27], as previously addressed by Wang et al. [26]. Such interaction was the consequence of the electrostatic attraction between the ring of imidazole and the backbone of DNA [27]. Aside from electrostatic attraction, the occurrence of T-shape π-π stacking between the bases and the aromatic ring of imidazole was also observed in the stable docking conformation [27]. Preferable binding of smaller cations of [$C_n$bim]Br ILs to the minor groove of ct-DNA changed the microenvironment to be more hydrophobic by expelling water from immediate contact with DNA [27].

This dehydration effect can extend the lifetime of ct-DNA by circumventing a hydrolytic attack that instigates rapid deamination and depurination while increasing the thermal stability of the base pairs [27]. Imidazolium-based ILs were further investigated by Rossi et al. [28], where DNA sodium salt from salmon testes (sDNA) was added to 62 mM of the ILs 1-methylimidazolium chloride ([mim]Cl), 1-ethyl-3-methylimidazolium chloride ([C_2mim]Cl), 1-butyl-3-methylimidazolium chloride ([C_4mim]Cl), and 1-butyl-3-methylimidazolium bromide ([C_4mim]Br) solutions and [TRIS] buffer. Multi-wavelength Ultra-violet Resonance Raman (UVRR) spectroscopy was used for selectively detecting heat-induced structural transitions of DNA localized on specific base tracts [28]. This study reported the establishment of preferential interactions between the imidazole cations of ILs and the guanine bases in the DNA groove that led to more effective stacking between the guanine bases even at high temperatures (highest temperature registered was around 70ºC) [28]. The authors also observed that the trend for ILs sharing the same chloride anion was further enhanced as the alkyl chain on the imidazolium cation gets shorter [28].

A study conducted by Bottari et al. [29] investigated the molecular interactions and the effect of the imidazolium-based ILs 1-butyl-3-methylimidazolium triflate ([C_4mim][TfO]), 1-methylimidazolium triflate ([mim][TfO]) and 1-butyl-3-methylimidazolium methyl sulfate ([C_4mim][$MeSO_4$]) on the thermal unfolding process of sDNA UVRR and Circular Dichroism (CD) spectrometry analysis revealed that both the cation and anion of the ILs strongly interact with the structure of dsDNA, thus affecting the melting process but not perturbing the pre-melting transition that precedes the complete separation of the strands of DNA [29]. Overall, the results suggested that the dominant interaction occurred between the imidazole cation and the bases of guanine and thymine in the structure of DNA [29].

Toxicity Concerns of Imidazolium-Based ILs

The attractive properties of ILs, namely the negligible vapor pressure, classifies these compounds as green solvents, being unharmful air pollutants [1]. However, the idea that all ILs are completely green solvents has been refuted by some studies [31]. Due to their significant solubility in water, ILs could become as persistent pollutants in wastewater and endangered environmental systems [1].

In this context, Pham et al. [32] have reviewed the toxic effect of ILs on the environment and biological systems in a comprehensive way. Literature reports that ILs with imidazolium, pyridinium and pyrrolidinium cations with different anions (e.g., chloride, bromide, nitrate, fluorine) cause significant negative effects on fishes [32]. Furthermore, the literature also reveals that increased alkyl chain length in the cation of ILs showed higher toxicities on biological systems [31, 33–35]. It is suggested that the lipophilicity of ILs, which is increased by higher alkyl chain, dominates uptake into the cells, raising the toxicity [31]. Even though this is not a feature shared by all the ILs, it is a strong feature of imidazolium- and phosphonium-based ILs [35].

Recent studies (2020 and 2021) were gathered to corroborate these results [36–41]. For instance, the cytotoxicity of [$C_n$mim] (n=2, 6, 8 and 10) was investigated by Tarek M. et al. [36] in human hepatocytes. In agreement with the previous studies [31, 33–35], the authors observed that longer chain methylimidazolium ILs were toxic to cells since they readily integrated within the inner mitochondrial membrane, interfering with the electron chain leading to oxidative stress. In addition, oxygen consumption was inhibited, caspase expression was induced and DNA was fragmentation was increased with the increasing of IL alkyl chain length [36]. The cytotoxicity of 1-decyl-3-methylimidazolium chloride ([C_{10}mim]Cl) in a human cervical cancer cell line (HeLa) was also investigated by LX et al. [37], which led to increased reactive oxygen species (ROS) production, decreased membrane potential, induced cell apoptosis and cell cycle arrest. The phytotoxicity of the IL 1-allyl-3-methylimidazolium chloride ([Amim]Cl) and the ILs 1-hexyl-3-methylimidazolium bromide ([C_6mim]Br), 1-hexyl-3-methylimidazolium nitrate ([C_6mim][NO_3]) and 1-hexyl-3-methylimidazolium tetrafluoroborate ([C_6mim][BF4]) against maize and *Vicia faba* seedling was assessed by Xiaohong et al. [38] and Yaqi et al. [39], respectively.

Results demonstrated the IL [Amim]Cl induced the inhibition of maize seedling growth, which was associated with changes in the gene expression profiles of senescence-related hormones as well as inactivation of growth-promoting hormones metabolic pathways [38]. Yaki et al. [39] calculated the integrated biomarker response (IBR) to compare the toxicity of the three ILs on *Vicia faba* and the toxicity order was: [C_6mim][NO_3]$<$[C_6mim]Br$<$ [C_6mim][BF_4]. These ILs led to the generation of ROS and then caused oxidative damage, including DNA damage, which triggered an increase in antioxidant content and enzyme activity [39].

Moreover, *in vivo* studies about the toxicity of ILs was reported by Guilherme et al. [40], that evaluated the ecotoxicity of ILs exposed to *Daphnia magna*, and Wenxiu et al. [41] investigated the toxicity in zebrafish. Guilherme et al. [40] highlighted, once again, the fact that the IL with longer alkyl chain, 1-dodecyl-3-methylimidazolium chloride ([C_{12}mim]Cl), had a more toxic effect towards this organism, including cellular membrane and cytoskeleton damage, oxidative stress, inhibition of antioxidant enzymes, mitochondrial impairment, changes in protein biosynthesis and energy production, and DNA damage.

Wenxiu et al. [41] determined the lethal 96h median lethal concentration (96h LC50) of the three ILs 1-aminoethyl-3-methylimidazolium tetrafluoroborate ([C_2NH_2mim][BF_4]), 1-methoxyethyl-3-methylimidazolium tetrafluoroborate ([MOC_2mim][BF_4]) and 1-hydroxy-ethyl-3-methylimidazolium tetrafluoroborate ([HOmim][BF_4]), which were 143.8 mg/L, 2492.5 mg/L and 3086.7 mg/L, respectively. During the experiment, the authors reported increased contents of ROS and Malondialdehyde (MDA) detected; enzymatic activities especially superoxide dismutase (SOD) were inhibited; and DNA damage occurred in zebrafish [41].

Despite the good results concerning the stability of DNA in the imidazolium-based ILs investigated [26–29], and based on the previously mentioned concerns on ILs toxicity, new formulations with more benign and biocompatible ILs have emerged [21]. A relevant example is cholinium-based ILs, biocompatible ILs that are potential candidates to overcome the drawbacks of ILs toxicity [25]. Nonetheless, toxic concentrations of cholinium-based ILs should also be taken in consideration, as reported by Guilherme et al. [40], that reported the ecotoxicity of cholinium chloride ([Ch]Cl) in *Daphnia magna*.

Cholinium-Based ILs: Introduction of More Biocompatible Formulations of ILs

McFarlane and coworkers [42] investigated the behavior of long DNA duplexes from salmon solubilized in the hydrated ILs cholinium lactate ([Ch][Lac]), cholinium dihydrogen phosphate ([Ch][DHP]) and cholinium nitrate ([Ch][NO_3]). CD spectra were recorded for DNA dissolved in these three ILs and in aqueous solution controls, stored at different temperatures (room temperature (RT) and 4ºC) and time intervals (1 week, 1 month and 6 months). It was observed that the DNA duplex formed a B-form in the presence of 80 wt% of IL, and no structural and chemical degradation of the DNA double-helix was observed for 6 months of storage for all the ILs studied. On the other hand, DNA in an aqueous solution was substantially denatured after 1 month under RT conditions [42].

Fluorescence spectroscopy and differential scanning calorimetry analyses were also carried by the authors [42] in order to evaluate the stability of DNA (already stored for 1 year at RT) dissolved in 50 wt% [Ch][DHP] at high temperatures. Results show that no denaturation was observed below 100ºC, while an aqueous sample would be substantially denatured by this excursion, suggesting that the DNA sample is stabilized at higher temperatures by the hydrated IL medium [42]. To determine if any degradation was taking place during storage, gel electrophoresis studies were carried out in DNA samples stored in buffer and in the hydrated ILs stored at RT for 6 months, but no degradation was observed in the DNA/IL samples. Overall, DNA was found to be soluble and exhibits exceptional long-term stability in hydrated ILs based on [Ch][Lac], [Ch][NO_3], or [Ch][DHP] [42].

These findings on the enhanced storage of nucleic acids at RT for long periods of time in ILs is a major progress, since it solves some of the drawbacks associated to conventional sample freezing. In addition to the fact that there is no risk of damaging the sample with repeated freeze-thaw, the energy costs associated with refrigeration are no longer required, and the transport of these samples becomes easier since the transport of refrigerated samples is usually quite complicated. The enhanced stability offered by cholinium-based ILs opened the door for several applications of nucleic acids in the biotechnological and biomedical field.

A relevant example is a study conducted by Morimitsu et al. [43], where an improved stability of DNA solid films was attained with the introduction of 10-40 wt% of the IL [Ch][DHP]. The addition of the IL enhanced several parameters such as, water retentivity of the films at RT, packing density of DNA helical chains, suppression of conformational transitions and cooperative motion of the DNA duplex in the film [43]. Moreover, when the amount of hydrated [Ch][DHP] in the film increased, the mechanical response of the film changed from glassy-like to rubbery-like via a semicrystalline-like state, which is useful for the design of further functional and sustainable devices using DNA from natural sources instead of synthetic polymers [43].

A study by EEL et al. [44] represents how the IL cholinium octanoate ([Ch][Oct]) was able to maintain the framework nucleic acids (FNAs) integrity under RT storage conditions for 10 days [44]. The FNAs stored in 50% IL/buffer exhibit hydrodynamic radios much closer to the freshly prepared stock compared with those stored in the TM buffer (50 mM Tris HCl, 10 mM magnesium sulfate at pH 7.5) [44]. Gel electrophoresis was subsequently performed to confirm retention of FNA integrity. The band appearing for the freshly made stock, appeared in both the TM buffer and in the IL, although a tighter band was shown in the IL samples, confirming the stability of the FNA structures in [Ch][Oct] and hinting at its superiority over the conventional storage buffer [44].

Finally, the FNAs in IL [Ch][Oct] formulation were successfully delivered, intact, *ex vivo* on porcine skin, to the lower dermis region, paving the way for future progress in the area of transdermal delivery of nanoscale entities [44]. Another study that illustrates the enhanced stability of nucleic acids in cholinium-based ILs was reported by PR et al. [45] which investigated the ILs [Ch][Lac], cholinium levulinate ([Ch][Lev]) and [Ch][DHP] ability to fix ocular tissues as alternative to VOCs. The three ILs at 20 wt% were able to fix scleral tissues and in contrast with the tissues fixed with formalin, while maintaining the integrity and purity of DNA, as revealed by gel electrophoresis and absorbance measurements [43]. Moreover, [Ch][DHP]-fixed tissue resulted in relatively higher yield of DNA, compared to the other ILs [43].

In order to compare and understand the mechanisms underlying the stabilization of nucleic acids in imidazolium and cholinium-based ILs, Chandran et al. [46] carried out molecular dynamic (MD) simulations and spectroscopic experiments (CD, ultraviolet-visible (UV-Vis) and fluorescence) to evaluate the stability of the model DNA (Dickerson−Drew dodecamer) in several hydrated ILs. They confirmed that the DNA maintained the B-conformation in the hydrated ILs [C_4mim]Cl, 1-butyl-3-methyl imidazolium nitrate ([C_4mim][NO_3]), 1-butyl-3-methyl imidazolium 1-butyl-3-methyl imidazolium lactate ([C_4mim][Lac]), [Ch][NO_3] and [Ch][Lac] at concentrations up to 80 wt% of each IL.

Results demonstrated that [C_4mim]Cl and [C_4mim][NO_3] bind to DNA more strongly than [C_4mim][Lac]. In the presence of 80 wt% of both ILs, 55−75% of the water molecules could be stripped from the solvation shell of DNA, [C_4mim]$^+$ cations seem to disrupt the spine of hydration by intruding into the DNA minor groove and bind to DNA minor grooves more effectively than cholinium cations [46]. Garai et al. [47] presented a recent (2018) and systematic study, in order to determine meticulously the mechanical properties of double stranded (dsDNA) in the hydrated ILs 1-ethyl-3-methyl imidazolium acetate ([C_2mim][Ac]), 1-butyl-3-methyl imidazolium acetate ([C_4mim][Ac]), [C_4mim][Lac], [C_4mim][NO_3], cholinium acetate ([Ch][Ac]), [Ch][Lac], and [Ch][NO_3]. MD simulations and theoretical models were carried out and results demonstrated that dsDNA in 50 wt% ILs have lower persistence length and stretch modulus in comparison to 80 wt% ILs [47]. Therefore, both the persistence length and stretch modulus of dsDNA increase with the increase of the ILs concentration, which supports the predictions of the macroscopic elastic theory, in contrast to the behavior exhibited by dsDNA in monovalent salts.

The authors further suggested the preferable ILs that can be used for maintaining DNA stability during long-term storage [47]. The planar imidazolium head group and long hydrophobic tail of the [C_4mim]$^+$ cation enable easy accessibility into the depths of the minor groove where the cation establishes electrostatic and van der Waals interactions with the base atoms [47]. Conversely, the bulky trimethylamine (NMe_3) headgroup of cholinium did not allow it to penetrate deep into the groove, resulting in comparatively weaker binding. The shorter ethyl chain of the $[C_2mim]^+$ cation also weakens its interaction with the DNA groove. Thus, the stronger the interactions of DNA with the IL are, the more rigid and stable the DNA duplex becomes [47]. However, the persistence length of DNA solvated in $[C_4mim][NO_3]$/ water mixtures exhibited values close to the control water system irrespective of the IL concentration, and, among all ILs, $[C_4mim][NO_3]$ was found to strip the least amount of water from the DNA solvation shell [47]. This is plausibly because $[C_4mim]^+$ remains strongly paired with the $[NO_3]^-$ anion, preventing the cation from penetrating into the narrow minor groove of DNA [47].

Therefore, it can be clearly surmised that the level of hydration of DNA is inversely related to its persistence length in IL/water binary mixtures [47]. Opposite to the higher persistence length of DNA observed in $[C_4mim]^+$ containing binary mixtures, cholinium-based ILs interact with greater affinity with the DNA backbone owing to the localized positive charge on the NMe_3 headgroup of the cholinium cation [47]. As a result, it requires larger force to stretch cholinium-bound DNA, leading to higher values of stretch moduli in these systems. Moreover, in the presence of cholinium based ILs, a higher amount of water is retained in the DNA solvation shell compared to $[C_4mim]^+$ based ILs [47].

The hydroxyl (OH) group of the cholinium cation is known to interact with water and form a complex cation-water hydrogen bonded network. Such interactions in the solvation shell of DNA dissolved in cholinium based ILs plausibly makes the hydration layer quite rigid, thereby making it difficult to stretch the DNA [47]. It is known that flexibility of the DNA substrate is key in mediating appropriate DNase I binding and cleavage. Thus, increased rigidity of DNA in ILs might prevent DNase I binding to DNA, thereby conferring exceptional long-term stability of DNA during storage [47]. Overall, enhanced stability of DNA in hydrated ILs can be explained by the predominant anionic character of DNA. The negatively charged phosphate groups cause an electrostatic association of the organic cations in ILs with the DNA backbone, expelling the water molecules from immediate contact with DNA and disrupting the water cage around it [19, 47]. This partial dehydration of DNA prevents hydrolytic attack and increases thermal stability, stabilizing DNA in the long run [19]. In addition, groove binding by IL cations via hydrophobic and polar interactions also contributes significantly to DNA stability [22].

The IL cation choice between binding to the major or minor groove of DNA is affected by valence, size, ambient pH, and also depends on the DNA sequence [19]. Furthermore, solutions with 80 wt% of IL have unique properties, such as decreased water activity, lower dielectric constant and altered ion networks relative to aqueous buffers [48]. A comprehensive analysis and stability comparison of the dsDNA in imidazolium, phosphonium, ammonium and cholinium-based ILs was reported by Freire and co-workers [49]. Aqueous solutions of the ILs at different concentrations (5, 15 and 30 wt%) at pH 7.2 controlled by the use of Tris-HCl buffer were used for the experiments [49]. CD analysis demonstrated that the dsDNA maintained its B-conformational structure in all the ILs at the three different concentrations, with the exception of tetrabutylphosphonium bromide ($[P_{4444}]Br$), which caused complete precipitation of DNA at 30 wt% and [Ch][DHP] that also induced significant perturbations in the dsDNA native conformations [49].

The authors concluded that $[P_{4444}][Br]$ and [Ch][DHP] provided highly acid conditions, due to impurities and the IL anion, respectively [49]. Moreover, dsDNA stability was attained in the buffered [Ch][DHP] through the addition of cholinium hydroxide up to pH 7, and aqueous solutions composed of [P_{4444}][Br] at 30 wt% in 1000 mM of Tris–HCl (pH = 7.2) it was observed that the solubility and stability of DNA increased [49]. In order to investigate the interaction patterns between DNA and the ILs, UV-Vis absorption measurements were carried to detect the unstacking of nucleobases of the dsDNA, as a consequence of uncoiling or denaturation process [49].

Results demonstrated a specific trend of the cation structure independently of the IL concentration, the hyperchromicity increased in the following order: cholinium bromide ([Ch]Br) < [C_2mim]Br < tetrabutylammonium bromide ([N_{4444}]Br) < [P_{4444}]Br; and amongst the cholinium-based ILs investigated, cholinium glycolate ([Ch][Gly]) led to a higher unstacking of nucleobases in DNA [49]. Thus, [Ch]Br and [Ch][Ac] were classified as the most appropriate ILs to keep the DNA structure at the two concentrations investigated (15 and 30 wt%) [49]. Phosphorus-31 Nuclear Magnetic Resonance (^{31}P NMR) experiments supported the previous CD and UV-Vis results, since [Ch]Br presented the lowest value of ellipticity and the lowest intensity peak of DNA phosphate groups, supporting the possibility of the IL cation being more strongly interacting with the phosphate backbone of dsDNA, thus lowering the phosphorous exposition to the aqueous environment [49]. At 30 wt%, there is a general trend of the cholinium-based ILs to display lower ^{31}P NMR peaks intensity, reinforcing that the cholinium cation preferentially interacts with the DNA phosphate groups when compared with imidazolium and tetrabutylphosphonium/ammonium cations [49].

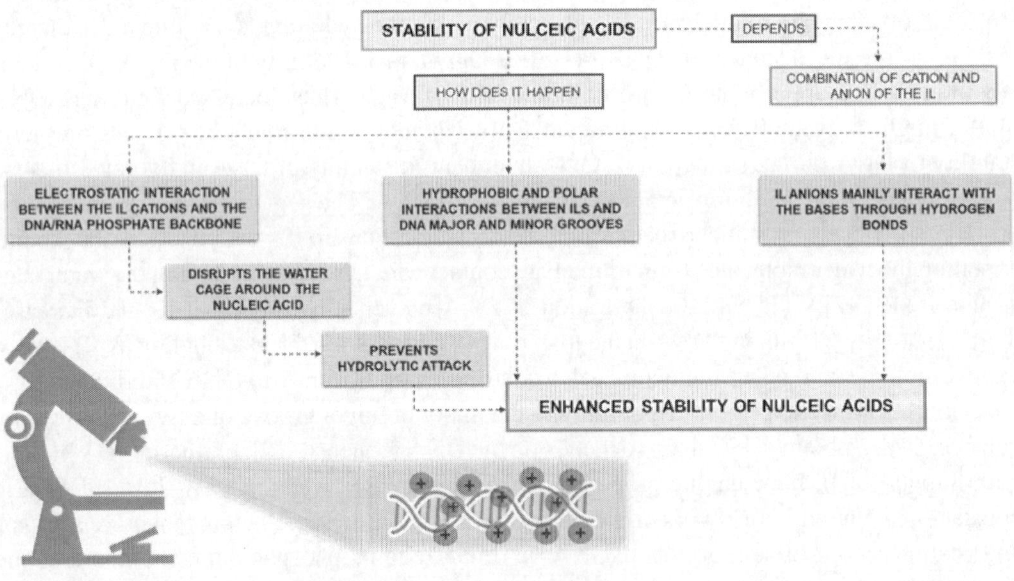

Figure 5. Scheme summarizing the main interactions and properties involved in the stability of nucleic acids in the presence of ionic liquids (ILs).

The main molecular mechanisms proposed by the authors underlying the stabilization of dsDNA in these ILs point that at low IL concentrations (5 wt%), there are less electrostatic interaction established between the IL cation and the DNA phosphate groups, which may be due to difficulties in "breaking" the dsDNA hydration shell [49]. However, at a higher IL concentration (30 wt%), the IL cations compete stronger with the water molecules and bind to the DNA phosphate backbone, stabilizing the biomolecule [49]. In addition, the hydrophobicity of the IL cation also plays a main role since an increase in DNA denaturation and loss of native conformation was observed with more hydrophobic cations [49].

Finally, the authors concluded that the IL anion has a weaker impact when compared to the cation on interacting with the DNA molecules. Nonetheless, it can influence other important parameters, such as the IL solution pH [49]. In summary, the main interactions and properties involved in the stabilization of nucleic acids in the presence of ILs are represented in Figure 5.

ILs Reverse the Stability of Nucleic Acids Base Pairs

It is well known that under physiological conditions, G-C base pairs are more stable than A-T base pairs, since these bases are bonded by three and two hydrogen bonds, respectively [50, 51]. Certain ions are known to preferentially stabilize certain base pairs. For instance, alkylammonium ions bind to A–T base pairs in the minor groove of a DNA duplex, stabilizing regions of A–T base pairs [50]. Some osmolytes, such as glycine and betaine which have alkylammonium derivative ions, destabilize G–C rich DNA duplexes by binding to unpaired guanines, at high salt concentrations [50].

Considering that [Ch][DHP] dissolved in a small amount of water provided long-term stability to nucleic acids and also contains alkylammonium ions [46, 47, 50], Karimata et al. [50] showed that this hydrated IL reversed the stabilities of the base pairs. Quantitative thermodynamic analyses demonstrated that A–T base pairs are more stable than G–C base pairs in a solution containing cholinium cations, which is the reverse base pair stability of DNA duplexes [50]. Later, Karimata and co-workers [51] presented a microscopic view that explained this unique stabilization. MD simulations showed that the affinity of cholinium ion for the minor groove of A-T base pairs is more than 2 times that for other groove areas.

In the narrow A-T minor groove, cholinium ion has high affinity for the ribose atoms of thymine [51]. Cholinium ions also destabilize the formation of hydrogen bonds between G-C base pairs by binding to base atoms preferentially for both of duplex and single-strand DNA, which are associated with the bonds between G-C base pairs [51]. Portella et al. [52] also studied the origin of these stabilization phenomena using a combination of MD simulations and NMR spectroscopy. They demonstrated that cholinium and tetramethyl ammonium (TMA) ions were preferentially localized in the minor groove of DNA duplexes at A-T base pairs and these same ions show less pronounced localization in the major groove. Furthermore, free energy calculations showed that single-stranded GC-rich sequences exhibited more favorable solvation by cholinium than single-stranded AT-rich sequences [52].

The sequence-specific nature of cholinium and TMA binding provides a rationale for the enhanced stability of AT-rich sequences when alkyl-ammonium ions are used as the counterions of DNA [52]. These findings, not only lead to better control of DNA stability for use in DNA nanodevices, but also provide new insight into the stability of DNA duplexes under crowding conditions found in living cells [50–52]. The role of the hydrogen bond of ILs in the stabilization/destabilization of Watson-Crick base pairs of DNA was further investigated by Sarkar and co-workers [53]. Spectroscopic and MD simulation methods were combined to study the effect of the ILs 1,1,3,3- tetramethylguanidinium acetate (TMG) and 2,2-diethyl-1,1,3,3-tetramethylguanidinium acetate (DETMG) on ct-DNA [53]. Thermal denaturation experiments of DNA were carried out at various temperatures ranging from 25 to 90°C and results suggested that TMG and DETMG both intrude into the minor groove region of DNA, however, their mode of interaction with DNA was different [53].

The authors reported that the hydrogen bonding capability of TMG enhanced the hydrogen bonding and stacking between the Watson-Crick base pairs of DNA resulting to its stabilization [53]. In contrast, the bigger size as well as absence of hydrogen bonding site of DETMG perturbed the minor groove width as well as base pair step parameters of DNA during the intrusion in the minor groove, which decreased the hydrogen bond and stacking between the Watson Crick base pairs of DNA leading to destabilization [53].

Conformational Preferences of DNA in ILs

As already mentioned before, ultra-low temperatures are used for long-term storage of DNA [20]. However, not only refrigeration is limited to small number of samples, but repeated freeze-thaw of frozen samples can also cause structural damage to the DNA [20, 54]. As a result, techniques employing RT storage of DNA are rapidly gaining recognition [42, 55–57]. Following nature's lead, most RT storage methods employ the strategy of 'anhydrobiosis' or storage under a state of extreme dehydration [56, 57]. Nevertheless, using artificial agents to induce dehydration severely perturbs the DNA conformation, as water is essential for stabilizing the physiological B-form of dsDNA [54]. B-DNA succumbs to A-form DNA upon dehydration and can further be denatured [54]. Considering that ILs have been reported to maintain the chemical and structural stability of B-DNA at RT for long periods of time [42], Ghoshdastidar et al. [54] explored the conformational preferences of dehydrated DNA in the presence of ILs, and attempted to understand the underlying mechanism of conformational transitions in DNA (Figure 6).

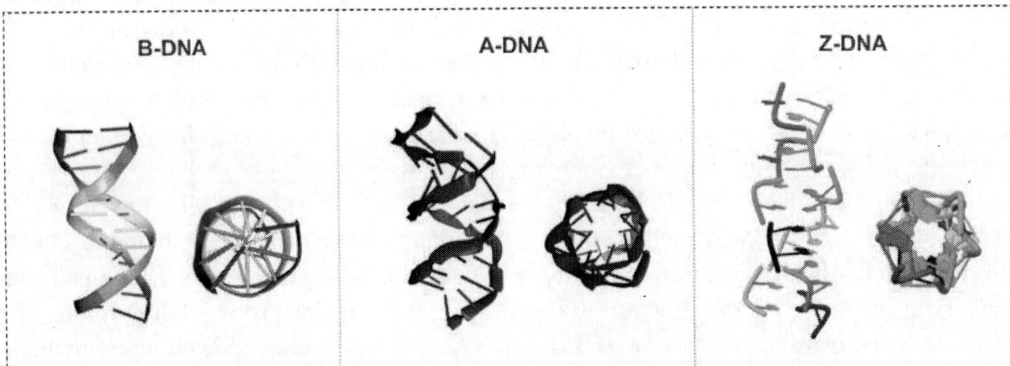

Figure 6. Common conformations adopted by DNA duplexes. B-DNA is the physiological conformation of ds-DNA, A-DNA forms upon dehydration and can easily be denatured. Z-DNA, also known as left-handed DNA, can form under physiological conditions playing a role in gene regulation or in vitro under high salt concentrations.

To emulate the dehydrated condition, the authors [54] selected the ethanol/water mixture (>78% vol/vol ethanol/water), known to present a low water activity (aw) (aw ≤ 74.4%), which induces B-DNA to undergo a spontaneous conformational transition to the A-form [54, 58]. The B conformation can be restored upon addition of water up to aw ≥ 80% relative humidity (r.h), which restores the DNA hydration shell that is critical in maintaining the B-form DNA (B-DNA) [54, 59, 60].

A series of ILs, [C_2mim][Ac], [C_4mim][Ac], [Ch][Ac], 1-ethyl-3-methylimidazolium lactate ([C_2mim][Lac]) and [Ch][Lac]), were added into to the dehydration environment of 78%

ethanol/water containing the A-form of DNA [54]. Interestingly, CD spectra showed a clear and completed transition of A-DNA to the B-form with the addition of a maximum 1.5 mM concentration of all ILs studied [54]. MD simulations were carried to understand the mechanism of IL-induced A to B-DNA transition under dehydration by calculating the root mean square deviation (RMSD) from the B-DNA crystal structure [54]. Thus, a set of mechanisms were presented by the authors that explain this phenomenon [54].

The A-DNA is stabilized at low water activity due to its compact architecture: the major groove of A-DNA harbors dense negative charge, which becomes a hub for increased condensation of Na^+ counter-ions promoted by the low-dielectric ethanol medium [61, 62]. These Na^+ counter-ions bridge between phosphate oxygens on complementary strands of A-DNA either directly or through water molecules, thereby lowering inter-phosphate repulsions and electrostatically stabilizing the A-form [54, 63]. The IL cations disrupted the Na^+ ion condensation at the major groove and formed electrostatic interactions with the DNA backbone, mimicking the so-called 'cone-of-hydration' - the shell of water molecules solvating the B-DNA backbone in water, thus the IL cations came in rescue to stabilize the B-DNA backbone [54].

Apart from this, IL cations also interacted very strongly with the DNA minor groove [54]. Randomly distributed water molecules in the minor groove of A-DNA were replaced by an ordered network of IL cations upon transiting to B-DNA [54]. The intruding of IL cations did not only reside in the minor groove of DNA, but also formed H-bonding interactions with the electronegative base atoms forming an IL-water spine that stabilized B-DNA [54]. Furthermore, Ghoshdastidar et al. [54] concluded that the nature of the IL cation influenced the transition of A-DNA to B-DNA. The positive charge on the cholinium head group is more localized compared to the imidazolium ring, in which the charge is distributed over five different sites [54]. As a consequence, cholinium cations experienced a greater affinity for the negatively charged DNA backbone and grooves that plausibly brought about a more rapid A→B transition [54]. Moreover, the water-assisted [Ch]-spine formed at the minor groove contained more water than [C_2mim]-spine, mainly because the small chain length of the cholinium moiety leaves more room for water and secondly, the OH group of cholinium facilitates Hydrogen bonding interactions with water [54].

In addition to prompting the transition of A-DNA to B-DNA in a dehydrated media [54], these ILs also highly stabilized the B conformation in aqueous solutions with a gradual addition of ethanol [54]. The resistance of the IL-bound B-DNA was attributed to the IL-water spine persisting in the minor groove, which locks DNA in its B-conformation [54]. The most effective IL was [C_4mim][Ac] since the planar imidazolium ring of the cation fits effectively into the minor groove pocket making greater number of contacts with the DNA base atoms, as already described by Chandran and co-workers [46, 47, 54].

Stabilization of Nucleic Acids Structural Motifs by Ionic Liquids

In addition to enhanced stability of dsDNA in hydrated cholinium-based-ILs, namely [Ch][DHP] [42, 46, 47, 50–52], this biocompatible solvent was also reported to stabilize several structural motifs of DNA, whose stability is highly limited [23, 48, 64]. For instance, the instability of Hoogsteen base pairs limits biological applications of triplex-forming oligonucleotides [23], however, Karimata et al. [23] found that the DNA triplex formation was significantly stabilized in aqueous solutions of [Ch][DHP] 80 wt% as compared with an aqueous buffer at neutral pH [23]. Interestingly, the stability of Hoogsteen base pairs was found to be comparable with that of Watson–Crick base pairs in the hydrated [Ch][DHP] [23].

MD simulations of a DNA triplex in the presence of cholinium cations revealed that the DNA triplex was stabilized because of the binding of cholinium cations around the third strand in the grooves [23]. The stabilization of DNA triplex in hydrated ILs facilitates the development of new DNA materials and helps to better understand its biological function, since triplex formation may be stabilized inside cells where cholinium ions and their derivatives are abundant in vivo [23]. Stable G-quadruplexes were also formed in [Ch][DHP] in a study reported by Fujita et al. [64], in which cholinium cations existed in the core of the quadruplex structure similar to that observed for Na^+ or K^+ ions [64].

Interestingly, a later study by Karimata et al. [48] demonstrated that i-motifs in 80 wt% [Ch][DHP] were more stable than G-quadruplexes, as cholinium ions bound to loop regions and grooves of the i-motifs 48. Usually, i-motifs are not formed at physiological pH [48]; however, cholinium ion binding dramatically induced the formation of i-motifs at physiological pH [48]. Therefore, [Ch][DHP] seems to be an excellent candidate for the stabilization of DNA and its several structural motifs, as represented in Figure 7.

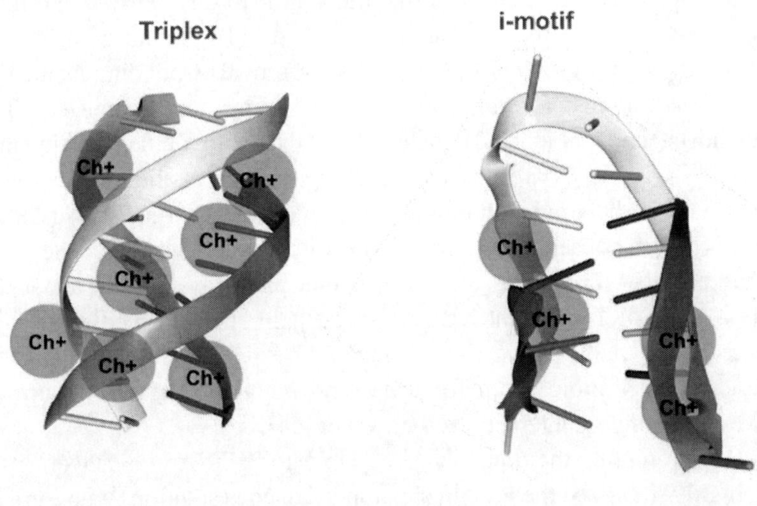

Figure 7. Triplex and i-motif stabilized by the IL [Ch][DHP]. The triplex was stabilized by binding of cholinium cations around the third strand in the grooves, and i-motifs stabilized due to cholinium cations bound to loop regions and grooves.

DNA Stabilization

For safe storage of genetic information, compaction of DNA is extremally relevant [65]. DNA comprises a total distance of 2 meters per human cell so an unbelievable length of 100 trillion meters of DNA per human appears in picture [65]. Somehow, nature manages to keep this amount of DNA stable and resistant to UV radiation and chemical, biochemical and mechanical stresses through its compaction in the nucleus by the formation of a complex named chromatin [65]. Therefore, storage and long-term stabilization of DNA can be achieved through artificial DNA compaction in an efficient way [65].

Studies have started applying these compaction tricks using polyamines [66, 67], surfactants [68, 69], nanoparticles [70, 71], polyethylene glycol (PEG) [72], dendrimers [73, 74], multivalent ions [75] and proteins [76]. Considering their low cytotoxicity and high biodegradability, ILs could play an important role in DNA compaction, since they are composed of a cation, which can compensate DNA backbone negative charges through electrostatic interaction and may lead to the formation of self-aggregated polyvalent structures [65]. Satpathi et al. [65] reported how the highly biodegradable and eco-friendly guanidinium tris(pentafluoroethyl)trifluorophosphate IL (Gua-IL), is found to be a very effective DNA compaction agent for ct-DNA and plasmid DNA. Compaction of DNA needs the neutralization of its negatively charged phosphate groups, and for this neutralization process, counter ion valency (Z) of the compacting agent should be greater or equal to 3 ($Z \geq 3$).

Through the data obtained from several spectroscopies (UV-Vis, fluorescence, CD, dynamic lights scattering (DLS), scanning electron microscopy (SEM), scanning tunnelling microscopy (STM), it was possible to understand that for a concentration equal or higher of 1mM, Gua-IL acts as a multivalent ion ($Z \geq 3$) [65], forming micellar like aggregates. Guanidium cations have two positively charged centers, one positively charged part neutralizes the negatively charged phosphate groups of DNA and the other part is involved in electrostatic interaction with the Gua-IL anion. When the Gua-IL hits its critical aggregation concentration (CAC= 1 mM), the anions form a micellar kind of aggregate [65], then the counter cations situated at the micellar surface drag the negatively charged DNA strands to the micellar surface due to electrostatic attraction. This process leads to the wrapping of several parts of DNA around this micellar kind of aggregate (Figure 8).

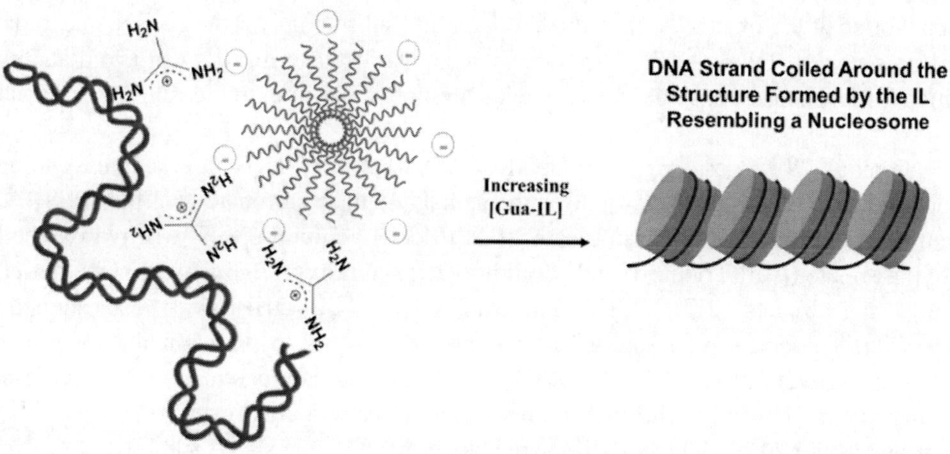

Figure 8. Schematic representation of how Gua-IL is able to stabilize DNA through its packaging. The anions of the Gua-IL form micellar-like aggregates. The cations have two positively charged centers, one is attracted to the negative charges of the anions and the other is electrostatically attracted to the phosphate backbone of DNA. The DNA molecule coils around these structures and, by increasing the concentration of the Gua-IL, the coiled DNA transitions to a globule-like state resembling a nucleosome.

Furthermore, by increasing Gua-IL concentration, the number of loops increases and the intra/interstrand repulsion between DNA strands ends up collapsing. Thus, the biomolecule transitions from coil to globule state showing that indeed Gua-IL has a remarkable ability for DNA compaction, reaching a state that seems to be analogue of a Manning-Oosawa condensate [19], comparable to the DNA compaction by histone protein, nanoparticles, dendrimers and PEG [65].

A further example of DNA packaging using [C_4mim]Cl is reported by Ding and co-authors [77] that demonstrated, for an IL concentration above 0.06 M, the cationic headgroups stay near DNA phosphates, and the hydrocarbon chains stay perpendicularly attached to the DNA surface through strong hydrophobic interactions, inducing a coil-to-globule transition of DNA [77]. Therefore, both studies prove that DNA can be stabilized through its packaging with ILs.

RNA Stabilization

RNA is intrinsically less stable than DNA [19], and the literature on ILs and RNA is far from abundant. Studies concerning the development of new strategies and protocols to maintain or increase the RNA stability and biological activity, ideally using less extreme temperatures and more benign solvents, are extremely relevant since RNA is becoming an increasing component of the pharmaceutical industry.

General approaches to maintain RNA stability focus on the inactivation of ribonucleases, and alternative strategies comprise chemical modifications and the use of cations or polycationic amines that increase RNA stability by shielding its negative phosphate group in the backbone [20]. Therefore, ILs can be also foreseen as alternative solvents for RNA. As already discussed in this chapter, cholinium-based ILs usually present high biodegradability, low toxicity and are able to maintain the structure of some biomolecules, including RNA.

Among the few reports comprising ILs and RNA, the work of Mazid et al. [78] can be highlighted since they proved that aqueous solutions up to 50 wt% of [Ch][DHP] are able to prolong the shelf life of siRNA up to three months and substantially reduce siRNA degradation in the presence of RNase A. Furthermore, the use of self-buffering ILs can avoid adding an addition buffer, considering that nucleic acids are not stable in extreme ranges of pH values [20].

Pedro et al. [20] investigated four Good's buffers-ILs (GB-ILs) comprising the cholinium cation: cholinium 2-[4-(2-hydroxyethyl)piperazin-1-yl]ethanesulfonate ([Ch][HEPES]), cholinium 2-(N-morpholino)ethanesulfonate ([Ch][MES]), cholinium N-[tris(hydroxymethyl) methyl]glycinate ([Ch][Tricine]), and cholinium 2-[(2-hydroxy-1,1-bis(hydroxymethyl)ethyl) amino]ethanesulfonate ([Ch][TES]). The well-known [Ch][DHP] was also studied for comparison. These ILs were studied as alternative solvents to maintain the integrity and improve the stability of small RNA (sRNA) containing the therapeutically relevant pre-miR-29. Analysis of CD spectra and thermal denaturation experiments revealed that the GB-ILs, with the exception of [Ch][TRICINE], were able to maintain the sRNA stability, at least up to 30 days with concentrations up to 50 wt% [20].

The authors also provided a closer look on the ILs anions effect on the stability of RNA, being very interesting and innovative since the role of anions is left unspecified by most of the studies found on the literature [20], although thermodynamic properties such as the free energy of binding of IL ion pairs to RNA are known to depend on the anion choice [42]. In fact, MD simulations allowed to better understand the molecular-level phenomena responsible for the observed improved stability of RNA, in which the overall charge of the biopolymer first solvation sphere plays a primary role [20]. The negatively charged sulfonate group of the GB-ILs anions, has similar size to that of [DHP], yet fails to be accommodated in an analogous way. Therefore, the stability of the sRNA chain is higher in the presence of the GB-ILs because the global charge of the overall complex is less negative than in presence of [Ch][DHP].

The exception in the RNA stability and integrity for the GB-IL [Ch][TRICINE] can be explained by the presence of a higher number of $[TRICINE]^-$ anions in close proximity to RNA, because these anions have a terminal carboxylic group, in contrast with all the remaining GB-ILs that present a negatively charged sulfonate group, which seems to be of high relevance to improve the RNA stability. Electrophoretic runs revealed that [Ch][HEPES] selectively stabilizes pre-miR-29 over the other host RNAs, this behavior may be due to the stem-loop structure of pre-miR-29 contrasting with the linear structure of other host RNAs. Amongst the diverse ILs anions investigated, $[HEPES]^-$ has been described as the most polar anion according to its dipole moment values, which can further justify its higher polarity and ability to establish hydrogen bonds with the RNA nitrogen bases and that may be responsible for the enhanced stabilization of stem-loop RNAs [20].

Amongst the most suitable ILs that improve the sRNA stability, [Ch][HEPES] and [Ch][MES], with concentrations up to 50 wt%, were additionally identified as non-cytotoxic to human cell lines (human fibroblasts and HeLa), which is particularly relevant when envisaging the use of RNA in therapeutic applications. The sustainability of GB-ILs solvents and preservation approach to RNA is reinforced by their low environmental impact, low cytotoxicity, but also by the possibility of the ILs recycling without compromising the sRNA integrity [20]. This study demonstrated that aqueous solutions of [Ch][MES] and [Ch][HEPES] are remarkable stabilizers and storage media of recombinant sRNAs at RT, without requiring the typically carried out samples freezing, thus reinforcing their effectiveness in RNA bioprocessing [20].

The stability and integrity of the same type of sRNA in aqueous solutions of nontoxic, biodegradable, and biocompatible amino-acid-based ILs (AA-ILs) was investigated by Quental et al. [79]. The stability of RNA at 25°C, in the presence of aqueous solutions of ILs at 20 wt % with the cholinium cation and anions derived from L-lysine ([Ch][Lys]), L-arginine ([Ch][Arg]), L-glutamic acid ([Ch][Glu]), and L-aspartic acid ([Ch][Asp]), were carried out by agarose gel electrophoresis and CD melting temperatures, respectively [79]. Overall, the authors found that all the hydrated AA-ILs maintained the RNA integrity.

Furthermore, a remarkable enhancement in the melting temperature (Tm) of RNA of about 15ºC after 1h and 10°C after 15 days was observed for both [Ch][Glu] and [Ch][Asp] [79]. Zeta potential assays demonstrated fewer negative values when the sRNA was in the presence of these ILs, indicating that when in aqueous solution the cholinium cation forms an outer shell around RNA, in agreement with previously reported MD simulations results [20, 78]. The ILs with $[Asp]^-$ and $[Glu]^-$ anions were more effective stabilizing RNA, demonstrating that the anionic counterpart of ILs also has a significant influence on the RNA structure and stability [79]. These differences may be related to the polarity gap between the different anions of the AA-ILs. Cytotoxicity studies in two different human cell lines (human fibroblasts and HeLa) revealed that all AA-ILs, with the exception of [Ch][Lys], are harmless to both cell lines, thereby confirming their noncytotoxic character at the concentration under study (20 wt%) [79].

The major factor affecting the RNA stability, and thereby causing its degradation, is the ubiquitous presence of RNases (active nucleases). Therefore, the protective effect of both [Ch][Glu] and [Ch][Asp] aqueous solutions against fetal-bovine-serum (FBS)-derived ribonucleases was also investigated by the authors [79]. In comparison with RNA in high-purity water, CD analyses showed that the ellipticity of RNA is slightly higher in the presence of [Ch][Asp], meaning that this IL is more effective on RNA protection from degradation by RNases, reinforcing the idea that this IL is particularly relevant to be used as preservation and stabilization media for RNA, as example, for storage or to be used as an additive in biological assays [79].

Quental el al. [79] also investigated the best identified AA-ILs ([Ch][Glu] and [Ch][Asp]) as phase-forming components of aqueous biphasic systems (ABS) in order to develop an integrated extraction and purification platform for RNA. Both with pure RNA and bacterial lysate samples, RNA was successfully extracted to the IL-rich phase without compromising its integrity and stability. The IL-rich phase can be used as the preservation medium of RNA up to its use; then RNA can be then recovered from the IL-rich phase by ethanol precipitation, and the ABS phase-forming components can be recovered and reused [79]. The AA-ILs [Ch][Glu] and [Ch][Asp] were further investigated by Tulsiyan et al. [80] for the structural stabilization of RNA sourced from yeast. Steady-state fluorescence, UV−vis absorbance, CD spectroscopy, and isothermal titration calorimetry (ITC) experiments revealed enhanced stabilization of RNA in these AA- ILs, adding evidence that AA-ILs display great as solvent medium for RNA.

Inspired by the work of Quental et al. [79], Chenghui Zhu and co-workers [81] incorporated amino acids in the structure of the magnetic ILs tetra(1-octylimidazole)cobaltate(II) di(L-glutamate) ([Co(OIM)$_4$ $^{2+}$][Glu$^-$]$_2$) and tetra(1-octylimidazole)cobaltate- (II) di(L-aspartate) ([Co(OIM)$_4$ $^{2+}$][Asp$^-$]$_2$) for the extraction and protection of yeast total RNA from RNase A degradation. To facilitate RNA recover, ILs were added to polypropylene glycol (PPG)-2000 in a 1:5 (v/v) ratio [81]. In addition to the ability of the ILs/PPG-2000 system to extract DNA, high performance liquid chromatography (HPLC) and gel electrophoresis revealed that the recovered RNA could be detected even after 24h incubation in the presence of RNase A [81].

Moreover, spiking RNase A into the IL/PPG-2000 system demonstrated that the extraction efficiency of RNase A was lower than RNA, clarifying that these ILs provided an anhydrous microenvironment to prevent RNA from interacting with RNase A [81]. Although improvements in the purity level of RNA are still required, these pioneer approaches reported by the authors [79, 81], represent a step forward in the development of sustainable processes to overcome the critical demand of high-quality/high-purity RNA to be used as biotherapeutics.

The studies reviewed here reporting the stabilization of nucleic acids in hydrated ILs are summarized in Table 1.

Table 1. Summary of the studies reporting the stabilization of nucleic acids in hydrated ILs

IL	DNA/RNA source	Remarks	Ref.
[$C_n$mim]Br (n=4, 6, 8, 10 and 12)	Calf thymus DNA (ct-DNA)	Shorter alkyl chains favored the binding between the ILs and the DNA.	[26]
[$C_n$bim]Br (n = 2, 4, 6)	ct-DNA	Shorter alkyl chains favored the binding between the ILs and the DNA. IL cations preferably bind to the minor groove of DNA by electrostatic interactions. π–π stacking between the bases and the aromatic ring 0.5 M IL solution at 25°C enabled a hydrophobic microenvironment by expelling water from immediate contact with DNA.	[27]
[Cmim]Cl, [C_2mim]Cl, [C_4mim]Cl and [C_4mim]Br	DNA duplex from salmon testes (sDNA)	High thermal stability of DNA in 62mM solutions of ILs: base stacking at 70°C. Establishment of interactions between the imidazole and guanine bases led to effective stacking between bases. Shorter alkyl chains in the IL with Cl anion favor DNA stability.	[28]

IL	DNA/RNA source	Remarks	Ref.
$[C_4mim][TfO]$), $[C_1mim][TfO]$), and $[C_4mim][MeSO_4]$	sDNA	Both the cation and anion of the ILs interacted with the DNA structure. Affected the thermal unfolding process but not the pre-melting transition. Dominant interaction between the imidazole cation and guanine and thymine bases.	[29]
[Ch][DHP], [Ch][Lac] and $[Ch][NO_3]$	sDNA	DNA preservation up to 6 months in 80 wt% of all ILs.	[42]
[Ch][DHP]	DNA solid films	Introduction of 10-40 wt% of IL into DNA solid films led to water retentivity at RT and DNA structural stability.	[43]
Cholinium octanoate	Framework nucleic acids (FNAs)	FNAs stored in 50% IL/buffer, maintained integrity under RT for 10 days. FNAs in the IL formulation successfully delivered *ex vivo* on porcine skin.	[44]
[Ch][Lac], [Ch][Lev], and [Ch][DHP]	DNA from fixed scleral tissues	All IL at 20 wt% were able to fix the scleral tissue while maintaining the integrity and purity of DNA.	[45]
$[C_4mim]Cl$, $[C_4mim][NO_3]$, $[C_4mim][Lac]$, [Ch][Lac], and $[Ch][NO_3]$	Model DNA (Dickerson–Drew dodecamer)	DNA maintained the B-conformation in all the ILs. $[C_4mim]^+$ cations bound to DNA minor grooves more effectively than $[Ch]^+$ cations. 55-75% of the water molecules could be stripped from the solvation shell with 80 wt% of IL.	[46]
$[C_4mim]Cl$, $[C_4mim][NO_3]$, $[C_4mim][Lac]$, [Ch][Lac] and $[Ch][NO_3]$	Model DNA (Dickerson–Drew dodecamer)	Persistence length and stretch modulus of dsDNA increased with IL concentration (80 wt%). $[C_4mim]$-based ILs, with the exception of $[C_4mim][NO_3]$, bound strongly to the DNA minor groove. Cholinium-based ILs interacted with greater affinity with the DNA backbone.	[47]
$[C_2mim][Ac]$, $[C_2mim][Lac]$, $[C_4mim][Ac]$, [Ch][Ac] and [Ch][Lac]	DNA duplexes	ILs induce conformational transitions in DNA (A to B-DNA). ILs to disrupted Na+ ion condensation and formed 'IL spine' in DNA minor groove. The nature of the IL cation influenced the transition of A-DNA to B-DNA.	[54]
[Ch][DHP], [Ch]Br, [Ch][Ac], [Ch][Gly], $[C_2mim]Br$, $[N_{4444}]Br$, and $[P_{4444}]Br$.	sDNA	dsDNA maintained its B-conformation in all ILs, with the exception of [Ch][DHP] and $[P_{4444}]Br$, due to highly acidic conditions provided by these ILs. More hydrophobic cations increased DNA denaturation and conformational loss. The best ILs to stabilize DNA were [Ch]Br and [Ch][Ac] at 15 and 30 wt% IL. Electrostatic interaction between IL cation and the DNA phosphate groups stabilizes the biomolecule by breaking the hydration shell of DNA.	[49]
[Ch][DHP]	DNA Triplexes	Cholinium cations bound around the third strand in the grooves and stabilized the triplex DNA.	[23]
[Ch][CHP]	G-quadruplexes	Cholinium cations existed in the core of the G-quadruplex stabilizing its structure.	[64]
[Ch][DHP]	i-motifs	Cholinium cations bound to loop regions and grooves of the i-motifs enabling its formation at physiological pH.	[48]
TMG and DETMG	ct-DNA	Both ILs intrude into the minor groove region of DNA. TMG is able to establish hydrogen bonds and stacking between the base pairs, stabilizing DNA. The bigger size and the absence of hydrogen bonding site of DETMG led to DNA destabilization.	[53]

Table 1. (Continued)

IL	DNA/RNA source	Remarks	Ref.
Gua-IL	Plasmid and ct-DNA	The IL induced a coil to globule transition of DNA, acting as a compaction agent of the biomolecule. CAC=1mM	[65]
[C_4mim]Cl	ct-DNA	The IL induces a coil to globule transition of DNA, acting as a compaction agent of the biomolecule. CAC=0.06M	[77]
[Ch][DHP]	siRNA	The IL (50 wt%) prolongs the shelf life of siRNA up to 3 months. Reduced siRNA degradation in the presence of RNase A.	[78]
[Ch][DHP], [Ch][HEPES], [Ch][MES], [Ch][TRICINE], and [Ch][TES]	ncRNA (human pre-miR-29b, tRNA and 6S RNA)	[Ch][MES] and [Ch][HEPES] are remarkable stabilizers and are non-cytotoxic to human cell lines with concentrations up to 50 wt%. Possible to recycle the ILs without comprising the RNA integrity.	[20]
[Ch][Lys], [Ch][Arg], [Ch][Glu] and [Ch][Asp]	ncRNA (human pre-miR-29b, tRNA and 6S RNA)	[Ch][Glu] and [Ch][Asp] enhance RNA stability up to 15 days of incubation and are non-cytotoxic to human cell lines with concentration up to 20 wt%. [Ch][Asp] protects RNA from degradation by RNases. Possible to recycle the ILs without comprising the RNA integrity.	[79]
[Ch][Glu] and [Ch][Asp]	Yeast RNA	Structural stabilization of RNA in these AA-ILs.	[80]
[Co(OIM)$_4$ $^{2+}$][Glu$^-$]$_2$ and [Co(OIM)$_4$ $^{2+}$][Asp$^-$]$_2$)	Yeast RNA	ILs/PPG-2000 systems 1:5 (v/v) were able to recover RNA and protect it for 24h in the presence of RNase A.	[81]

Neat Ionic Liquids

As previously mentioned, there is an emerging need to find fully non-aqueous media for nucleic acids that do not involve the use of organic solvents. ILs emerge as potential alternative solvents, however, the low solubility of DNA in some aqueous solutions of ILs and very lengthy dissolution time limits the application of the biomolecule when it's necessary to have a fully non-aqueous environment [82]. Therefore, the number of studies reporting the stabilization of nucleic acids in neat ILs are significantly less than the ones reporting hydrated ILs.

Nevertheless, diversified studies reporting neat ILs that can successfully solubilize and stabilize nucleic acids are herein reviewed. Cardoso et al. [83] examined the molecular solvation of dsDNA and single stranded DNA (ssDNA) in neat RT ILs based on imidazolium, oxazolium, pyrrolidinium, pyridinium and quaternary ammonium, composed of the well-studied tetrafluoroborate ([BF_4]$^-$) and hexafluorophosphate ([PF_6]$^-$) anions. The authors suggested that cations tend to move close to the DNA main chain due to strong electrostatic interactions with the phosphate groups as well as hydrogen bonding and π interactions with the DNA bases, while anions mainly form hydrogen bonds with DNA bases [83]. In addition, ssDNA bases are preferentially solvated by the anions via hydrogen bonding with the fluorine atoms [83].

MD simulation studies indicated that the DNA structure is highly stable in these ILs, retaining the native B-form dsDNA crystallographic structure [83]. The best ILs to stabilize the native form of dsDNA were found to be based on quaternary ammonium and pyridinium cations combined with $[PF_6]^-$ anions. The change of anion from $[PF_6]^-$ to $[BF_4]^-$ seems to have a slight detrimental effect on the dsDNA structure since $[BF_4]^-$ anions are able to establish more hydrogen bonds with DNA bases, which affects the double helix integrity [83]. This fact enables ssDNA to form a stable conformation in ILs based on $[BF_4]^-$ anions [83].

The effect of the neat IL [C_4mim][Ac] on the long-term structural and chemical stability of basidiomycetes DNAs integrated within Schottky like junctions was investigated by Rizan et al. [84]. Significant improvement in the rectifying profiles was reported by the authors for DNA dissolved in [C_4mim][Ac] compared to DNA in aqueous solution. DNA with and without the addition of IL was stored at RT (28ºC) at different time intervals (1 day, 1 week, and 1, 2 and 3 months) [84]. The addition of [C_4mim][Ac] in the DNA solution led to long-term stability of DNA up to 4 months, which was limited to the period of this current experiment [84]. Therefore, these studies help to clarify the compatibility of ILs with DNA and to rationally design ILs for DNA-based applications [83, 84].

Prasad and co-workers [82] also reported the stabilization of DNA in neat ILs. The authors were able to dissolve in 6 h, up to 3.5 wt% DNA (from salmon testes) in two bio-based ILs (cholinium indole-3-acetate ([Ch][IAA]) and cholinium indole-3-butyrate ([Ch][IBA]). DNA maintained B-form structures and long-term stability in the first IL after 6 months of storage but is denatured by the second one during the dissolution process [82]. The IL that led to the best results ([Ch][IAA]) was proven to be recycled with very high yield (90–95%) for consecutive reuse in the redissolution of DNA [82]. The dissolution of DNA in different neat cholinium-based ILs was further assessed by Sharma et al. [85]. The authors reported that at RT the DNA dissolution in the different ILs was: [Ch][Ac]: 5 wt%, cholinium formate ([Ch][For]): 12 wt%, cholinium glycine ([Ch][Gly]): 8 wt% and cholinium pyruvate ([Ch][Pyr]): 2 wt% [85].

In addition to the A and B conformation, ds-DNA can also adopt a left-handed conformation known as Z-DNA, which has attracted the attention of the scientific community due to its characteristic features such as its possible role in regulation of gene expression and genetic instability [86]. Right-handed B-DNA can be transformed into left-handed Z-DNA in vitro at high salt concentrations or in vivo under physiological conditions [86]. Bhatt et al. [86] developed a study in which DNA from salmon testes was solubilized in neat [Ch][For] up to 1 wt% and was found to reduce Ag(I) salt into silver nanoparticles (AgNPs), whose presence was confirmed by UV-Vis spectrophotometer and TEM [86]. During this process, CD spectroscopic measurements demonstrated that the interaction of DNA with the IL induced a B-Z transition in the DNA secondary structure [86].

According to docking simulations, the IL presented higher preference for minor-groove binding on the DNA structure and a high ability to promote hydrogen bonds with the nucleic acids [86]. [Ch][For] presented different binding sites for IL cation and anion in comparison with earlier studies describing the interactions between ILs and nucleic acids [86]. This IL displayed lower ability to stabilize the DNA structure and was more willing to promote torsions on DNA structure leading to possible changes in DNA secondary structure [86]. Bhatt et al. [86] suggested that the induced torsion is more prominent in the presence of a metal salt, resulting in a conformational transformation of B to Z-DNA.

Another work regarding the DNA stabilization in neat ILs is reported by Soni et al. [87] that demonstrated how 1-butyl-3-methyl imidazolium hexafluorophosphate ([C_4mim][PF_6]) can form self-assembled nanostructures by interacting with plasmid DNA, increasing the efficiency of the bacterial green fluorescent protein plasmid (pGFP) gene transformation in cells. Bacterial gene transformation is a key technique in molecular biology, being extremely important developing plasmid delivery vectors that efficiently transfer these genetic materials across the bilayer membrane and prevent any enzymatic damage to the DNA molecules [87]. Currently, calcium chloride method followed by heat shock treatment has been used and it proposes that divalent cations (Ca^{2+} and Mg^{2+}) bind with anionic DNA due to their opposite charges and then it is applied the heat shock treatment in order to push the DNA molecules through the cell wall [87]. However, these metal inorganic cations do not direct any self-assembly due to their hydrophilic character, they also neutralize the DNA negative charge which then requires heat shock for the transformation [87].

In this context, using an IL (that is composed by an organic cation and inorganic anion, by definition) as nonviral synthetic vectors may be considered to enhance the efficiency of the bacterial gene transformation process [87, 88]. In fact, Soni et al. [87] characterized the functional nanostructure of IL-DNA by multiple techniques. Agarose gel electrophoresis showed that the IL-DNA complex migrates to the same position as pure DNA. Thus, the electrostatic interactions between DNA and the $[C_4mim]^+$ cation do not neutralize the negative charge, so the heat shock is not necessary for the transformation as it is in the calcium chloride method. CD analyses indicated that the B-form of DNA was retained within the nanostructures, which is essential for efficient transformation. NMR, Fourier-transform infrared spectroscopy (FTIR), and X-ray photoelectron spectroscopy revealed the nature of the interaction between the $[C_4mim]^+$ cations and the negatively charged phosphate groups, and thermogravimetric analyses showed these nanostructures consist up to 16 wt% of IL molecules, which provides stability and protects DNA against enzymatic degradation while providing enough charge density to transfer the molecule across the cell [87].

Overall, these DNA-IL nanostructures enhanced the transformation efficiency up to 300-400% as compared to pure plasmid DNA, keeping a constant volume of 1μL of IL [87]. Cytotoxicity studies showed no toxicity in bacterial cells for IL quantities up to 5μL (much more than the 1μL of IL required for the maximum transformation rates) [87]. Zhang et al. [88] also found that poly[3-butyl-1-vinylimidazolium L-proline salt] could condense plasmid DNA to form stable complexes against enzymatic degradation by deoxyribonuclease I, as suggested by the agarose gel electrophoresis results [88]. DNA could be released from the complexes by competitive exchange with sodium polyacrylate without losing its structures. The cell viability was determined through propidium iodine (PI) staining and flow cytometry, showing marginal toxicity toward the cells examined [88]. Therefore, the self-assembled nanostructures of plasmid DNA and these ILs demonstrated to be promising nonviral vectors for gene transformation that successfully mediate the gene expression without the aid of an additional agent.

Shi et al. [89] observed that a bicyclic imidazolium IL, was highly effective, not only for promoting polymerase chain reactions (PCR) of GC-rich DNA by minimizing non-specific amplification, but also for facilitating PCR of normal-GC DNA under mild conditions because of destabilization of DNA duplexes [89]. Thus, formulations of ILs can be used both to storage the DNA until its use, and when used, presenting an improved PCR amplification.

It is possible to notice from the papers reviewed up to now, that there are significantly more studies concerning the stability of nucleic acids in hydrated ILs when compared to neat ILs. On one hand, it is hard to dissolve large amounts of nucleic acids in a completely neat IL, and on the other hand, aqueous solutions of ILs are economically and environmentally advantageous since a smaller amount of IL is necessary and the universal greenest solvent is used – water. In addition, most studies reporting neat ILs are usually based in fluorinated anions, that are more expensive, less benign and hence less attractive. Moreover, it is important to highlight that aqueous solutions of ILs are also very advantageous foreseeing systemic drug administration.

Considering that one of the most important physicochemical properties of bioformulations is the viscosity, the use of aqueous solutions enables viscosity reduction, and thus hydrated ILs appears as more adequate formulations for biopharmaceuticals. Nevertheless, there is a lack of studies addressing the viscosity of ILs, which is an extremely relevant issue since these compounds are being perused as alternative solvents in the field of biopharmaceuticals. The studies reviewed here reporting the stabilization of nucleic acids in neat ILs are summarized in Table 2.

Table 2. Summary of the studies reporting the stabilization of nucleic acids in neat ionic liquids (ILs)

IL	DNA/RNA source	Remarks	Ref.
[BOXA][PF_6], [BIM][PF_6], [BPYR][PF_6], [BPYRR][PF_6], [N111ETOH][PF_6], [BOXA][BF_4] [BIM][BF_4] [BPYR][BF_4], [BPYRR][BF_4] and [N_{111}ETOH][BF_4]	ssDNA and dsDNA	[BPYR][PF_6] and [N_{111}ETOH][PF_6] are the ILs that perform best in retaining the native structure of dsDNA. ssDNA structure is more stable in ILs based on [BF_4]$^-$ anions.	[83]
[C_4mim][Ac]	Basidiomycetes DNAs integrated within Schottky like junctions	Long-term stability of DNA up to 4 months at RT.	[84]
[Ch][IAA], [Ch][IBA]	sDNA	3.5 wt% of DNA was dissolved in both ILs in a short period of time. DNA presented stability and integrity in [Ch][IAA] up to 6 months, but was degraded in [Ch][IBA]. Possible to recycle [Ch][IAA] with very high yield.	[82]
[Ch][Ac], [Ch][For], [Ch][Gly] and [Ch][Pyr]	DNA duplexes	Dissolution of DNA in the neat ILs at RT was: 5 wt%, 12 wt%, 8 wt% and 2 wt%, respectively.	[85]
[Ch][HCOO]	sDNA	Interaction of DNA with the IL induced a B-Z transition in the DNA secondary structure.	[86]
[C_4mim][PF_6]	Plasmid DNA	Stable self-assembled nanostructures of DNA and IL that act as efficient nonviral vectors for bacterial gene transformation. Protection against enzymatic degradation. The IL is non-cytotoxic for bacterial cells.	[87]
Poly[3-butyl-1-vinylimidazolium L-proline] salt	Plasmid DNA	Stable self-assembled nanostructures of DNA and IL that act as efficient nonviral vectors for bacterial gene transformation. Protection against enzymatic degradation. The IL is non-cytotoxic for bacterial cells.	[88]

Deep Eutetic Solvents

Since DESs share many common properties with ILs, the possibility to use them as alternative solvents for nucleic acids is also of upmost importance [10, 12–14]. Several studies report the behavior of nucleic acids in a typical DES known as cholinium chloride:urea (ChCl:urea) at

1:2 [90–94]. Although DNA duplexes were reported to form in this DES by Mamajanov et al. [90], CD analyses and melting transition midpoints of dsDNAs in the DES, showed altered helical structures and low duplex stability [90]. It is perhaps not surprising that duplex stability is lower in this DES than in an aqueous solution, as urea is commonly used as a denaturant [90].

However, Tortora et al. [94] managed to demonstrate excellent physical–chemical properties of ChCl:urea DES as a co-solvent for dsDNA, able to not only ensure sDNA solubility but also providing increased stability to temperature-induced conformational alterations. For this matter, the authors mixed dsDNA from salmon testes in Tris solution with the hydrated DES ChCl:urea (1:2), reaching the final two molar rations of 158 and 474 DES molecules for each DNA base pair [94]. UVRR and CD analysis demonstrated that both DNA samples show higher transition temperatures than in buffer solution [94]. The increased interaction between the DES and DNA was attributed to the preferential hydrogen binding between the cholinium ions and the bases' proton-acceptor sites [94].

Urea played a minor role in the solvation of double strands, taking part as an enhancer of hydrogen bonding [94]. Moreover, the authors reported that as the ChCl:urea concentration increased, DES-DES interactions prevailed on DES-DNA ones, minimizing DES-induced structural stabilization [94]. Thus, hydrated solutions of the DES ChCl:urea are a better alternative for the stabilization of DNA duplexes [94]. Although neat ChCl:urea DES wasn't a very efficient solvent for DNA duplexes [90], G-quadruplex structures in neat ChCl:urea become more stable than in aqueous media; some G-quadruplex DNA molecules are even thermally stable at temperature higher than 110ºC [90–92].

In order to understand the enhanced stabilization of G-quadruplexes in ChCl:urea (1:2), Pal and Paul [95] explored the temperature-mediated conformational dynamics of c-kit oncogene promoter G-quadruplex DNA in this natural DES in the temperature range of 27-227ºC [95]. They observed that the c-kit DNA was stable at a temperature of 177ºC in ChCl:urea medium [95]. However, it unfolded when the water was used as solvent at the same temperature [95] which led to conclude that reline provided the ultrastability to c-kit G-quadruplex DNA at elevated temperature in comparison with water medium, previously validated by other studies [91–93, 95].

The authors found that the hydrogen bonding interactions between c-kit G-quadruplex DNA and ChCl:urea played a key role in the stabilization of the G-quadruplex DNA even at high temperature (up to 176.85ºC) [95]. Moreover, they also observed that π–π stacking interactions of tetrads (crucial in the stabilization of G-quadruplex formation) were very stable in DES medium up to 176.85ºC, but in water medium no such π–π stacking interactions were observed at high temperature [95]. When the temperature was elevated to 227ºC, the intratetrad hydrogen bounds of tetrad-1 were broken [95]. Nevertheless, other tetrads were stable for a long time, although they too collapsed eventually [95].

Hence, ChCl:urea not only protected the c-kit G-quadruplex but it also slowed the thermal denaturing process [95]. Furthermore, this study provided information not yet reported about the central cation K^+ oscillating its position in between tetrads at 176.85 and 226.85ºC in ChCl:urea medium with an increasing rate of oscillations as the temperature was raised (Figure 9) [95]. This distinctive "up down" phenomenon of the central cation was absent in water medium at 176.85ºC and it can play a major role in the stabilization of G-quadruplex DNA at high temperature [95].

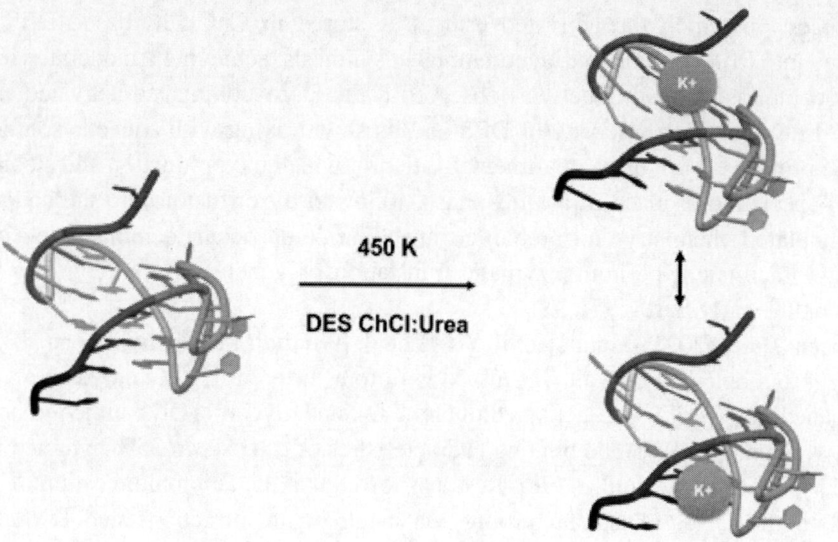

Figure 9. G-quadruplex present enhanced thermal stability in the DES ChCl:urea. The molecular mechanisms underlying its stabilization are hydrogen bonding interactions between G-quadruplex DNA and ChCl:urea, π–π stacking interactions of tetrads, and a distinctive "up adown" phenomenon of the central cation K^+ oscillating its position in between tetrads.

Prasad and co-workers [93] demonstrated that two DESs cholinium chloride: ethylene glycol (ChCl:EtG) at 1:2 and cholinium chloride: glycerol (ChCl:Gly) at 1:2, could dissolve up to 5.5 wt% and 2.5 wt% of DNA from salmon testes, respectively. They observed that DNA maintains B-form helical structures in these DESs and the regenerated DNA shows a high thermal and pH stability. The authors also proposed a likely DNA solvation mechanism being the interaction of cholinium cations with DNA phosphate on its basis. DNA duplex stability was also demonstrated by Gállego et al. [96] in the DES ChCl:Gly (1:4). In addition, it was also reported in this study the folding of 2D and 3D DNA nanostructures in this neat DES under isothermal conditions [96].

Furthermore, they observed that tuning the solvent composition with water leads to more rapid folding of the same structures and relatively fast folding of more complex ones [96]. These results indicated that ChCl:Gly (1:4) reduces the problem of kinetic traps typically encountered during folding in aqueous buffer, and that tuning the viscosity of water/DES mixtures provides a means to control the kinetics of DNA nanostructure self-assembly [96]. This study is innovative because it exploits the properties of ChCl:Gly (1:4) and its hydrated forms for the synthesis of more complex DNA structures with fewer local and global defects, representing a key challenge for DNA nanotechnology where the use of a complex DNA structure in anhydrous solvent is required [96]. However, it would be possible to perform accelerated folding in hydrated glycholinium followed by rendering the solvent anhydrous simply by the removal of water from ChCl:Gly (1:4) under vacuum [96].

Núñez-Pertíñez and Wilks [97] also explored ChCl:Gly-based DES as alternative solvents. In this study, the authors found that these DES have a positive impact on the stability of DNA-conjugated activated esters when compared with buffered aqueous solutions. The results highlighted that it is very challenging to produce DNA-conjugated activated esters in aqueous media for subsequent templated reactions, as they quickly decompose [97]. However, a two-stage process in which the activated ester was stored in ChCl:Gly-based DES and then transferred into PBS solution to trigger templated synthesis, achieving a moderate yield of 30% [97]. By comparison, no product was observed if the DNA-conjugated activated esters were stored in PBS. The high viscosity of DES and its slow mixing with aqueous solutions might allowed a primitive form of compartmentalization in which the production and stabilization of activated species took place in a DES phase, followed by diffusion into the aqueous phase where templated chemistry and prebiotic translation could occur, demonstrating how DESs could have facilitated the non-enzymatic translation of genetic information into functional peptides on the early Earth [97].

A recent study (2021) conducted by Yusof et al. [98] didn't use [Ch]Cl as an HBA. Instead tetrabutyl bromide (TBABr) salt was mixed with four different HBDs: ethylene glycol (EtG); 1,3-propanediol (1,3-PD); 1,5-pentanediol (1,5-PD), and glycerol (Gly) at the molar rations of 1:2, 1:3, 1:4, and 1:5 [98]. The binding characteristics of ct-DNA in 25% hydrated DESs was assessed by fluorescence studies [98]. Results showed that the ammonium cation of TBA^+ was attracted to the DNA phosphate groups via electrostatic attractions and DNA bases via hydrophobic interactions, being the two main interactions between the DESs and ct-DNA [98]. In addition, DES also groove bound via hydrogen bonding to the DNA's minor groove [98]. The authors also reported that the shorter the alkyl chain length (2 carbons) and higher EtG ratios of TBABr:EtG (1:5) increased the binding constant (K_b) between DES and DNA and decreased the binding of Gibbs energy [98]. Moreover, the higher number of OH groups caused the TBABr:Gly to form more hydrogen bonds with DNA bases and had the highest ability to quench DAPI from DNA [98].

Although DES are commonly describer as safe, the cytotoxicity and genotoxicity of DESs made of compounds commonly used for the stabilization of nucleic acids (ChCl:oxalic acid (OX):EtG; ChCl:OX:Gly; ChCl:citric acid (CA):EtG; and ChCl:CA:Gly at varying molar ratios was reported by Jangir et al. [99]. These DESs exhibited high toxicity towards HeLa cancer cell line, which the authors attributed to the presence of secondary HBDs [99]. The highest inhibitory effect was observed for ChCl:OX:EtG, while the lowest was observed for ChCl:CA:Gly. The DNA fragmentation activity was examined after exposure of HeLa cells to different DESs concentrations (50 mL and 200 mL) and results indicated that DNA impairment increased with the amount of solvent [99]. This study reinforces the idea that; however, the alternative solvents herein reviewed are commonly described as green and non-toxic, it is essential to include toxicity assays when using ILs or DESs as solvents for biomolecules with biotechnological and therapeutic applications.

Overall, it is possible to observe that the studies addressing the stabilization of nucleic acids in DES report mostly neat DES. Considering that these alternative solvents are eutectic mixtures of compounds; the addition of water may alter the properties of the mixture and lead to the loss of the nonconventional character of the DES. Moreover, the exploration of DES as alternative media for biomolecules is a recent subject, especially for nucleic acids. These seem to be the most likely explanations for which there are only few studies about nucleic acids in aqueous solutions of DES. Using neat DES as media for biomolecules may raise some economic concerns, if the compounds used to make the DES are expensive.

In addition, the viscosity of a neat DES can also represent a major problem, making it unsuitable for several applications in the pharmaceutical industry, such as the systemic administration of drugs. Therefore, the stabilization of nucleic acids in DES with small amounts of water added should be investigated in future studies. Nevertheless, the use of a fully non-aqueous media, such neat DES or ILs, can also be advantageous for other purposes. For instance, to acquire experimental results at very high temperatures or to promote reactions involving nucleic acids as hybrid catalysts.

The studies reviewed here reporting the stabilization of nucleic acids in DES are summarized in Table 3.

Table 3. Summary of the studies reviewed in this work reporting the stabilization of nucleic acids in DES

DES	DNA/RNA source	Remarks	Ref.
ChCl:urea (1:2)	Duplex, Triplex, and G-Quadruplex	DNA duplexes presented altered helical structures and low stability. G-quadruplexes presented enhanced stability in this neat DES.	[90]
ChCl:urea (1:2)	DNA duplexes	DNA duplexes were dissolved and presented enhanced thermal stability in a DES/buffer solution. Interaction between the cholinium ions the bases' proton-acceptor sites. Increased DES concentrations minimizes DES-DNA interaction and subsequent stabilization.	[94]
ChCl:urea (1:2)	Human telomere sequence (HTS) DNA	G-quadruplexes presented enhanced stability in this neat DES.	[91]
ChCl:urea (1:2)	G-quadruplexes	G-quadruplexes presented enhanced stability in this neat DESs.	[92]
ChCl:urea (1:2)	c-kit oncogene promoter G-quadruplex DNA	Stable G-quadruplex in this neat DES at temperature up to 450K. Hydrogen bonding interactions between G-quadruplex and reline played key role in the stabilization. The rate of oscillation of the G-quadruplex central cation was increased when temperature was increased.	[95]
ChCl:EtG (1:2) and ChCl:Gly (1:2)	DNA duplex from salmon testes	High stability of DNA duplex in both DES. DNA was more soluble in the ChCl:EtG mixture.	[93]
ChCl:Gly (1:4)	dsDNA	Stable duplex in this DES. Folding of 2D and 3D DNA nanostructures in this neat DES under isothermal conditions.	[96]
ChCl:Gly-based DES	DNA-conjugated activated esters	Enhanced stabilization of DNA-conjugated activated esters when compared with buffered aqueous solutions	[97]
TBABr:EG; TBABr:1,3-PD; TBABr:1,5-PD; and TBABr:Gly at (1:2; 1:3; 1:4; and 1:5)	ct-DNA	TBA^+ was attracted to the DNA phosphate groups via electrostatic attractions and DNA bases via hydrophobic interactions. The DES TBABr:EG was the one with highest binding constant and lower binding Gibbs energy, especially at higher EG ratios and due to the shorter alkyl chain. The DES TBABr:Gly had the highest ability to bind to DNA due to the large number of OH groups leading to the formation of more hydrogen bonds.	[98]

Conclusion

This chapter discussed the recent works that deals with the stability of nucleic acids in alternative solvents, for instance in ILs and DESs, while giving an overview on the main molecular mechanisms and interactions underlying these phenomena. Moreover, the main fields and applications that could take advantage of the increased stability and long-term storage of these biomolecules in these solvents were also highlighted in this chapter and summarized in Figure 10.

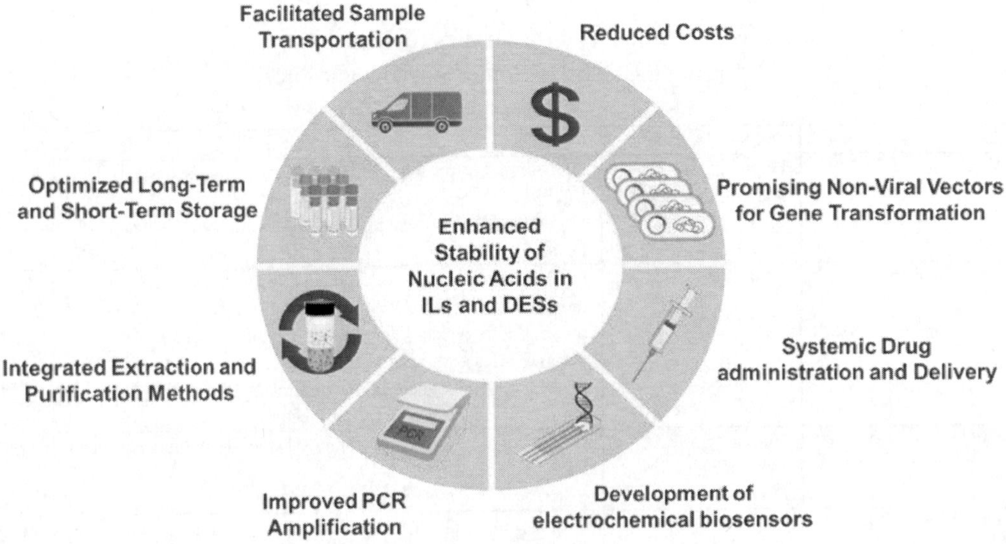

Figure 10. Schematic summary of the main advantages and applications due to the enhanced stability of nucleic acids in ILs and DESs.

The extent of stabilization ability of ILs and DES for nucleic acids vary and usually depend upon the combination of ions (both cation/anion), type of structure (of the biomacromolecule), water content and molar ratio (in the case of DES). Nevertheless, it is interesting to notice that for the main groups of biomolecules focused on this chapter, ILs and DES were found to be effective in maintaining/improving their structure and/or activity, while simultaneously contributing for the development of more sustainable strategies for nucleic acids handling/ processing/storage.

Having in mind the environmental and health concerns on the use of new alternative solvents, some studies were found to have already integrated ecotoxicity and cytotoxicity studied of the developed formulations in their investigations, showing for the majority of the cases, that low toxicities are associated with ILs/DES. However, these ecotoxicity and cytotoxicity credentials are not always ascertained in the majority of the studies; thus, due to the upmost importance of these features, they should always be considered in future studies.

There is also an increasing tendency observed in the last few years, in the replacement of the typical and well-studied imidazolium-based ILs by new more sustainable, biodegradable and biocompatible ILs, obtained from natural and renewable sources. Thus, efforts should continue to be made to develop novel formulations of ILs and DES based on natural and renewable sources in order to enhance the stability of nucleic acids through strategies that can be faithfully considered fully sustainable and biocompatible, allowing further improvements in several applications fields.

Recently, few studies were still found regarding the use of DES in the stability of nucleic acids, probably due to their novelty when comparing with ILs that have been deeply studied for the last decades. Therefore, it is expected that, in the future, more studies using this class of solvents could be reported for the processing/stability of a wider spectra of (bio)molecules. Finally, the possibility of reusing and recycling the ILs/DES after the biomolecule processing should also be taken into consideration in future studies on this field, in order to fully accomplish the green credentials of the reported strategies.

Acknowledgments

This work was developed within the scope of the project CICECO-Aveiro Institute of Materials, UIDB/50011/2020, UIDP/50011/2020 & LA/P/0006/2020, financed by national funds through the FCT/MCTES (PIDDAC). Ana P. M. Tavares acknowledges FCT for the research contract CEECIND/2020/01867.

References

[1] Kumar, A.; Bisht, M.; Venkatesu, P. Biocompatibility of Ionic Liquids towards Protein Stability: A Comprehensive Overview on the Current Understanding and Their Implications. *Int. J. Biol. Macromol.* 2017, *96*, 611–651. https://doi.org/10.1016/j.ijbiomac.2016.12.005.

[2] Cardoso, I. S.; Pedro, A. Q.; Silvestre, A. J. D.; Freire, M. G. Ionic Liquids and Deep Eutectic Solvents in the Field of Environmental Monitoring. *Green Anal. Chem.* 2019, 203–240. https://doi.org/10.1007/978-981-13-9105-7_8.

[3] Freire, M. G.; Coutinho, J. A. P. Editorial Overview: A Closer Look on Sustainable Solvents and Processes. *Curr. Opin. Green Sustain. Chem.* 2019, *18*, A4–A6. https://doi.org/10.1016/J.COGSC.2019.07.003.

[4] Pedro, S. N.; Freire, C. S. R.; Silvestre, A. J. D.; Freire, M. G. The Role of Ionic Liquids in the Pharmaceutical Field: An Overview of Relevant Applications. *Int. J. Mol. Sci.* 2020, *21* (21), 8298. https://doi.org/10.3390/IJMS21218298.

[5] Park, S.; Kazlauskas, R. J. Biocatalysis in Ionic Liquids – Advantages beyond Green Technology. *Curr. Opin. Biotechnol.* 2003, *14* (4), 432–437. https://doi.org/https://doi.org/10.1016/S0958-1669(03)00100-9.

[6] Freire, M. G.; Cláudio, A. F. M.; Araújo, J. M. M.; Coutinho, J. A. P.; Marrucho, I. M.; Lopes, J. N. C.; Rebelo, L. P. N. Aqueous Biphasic Systems: A Boost Brought about by Using Ionic Liquids. *Chem. Soc. Rev.* 2012, *41* (14), 4966–4995. https://doi.org/10.1039/C2CS35151J.

[7] Magina, S.; Barros-Timmons, A.; Ventura, S. P. M.; Evtuguin, D. V. Evaluating the Hazardous Impact of Ionic Liquids – Challenges and Opportunities. *J. Hazard. Mater.* 2021, *412*, 125215. https://doi.org/10.1016/J.JHAZMAT.2021.125215.

[8] Zhao, H. DNA Stability in Ionic Liquids and Deep Eutectic Solvents. *J. Chem. Technol. Biotechnol.* 2015, *90* (1), 19–25. https://doi.org/https://doi.org/10.1002/jctb.4511.

[9] Capela, E. V; Valente, A. I.; Nunes, J. C. F.; Magalhães, F. F.; Rodríguez, O.; Soto, A.; Freire, M. G.; Tavares, A. P. M. Insights on the Laccase Extraction and Activity in Ionic-Liquid-Based Aqueous Biphasic Systems. *Sep. Purif. Technol.* 2020, *248*, 117052. https://doi.org/https://doi.org/10.1016/j.seppur.2020.117052.

[10] Pätzold, M.; Siebenhaller, S.; Kara, S.; Liese, A.; Syldatk, C.; Holtmann, D. Deep Eutectic Solvents as Efficient Solvents in Biocatalysis. *Trends Biotechnol.* 2019, *37* (9), 943–959. https://doi.org/10.1016/j.tibtech.2019.03.007.

[11] Capela, E. V.; Santiago, A. E.; Rufino, A. F. C. S.; Tavares, A. P. M.; Pereira, M. M.; Mohamadou, A.; Aires-Barros, M. R.; Coutinho, J. A. P.; Azevedo, A. M.; Freire, M. G. Sustainable Strategies Based on

Glycine-Betaine Analogue Ionic Liquids for the Recovery of Monoclonal Antibodies from Cell Culture Supernatants. *Green Chem.* 2019, *21* (20), 5671–5682. https://doi.org/10.1039/c9gc02733e.

[12] Macário, I. P. E.; Jesus, F.; Pereira, J. L.; Ventura, S. P. M.; Gonçalves, A. M. M.; Coutinho, J. A. P.; Gonçalves, F. J. M. Unraveling the Ecotoxicity of Deep Eutectic Solvents Using the Mixture Toxicity Theory. *Chemosphere* 2018, *212*, 890–897. https://doi.org/10.1016/J.CHEMOSPHERE.2018.08.153.

[13] Macário, I. P. E.; Oliveira, H.; Menezes, A. C.; Ventura, S. P. M.; Pereira, J. L.; Gonçalves, A. M. M.; Coutinho, J. A. P.; Gonçalves, F. J. M. Cytotoxicity Profiling of Deep Eutectic Solvents to Human Skin Cells. *Sci. Reports* 2019, *9* (1), 1–9. https://doi.org/10.1038/s41598-019-39910-y.

[14] Pedro, S. N.; Freire, M. G.; Freire, C. S. R.; Silvestre, A. J. D. Deep Eutectic Solvents Comprising Active Pharmaceutical Ingredients in the Development of Drug Delivery Systems. 2019, *16* (5), 497–506. https://doi.org/10.1080/17425247.2019.1604680.

[15] Liu, P.; Hao, J. W.; Mo, L. P.; Zhang, Z. H. Recent Advances in the Application of Deep Eutectic Solvents as Sustainable Media as Well as Catalysts in Organic Reactions. *RSC Advances*. 2015, 48675–48704. https://doi.org/10.1039/c5ra05746a.

[16] Teles, A. R. R.; Capela, E. V; Carmo, R. S.; Coutinho, J. A. P.; Silvestre, A. J. D.; Freire, M. G. Solvatochromic Parameters of Deep Eutectic Solvents Formed by Ammonium-Based Salts and Carboxylic Acids. *Fluid Phase Equilib.* 2017, *448*, 15–21. https://doi.org/10.1016/j.fluid.2017.04.020.

[17] Pena-Pereira, F.; Namieśnik, J. Ionic Liquids and Deep Eutectic Mixtures: Sustainable Solvents for Extraction Processes. *ChemSusChem* 2014, *7* (7), 1784–1800. https://doi.org/10.1002/cssc.201301192.

[18] Choi, Y. H.; van Spronsen, J.; Dai, Y.; Verberne, M.; Hollmann, F.; Arends, I. W. C. E.; Witkamp, G.-J.; Verpoorte, R. Are Natural Deep Eutectic Solvents the Missing Link in Understanding Cellular Metabolism and Physiology? *Plant Physiol.* 2011, *156* (4), 1701–1705. https://doi.org/10.1104/pp.111.178426.

[19] Benedetto, A.; Ballone, P. Room Temperature Ionic Liquids Meet Biomolecules: A Microscopic View of Structure and Dynamics. *ACS Sustain. Chem. Eng.* 2016, *4* (2), 392–412. https://doi.org/10.1021/acssuschemeng.5b01385.

[20] Pedro, A. Q.; Pereira, P.; Quental, M. J.; Carvalho, A. P.; Santos, S. M.; Queiroz, J. A.; Sousa, F.; Freire, M. G. Cholinium-Based Good's Buffers Ionic Liquids as Remarkable Stabilizers and Recyclable Preservation Media for Recombinant Small RNAs. *ACS Sustain. Chem. Eng.* 2018, *6* (12), 16645–16656. https://doi.org/10.1021/acssuschemeng.8b03900.

[21] Duan, J.; Wang, X.; Kizer, M. E. Biotechnological and Therapeutic Applications of Natural Nucleic Acid Structural Motifs. *Top. Curr. Chem.* 2020, *378* (2), 26. https://doi.org/10.1007/s41061-020-0290-z.

[22] Tateishi-Karimata, H.; Sugimoto, N. Biological and Nanotechnological Applications Using Interactions between Ionic Liquids and Nucleic Acids. *Biophys. Rev.* 2018, *10* (3), 931–940. https://doi.org/10.1007/s12551-018-0422-7.

[23] Tateishi-Karimata, H.; Nakano, M.; Sugimoto, N. Comparable Stability of Hoogsteen and Watson-Crick Base Pairs in Ionic Liquid Choline Dihydrogen Phosphate. *Sci. Rep.* 2014, *4*, 3593. https://doi.org/10.1038/srep03593.

[24] Coutinho, M. F.; Matos, L.; Santos, J. I.; Alves, S. RNA Therapeutics: How Far Have We Gone? *Adv. Exp. Med. Biol.* 2019, *1157*, 133–177. https://doi.org/10.1007/978-3-030-19966-1_7.

[25] Lieberman, J. Tapping the RNA World for Therapeutics. *Nat. Struct. Mol. Biol.* 2018, *25* (5), 357–364. https://doi.org/10.1038/s41594-018-0054-4.

[26] Wang, H.; Wang, J.; Zhang, S. Binding Gibbs Energy of Ionic Liquids to Calf Thymus DNA: A Fluorescence Spectroscopy Study. *Phys. Chem. Chem. Phys.* 2011, *13* (9), 3906–3910. https://doi.org/10.1039/c0cp01815e.

[27] Jumbri, K.; MA, K.; NM, Y.; MBA, R.; Ahmad, H.; RA, W. Fluorescence and Molecular Simulation Studies on the Interaction between Imidazolium-Based Ionic Liquids and Calf Thymus DNA. *Processes* 2020, *8* (1). https://doi.org/10.3390/pr8010013.

[28] Rossi, B.; Tortora, M.; Catalini, S.; Vigna, J.; Mancini, I.; Gessini, A.; Masciovecchio, C.; Mele, A. Insight into the Thermal Stability of DNA in Hydrated Ionic Liquids from Multi-Wavelength UV Resonance Raman Experiments. *Phys. Chem. Chem. Phys.* 2021, *23* (30), 15980–15988. https://doi.org/10.1039/d1cp01970h.

[29] Bottari, C.; Catalini, S.; Foggi, P.; Mancini, I.; Mele, A.; DR, P.; Paciaroni, A.; Gessini, A.; Masciovecchio, C.; Rossi, B. Base-Specific Pre-Melting and Melting Transitions of DNA in Presence of Ionic Liquids Probed by Synchrotron-Based UV Resonance Raman Scattering. *J. Mol. Liq.* 2021, *330*. https://doi.org/10.1016/j.molliq.2021.115433.

[30] An Introduction to Ionic Liquids. *Choice Reviews Online* 2010, 47-6874-47–6874. https://doi.org/10.5860/choice.47-6874.

[31] Zhao, D.; Liao, Y.; Zhang, Z. Toxicity of Ionic Liquids. *CLEAN – Soil, Air, Water* 2007, *35* (1), 42–48. https://doi.org/https://doi.org/10.1002/clen.200600015.

[32] Thuy Pham, T. P.; Cho, C. W.; Yun, Y. S. Environmental Fate and Toxicity of Ionic Liquids: A Review. *Water Res.* 2010, *44* (2), 352–372. https://doi.org/10.1016/j.watres.2009.09.030.

[33] Docherty, K. M.; Kulpa, C. F. Toxicity and Antimicrobial Activity of Imidazolium and Pyridinium Ionic Liquids. *Green Chem.* 2005, *7* (4), 185–189. https://doi.org/10.1039/b419172b.

[34] Stepnowski, P.; Skladanowski, A. C.; Ludwiczak, A.; Laczyńska, E. Evaluating the Cytotoxicity of Ionic Liquids Using Human Cell Line HeLa. *Hum. Exp. Toxicol.* 2004, *23* (11), 513–517. https://doi.org/10.1191/0960327104ht480oa.

[35] Ventura, S. P. M.; Marques, C. S.; Rosatella, A. A.; Afonso, C. A. M.; Gonçalves, F.; Coutinho, J. A. P. Toxicity Assessment of Various Ionic Liquid Families towards Vibrio Fischeri Marine Bacteria. *Ecotoxicol. Environ. Saf.* 2012, *76* (2), 162–168. https://doi.org/10.1016/j.ecoenv.2011.10.006.

[36] M, A. T.; C, L. A.; Irena, N.; Ibrahim, I.; Satomi, M.; Colin, W.; Sandrine, H.; C, W. M. Emerging Risk from "Environmentally-Friendly" Solvents: Interaction of Methylimidazolium Ionic Liquids with the Mitochondrial Electron Transport Chain Is a Key Initiation Event in Their Mammalian Toxicity. *Food Chem. Toxicol.* 2020, *145*, 111593. https://doi.org/https://doi.org/10.1016/j.fct.2020.111593.

[37] LX, H.; Xiong, Q.; WJ, S.; GY, H.; YS, L.; GG, Y. New Insight into the Negative Impact of Imidazolium-Based Ionic Liquid [C(10)Mim]Cl on Hela Cells: From Membrane Damage to Biochemical Alterations. *Ecotoxicol. Environ. Saf.* 2021, *208*. https://doi.org/10.1016/j.ecoenv.2020.111629.

[38] Xiaohong, C.; Hua, Y.; Chunming, G.; Ruxia, Y.; Zhaoxue, H.; Yajun, L. Transcriptomic Analysis of the Phytotoxic Effects of 1-Allyl-3-Methylimidazolium Chloride on the Growth and Plant Hormone Metabolic Pathways of Maize (Zea Mays L.) Seedlings. *Chemosphere* 2020, *241*, 125013. https://doi.org/https://doi.org/10.1016/j.chemosphere.2019.125013.

[39] Yaqi, X.; Jinhua, W.; Zhongkun, D.; Bing, L.; Albert, J.; Meiying, T.; Lusheng, Z.; Jun, W. Toxicity Evaluation of Three Imidazolium-Based Ionic Liquids ([C6mim]R) on Vicia Faba Seedlings Using an Integrated Biomarker Response (IBR) Index. *Chemosphere* 2020, *240*, 124919. https://doi.org/10.1016/j.chemosphere.2019.124919.

[40] Guilherme, J.; Fátima, J.; P.M, V. S.; J.M, G. F.; Jana, A.; L, P. J. New Insights on the Effects of Ionic Liquid Structural Changes at the Gene Expression Level: Molecular Mechanisms of Toxicity in Daphnia Magna. *J. Hazard. Mater.* 2021, *409*, 124517. https://doi.org/10.1016/j.jhazmat.2020.124517.

[41] Wenxiu, L.; Lei, Z.; Zhongkun, D.; Bing, L.; Jinhua, W.; Jun, W.; Cheng, Z.; Lusheng, Z. Acute Toxicity, Oxidative Stress and DNA Damage of Three Task-Specific Ionic Liquids ([C2NH2MIm]BF4, [MOEMIm]BF4, and [HOEMIm]BF4) to Zebrafish (Danio Rerio). *Chemosphere* 2020, *249*, 126119. https://doi.org/https://doi.org/10.1016/j.chemosphere.2020.126119.

[42] Vijayaraghavan, R.; Izgorodin, A.; Ganesh, V.; Surianarayanan, M.; MacFarlane, D. R. Long-Term Structural and Chemical Stability of DNA in Hydrated Ionic Liquids. *Angew. Chemie - Int. Ed.* 2010, *49* (9), 1631–1633. https://doi.org/10.1002/anie.200906610.

[43] Morimitsu, Y.; Matsuno, H.; Ohta, N.; Sekiguchi, H.; Takahara, A.; Tanaka, K. Mechanical Stabilization of Deoxyribonucleic Acid Solid Films Based on Hydrated Ionic Liquid. *Biomacromolecules* 2020, *21* (2), 464–471. https://doi.org/10.1021/acs.biomac.9b01207.

[44] EEL, T.; Wiraja, C.; CA, C.; CJ, X.; Mitragotri, S. Stabilization and Topical Skin Delivery of Framework Nucleic Acids Using Ionic Liquids. *Adv. Ther.* 2020, *3* (7). https://doi.org/10.1002/adtp.202000041.

[45] PR, D.; Nalini, V.; Surianarayanan, M.; Krishnakumar, S. Towards Safer Non-Volatile Tissue Fixatives: Evaluation of Choline-Based Ionic Liquids for Fixing Ocular Tissues. *Ecotoxicol. Environ. Saf.* 2021, *209*. https://doi.org/10.1016/j.ecoenv.2020.111777.

[46] Chandran, A.; Ghoshdastidar, D.; Senapati, S. Groove Binding Mechanism of Ionic Liquids: A Key Factor in Long-Term Stability of DNA in Hydrated Ionic Liquids? *J. Am. Chem. Soc.* 2012, *134* (50), 20330–20339. https://doi.org/10.1021/ja304519d.

[47] Garai, A.; Ghoshdastidar, D.; Senapati, S.; Maiti, P. K. Ionic Liquids Make DNA Rigid. *J. Chem. Phys.* 2018, *149* (4), 45104. https://doi.org/10.1063/1.5026640.

[48] Tateishi-Karimata, H.; Nakano, M.; Pramanik, S.; Tanaka, S.; Sugimoto, N. I-Motifs Are More Stable than G-Quadruplexes in a Hydrated Ionic Liquid. *Chem. Commun.* 2015, *51* (32), 6909–6912. https://doi.org/10.1039/C5CC00666J.

[49] Dinis, T. B. V.; Sousa, F.; Freire, M. G. Insights on the DNA Stability in Aqueous Solutions of Ionic Liquids. *Front. Bioeng. Biotechnol.* 2020, *8*. https://doi.org/10.3389/fbioe.2020.547857.

[50] Tateishi-Karimata, H.; Sugimoto, N. A-T Base Pairs Are More Stable than G-C Base Pairs in a Hydrated Ionic Liquid. *Angew. Chemie - Int. Ed.* 2012, *51* (6), 1416–1419. https://doi.org/10.1002/anie.2011 06423.

[51] Nakano, M.; Tateishi-Karimata, H.; Tanaka, S.; Sugimoto, N. Choline Ion Interactions with DNA Atoms Explain Unique Stabilization of A-T Base Pairs in DNA Duplexes: A Microscopic View. *J. Phys. Chem. B* 2014, *118* (2), 379–389. https://doi.org/10.1021/jp406647b.

[52] Portella, G.; Germann, M. W.; Hud, N. V; Orozco, M. MD and NMR Analyses of Choline and TMA Binding to Duplex DNA: On the Origins of Aberrant Sequence-Dependent Stability by Alkyl Cations in Aqueous and Water-Free Solvents. *J. Am. Chem. Soc.* 2014, *136* (8), 3075–3086. https://doi.org/10. 1021/ja410698u.

[53] Sarkar, S.; Rajdev, P.; PC, S. Hydrogen Bonding of Ionic Liquids in the Groove Region of DNA Controls the Extent of Its Stabilization: Synthesis, Spectroscopic and Simulation Studies. *Phys. Chem. Chem. Phys.* 2020, *22* (27), 15582–15591. https://doi.org/10.1039/d0cp01548b.

[54] Ghoshdastidar, D.; Senapati, S. Dehydrated DNA in B-Form: Ionic Liquids in Rescue. *Nucleic Acids Res.* 2018, *46* (9), 4344–4353. https://doi.org/10.1093/nar/gky253.

[55] Bonnet, J.; Colotte, M.; Coudy, D.; Couallier, V.; Portier, J.; Morin, B.; Tuffet, S. Chain and Conformation Stability of Solid-State DNA: Implications for Room Temperature Storage. *Nucleic Acids Res.* 2010, *38* (5), 1531–1546. https://doi.org/10.1093/nar/gkp1060.

[56] Paunescu, D.; Fuhrer, R.; Grass, R. N. Protection and Deprotection of DNA—High-Temperature Stability of Nucleic Acid Barcodes for Polymer Labeling. *Angew. Chemie Int. Ed.* 2013, *52* (15), 4269–4272. https://doi.org/https://doi.org/10.1002/anie.201208135.

[57] Lou, J. J.; Mirsadraei, L.; Sanchez, D. E.; Wilson, R. W.; Shabihkhani, M.; Lucey, G. M.; Wei, B.; Singer, E. J.; Mareninov, S.; Yong, W. H. A Review of Room Temperature Storage of Biospecimen Tissue and Nucleic Acids for Anatomic Pathology Laboratories and Biorepositories. *Clin. Biochem.* 2014, *47* (4–5), 267–273. https://doi.org/10.1016/j.clinbiochem.2013.12.011.

[58] Ivanov, V. I.; DYu, K. A-DNA in Solution as Studied by Diverse Approaches. *Methods Enzymol.* 1992, *211*, 111–127. https://doi.org/10.1016/0076-6879(92)11008-7.

[59] Ivanov, V. I.; Minchenkova, L. E.; Minyat, E. E.; Frank-Kamenetskii, M. D.; Schyolkina, A. K. The B to A Transition of DNA in Solution. *J. Mol. Biol.* 1974, *87* (4), 817–833. https://doi.org/10.1016/0022-2836(74)90086-2.

[60] Malenkov, G.; Minchenkova, L.; Minyat, E.; Schyolkina, A.; Ivanov, V. The Nature of the B-A Transition of DNA in Solution. *FEBS Lett.* 1975, *51* (1), 38–42. https://doi.org/10.1016/0014-5793(75)80850-7.

[61] Cheatham, T. E.; Crowley, M. F.; Fox, T.; Kollman, P. A. A Molecular Level Picture of the Stabilization of A-DNA in Mixed Ethanol–Water Solutions. *Proc. Natl. Acad. Sci.* 1997, *94* (18), 9626 LP – 9630. https://doi.org/10.1073/pnas.94.18.9626.

[62] Sprous, D.; Young, M. A.; Beveridge, D. L. Molecular Dynamics Studies of the Conformational Preferences of a DNA Double Helix in Water and an Ethanol/Water Mixture: Theoretical Considerations of the A ⇔ B Transition. *J. Phys. Chem. B* 1998, *102* (23), 4658–4667. https://doi.org/ 10.1021/jp980548g.

[63] Jayaram, B.; Sprous, D.; Young, M. A.; Beveridge, D. L. Free Energy Analysis of the Conformational Preferences of A and B Forms of DNA in Solution. *J. Am. Chem. Soc.* 1998, *120* (41), 10629–10633. https://doi.org/10.1021/ja981307p.

[64] Fujita, K.; Ohno, H. Stable G-Quadruplex Structure in a Hydrated Ion Pair: Cholinium Cation and Dihydrogen Phosphate Anion. *Chem. Commun.* 2012, *48* (46), 5751–5753. https://doi.org/10.1039/C2CC30554B.

[65] Satpathi, S.; Sengupta, A.; Hridya, V. M.; Gavvala, K.; Koninti, R. K.; Roy, B.; Hazra, P. A Green Solvent Induced DNA Package. *Sci. Rep.* 2015, *5* (1), 9137. https://doi.org/10.1038/srep09137.

[66] Gosule, L. C.; Schellman, J. A. Compact Form of DNA Induced by Spermidine. *Nature* 1976, *259* (5541), 333–335. https://doi.org/10.1038/259333a0.

[67] Yoshikawa, Y.; Umezawa, N.; Imamura, Y.; Kanbe, T.; Kato, N.; Yoshikawa, K.; Imanaka, T.; Higuchi, T. Effective Chiral Discrimination of Tetravalent Polyamines on the Compaction of Single DNA Molecules. *Angew. Chemie Int. Ed.* 2013, *52* (13), 3712–3716. https://doi.org/10.1002/anie.201209144.

[68] Dias, R.; Mel'nikov, S.; Lindman, B.; Miguel, M. G. DNA Phase Behavior in the Presence of Oppositely Charged Surfactants. *Langmuir* 2000, *16* (24), 9577–9583. https://doi.org/10.1021/la000640f.

[69] Grueso, E.; Cerrillos, C.; Hidalgo, J.; Lopez-Cornejo, P. Compaction and Decompaction of DNA Induced by the Cationic Surfactant CTAB. *Langmuir* 2012, *28* (30), 10968–10979. https://doi.org/10.1021/la302373m.

[70] Ganguli, M.; Babu, J. V.; Maiti, S. Complex Formation between Cationically Modified Gold Nanoparticles and DNA: An Atomic Force Microscopic Study. *Langmuir* 2004, *20* (13), 5165–5170. https://doi.org/10.1021/la036049a.

[71] Zinchenko, A.; Tsumoto, K.; Murata, S.; Yoshikawa, K. Crowding by Anionic Nanoparticles Causes DNA Double-Strand Instability and Compaction. *J. Phys. Chem. B* 2014, *118* (5), 1256–1262. https://doi.org/10.1021/jp4107712.

[72] Yoshikawa, K.; Yoshikawa, Y.; Koyama, Y.; Kanbe, T. Highly Effective Compaction of Long Duplex DNA Induced by Polyethylene Glycol with Pendant Amino Groups. *J. Am. Chem. Soc.* 1997, *119* (28), 6473–6477. https://doi.org/10.1021/ja970445w.

[73] Fant, K.; Esbjörner, E. K.; Lincoln, P.; Nordén, B. DNA Condensation by PAMAM Dendrimers: Self-Assembly Characteristics and Effect on Transcription. *Biochemistry* 2008, *47* (6), 1732–1740. https://doi.org/10.1021/bi7017199.

[74] Su, C. J.; Chen, C. Y.; Lin, M. C.; Chen, H. L.; Iwase, H.; Koizumi, S.; Hashimoto, T. Nucleosome-like Structure from Dendrimer-Induced DNA Compaction. *Macromolecules* 2012, *45* (12), 5208–5217. https://doi.org/10.1021/ma300308y.

[75] Zinchenko, A. A.; Sergeyev, V. G.; Yamabe, K.; Murata, S.; Yoshikawa, K. DNA Compaction by Divalent Cations: Structural Specificity Revealed by the Potentiality of Designed Quaternary Diammonium Salts. *ChemBioChem* 2004, *5* (3), 360–368. https://doi.org/10.1002/cbic.200300797.

[76] Soufi, A.; Sawasdichai, A.; Shukla, A.; Noy, P.; Dafforn, T.; Smith, C.; Jayaraman, P.-S.; Gaston, K. DNA Compaction by the Higher-Order Assembly of PRH/Hex Homeodomain Protein Oligomers. *Nucleic Acids Res.* 2010, *38* (21), 7513–7525. https://doi.org/10.1093/nar/gkq659.

[77] Ding, Y.; Zhang, L.; Xie, J.; Guo, R. Binding Characteristics and Molecular Mechanism of Interaction between Ionic Liquid and DNA. *J. Phys. Chem. B* 2010, *114* (5), 2033–2043. https://doi.org/10.1021/jp9104757.

[78] Mazid, R. R.; Divisekera, U.; Yang, W.; Ranganathan, V.; MacFarlane, D. R.; Cortez-Jugo, C.; Cheng, W. Biological Stability and Activity of SiRNA in Ionic Liquids. *Chem. Commun.* 2014, *50* (88), 13457–13460. https://doi.org/10.1039/C4CC05086J.

[79] Quental, M. V. V.; Q. Pedro, A.; Pereira, P.; Sharma, M.; A. Queiroz, J.; A.P. Coutinho, J.; Sousa, F.; G. Freire, M.; Pedro, A. Q.; Pereira, P.; Sharma, M.; Queiroz, J. A.; Coutinho, J. A. P. P.; Sousa, F.; Freire, M. G.; Q. Pedro, A.; Pereira, P.; Sharma, M.; A. Queiroz, J.; A.P. Coutinho, J.; Sousa, F.; G. Freire, M.; Pedro, A. Q.; Pereira, P.; Sharma, M.; Queiroz, J. A.; Coutinho, J. A. P. P.; Sousa, F.; Freire, M. G. *Integrated Extraction-Preservation Strategies for RNA Using Biobased Ionic Liquids*; American Chemical Society 2019; 7, 9439–9448. https://doi.org/10.1021/acssuschemeng.9b00688.

[80] Tulsiyan, K. D.; Jena, S.; González-Viegas, M.; Kar, R. K.; Biswal, H. S. Structural Dynamics of RNA in the Presence of Choline Amino Acid Based Ionic Liquid: A Spectroscopic and Computational Outlook. *ACS Cent. Sci.* 2021. https://doi.org/10.1021/acscentsci.1c00768.

[81] CH, Z.; Varona, M.; Anderson, J. Magnetic Ionic Liquids as Solvents for RNA Extraction and Preservation. *ACS OMEGA* 2020, *5* (19), 11151–11159. https://doi.org/10.1021/acsomega.0c01098.

[82] Mukesh, C.; Mondal, D.; Sharma, M.; Prasad, K. Rapid Dissolution of DNA in a Novel Bio-Based Ionic Liquid with Long-Term Structural and Chemical Stability: Successful Recycling of the Ionic Liquid for Reuse in the Process. *Chem. Commun.* 2013, *49* (61), 6849–6851. https://doi.org/10.1039/C3CC42829J.

[83] Cardoso, L.; Micaelo, N. M. DNA Molecular Solvation in Neat Ionic Liquids. *Chemphyschem* 2011, *12* (2), 275–277. https://doi.org/10.1002/cphc.201000645.

[84] Rizan, N.; HA, T.; YS, T.; Abdullah, Z.; Idris, A.; Periasamy, V. Effect of Ionic Liquid on the Long-Term Structural and Chemical Stability of Basidiomycetes DNAs Integrated within Schottky-like Junctions. *Appl. Phys. A-MATERIALS Sci. Process.* 2021, *127* (2). https://doi.org/10.1007/s00339-021-04303-4.

[85] Sharma, G.; RA, S.; MM, P.; TK, M.; NA, C.; Prasad, K. Are Ionic Liquids and Deep Eutectic Solvents the Same?: Fundamental Investigation from DNA Dissolution Point of View. *J. Mol. Liq.* 2021, *328*. https://doi.org/10.1016/j.molliq.2021.115386.

[86] Bhatt, J.; Pereira, M. M.; Prasad, K. Simultaneous Morphological Transformation of Metal Salt and Conformations of DNA in a Bio-Based Ionic Liquid. *Int. J. Biol. Macromol.* 2019, *135*, 926–930. https://doi.org/10.1016/j.ijbiomac.2019.06.012.

[87] Soni, S. K.; Sarkar, S.; Mirzadeh, N.; Selvakannan, P. R.; Bhargava, S. K. Self-Assembled Functional Nanostructure of Plasmid DNA with Ionic Liquid [Bmim][PF6]: Enhanced Efficiency in Bacterial Gene Transformation. *Langmuir* 2015, *31* (16), 4722–4732. https://doi.org/10.1021/acs.langmuir.5b00402.

[88] Zhang, Y.; Chen, X.; Lan, J.; You, J.; Chen, L. Synthesis and Biological Applications of Imidazolium-Based Polymerized Ionic Liquid as a Gene Delivery Vector. *Chem. Biol. Drug Des.* 2009, *74* (3), 282–288. https://doi.org/https://doi.org/10.1111/j.1747-0285.2009.00858.x.

[89] Shi, Y.; Liu, Y. L.; Lai, P. Y.; Tseng, M. C.; Tseng, M. J.; Li, Y.; Chu, Y. H. Ionic Liquids Promote PCR Amplification of DNA. *Chem. Commun.* 2012, *48* (43), 5325–5327. https://doi.org/10.1039/c2cc31740k.

[90] Mamajanov, I.; Engelhart, A. E.; Bean, H. D.; Hud, N. V. DNA and RNA in Anhydrous Media: Duplex, Triplex, and G-Quadruplex Secondary Structures in a Deep Eutectic Solvent. *Angew. Chemie Int. Ed.* 2010, *49* (36), 6310–6314. https://doi.org/https://doi.org/10.1002/anie.201001561.

[91] Lannan, F. M.; Mamajanov, I.; Hud, N. V. Human Telomere Sequence DNA in Water-Free and High-Viscosity Solvents: G-Quadruplex Folding Governed by Kramers Rate Theory. *J. Am. Chem. Soc.* 2012, *134* (37), 15324–15330. https://doi.org/10.1021/ja303499m.

[92] Zhao, C.; Ren, J.; Qu, X. G-Quadruplexes Form Ultrastable Parallel Structures in Deep Eutectic Solvent. *Langmuir* 2013, *29* (4), 1183–1191. https://doi.org/10.1021/la3043186.

[93] Mondal, D.; Sharma, M.; Mukesh, C.; Gupta, V.; Prasad, K. Improved Solubility of DNA in Recyclable and Reusable Bio-Based Deep Eutectic Solvents with Long-Term Structural and Chemical Stability. *Chem. Commun.* 2013, *49* (83), 9606–9608. https://doi.org/10.1039/C3CC45849K.

[94] Tortora, M.; Vigna, J.; Mancini, I.; Mele, A.; Gessini, A.; Masciovecchio, C.; Rossi, B. Effect of Hydrated Deep Eutectic Solvents on the Thermal Stability of DNA. *CRYSTALS* 2021, *11* (9). https://doi.org/10.3390/cryst11091057.

[95] Pal, S.; Paul, S. Understanding The Role of Reline, a Natural DES, on Temperature-Induced Conformational Changes of C-Kit G-Quadruplex DNA: A Molecular Dynamics Study. *J. Phys. Chem. B* 2020, *124* (15), 3123–3136. https://doi.org/10.1021/acs.jpcb.0c00644.

[96] Gállego, I.; Grover, M. A.; Hud, N. V. Cover Picture: Folding and Imaging of DNA Nanostructures in Anhydrous and Hydrated Deep-Eutectic Solvents (Angew. Chem. Int. Ed. 23/2015). *Angew. Chemie Int. Ed.* 2015, *54* (23), 6653. https://doi.org/https://doi.org/10.1002/anie.201503959.

[97] Núñez-Pertíñez, S.; Wilks, T. R. Deep Eutectic Solvents as Media for the Prebiotic DNA-Templated Synthesis of Peptides. *Front. Chem.* 2020, *8*, 41. https://doi.org/10.3389/fchem.2020.00041.

[98] Yusof, R.; Jumbri, K.; Ahmad, H.; Abdulmalek, E.; MBA, R. Binding of Tetrabutylammonium Bromide Based Deep Eutectic Solvent to DNA by Spectroscopic Analysis. *Spectrochim. ACTA PART A-MOLECULAR Biomol. Spectrosc.* 2021, *253*. https://doi.org/10.1016/j.saa.2021.119543.

[99] Jangir, AK; Lad, B.; Dani, U.; Shah, N.; Kuperkar, K. *In Vitro* toxicity Assessment and Enhanced Drug Solubility Profile of Green Deep Eutectic Solvent Derivatives (DESDs) Combined with Theoretical Validation. *RSC Adv.* 2020, *10* (40), 24063–24072. https://doi.org/10.1039/c9ra10320a.

Chapter 8

The Exfoliation and Dispersion of Graphene in Ionic Liquids

Rima Biswas[*]
Process Simulation Research Group, School of Chemical Engineering,
Vellore Institute of Technology, Vellore, Tamil Nadu, India

Abstract

Graphene has attracted a lot of interest due to its exceptional physical and chemical properties as well as its wide range of applications. A lot of different techniques are used to produce graphene sheets. Liquid-phase exfoliation (LPE) is considered to be a very promising large-scale technique to obtain high-quality graphene sheets in certain solvents. Novel ionic liquids (ILs) are synthesized to contain aromatic groups on the imidazolium cation that interact non-covalently with the graphene surface.

In this chapter, we focus on the emerging use of ILs for the dispersion, exfoliation, and modification of graphene nanosheets. The potentials of mean force and free energies, characterizing steered separation of graphene sheets and their subsequent solvation by ILs are also reviewed in this chapter. The importance of solvation energy as an indicator of solvent quality for spontaneous exfoliation is demonstrated using a simple thermodynamic model.

Based on the established mechanism, one can predict and screen a suitable IL for spontaneous exfoliation of graphene. Density function theory (DFT) calculations correlate with experimental observations and provide a basis for predictive modeling of IL design for graphene dispersions. Computational and experimental results corroborate the development of new high-performance ILs for use in solvent-assisted graphene exfoliation.

Keywords: graphene, ionic liquids, spontaneous exfoliation, density functional theory

Abbreviations

AFM	Atomic Force Microscopy
DMF	Dimethylformamide

---

[*] Corresponding Author's Email: rima.biswas@vit.ac.in.

In: Properties and Applications of Ionic Liquids
Editors: Noureddine Abidi and Julia L. Shamshina
ISBN: 979-8-88697-635-9

FESEM	Field emission scanning electron microscopy
GNs	Graphene nanosheets
GO	Graphene oxide
HOPG	Highly ordered pyrolytic graphite
IGAs	Inorganic graphene analogues
ILCs	Ionic liquid crystals
NMR	Nuclear Magnetic Resonance
PANI	Poly(aniline)
PIL	Poly(ionic liquid)
PMMA	Poly(methyl methacrylate)
PU	Poly(urethane)
QC	Quantum calculations
RGO	Reduced graphene oxide
RTILs	Room-temperature ionic liquids
SWCNTs	Single-walled carbon nanotubes
TEM	Transmission electron microscopy
VDW	van der Waals force
XPS	X-ray photoelectron spectroscopy

Introduction

Graphene is the most fundamental component of all allotropic materials [1, 2] As shown in Figure 1, it can be wrapped into 0D fullerenes, rolled into 1D carbon nanotubes (CNTs), and stacked into 3D graphite [3, 4]. Graphene is the strongest material yet measured (fracture strength, 125 GPa) and has the highest Young's modulus (~1100 GPa) [5]. It can withstand current densities six orders of magnitude higher than copper, is impermeable to gases, has a large specific surface area (estimated value, 2630 $m^2 g^{-1}$) [6], has a record thermal conductivity (~5000 $W m^{-1} K^{-1}$) [7], and has an abnormal quantum Hall effect [8]. The charge carriers in this material have enormous inherent mobility, have no effective mass, and can travel for microns without dispersing at room temperature, making it a boon to next-generation electronics and devices.

Following the Scotch tape method [1] a plethora of methods for producing graphene nanosheets (GNs) have emerged, ranging from epitaxy to chemical vapor deposition [9], and liquid-phase exfoliation to separate GNs from a variety of graphitic starting materials (e.g., expanded graphite, graphene oxide (GO), highly ordered pyrolytic graphite (HOPG)) [10]. Among the several ways for producing GNs, liquid-phase exfoliation has a distinct advantage due to its ease of use, low variable cost, and enormous potential for large-scale production [10, 11]. However, there are certain disadvantages to this process, the most significant of which is the low quality of GNs produced as a result of residual solvents on the surface and the nature of the volatile organics utilized, which are often hazardous to the environment.

ILs are organic or partially inorganic salts that are liquid at temperatures lower than 100°C. They are utilized in numerous alternative synthetic approaches, including those connected to green chemistry, because of several of their features, such as compatibility with a wide range of organic and inorganic solvents, minimal risk of flammability, low volatility, and explosiveness [12, 13]. Furthermore, ILs can easily be separated from mixtures with other compounds via liquid-liquid extraction or solid-phase extraction. Interesting applications include conductive and stimuli-responsive materials, particularly in those circumstances when a non-volatile solvent is required [14, 15].

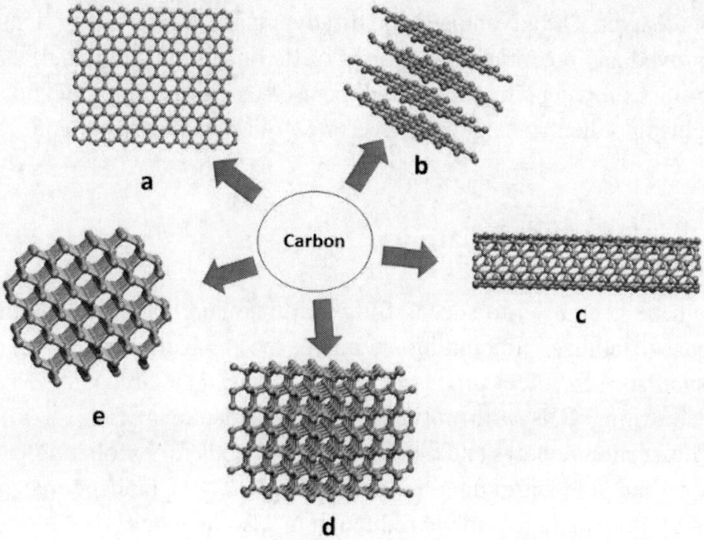

Figure 1. Different allotropes of carbon: (a) graphene, (b) graphite, (c) carbon nanotube, (d) lonsdaleite, and (e) diamond.

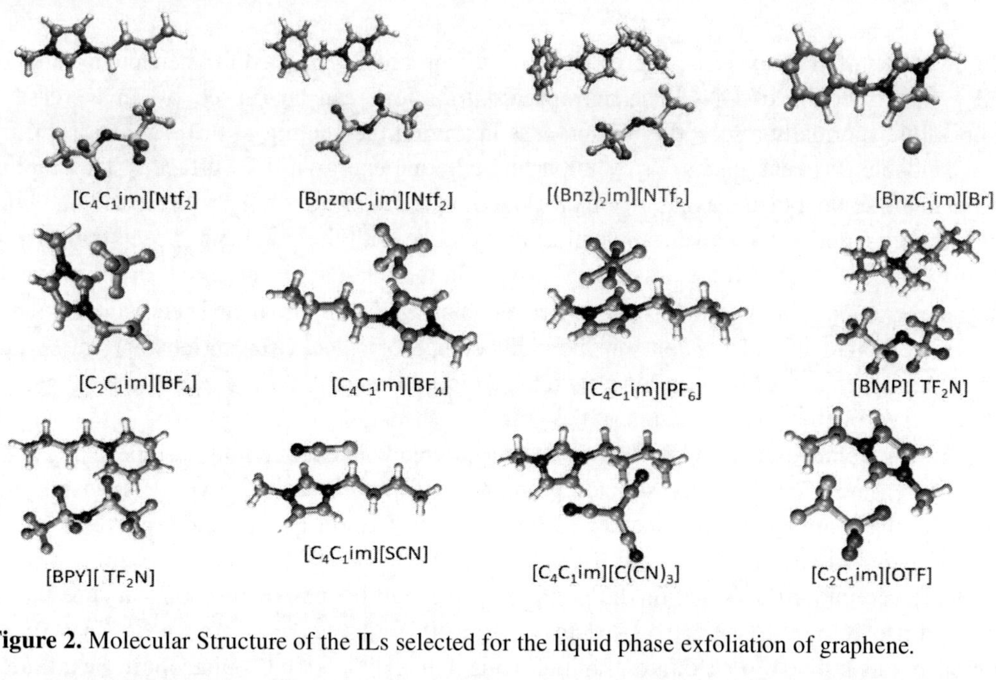

Figure 2. Molecular Structure of the ILs selected for the liquid phase exfoliation of graphene.

Imidazolium-based ILs have been used to successfully disperse single-walled carbon nanotubes (SWCNTs) and IL-based bucky gels, a gelatinous mixture of SWCNTs and ILs [16, 17]. ILs have surface tensions that closely match the surface energy of graphite. ILs can allow

GN dispersions by stabilizing against re-aggregation via Coulomb repulsion of the surface charges presented by the adsorbed IL. ILs are being used as an efficient solvent for the exfoliation and stabilization of GNs for their excellent characteristics. The different types are ILs are used for the exfoliation of graphene as shown in Figure 2.

The ability of ILs to disperse, exfoliate, and modify GNs, GO, and reduced GO (RGO), is examined in this chapter. The advantages and disadvantages of employing ILs will be reviewed, as well as the proved and potential applications of IL-functionalized materials, with an eye on industrial scale-up. Computer simulations will be used as a screening tool and a way to provide molecular insights into the mechanism of ILs for exfoliation and dispersion.

Methods of IL-Assisted Exfoliation

Many methods have been used to successfully exfoliate and disperse GNs into ILs and their mixes for IL-based stabilizers, including chemical, mechanical, microwave, electrochemical, hydrothermal, and sonochemical processes. Furthermore, IL-stabilized GNs provide a good substrate for fabricating GNs with polymer and metal nanoparticles for improved material applications. This section reviews and discusses the methods for exfoliating GNs with ILs. The usage of ILs for dispersion and reduction of GO into RGO, a graphene material having flaws left over from oxidation and incomplete reduction, is also included.

Grinding Methods and Sonication

Minimization of the Gibbs free energy (ΔG) for the process is required for efficient exfoliation and good dispersion of GNs. The entropic contributions can be caused by an increase in temperature (normally a solvothermal process involving the heating of the graphite, resulting in expandable graphene layers) or by imparting external energy to the surface of the graphite sheets in the solvent media using mechanical methods such as grinding or ultrasound, in which energy is imparted to the solvent molecules that collide with the graphene sheets, transferring energy. Through a shear force, this energy assists in the relative displacement of the graphene sheets, overcoming the π-π stabilizing interactions between the graphene sheets. Furthermore, efficient exfoliation is dependent on the right choice of solvent (or surfactant) via surface interactions such as surface matching, which balances the energy cost of graphite layer growth while preventing the re-aggregation of the graphene sheets.

Direct graphite exfoliation in ILs has been described with encouraging results. Wang et al., [18] were the first to investigate graphite exfoliation using two ILs: 1-butyl-3-methylimidazolium bis(trifluoromethanesulfonyl)imide [C_4C_1im][NTf_2] and 1-butyl-1-methyl-pyrrolidinium bis(trifluoromethanesulfonyl)imide [$Pyrr_{4,1}$][NTf_2]. Table 1 summarises the exfoliation conditions. Exfoliation did not lead to any significant oxidation of graphene flakes, according to X-ray photoelectron spectroscopy (XPS) data. In the same year, an acoustic approach was utilized to exfoliate graphite using 1-hexyl-3-methyl-imidazolium hexafluoro-phosphate ([C_6C_1im][PF_6]) [19]. Longer sonication times and a high beginning concentration were shown to be advantageous conditions for achieving higher exfoliated graphene concentrations, according to the authors. For the graphene exfoliation, new ILs have been synthesized.

Bari et al., [20] used one or two benzyl groups to functionalize an imidazolium cation in order to establish π-π interactions in contact with the graphene surface. The energy for the exfoliation of graphite into suspensions was supplied by ultrasound. UV-vis absorbance spectroscopy was used to determine the concentration of graphene in suspension. The viscosity of $[BnzC_1im]Br$ was determined to be too high to scatter graphite. With the author's setup, the concentration of suspended graphene in the $[C_4C_1im][NTf_2]$ was too low to be identified by absorbance. In comparison to $[BnzC_1im][NTf_2]$, $[(Bnz)_2im][NTf_2]$ stabilized a bigger amount of graphene (as shown in Table 1). *Ab initio* calculations revealed that both benzyl groups of $[(Bnz)_2im][NTf_2]$ were discovered to be orientated parallel to the surface of graphene, in contrast to the single benzyl substituent in $[BnzC_1im][NTf_2]$ [20].

Matsumoto et al., [21] established a quick and efficient method for exfoliating natural graphite. Microwaves were used to exfoliate a 25 mg mL^{-1} suspension of graphite powder in an oligomeric IL containing repeating units of $[(ethoxy)im][PF_6]$ for 0.5 hours. The temperature of the combination hit 443 K. The authors discovered that 10% of the IL was destroyed using proton NMR. There were no other structural flaws in graphene, and 95% of the flakes produced were single-layer graphene. In $[C_4C_1im][PF_6]$, the identical technique was employed, but with a reduced exfoliation rate.

Table 1. Comparison of ILs used for the liquid phase exfoliation of graphite

ILs	Exfoliation set up	Mass Yield	Size of flakes	No. of exfoliated layers
$[C_4C_1im][NTf_2]$ [18]	Ultrasound(750W)/1h+10,000rpm/1h	47.5%	≈ μm	1-5
$[Pyrr_{41}][NTf_2]$ [18]	Ultrasound(750W)/1h+10,000rpm/1h	47.5%	≈ μm	1-5
$[C_4C_1im][NTf_2]$ [20]	Ultrasound(10W)/1h+5,000rpm/6h	unstable	-	-
$[C_6C_1im][PF_6]$ [19]	Ultrasound(550W)/24h+4,000rpm/0.33h	8.17%	3-4 μm	1-5
$[BnzC_1im][NTf_2]$ [20]	Ultrasound(10W)/1h+5,000rpm/6h	0.81%	-	-
$[(Bnz)_2im][NTf_2]$ [20]	Ultrasound(10W)/1h+5,000rpm/6h	58%	-	2-5
$[C_4C_1im][PF_6]$ [22]	Grinding/4h+3,000rpm/0.5h	20%	9 nm	2-7
$[(ethoxy)im][PF_6]$ [21]	Microwaves(30W)/0.5h	92%	1-5 μm	1

Cavitation, the creation of tiny bubbles from dissolved gases or vapors as a result of pressure fluctuations and shear stresses given by sonication waves, occurs frequently during sonication. These bubbles have a high internal Laplace pressure, which can cause cyclic stress on the graphene surface, resulting in GNs fatigue and defects. Instead of sonication, methods that rely on shearing forces to exfoliate graphite, such as mechanical grinding [22] (ball milling, mortar, and pestle, etc.) or the use of a vortex fluidic device, can be used to avoid cavitation effects [23]. To exfoliate graphene from virgin graphitic flakes, Shang et al., [22] used mechanical grinding in combination with modest amounts of IL. Using a mortar and pestle, graphite flakes (< 20 μm^2) were combined and ground with $[C_4mim][PF_6]$ for 4 h to generate a gel. After that, dimethylformamide (DMF) and acetone were added, followed by centrifugation to remove any remaining IL, sedimentation, and drying. For the type of end product, grinding periods and beginning graphite: IL ratios were critical. When grinding periods were less than 30 min (with 1: 10 – 1: 4 ratios of graphite flakes (mg) to IL (mL)), graphene sheets with 2–5 layers were prominent. On the other hand, increasing the grinding duration and decreasing the amount of IL, resulted in the creation of graphene nanodots with diameters of 9–29 nm and heights of 1–16 nm, which were controlled by the grinding time.

Electrochemical Methods

Exfoliation of graphite precursors to create GNs has also been done using electrochemical techniques. Electrochemical methods have potential benefits such as a simple one-step process, ease of operation and control during the process, mass production of large sheets ranging from a few nanometres to micrometers in length under ambient conditions, and without introducing a large number of defects or oxygenation on the surface of the GNs [24]. The end product's high quality allows it to be used in a wide range of applications, including imaging and sensing, fuel cells, and electrocatalysts.

Liu et al., [25] were the first to report the electrochemical synthesis of GNs to produce IL-functionalized GNs. The electrolyte in their study was a mixture of imidazolium-based ILs (1-octyl-3-methylimidazolium hexafluorophosphate [C_8mim][PF_6], 1-octyl-3-methylimidazolium tetrafluoroborate [C_8mim][BF_4], 1-octyl-3-methylimidazolium chloride [C_8mim]Cl and [C_4mim][PF_6]). A dark precipitate of IL functionalized GNs was formed at the bottom of the reactor after applying a 15 V potential between two graphite rods. TEM and FESEM images revealed that the IL-functionalized GNs flakes were up to ~700 nm long and 500 nm wide, with an average thickness of 1.1 nm, which was validated by AFM tests. These IL-functionalized GNs may be distributed uniformly in polar solvents like DMF. The IL cation reducing on the cathode, generating a free radical at the C2 position of the IL cation, which subsequently coupled with the π-bond electron of a GNs, was thought to be the process of IL-functionalized GNs production [25].

Lu et al., [24] studied the mechanism for electrochemical GNs exfoliation and functionalization by ILs. They discovered that the amount of water in the ILs was crucial in determining the shape and type of carbon nanostructure that formed, which ranged from nanoparticles to nanoribbons and nanosheets. Carbon nanoparticles formed when the water content was greater than 10%, while nanoribbons or nanosheets formed when the water content was lower. Furthermore, IL-functionalization of the GNs was only detected at water concentrations of less than 10%, and larger water concentrations were linked to higher oxidation of the resultant structures. Exfoliation experiments with neat [C_4C_1im][BF_4] IL produced a highly viscous electrolyte from which a bucky gel composed of IL-functionalized GNs and carbon nanoparticles could be isolated after centrifugation with the supernatant containing soluble exfoliates consisting of fluorescent carbon structures of nanoparticles and nanoribbons could be isolated [24]. This "one-pot" synthesis has a lot of potential in terms of industrial scalability and mass production, as well as biological labelling and imaging applications.

Singh et al., [26] used a three-electrode system for exfoliation, rather than the traditional two-electrode technique, comprising of a graphite pencil and two platinum electrodes bathed in triethylsulfonium bis(trifluoromethylsulfonyl)imide $[S_{222}][Tf_2N]$ IL. Electrochemical exfoliation techniques are gaining popularity for GNs destined for battery applications due to their ability to generate controlled homogeneous size, shape, and increased specific energy capacity. In comparison to graphite commonly employed in batteries (200 mA h g^{-1}), Wei *et al.*,[27] used electrochemical exfoliation of graphite in combinations of $[C_4C_1im][BF_4]$, deionized water, and lithium salts to generate GNs with conductivities as high as 750 mA h g^{-1}.

IL as a Stabilizer or Surfactant

The solubility of resultant RGO in water and other organic solvents is frequently limited, resulting in unwanted RGO sheet aggregation. As surfactants/stabilizer additives, different salts and ILs have been utilized to overcome this aggregation. However, for such additives to be successful, their surface energy must be precisely aligned with that of RGO nanosheets. Zhou et al., [28] studied the Hummers method to scatter GO in water, then added poly(1-vinyl-3-butylimidazolium chloride), a polymer ionic liquid (PIL), followed by hydrazine for chemical reduction of GO. By engaging with the GO via non-covalent – interactions, the PIL worked as a stabilizer.

At high concentrations, Zhang et al., [29] observed the production of a persistent RGO dispersion in three hydrophilic ILs $[C_4C_1im][BF_4]$, 1-allyl-3-methylimidazolium chloride [Amim]Cl, and N-butylpyridinium ($[C_4py][BF_4]$. The IL was mixed with an aqueous GO solution prepared using Hummer's technique. Following the removal of the water, the GO was reduced at 80 °C for 12 h with 85% hydrazine monohydrate to generate RGO suspensions (easily up to 7.0 mg mL^{-1}) within ILs that remained stable for several months. The stabilising effect of ILs was attributed to an IL molecular ordering layer on the surface of RGO, which was likely caused via cation–π or π–π interactions.

Kim et al., [30] reported an anion exchange mechanism to reversibly phase transfer RGO between water and organic solvents, using PIL, poly(1-vinyl-3-ethylimidazolium) bromide to form a complex with GO, which was suspended in an aqueous solvent and further reduced to RGO, resulting in the [PIL]Br-G complex. When the lithium salt ($Li[Tf_2N]$) was introduced to the system, it caused an anion exchange between the Br^- and $[Tf_2N]^-$ anions, resulting in the [PIL]Br-G complex (water-soluble) generating the $[PIL][NTf_2]$-G complex (organic phase soluble). Yan et al., [31] also described a simple and successful approach for phase transfer of dispersed aqueous GO from the aqueous phase to a non-polar organic solvent such as ortho-dichlorobenzene in the presence of the IL 1,3-didodecylimidazolium bromide $[C_{12}C_{12}im]Br$.

Due to their better solubility in the organic solvent than other shorter alkyl chain imidazolium bromide ILs like $[C_4C_1im]Br$ and $[C_{12}C_1im]Br$, the two dodecyl groups $[C_{12}C_{12}im]Br$ were critical for facilitating the phase transfer process. The transfer of GO from the aqueous to the organic phase was also influenced by experimental conditions such as pH and the GO/IL ratio, with the best results obtained at pH = 9 and a weight ratio of 1: 8 GO: DDIB. By expanding the length of the cation alkyl chain to 12 or more carbons, ILs can change into a smectic liquid crystal state. The features of both ILs and liquid crystals are combined in these IL crystals (ILCs). Different forms of mesophases exist in ILCs, with the smectic A phase being the most prevalent. This is the least ordered phase, with molecules arranged in layers and the long alkyl chain axis perpendicular to the plane on average [32]. The viscosity of ILC can be lowered by thermal heating, helping in the intercalation of graphite to form GNs.

Acik et al., [33] studied three ammonium-based ILs as intercalating agents between GO sheets distributed in water or propylene carbonate (*N*-methyl-*N*,*N*,*N*-tris(2-hydroxyethyl) ammonium iodide, *N*-methyl-*N*,*N*,*N*-tris(2-hydroxyethyl) ammonium methane sulphate, and *N*,*N*,*N*-tributyl-*N*-octylammonium methane sulphate). While the aqueous GO-IL solutions were centrifuged for 30 min and stirred overnight, the GO-ILs solution in PC was tip sonicated for 10 min. The samples were subsequently thermally reduced at 500–900 °C, allowing interlayer separation and RGO exfoliation of carbon nanosheets to take place. Due to a lack of covalent functionalization of the GO by the other two ILs examined, the *N*,*N*,*N*-tributyl-*N*-octylammonium methane sulphate IL best promoted this expansion and exfoliation.

Microwaves, which are electromagnetic radiation with a frequency ranging from 0.3 to 300 GHz, are a quick and easy way to prepare exfoliated graphene sheets from a variety of precursors. When utilized as a microwave reaction medium, IL, in particular, is responsive to microwave heating via ionic conduction, resulting in favorable yields and reaction rates. Matsumoto et al., [21] used microwave irradiation to agitate graphitic layers, as well as the newly developed oligomeric ionic liquids $IL2PF_6$ and $IL4PF_6$ as fluid media. Microwaves are well absorbed by graphite and graphene as well as ILs [34-36]. They discovered that graphite can be exfoliated in 30 min, yielding high-purity pure graphene with a 93% yield based on the weight of loaded graphite and a 95% selectivity towards single-layer graphene (as shown in Figure 3). Furthermore, its structural integrity was only marginally poorer than that of the precursor graphite, as determined by Raman spectroscopy and X-ray photoelectron spectroscopy (XPS).

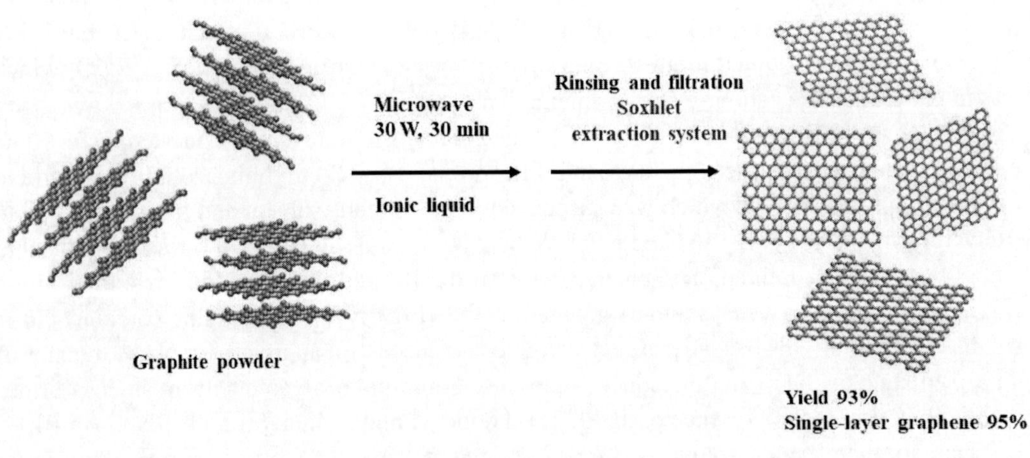

Figure 3. Microwave-assisted liquid-phase exfoliation of graphite in Ils [21].

Individual graphene sheets with polymer compatibility, good dispersity, and improved chemical and physical properties can be produced by functionalizing graphene sheets with polymers such as poly(methyl methacrylate) (PMMA), poly(aniline) (PANI), poly(urethane) (PU), and metal nanoparticles such as Ag, Au, TiO_2, and MnO_2 NPs, etc., using covalent or noncovalent methods involving ILs as a media. In such procedures, ILs operate as a surfactant, forming stable graphene nanocomposites by interacting with GO sheets and polymers via cation–π or π-π interactions. Yang et al., [37] demonstrated that using an amine-terminated IL,

1-(3-aminopropyl)-3-methylimidazolium bromide ($[NH_2C_3C_1im]Br$) to functionalize GO via a nucleophilic ring-opening reaction, a feasible approach for the poly-dispersion of chemically transformed graphene sheets can be achieved. Through electrostatic repulsion and increasing the distance between the GO sheets from 9.6 to 14.9 Å, the grafted IL on the GO sheets enhanced their stability and dispersion (in multiple solvents including water, DMSO, and DMF). This enhanced stability is beneficial in electrochemical applications.

Insights from Computational Methods

From a microscopic perspective, researchers believe that π–π interactions, also known as cation–π interactions, are the surface contacts that allow GNs to stabilize diverse solvents and ILs. Computer simulations provide a means of better understanding the factors that cause GNs to exfoliate and stabilize in various solvents, as well as an a priori screening tool for rating the exfoliation effectiveness of different solvents for GNs and IGAs. Since the 1980s, simulated studies on solvent interactions with graphite surfaces have been conducted, although the solvents used have been limited to water [38] and other organic solvents [39, 40]. The study of the interactions of graphite and other surfaces with ILs has grown in popularity over the last decade [41]. While these simulations show a static picture of distinct surfaces interacting with ILs (due to the fact that the sheets or surfaces were assumed to be static in nature), the energetics and dynamics of interactions during exfoliation and dispersion of 2-D materials in ILs are missing.

Towards this, I recently evaluated the solvation energies for graphene exfoliation using room-temperature ionic liquids (RTILs) from molecular dynamics (MD) simulations and quantum calculations (QC) [42]. The importance of the solvation energy as an indicator of solvent quality for spontaneous exfoliation is demonstrated using a simple thermodynamic model [42]. MD simulations were used to calculate the solvation energy of 10 ILs surrounding a single graphene sheet. Surface excess, sheet deformation, and surface tension are all found to be closely connected. QC was used to examine the properties of the interaction between ILs and graphene. The interaction of the researched ILs' components was investigated, and the interaction energy E_{int} between ILs and graphene is reported in Figure 4. The connection between a cation and graphene is much stronger than that between anions. Cations are more attracted to the graphene plane when they approach it from above via cation-π interactions. Cations approaching from the edge have less appealing interactions [42].

Bari *et. al.* [20] reported new ILs with regulated chemical functionality variations and show that they can be used to disperse graphene at high concentrations. ILs without these aromatic groups are less effective in graphene dispersion and the concentration of dispersed graphene increases as the contact between the cation and the graphene surface increases. The binding energies to graphene of the cations and anion of three ILs were computed: $[C_1C_1im]^+$, $[BnzC_1im]^+$, $[(Bnz)_2im]^+$, and $[NTf_2]^-$. The binding energies of the graphene–cation and graphene–anion complexes were calculated as the difference between the energies of the complex and the individual structures (*i.e.*, E = *E*(complex) [*E*(ion) + *E*(graphene)]) and are shown in Figure 5. The dimethylimidazolium ring adopts a parallel–displaced–stacking type of structure in the absence of aromatic substituents. According to the computational calculations, adding π–π stacking competent substituents will boost ionic liquid adsorption on the graphene surface, potentially improving exfoliation and stabilizing processes.

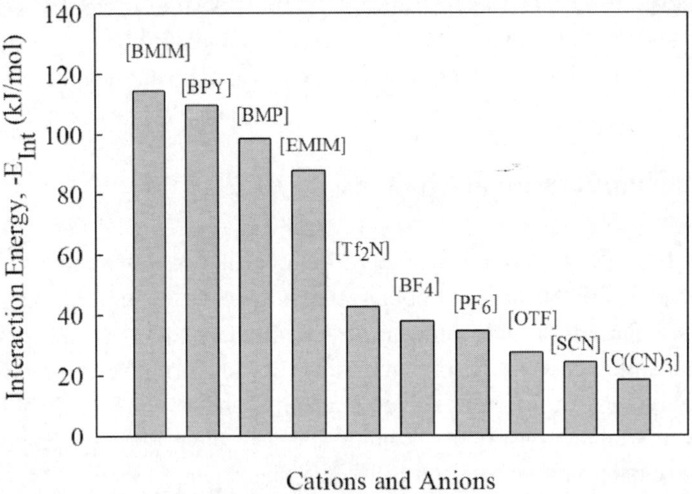

Figure 4. The interaction energies (-Eint) between cations or anions of IL and graphene were calculated at B3LYP/6–311 + g(d,p) level of theory [42].

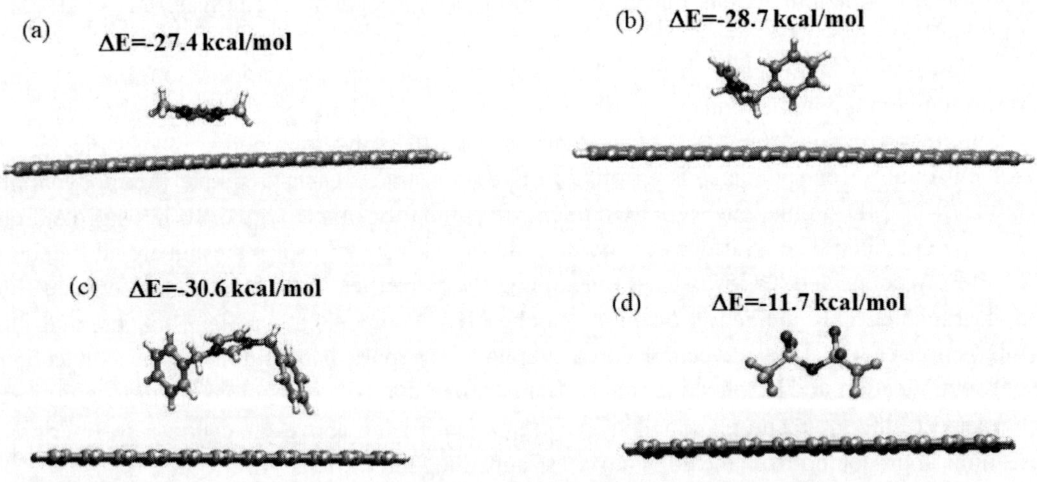

Figure 5. Interaction energy between Graphene and IL: (a) $[C_1C_1im]^+$ cation and graphene, (b) $[BnzC_1im]^+$ cation and graphene, (c) $[(Bnz)_2im]^+$ and graphene, and (d) $[NTf_2]^-$ anion and graphene [20].

For the imidazolium ring engaging with the graphene surface, the alkyl–π interactions are dominating. While this contradicts Bell and co-workers' [43] simulations (which showed a dominance of the cation ring parallel to the graphite surface, implying π–π interactions), it appears to be consistent with sum-frequency vibrational-spectroscopy experiments on imidazolium-based ILs on silica and graphene surfaces [44, 45]. The cations preferentially orient at an angle with the alkyl groups perpendicular to the surface in these experiments. Nonetheless, computational tools are proven to be an effective complement to atomic-scale research and analysis. In addition to the better understanding gained from the kinetic theory of colloid aggregation [39] and Flory Higgins theory adapted for the dispersion of nanotubes [46] and nanosheets, using normal-mode analysis and solvation-shell structure around the nanosheets could further enhance our understanding of the dynamics of exfoliation and stabilization processes. Molecular simulations can reveal the shape of the curve for the differential capacitance of the graphite–IL interface versus electrode potential, revealing the electrode's surface roughness and suggesting that controlling the electrode's topography could improve energy densities in electric double-layer capacitors [47].

Zhao and Hu [48] computed the free energy difference between a system of two graphene sheets. The work required to bring two graphene sheets from infinity to r is equivalent to the free energy difference between a system of two graphene sheets separated at a distance of r and a system of graphene separated at a distance of infinity ($\Delta G(r) = G(r) - G(\infty)$). Andersen [49] first devised this methodology for determining the potential of mean force in 1986, and it was recently utilized to explore the effect of solvents on the thermodynamics of graphene sheets [50]. Berne and co-workers [51] demonstrated that the addition of urea to water can disrupt the hydrophobic contact between two graphene plates by observing the difference in potential of mean force.

Blankschtein and co-workers [39, 50] calculated the potential of mean force between two graphene sheets coated with surfactant and placed in different polar organic solvents, confirming the presence of a metastable bilayer graphene structure provided the right solvents are used. In ILs, calculating the free energy difference as a function of distance for two graphene sheets has its own relevance and difficulty.

This computation serves three purposes. First, cation interactions have long been thought of as the fundamental interaction between positive-charged particles and molecules rich in -electrons, and have been widely employed to interpret experimental data, despite the fact that the underlying relationship is unclear. It will be fascinating to see if a molecular model that uses simply the VDW interaction to represent both short-ranged strong cation and long-ranged weak dispersion interactions can explain experimental results. Second, because of their versatility and green chemistry qualities, ILs have been considered as viable alternatives for typical organic solvents. Understanding graphene dispersion in ILs at various temperatures, charges, and chain lengths provides a thorough examination of the effects of various ionic solvents on graphene characteristics. Third, it is significantly more difficult to simulate IL systems than any other solvents due to the existence of unique spatial and energetic heterogeneity in ILs, which was initially identified by Voth and coworkers [52, 53] and Lopes and co-workers [54, 55]. Because of the sluggish dynamics of ILs, simulation results based on various initial configurations frequently differ dramatically. The properties of ILs are substantially more difficult to converge on than those of typical organic solvents.

Conclusion

The exfoliation of graphene in various ILs and the application of these IL-GNs composites are reviewed in this chapter. This IL-based exfoliation method will also allow for the adoption of low-cost solution processing techniques including electrostatic adsorption, filtration, drop-casting, spraying, dip- or spin-coating of dispersions to fabricate graphene-based devices that have a wider range of applications.

The benefits of employing ILs are highlighted, with a wide range of permutations and combinations available in the anions and cations, revealing enormous promise for exfoliated RGO, GNs, and IGA manufacturing capabilities. While still in its early stages, the use of ILs as solvents for liquid-phase exfoliation and colloidal stabilization performs better than several standard solvents. This demands for more research into the benefits of combining ILs, phase thermodynamics for optimum exfoliation, technique parameter optimization, and post-processing processes, among other things.

With the goal of creating GNs "by the pound," a better fundamental knowledge of the interactions between graphene and IL components is critical, as is a systematic search for novel functional cations/anions with the goal of improving intercalation and dispersion. The use of molecular dynamics and the potential of mean force simulations is critical, within silico feedback guiding experimental results. However, there is still an obvious need for more specialized experimental techniques and simulation methodologies to examine the interface. These fundamental methodologies and insights can be used to exfoliation of inorganic 2-D analogues such as MoS_2, BN, Bi_2Te_3, and WS_2, for example. Efforts along this road have undoubtedly already begun in several laboratories.

References

[1] Novoselov, K. S.; Geim, A. K.; Morozov, S. V.; Jiang, D.-e.; Zhang, Y.; Dubonos, S. V.; Grigorieva, I. V.; Firsov, A. A. Electric field effect in atomically thin carbon films. *Science* 2004, 306, 666-669. https://www.science.org/doi/10.1126/science.1102896.

[2]Geim, A. K.; Novoselov, K. S., The rise of graphene. In *Nanoscience and Technology: A Collection of Reviews From Nature Journals*, World Scientific, 2010, 11-19. https://www.worldscientific.com/doi/abs/10.1142/9789814287005_0002.

[3] Kroto, H. W.; Heath, J. R.; O'Brien, S. C.; Curl, R. F.; Smalley, R. E. C 60: buckminsterfullerene. *Nature* 1985, 318, 162-163.

[4] Iijima, S. Helical microtubules of graphitic carbon. *Nature* 1991, 354, 56-58. https://www.nature.com/articles/354056a0.

[5] Lee, C.; Wei, X.; Kysar, J. W.; Hone, J. Measurement of the elastic properties and intrinsic strength of monolayer graphene. *Science* 2008, 321, 385-388. https://www.science.org/doi/10.1126/science.1157996.

[6] Stoller, M. D.; Park, S.; Zhu, Y.; An, J.; Ruoff, R. S. Graphene-based ultracapacitors. *Nano Lett.* 2008, 8, 3498-3502. https://pubs.acs.org/doi/10.1021/nl802558y.

[7] Balandin, A. A.; Ghosh, S.; Bao, W.; Calizo, I.; Teweldebrhan, D.; Miao, F.; Lau, C. N. Superior thermal conductivity of single-layer graphene. *Nano Lett.* 2008, 8, 902-907. https://pubs.acs.org/doi/10.1021/nl0731872.

[8] Zhang, Y.; Tan, Y.-W.; Stormer, H. L.; Kim, P. Experimental observation of the quantum Hall effect and Berry's phase in graphene. *Nature* 2005, 438, 201-204. https://www.nature.com/articles/nature04235/.

[9] Li, X.; Magnuson, C. W.; Venugopal, A.; An, J.; Suk, J. W.; Han, B.; Borysiak, M.; Cai, W.; Velamakanni, A.; Zhu, Y. Graphene films with large domain size by a two-step chemical vapor deposition process. *Nano Lett.* 2010, 10, 4328-4334. https://pubs.acs.org/doi/10.1021/nl101629g

[10] Hernandez, Y.; Nicolosi, V.; Lotya, M.; Blighe, F. M.; Sun, Z.; De, S.; McGovern, I.; Holland, B.; Byrne, M.; Gun'Ko, Y. K. High-yield production of graphene by liquid-phase exfoliation of graphite. *Nat. Nanotechnol.* 2008, 3, 563-568. https://pubmed.ncbi.nlm.nih.gov/18772919/.

[11] Cui, X.; Zhang, C.; Hao, R.; Hou, Y. Liquid-phase exfoliation, functionalization and applications of graphene. *Nanoscale* 2011, 3, 2118-2126. https://pubs.rsc.org/en/content/articlelanding/2011/nr/c1nr10127g.

[12] Lu, J.; Yan, F.; Texter, J. Advanced applications of ionic liquids in polymer science. *Prog. Polym. Sci.* 2009, 34, 431-448. https://www.sciencedirect.com/science/article/pii/S0079670008001226?via%3Dihub.

[13] Rogers, R. D. In *Ionic Liquids: Industrial Applications to Green Chemistry*, ACS symposium series, American Chemical Society, 2002.

[14] Ohno, H.; Yoshizawa, M. Ion conductive characteristics of ionic liquids prepared by neutralization of alkylimidazoles. *Solid State Ion.* 2002, 154, 303-309. https://www.sciencedirect.com/science/article/abs/pii/S016727380200526X?via%3Dihub.

[15] Ueki, T.; Watanabe, M. Upper critical solution temperature behavior of poly (N-isopropylacrylamide) in an ionic liquid and preparation of thermo-sensitive nonvolatile gels. *Chem. Lett.* 2006, 35, 964-965. https://www.journal.csj.jp/doi/10.1246/cl.2006.964.

[16] Fukushima, T.; Kosaka, A.; Ishimura, Y.; Yamamoto, T.; Takigawa, T.; Ishii, N.; Aida, T. Molecular ordering of organic molten salts triggered by single-walled carbon nanotubes. *Science* 2003, 300, 2072-2074. https://www.science.org/doi/10.1126/science.1082289.

[17] Lee, J.; Aida, T. "Bucky gels" for tailoring electroactive materials and devices: the composites of carbon materials with ionic liquids. *ChemComm.* 2011, 47, 6757-6762. https://pubs.rsc.org/en/content/articlelanding/2011/cc/c1cc00043h#!.

[18] Wang, X.; Fulvio, P. F.; Baker, G. A.; Veith, G. M.; Unocic, R. R.; Mahurin, S. M.; Chi, M.; Dai, S. Direct exfoliation of natural graphite into micrometre size few layers graphene sheets using ionic liquids. *ChemComm.* 2010, 46, 4487-4489. https://pubs.rsc.org/en/content/articlelanding/2010/cc/c0cc00799d#!.

[19] Nuvoli, D.; Valentini, L.; Alzari, V.; Scognamillo, S.; Bon, S. B.; Piccinini, M.; Illescas, J.; Mariani, A. High concentration few-layer graphene sheets obtained by liquid phase exfoliation of graphite in ionic liquid. *J. Mater. Chem.* 2011, 21, 3428-3431. https://pubs.rsc.org/en/content/articlelanding/2011/jm/c0jm02461a#!.

[20] Bari, R.; Tamas, G.; Irin, F.; Aquino, A. J.; Green, M. J.; Quitevis, E. L. Direct exfoliation of graphene in ionic liquids with aromatic groups. *Colloids Surf. A: Physicochem. Eng. Asp.* 2014, 463, 63-69. https://www.sciencedirect.com/science/article/abs/pii/S0927775714007390?via%3Dihub.

[21] Matsumoto, M.; Saito, Y.; Park, C.; Fukushima, T.; Aida, T. Ultrahigh-throughput exfoliation of graphite into pristine 'single-layer'graphene using microwaves and molecularly engineered ionic liquids. *Nat. Chem.* 2015, 7, 730-736. https://www.nature.com/articles/nchem.2315

[22] Shang, N. G.; Papakonstantinou, P.; Sharma, S.; Lubarsky, G.; Li, M.; McNeill, D. W.; Quinn, A. J.; Zhou, W.; Blackley, R. Controllable selective exfoliation of high-quality graphene nanosheets and nanodots by ionic liquid assisted grinding. *ChemComm.* 2012, 48, 1877-1879. https://pubs.rsc.org/en/content/articlelanding/2012/cc/c2cc17185f#!.

[23] Chen, X.; Dobson, J. F.; Raston, C. L. Vortex fluidic exfoliation of graphite and boron nitride. *ChemComm.* 2012, 48, 3703-3705. https://pubs.rsc.org/en/content/articlelanding/2012/cc/c2cc17611d.

[24] Lu, J.; Yang, J.-x.; Wang, J.; Lim, A.; Wang, S.; Loh, K. P. One-pot synthesis of fluorescent carbon nanoribbons, nanoparticles, and graphene by the exfoliation of graphite in ionic liquids. *ACS Nano* 2009, 3, 2367-2375. https://pubs.acs.org/doi/10.1021/nn900546b.

[25] Liu, N.; Luo, F.; Wu, H.; Liu, Y.; Zhang, C.; Chen, J. One-step ionic-liquid-assisted electrochemical synthesis of ionic-liquid-functionalized graphene sheets directly from graphite. *Adv. Funct. Mater.* 2008, 18, 1518-1525. https://onlinelibrary.wiley.com/doi/10.1002/adfm.200700797.

[26] Singh, V. V.; Gupta, G.; Batra, A.; Nigam, A. K.; Boopathi, M.; Gutch, P. K.; Tripathi, B. K.; Srivastava, A.; Samuel, M.; Agarwal, G. S. Greener electrochemical synthesis of high quality graphene nanosheets directly from pencil and its SPR sensing application. *Adv. Funct. Mater.* 2012, 22, 2352-2362. https://onlinelibrary.wiley.com/doi/abs/10.1002/adfm.201102525.

[27] Wei, D.; Grande, L.; Chundi, V.; White, R.; Bower, C.; Andrew, P.; Ryhänen, T. Graphene from electrochemical exfoliation and its direct applications in enhanced energy storage devices. *ChemComm.* 2012, 48, 1239-1241. https://pubs.rsc.org/en/content/articlelanding/2012/cc/c2cc16859f.

[28] Zhou, X.; Wu, T.; Ding, K.; Hu, B.; Hou, M.; Han, B. Dispersion of graphene sheets in ionic liquid [bmim][PF6] stabilized by an ionic liquid polymer. *ChemComm.* 2010, 46, 386-388. https://pubs.rsc.org/en/content/articlelanding/2010/cc/b914763b#!.

[29] Zhang, B.; Ning, W.; Zhang, J.; Qiao, X.; Zhang, J.; He, J.; Liu, C.-Y. Stable dispersions of reduced graphene oxide in ionic liquids. *J. Mater. Chem.* 2010, 20, 5401-5403. https://pubs.rsc.org/en/content/articlelanding/2010/jm/c0jm01029d.

[30] Kim, T.; Lee, H.; Kim, J.; Suh, K. S. Synthesis of phase transferable graphene sheets using ionic liquid polymers. *Acs Nano* 2010, 4, 1612-1618. https://pubs.acs.org/doi/10.1021/nn901525e.

[31] Yan, W.; Huang, Y.; Xu, Y.; Huang, L.; Chen, Y. Rapid and effective functionalization of graphene oxide by ionic liquid. J. Nanosci. Nanotechnol. 2012, 12, 2270-2277. https://www.ingentaconnect.com/content/asp/jnn/2012/00000012/00000003/art00074;jsessionid=30ch69kjbggrt.x-ic-live-03.

[32] Binnemans, K. Ionic liquid crystals. *Chem. Rev.* 2005, 105, 4148-4204. https://pubs.acs.org/doi/10.1021/cr0400919.

[33] Acik, M.; Dreyer, D. R.; Bielawski, C. W.; Chabal, Y. J. Impact of ionic liquids on the exfoliation of graphite oxide. *J. Phys. Chem.* C 2012, 116, 7867-7873. https://pubs.acs.org/doi/10.1021/jp300772m.

[34] Qin, F.; Brosseau, C. A review and analysis of microwave absorption in polymer composites filled with carbonaceous particles. *J. Appl. Phys.* 2012, 111, 4. https://aip.scitation.org/doi/10.1063/1.3688435.

[35] Wang, C.; Han, X.; Xu, P.; Zhang, X.; Du, Y.; Hu, S.; Wang, J.; Wang, X. The electromagnetic property of chemically reduced graphene oxide and its application as microwave absorbing material. *Appl. Phys. Lett.* 2011, 98, 072906. https://aip.scitation.org/doi/10.1063/1.3555436.

[36] Tang, J.; Radosz, M.; Shen, Y. Poly (ionic liquid) s as optically transparent microwave-absorbing materials. *Macromolecules* 2008, 41, 493-496. https://pubs.acs.org/doi/10.1021/ma071762i.

[37] Yang, H.; Shan, C.; Li, F.; Han, D.; Zhang, Q.; Niu, L. Covalent functionalization of polydisperse chemically-converted graphene sheets with amine-terminated ionic liquid. *ChemComm.* 2009, 3880-3882. https://pubs.rsc.org/en/content/articlelanding/2009/cc/b905085j#!.

[38] Lee, C. Y.; McCammon, J. A.; Rossky, P. The structure of liquid water at an extended hydrophobic surface. *J. Chem. Phys.* 1984, 80, 4448-4455. https://aip.scitation.org/doi/10.1063/1.447226.

[39] Shih, C.-J.; Lin, S.; Strano, M. S.; Blankschtein, D. Understanding the stabilization of liquid-phase-exfoliated graphene in polar solvents: molecular dynamics simulations and kinetic theory of colloid aggregation. *J. Am. Chem. Soc.* 2010, 132, 14638-14648. https://pubs.acs.org/doi/10.1021/ja1064284.

[40] Fu, C.; Yang, X. Molecular simulation of interfacial mechanics for solvent exfoliation of graphene from graphite. *Carbon* 2013, 55, 350-360. https://www.sciencedirect.com/science/article/abs/pii/S0008622312010500?via%3Dihub.

[41] Fedorov, M. V.; Lynden-Bell, R. Probing the neutral graphene–ionic liquid interface: insights from molecular dynamics simulations. *Phys. Chem. Chem. Phys.* 2012, 14, 2552-2556. https://pubs.rsc.org/en/content/articlelanding/2012/cp/c2cp22730d.

[42] Biswas, R. Molecular dynamics studies on the exfoliation of graphene in room temperature ionic liquids. *J. Mol. Liq.* 2021, 337, 116592. https://www.sciencedirect.com/science/article/abs/pii/S0167732221013167?via%3Dihub.

[43] Pinilla, C.; Del Pópolo, M. G.; Lynden-Bell, R. M.; Kohanoff, J. Structure and dynamics of a confined ionic liquid. Topics of relevance to dye-sensitized solar cells. *J. Phys. Chem. B* 2005, 109, 17922-17927. https://pubs.acs.org/doi/10.1021/jp052999o

[44] Fitchett, B. D.; Conboy, J. C. Structure of the room-temperature ionic liquid/SiO2 interface studied by sum-frequency vibrational spectroscopy. *J. Phys. Chem. B* 2004, 108, 20255-20262. https://pubs.acs.org/doi/10.1021/jp0471251.

[45] Baldelli, S.; Bao, J.; Wu, W.; Pei, S.-s. Sum frequency generation study on the orientation of room-temperature ionic liquid at the graphene–ionic liquid interface. *Chem. Phys. Lett.* 2011, 516, 171-173. https://www.sciencedirect.com/science/article/abs/pii/S0009261411012267?via%3Dihub.

[46] Bergin, S. D.; Nicolosi, V.; Streich, P. V.; Giordani, S.; Sun, Z.; Windle, A. H.; Ryan, P.; Niraj, N. P. P.; Wang, Z. T. T.; Carpenter, L. Towards solutions of single-walled carbon nanotubes in common solvents. *Adv. Mater.* 2008, 20, 1876-1881. https://onlinelibrary.wiley.com/doi/10.1002/adma.200702451.

[47] Vatamanu, J.; Cao, L.; Borodin, O.; Bedrov, D.; Smith, G. D. On the influence of surface topography on the electric double layer structure and differential capacitance of graphite/ionic liquid interfaces. *J. Phys. Chem. Lett.* 2011, 2, 2267-2272. https://pubs.acs.org/doi/10.1021/jz200879a.

[48] Zhao, Y.; Hu, Z. Graphene in ionic liquids: collective van der Waals interaction and hindrance of self-assembly pathway. *J. Phys. Chem. B* 2013, 117, 10540-10547. https://pubs.acs.org/doi/10.1021/jp405660d.

[49] Watanabe, K.; Andersen, H. C. Molecular dynamics study of the hydrophobic interaction in an aqueous solution of krypton. *J. Phys. Chem.* 1986, 90, 795-802. https://pubs.acs.org/doi/10.1021/j100277a019.

[50] Lin, S.; Shih, C.-J.; Strano, M. S.; Blankschtein, D. Molecular insights into the surface morphology, layering structure, and aggregation kinetics of surfactant-stabilized graphene dispersions. *J. Am. Chem. Soc.* 2011, 133, 12810-12823. https://pubs.acs.org/doi/10.1021/ja2048013.

[51] Zangi, R.; Zhou, R.; Berne, B. Urea's action on hydrophobic interactions. *J. Am. Chem. Soc.* 2009, 131, 1535-1541. https://pubs.acs.org/doi/10.1021/ja807887g.

[52] Wang, Y.; Voth, G. A. Unique spatial heterogeneity in ionic liquids. *J. Am. Chem. Soc.* 2005, 127, 12192-12193. https://pubs.acs.org/doi/10.1021/ja053796g.

[53] Wang, Y.; Voth, G. A. Tail aggregation and domain diffusion in ionic liquids. *J. Phys. Chem. B* 2006, 110, 18601-18608. https://pubs.acs.org/doi/10.1021/jp063199w

[54] Canongia Lopes, J. N.; Padua, A. A. Nanostructural organization in ionic liquids. *J. Phys. Chem. B* 2006, 110, 3330-3335. https://pubs.acs.org/doi/10.1021/jp056006y.

[55] Canongia Lopes, J. N.; Costa Gomes, M. F.; Pádua, A. A. Nonpolar, polar, and associating solutes in ionic liquids. *J. Phys. Chem. B* 2006, 110, 16816-16818. https://pubs.acs.org/doi/10.1021/jp063603r.

Chapter 9

Room-Temperature Ionic Liquids for Heavy Metal Ion Extraction

Rima Biswas*
Process Simulation Research Group, School of Chemical Engineering,
Vellore Institute of Technology, Vellore, Tamil Nadu, India

Abstract

Room-temperature ionic liquids (RTILs) have demonstrated promising applications for the removal of metal ions as green solvents and extractants. With water, RTILs create biphasic systems and are frequently employed in liquid-liquid extraction. Metal ions are extracted using crown ethers, phosphoryl derivatives, and calixarenes in traditional liquid-liquid extraction.

Studies of the behavior of alkali cations in systems like nitric acid solutions and water-immiscible ionic liquids (ILs) have shown patterns in the main way that these cations move from one phase to the next. Density functional theoretical (DFT) analysis was used to complement the dual mode of metal ion extraction by crown ether found in the experimental work.

Molecular dynamic (MD) simulations are used to examine the interfacial behavior of heavy metal ions extraction in IL-water biphasic systems. The existing uses of RTILs as solvents and carriers in liquid-liquid extraction and biphasic systems for heavy metal ions are presented in this chapter, along with a discussion of their potential future use.

Keywords: room temperature ionic liquids, heavy metal ions extraction, crown ether, solvation effect, molecular dynamics, quantum calculations

Abbreviations

$[C_2mim][Tf_2N]$	1-Ethyl-3-methylimidazolium bis(trifluoromethanesulfonyl)imide
$[C_4mim][Tf_2N]$	1-Butyl-3-methylimidazolium bis(trifluoromethanesulfonyl)imide
$[C_8mim][Tf_2N]$	1-Octyl-3-methylimidazolium bis(trifluoromethanesulfonyl)imide
$[C_4mim][PF_6]$	1-Butyl-3-methylimidazolium hexafluorophosphate

* Corresponding Author's Email: rima.biswas@vit.ac.in.

In: Properties and Applications of Ionic Liquids
Editors: Noureddine Abidi and Julia L. Shamshina
ISBN: 979-8-88697-635-9

$[C_6mim][PF_6]$	1-Hexyl-3-methylimidazolium hexafluorophosphate
$[C_8mim][PF_6]$	1-Octyl-3-methylimidazolium hexafluorophosphate
$[C_8mim][BF_4]$	1-Octyl-3-methylimidazolium tetrafluoroborate
[MTOA]Cl	Methyltrioctylammonium chloride
[A336]Cl	Tricaprylylmethylammonium chloride
[A336][SCN]	Tricaprylylmethylammonium thiocyanate
[A336][TS]	Tricaprylylmethylammonium thiosalicylate
[A336][Sal]	Tricaprylylmethylammonium salicylate
$[P_{6,6,6,14}]Cl$	Trihexyl(tetradecyl)phosphonium chloride
DCE	Dichloroethane
$CHCl_3$	Chloroform
$C_6H_5NO_2$	Nitrobenzene
$[C_2mim][BETI]$	Bis(pentafluoroethylsulfonyl)imide
DC18C6	Dicyclohexane-18-crown-6
DB18C6	Dibenzo-18-crown-6
DB21C7	Dibenzo-21-crown-7
18C6	18-crown-6
DB24C8	Dibenzo-24-crown-8
DCH18C6	Dicyclohexano-18-crown-6
BPC6	Bis(2-propyloxy)calix[4]crown-6

Introduction

Increased metal emissions into aquatic ecosystems and soils, particularly in emerging nations, were an unavoidable consequence of the rapid growth of heavy metal industrial use, such as in mining, batteries, tanneries, metal plating facilities, and fertilizer. Heavy metals (such as chromium, cadmium, lead, nickel zinc, nickel, copper, etc.) are non-biodegradable and tend to accumulate in diverse environmental compartments, notably in sediments, unlike many organic pollutants that are easily degradable [1]. It is also well-recognized that many heavy metal ions are toxic or carcinogenic [2, 3]. Meanwhile, heavy metal poisoning of ground and surface water bodies is now recognized as one of the most important environmental hazards to humans and biota [4-7].

This growing concern has led to a slew of international and national laws governing metal pollution monitoring and forecasting critical levels and loads [8]. As a result, using today's wastewater treatment technology, removing heavy metals from municipal and industrial wastewater streams remains a difficult task. A variety of methods including ion exchange, chemical precipitation, depletion from the aqueous phase and adsorption onto activated sludge, electrolysis, membrane technology, and solvent extraction were suggested for the removal of heavy metals [9-11]. Solvent extraction uses an extraction agent dissolved in organic solvents (toluene, kerosene, etc.) and has a negative environmental impact because the organic solvent volatilizes. Due to their great economic benefit, metal separation and recovery involve not only environmental but also economical benefits, making effective separation processes quite appealing.

Ionic liquids (ILs) at room temperature have shown to be a more effective class of environmentally friendly solvents for liquid-liquid extraction [12, 13], with higher extraction efficiency than traditional organic solvents [12, 14]. They are made up of cations (such as imidazolium derivatives, pyridinium, and ammonium) and anions and have lower melting temperatures than molten salts (usually less than 100°C). Many ILs have almost zero vapor pressure, great thermal stability, high ionic conductivity, low viscosity, and have the ability to remain liquid over a wide temperature range. ILs have attracted a lot of attention in chemical engineering [15, 16], chemistry [17, 18], and materials research [16, 19, 20] because of their unusual features [21]. As a result, efforts are being made to advance the use of IL in the field of separations, notably the employment of these materials as next-generation diluents in the solvent extraction of metal ions [22-25]. It is believed that the extraction mechanism connected to these hydrophobic ionic solvents involves the typical complexation of metal ions via an extractant as well as the exchange of aqueous metal ions with the cation of the ILs [26].

Crown ethers have excellent complexation ability with heavy metal ions and high binding selectivity with other alkali cations (hereafter M^+). This exceptional affinity for Cs^+ is seen during complexation in a homogeneous phase (e.g., acetonitrile, methanol), as well as during liquid-liquid extraction of M^+ from aqueous to organic phases (e.g., aromatic derivatives, halogenated alkanes), enabling significant improvements in Cs^+ partitioning from acidic liquid waste solutions [27, 28]. The use of ILs as a heavy metal ion extraction agent has previously been documented [24, 29, 30]. Investigations into the extraction of different metal ions have been conducted using Imidazolium-based ILs [31]. The extraction of metal ions into the hydrophobic ILs phase was demonstrated to be minimal in the majority of studies because metal ions recovered from the aqueous phase are strongly hydrated. Therefore, an extractant should be used that forms hydrophobic interactions with the metal ions and facilitates their transfer from the aqueous to the ILs phase.

Luo et al. [32] discovered extraordinarily high values of the distribution constant for Sr^{2+} and Cs^+ ions using DCH18C6 as an extractant in protic amide-based ILs. Additionally, it was found that BOBCalix6 (calix[4]arene-bis(tert-octylbenzo-crown-6) is a more efficient extractant for Cs^+ ion in ILs than straightforward organic solvents like chloroform and 1,2-dichloroethane [12]. The Cs^+ ion was discovered to be partitioned in ILs even in the absence of extractant molecules, whereas no such partitioning was seen in conventional organic molecular solvents [33]. Additionally, it was shown that the extraction mechanism for metal ions transfer from the aqueous phase to the ILs phase is different as compared to a molecular organic solvent. It is generally believed that when a neutral extractant is used, the exchange of metal ions from the aqueous phase with the cation of ILs takes place without the co-extraction of the counter anions, whereas in common molecular organic solvents, the co-anions are typically present when the metal ions are transferred. The mechanism and impact of radiation on the extraction of Cs^+ ion in ILs using calix-crown ether have been previously described [34]. It was demonstrated that calix-crown had a substantially higher extraction capacity than crown ether ligands.

Further experimental and theoretical studies should be conducted to learn more about this extraordinary and intriguing situation. Several investigations [35, 36] have described the separation of metal ions using ILs. Thermophysical characteristics of ILs [37, 38] or metal ion-crown ligand/calix-crown ligands [39, 40] have been the subject of theoretical studies. Theoretical investigations involving ILs and metal ion–ligand complexes, on the other hand, are uncommon. Sieffert et al. [41] performed MD simulation on Cs^+/Na^+ extraction in $[C_4mim][Tf_2N]$ among theoretical investigations. They discovered that the metal ion exchanges with the IL $[C_4mim]$-cation. Pouramini et al. [42] reported the dual exchange mechanism of ion transport from the aqueous to the ILs phase using MD simulation. The adsorption of cadmium ions at the interface can lead to some of them being diluted into the IL phase via an anion exchange or ion-pair mechanism. The diffusion of $[PF_6]^-$ anions to the centre of bulk water supports an anion exchange process [42]. The interactions between IL and lanthanides compounds in the organic phase was studied by Hunter et al. [43]. They reported the IL aggregation around a microhydrated La^{3+} with nitrates completing the inner coordination sphere using MD simulations. This chapter serves as a summary of data on the extraction of heavy metals using RTILs and crown ethers, which are relevant to separation processes.

Heavy Metal Ions Extraction in RTILs

Several research groups have investigated the behavior of "traditional" hydrophobic RTILs based on imidazolium cations in aqueous solutions for various metals (e.g., Mn, Zn, Co, and Ni) [44-47]. Table 1 lists the RTILs that were tested as extracting agents. In addition to the conventional physical adsorption process, the experiment findings suggest a cation exchange mechanism, and the study of anionic species in the aqueous phase suggests a concurrent partial transfer of non-charged metal species. Several 1-alkyl-3-methylimidazolium hexafluoro-phosphate ILs were examined as mercury extraction agents in aqueous solutions by Germani et al. [48]. Their results strongly imply that the hydrophilic/hydrophobic characteristics of the system as a whole, along with the length of the alkyl chain, affect the association and extraction of metal ions. De los Rios et al. [49] reported comparable results when testing a range of RTILs for the separation of Cd^{2+}, Zn^{2+}, and Fe^{3+} from hydrochloric acid solutions.

It is evident that the sequence $[Tf_2N] < [PF_6] < [BF_4]$ enhanced the extraction efficiency of ILs for Cd^{2+} and Zn^{2+}, which is compatible with the observed anions' hydrophilicity [49, 50]. On the other hand, increasing the length of the cation's alkyl chain improved the extraction performance of imidazolium-based ILs. A cationic ester derivative of glycine betaine was combined with bis(trifluoromethylsulfonyl)imide $[Tf_2N]$ or dicyanamide [Dca] anions to produce task-specific ionic liquids (TSILs) [51]. These ILs were used to extract Cu^{2+} ions from water at room temperature. The use of [Dca] substantially improves the efficiency of metal ion extraction. Understanding the metal ion transport mechanism in ILs is critical for developing predictive models and, as a result, enabling realistic operations. Furthermore, the fluorine-containing anions of RTILs are hydrophobic, one disadvantage of these systems is their instability towards hydrolysis. The potential development of hydrofluoric acid poses significant environmental dangers [52]. ILs containing hydrophobic, long-chain quaternary ammonium and phosphonium cations can also be used in extraction operations to avoid fluorine-containing anions. Rout and Binnemans [53] reported the dual mechanism (i.e., ion-pair extraction and anion exchange) for europium ion (Eu^{3+}) extraction using a non-fluorinated functionalized IL. They claimed that at low salt concentrations, Eu^{3+} was removed by the complex formation mechanism, whereas at high nitrate concentrations, the anion exchange mechanism was implicated. Because the aqueous phase contains insufficient nitrate ions to extract Eu^{3+} at low nitrate concentrations, the extraction is exclusively due to the complex formation mechanism.

Another way to prevent fluorine-containing anions in the extraction process is to use ILs that contain hydrophobic, long-chain quaternary ammonium and phosphonium cations. Cations with various alkyl chain lengths and compositions, including C_6, C_8, and C_{10}. According to Seeley and Crouse, Aliquat® 336 outperformed other amine extractants like Alamine®336 or

Amberlite LA-1 as a liquid ion exchanger for extracting various metal species from hydrochloric acid solutions [54]. Since then, Aliquat® 336 has undergone testing as an extractant for a number of metals from acidic aqueous solutions, including Fe^{3+}, Cd^{2+}, Pt^{2+}, and Hg^{2+} from hydrochloric acid solutions [55, 56]. Nayl [57] removed Co(II) and Ni(II) effectively from acidic sulfate solutions using kerosene- and Aliquat® 336-based ILs. Given that Aliquat® 336 was first used as an extractant in the mining sector, it is no surprise that various studies have been conducted to assess its potential to extract uranium, thorium, and a variety of lanthanides [58, 59]. In general, the extraction process with Aliquat® 336 is based on ion association/ anion exchange of sulfate species/ metal chloride; as a result, it is highly dependent on the aqueous phase composition and pH.

Rios et al. [49] demonstrated that the comparable chemical methyltrioctylammonium chloride, [MTOA]Cl, may be used to successfully extract Fe^{3+}, Zn^{2+}, and Cd^{2+} from hydrochloric acid solutions. The use of phosphonium-based ILs, such as trihexyl (tetradecyl)phosphonium chloride, Cyphos® IL 101 (Cytec Ind.), has recently drawn more attention due to their improved hydrophobicity and thermal stability. The use of phosphonium-based ILs, such as trihexyl(tetradecyl)phosphonium chloride, Cyphos® IL 101 (Cytec Ind.), has recently gained increased attention due to their increased hydrophobicity and thermal stability. Cyphos1 IL 101 has been thoroughly investigated as a new IL extracting agent dissolved in toluene for the removal of zinc and palladium from aqueous phases [60]. Alguacil et al. [61] investigated the transport of Cr^{6+} through a pseudo-emulsion membrane from feeding to an appropriate stripping phase using Cyphos® IL 101 as a metal carrier. There are several disadvantages to using Aliquat1 336 or Cyphos1 IL101 in liquid-liquid extraction methods. First, Aliquat1 336 and Cyphos1 IL101 are generally dissolved in organic solvents due to their high viscosity, which has an impact on both the ecological and economic sides of the process. Second, because the extraction mechanism is based on anion exchange, it is very sensitive to the aqueous phase composition. In addition, the aqueous phase leachability of both ILs and organic solvents could represent an additional environmental danger. Because long-chain quaternary ammonium- and phosphonium ILs have a high level of aquatic toxicity [61], their release into the environment should be strictly controlled.

To increase metal ion selectivity, other techniques involve adding a complexing agent to the aqueous or IL phase [62]. Introducing an organic coordinating molecule directly to the IL-phase, such as calixarenes, quinoline derivate, thenoyltrifluoroacetone, 2-aminothiophenol, dithiazone, and crown-ethers (CE), considerably enhanced the distribution ratios of metal ions between the IL- and aqueous phases [31, 63, 64]. However, adding a water-soluble complexing agent to the aqueous phase is not ideal for the selective removal of toxic metals from real water matrices. Despite numerous studies on metal extraction using organic ligands, including ILs, little is known about the extraction mechanism of these IL systems. Dietz et al. [31] showed that the exchange of a cationic Sr^{2+}-CE complex with the IL cation occurs during the extraction of Sr^{2+} from acidic environments by the macrocyclic dicyclohexano-18- crown-6 polyether into different imidazolium-based RTILs [35]. Shen et al. [65] examined the extraction of UO_2^{2+} into $[C_nmim][PF_6]$ ILs containing diglycolamide under low acidic conditions. They proposed that stable 2:1 complexes develop and that uranyl is transferred to the IL phase via a cation exchange mechanism.

The extraction technique, which entails the loss of an IL cation into the aquatic environment, casts doubt on the process's "greenness." While ILs may ease air pollution problems due to their low volatility, IL cations in the aqueous phase may represent a serious threat to aquatic life [61]. Furthermore, cation exchange causes ILs to decompose, reducing the number of viable extraction cycles and increasing the cost of using ILs in liquid-liquid extractions. An anion-exchange process employing fluorinated IL anions has the same difficulty. Ion exchange could be suppressed in favor of traditional extraction as a possible solution to this problem. Dietz [35] proposed the following approaches: enhancing the hydrophobicity of ILs by lengthening the alkyl chain of imidazolium-based ILs or fluorinating the alkyl chain, and using a cation exchanger, which, of course, adds to the method's cost and complexity.

Table 1. List of RTILs used in heavy metals extraction

IL	Molecular Structure	Ligand	Metal ion	Reference
$[C_2mim][Tf_2N]$	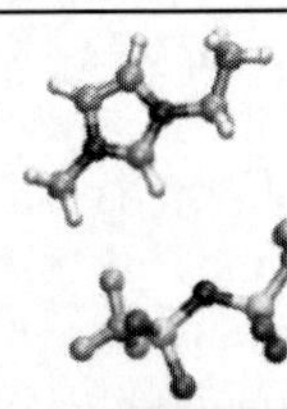	-	Zn^{2+}, Co^{2+}, Ni^{2+}, Mn^{2+}, and Fe^{3+}	[44, 66, 67]
$[C_4mim][Tf_2N]$	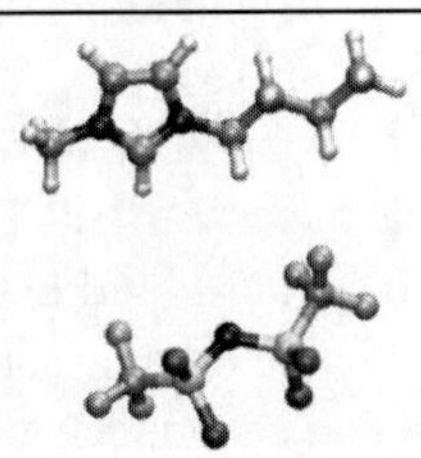	Calixarene, TBP, and thenoyltrifluoroacetone	Cs^{+}, Na^{+}, Zn^{2+}, Cd^{2+}, Fe^{3+} and UO_2^{2+}	[49, 68, 69]
$[C_8mim][Tf_2N]$	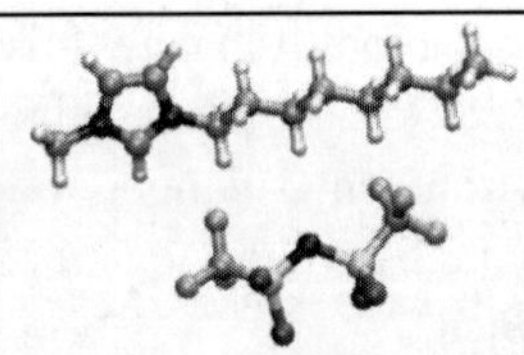	TBP, thenoyltrifluoroacetone	Zn^{2+}, Cd^{2+}, Fe^{3+},UO_2^{2+} Co^{2+}, Ni^{2+}, Mn^{2+}, and Cu^{2+}	[49, 68, 69]
$[C_4mim][PF_6]$	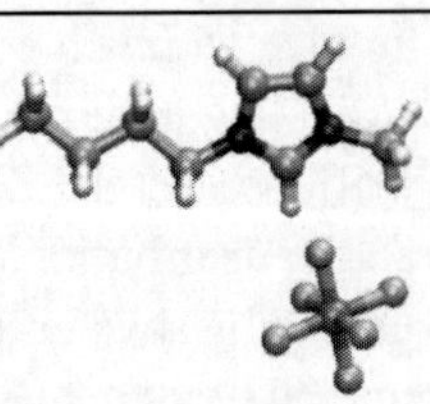	Dithizone 2-aminothiophenol quinoline derivate calixarene PAN, TAN, halides, SCN, CN, OCN Diglycolamide thenoyltrifluoroacetone, htta	Pb^{2+}, Ni^{2+}, Cu^{2+} Co^{2+}, Cd^{2+}, Cu^{2+}, Zn^{2+}, Ag^{+} Hg^{2+}, Cd^{2+}, Co^{2+}, Ni^{2+}, Fe^{3+}, and UO_2^{2+}	[48 49, 63-65, 70-74]
$[C_6mim][PF_6]$	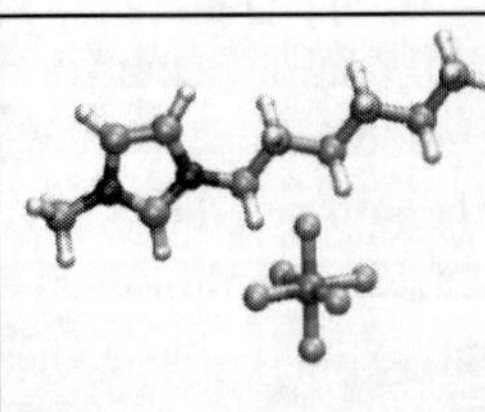	Calixarene PAN, TAN, halides, SCN, CN, OCN dithizone diglycolamide thenoyltrifluoroacetone, htta	Ag^{+}, Hg^{2+}, Cd^{2+}, Co^{2+}, Ni^{2+}, Fe^{3+}, Pb^{2+}, UO_2^{2+}, Cu^{2+}, Zn^{2+}, and Mn^{2+}	[48, 64, 65, 72-74]

IL	Molecular Structure	Ligand	Metal ion	Reference
$[C_8mim][PF_6]$	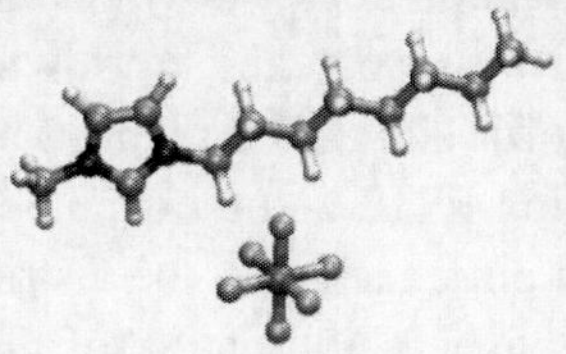	calixarene diglycolamide thenoyltrifluoroacetone, htta	Ag^+, UO_2^{2+}, Co^{2+}, Cd^{2+}, Cu^{2+}, Zn^{2+}, Pb^{2+}, and Mn^{2+}	[49, 65, 72, 74]
$[C_8mim][BF_4]$	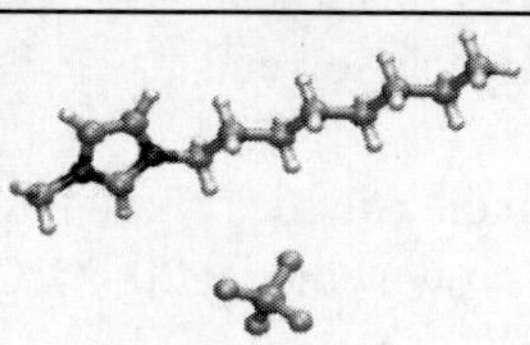		Zn^{2+}, Cd^{2+}, and Fe^{3+}	[49]
[MTOA]Cl	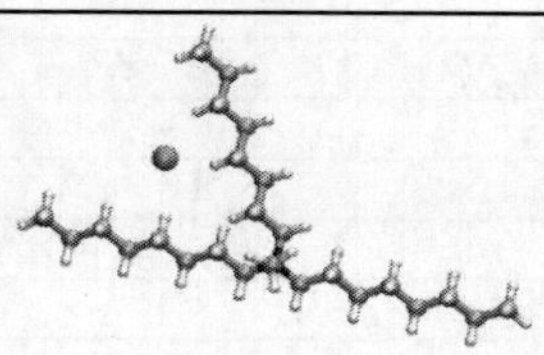		Zn^{2+}, Cd^{2+}, and Fe^{3+}	[49]
[A336]Cl	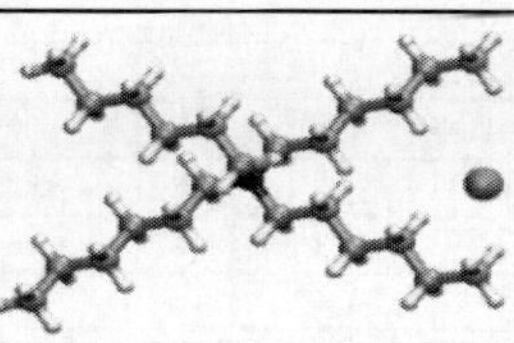		Zn^{2+}, Co^{2+}, Ni^{2+}, Fe^{3+}, Pt^{2+}, Hg^{2+}, Pt^{2+}, and UO_2^{2+}	[56, 57, 75-77]
[A336][SCN]	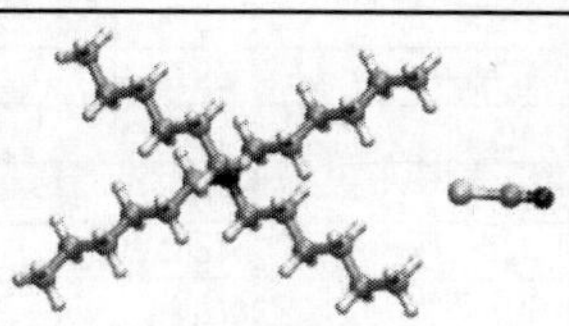		Co^{2+}, and Ni^{2+}	[57]
$[P_{6,6,6,14}]Cl$	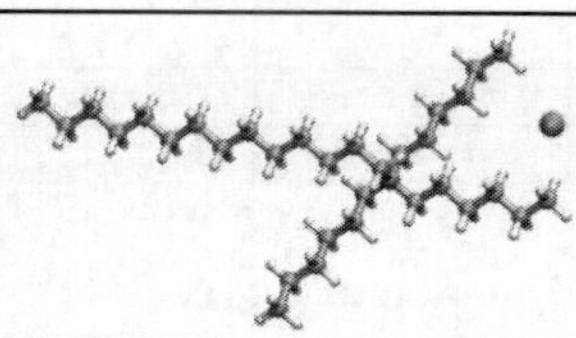		Zn^{2+}, Pd^{2+}, Cr^{6+}, Fe^{3+}, and Ni^{2+}	[60, 78-80]

Crown Ether-Assisted Metal Ion Extraction

Solvent extraction is a popular separation technique for removing metal ions and organic molecules, and it has a number of advantages over other methods. However, liquid-liquid extraction of volatile organic compounds typically uses water-immiscible organic solvents, many of which are volatile, combustible, and, poisonous [35]. ILs have a number of qualities that make them desirable as substitutes for conventional organic solvents [35]. For the separation of organic molecules and biomolecules, the distribution coefficients D have been shown to be higher than those of conventional solvents. The situation with inorganic cation extraction, on the other hand, is less apparent. Dai et al. [81] observed log $D = 4.04$ for extraction of Sr^{2+} from water into a series of dialkylimidazolium-based RTILs in the presence of DCH18C6, although other molecular solvents had very low distribution coefficients under the same conditions.

Furthermore, this research showed that the extraction degree may be adjusted by changing the RTIL anion or cation. Since the publication of this research, a worldwide investigation of the selectivity, ability, and extraction mechanism of RTIL for inorganic cation separation has been conducted. Table 2 lists the advantages of using RTIL and crown ether in Cs^+ extraction over the use of traditional organic solvents. In other circumstances, the distribution coefficients D_{Cs} grow by 2–4 orders of magnitude. Simultaneously, acidification of the feeding aqueous solution reduces RTIL efficiency, but it rises for molecular solvents.

Table 2. Efficiency of Cs^+ extraction from the aqueous phase into RTIL and organic solvents in presence of crown ether at 20–25°C

Organic phase	Crown ether, concentration, mol × dm^{-3}	log D_{Cs}	References
DCE	DC18C6, 0.4	−1.1	[82]
$CHCl_3$	DC18C6, 0.05	−1.54	[83]
$CHCl_3$	DB18C6, 0.05	−2.0	[83]
$CHCl_3$	18C6, 0.05	−1.89	[83]
$CHCl_3$	DB21C7, 0.1	−1.96	[84]
$C_6H_5NO_2$	DB21C7, 0.1	2.03	[84]
Octanol	DC18C6, 0.05	−0.89	[85]
[C_4mim][Tf_2N]	DC18C6, 0.01	1.9	[82]
[C_4mim][Tf_2N]	DC18C6, 0.1	2.59	[86]
[C_2mim][BETI]	DC18C6, 0.1	3.06	[87]
[C_4mim][Tf_2N]	DC18C6, 0.05	1.97	[85]
[C_4mim][PF_6]	DC18C6, 0.1	1.76	[87]
[C_4mim][Tf_2N]	18C6, 0.15	1.56	[88, 89]
[C_4mim][Tf_2N]	18C6, 0.15	−0.23	[88]
[C_4mim][Tf_2N]	DB18C6, 0.15	2.28	[88, 89]
[C_4mim][Tf_2N]	DB18C6, 0.15	-0.47	[88]
[C_4 mim][Tf_2N]	DB21C7, 0.15	1.67	[89, 90]
[C_4mim][Tf_2N]	DB24C8, 0.15	0.94	[89, 90]
[C_4mim][Tf_2N]	15C5, 0.15	0.30	[89]

Luo et al. [87] used DC18C6 in [C_2C_1im][BETI] to achieve a result for strontium that was higher than that studied by Dai et al. [81]: log D_{Sr} = 4.9. When RTIL and calix[4]arene were used together, the extraction of silver ions was shown to be significantly improved over that of chloroform [72]. In comparison to ordinary organic solvents, tri-n-octylphosphine oxide, octyl(phenyl)-*N,N*-diisobutylcarbamoylmethylphosphine oxide, and relative complexing agents dissolved in RTILs dramatically improve the extractability of actinides and lanthanides. Chelating agent-assisted distribution ratios in RTIL for Pu^{4+}, Am^{3+}, UO_2^{2+}, and Th^{4+} have been reported to be one order of magnitude more than those for VOCs [91, 92]. In the RTIL phase, Cs^+ extraction efficiency [93] and Am^{3+}/Eu^{3+} extraction selectivity [94] were found to be highly correlated with complex stability constant values. It is important to note that the high extraction efficiency would greatly increase cost-effectiveness for industrial separation applications. Figure 1 shows the different types of crown ether used as an extractant in heavy metal ion extraction from aqueous medium.

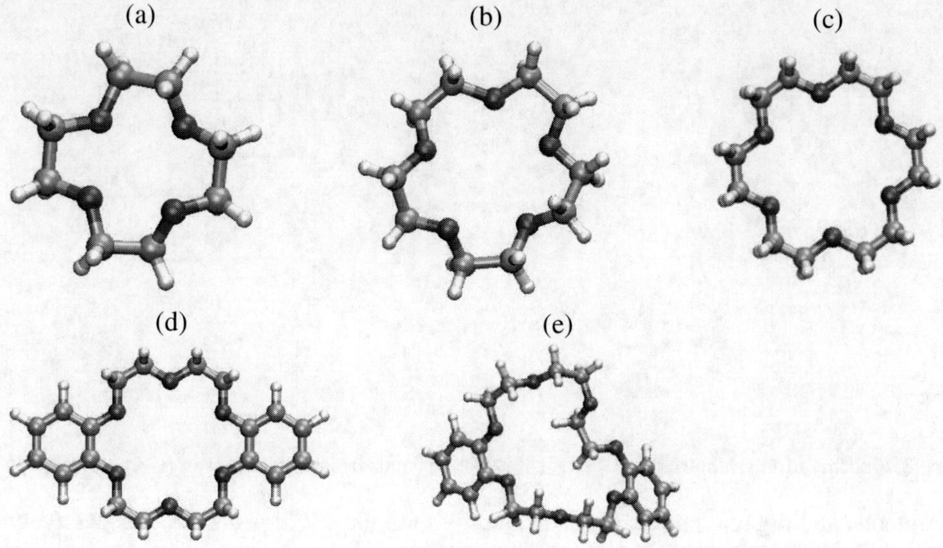

Figure 1. Molecular structure of (a) 12C4, (b) 15C5, (c) 18C6, (d) DB18C6, and (e) DB21C7 commonly used as ligand in heavy metal ion extraction from aqueous medium.

Compared to molecular solvents, the complex formation scheme in RTILs is more complex [95]. Both ions that make up the IL (cation Z^+ and anion X^-) can interact with cesium complex components in general. Z^+ competes with cesium for the ligand, but X^- solvates cesium, preventing the formation of complexes:

$$CsX_n^{1-n} + Z_m L^{m+} \rightleftarrows CsLX_{n-p}^{1-n+p} + mZ^+ + pX^- \quad (1)$$

NMR chemical shift data shows that even in triflimide-based RTILs, the crown ether does not completely substitute all RTIL anions X^- in the coordination sphere of cesium in $[Cs(18C6)]^+$ complexes (Figure 2a) [96]. Cesium is placed above the mean oxygen plane of the crown ether ring in all of these structures because its cavity size is bigger than that of 18C6 [97]. As a result, the exposed region of Cs^+ might interact with RTIL anions, resulting in a coordination number of 8 or 9, as seen crystallographically for complexes in molecular solvents. It was noted that the chemical shifts of $[Cs(18C6)_2]^+$ in ^{133}Cs NMR are virtually independent of the solvent's composition, yielding nearly identical values for RTILs and molecular solvents. This finding suggests that in a "sandwich" combination, the two crown ether molecules effectively screen the cesium ion from solvent contact (Figure 2b). In RTILs and molecular solvents, however, the chemical changes of $[Cs(18C6)]^+$ are substantially solvent-dependent.

It was noted that the extraction methods of hydrophobic RTILs and organic solvents are quite different. Extraction of neutral species of a cation Mn^+ is necessary for traditional solvents; for example, if M is present in an aqueous solution as nitrate, then a neutral (nonionized) complexing agent L promotes M's partitioning into the organic phase. A general equation can be used to convey this:

$$M_{aq}^{n+} + nNO_3^- + L_o \rightleftarrows [M(NO_3)_n L]_o \quad (2)$$

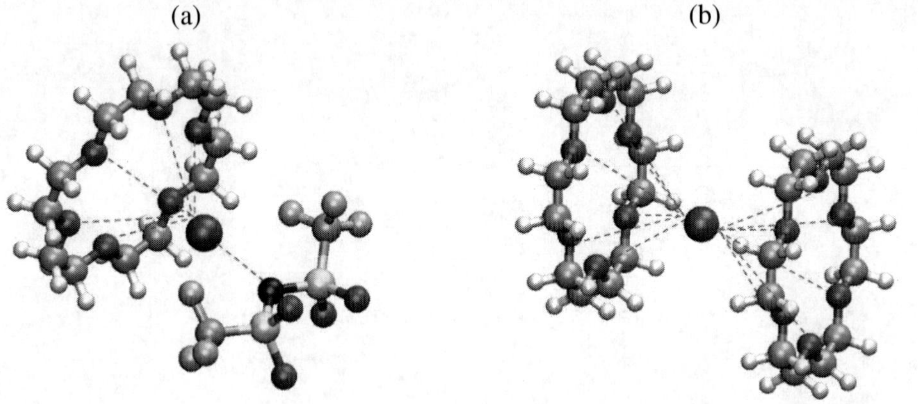

Figure 2. Optimization structures of (a) $[Cs(18C6)Tf_2N]^0$ and (b) $[Cs(18C6)_2]^+$ in $[C_4C_1im][Tf_2N]$.

Aqueous and organic phases are indicated by the indexes "*aq*" and "*o*," respectively. For the extraction of metal ion nitrates from water into conventional RTILs, such as $[C_4C_1im][PF_6]$ and $[C_4C_1im][Tf_2N]$ [35, 91], the ion exchange scenario was somewhat different and more complicated:

$$M_{aq}^{n+} + nNO_3^- + L_o + n[ZX]_o \rightleftarrows [ML][X_n]_o + nNO_3^-{}_{aq} + nZ_{aq}^+ \tag{3}$$

The cation and anion of an RTIL are denoted by Z^+ and X^-, respectively. A partitioning process in which the loss of an RTIL cation to the aqueous phase occurs after metal ion transfer raises serious concerns for potential liquid-liquid extraction applications of ILs. To begin with, this mechanism calls into doubt the status of RTILs as "green" extractants. Indeed, while RTILs can help to tackle the problem of air pollution produced by conventional solvents, their cations can create an increase in aqueous phase pollution. Extraction of metal ions such as Pb^{2+}, Ni^{2+}, and Cu^{2+} by RTILs in the presence of b-diketonates proceeds in the same way as in toluene, nitrobenzene, or chloroform – as a neutral hydrated complex $[ML_2(H_2O)n]°$, where n = 1, 2 [74]. Nonchelating inorganic ligands such as OCN^-, SCN^-, CN^-, and halide ions have all been observed in a similar way [98].

Molecular Dynamics and Quantum Calculations on Metal Ion Extraction

The mutations of the free and complexed cations in the three liquids (i.e., $[C_4mim][PF_6]$-dry, $[C_4mim][PF_6]$-humid, and $[C_4mim][Tf_2N]$-dry) are reviewed in this section [99]. Table 4 displays the results of using thermodynamic cycles to determine the relative free energy of complexation in a specific liquid, M^+ transfer from water to the IL, and liquid-liquid extraction of M^+ by the calixarene L. According to FEP calculations, Na^+ is the best solvated cation, and in the three liquids, the order of the cations is $Na^+ > K^+ > Rb^+ > Cs^+$. The computed ΔG_3/IL values for the $K^+ \leftrightarrow Rb^+$ and $Rb^+ \leftrightarrow Cs^+$ mutations are similar in all three liquids, whereas ΔG_3/IL for the $Na^+ \leftrightarrow K^+$ mutations is roughly 4 kcal/mol lower in the dry than in the humid $[C_4mim][PF_6]$ and dry $[C_4mim][Tf_2N]$ liquids.

In the three ILs investigated, the results demonstrate that L complexes primarily Cs^+, in the order $Cs^+ > Rb^+ > K^+ > Na^+$. The contributions of K^+/Na^+, Rb^+/K^+, and Cs^+/Rb^+ in dry and humid [C_4mim][PF_6] were fairly similar (approximately 1.5 kcal/mol), but the Cs^+/Na^+ binding selectivity in [C_4mim][Tf_2N] is primarily due to the K^+/Na^+ contribution (7.1 kcal/mol). The $\Delta\Delta G_c$ results (Table 3) show that the three ILs have a good Cs^+/Na^+ extraction selectivity by L.

Table 3. Free energy differences for cation complexation ($\Delta\Delta G_c$), cation transfer from water to ILs (without L) ($\Delta\Delta G_t$), and cation extraction from water to ILs (by L) ($\Delta\Delta G_e$) calculated by FEP calculations [99]

IL	$Na^+ \rightarrow K^+$	$K^+ \rightarrow Rb^+$	$Rb^+ \rightarrow Cs^+$	$Na^+ \rightarrow Cs^+$ [a]
Cation complexation by L in ILs				
[C_4mim][PF_6]-dry	1.9	1.3	1.6	4.8
[C_4mim][PF_6]-humid	1.7	1.9	0.3	3.9
[C_4mim][Tf_2N]-dry	7.1	1.6	1.1	9.8
Cation transfer from water to ILs (in absence of L)				
[C_4mim][PF_6]-dry	5.7	−0.6	−0.2	4.9
[C_4mim][PF_6]-humid	1.8	−1.8	−0.2	−0.2
[C_4mim][Tf_2N]-dry	1.1	−1.0	−0.1	0.0
Cation extraction from water to ILs (in presence of L)				
[C_4mim][PF_6]-dry	7.6	0.7	1.4	9.7
[C_4mim][PF_6]-humid	4.1	0.2	0.1	4.4
[C_4mim][Tf_2N]-dry	8.2	0.6	1.0	9.8

[a] $\Delta G_{(Na^+\rightarrow Cs^+)} = \Delta G_{(Na^+\rightarrow K^+)} = \Delta G_{(K^+\rightarrow Rb^+)} = \Delta G_{(Rb^+\rightarrow Cs^+)}$.

As demonstrated in Table 4, the LCs^+ complex interacts equally with the interface and the bulk IL, whereas the uncomplexed calixarene L interacts more strongly with its liquid environment at the interface than the bulk IL. These energies indicate that the complex is less surface-active at the IL interface than the L ligand, which is consistent with its high IL extractability. As opposed to bulk chloroform, both L and LCs^+ favor the interface, with LCs^+ being more surface-active than L due to its higher polarity and attraction to water.

Further comparisons with the two oil monophasic systems reveal that LCs^+ interacts with the IL significantly better than chloroform and is thus solvated by the IL. Contacts with IL and chloroform differ less in the case of the L ligand itself, and these are weaker than interactions at comparable interfaces. In general, complexes should be less attracted to IL interfaces than to classical aqueous interfaces, allowing for easier transfer to the IL receiving phase. This is in line with the fact that hydrophobic ILs extract more efficiently than organic solvents.

Table 4. Solute-solvent interaction energies (kcal/mol) in bulk phase or at the interface [99]

Phase	Species	IL	Chloroform	Water	Total
In bulk phase					
Bulk IL	L	−86	–	–	−86
	LCs^+	−136	-	-	−136
Bulk chloroform	L	-	−68	-	−68
	LCs^+	-	−77	-	−77
Bulk water	L	-	-	−84	−84
	LCs^+	-	-	−106	−106

Table 4. (Continued)

Phase	Species	IL	Chloroform	Water	Total
At the interfaces (biphasic systems)					
IL-water interface	L	−65	-	−38	−103
	LCs^+	−128	-	−11	−139
Chloroform-water interface	L	-	−57	−36	-93
	LCs^+	-	−65	−43	−108

The binding energies for Cs^+ and Na^+ ions with DCH18C6 in the gas phase are displayed in Table 5 [85]. In contrast, the calculated binding energy using the BP86 functional is lower than the value predicted from the B3LYP and TPSSH functional with the SVP basis set. The calculated binding energy of the Cs^+ with DCH18C6 using the B3LYP and TPSSH functional is equivalent with both the SVP and TZVP basis sets. The basis set has very little impact on the Cs^+ ion binding energy. For all DFT functional basis set combinations, the binding energy for Na^+ ion in the gas phase is always greater than Cs^+ ion by 28–32 kcal/mol, showing that Na^+ ion has stronger selectivity than Cs^+ ion.

The binding energy of Cs^+ and Na^+ ions using BPC6 was reported to be significantly higher than that of DCH18C6. Table 5 displays the computed BE for both Cs^+ and Na^+ ions using DCH18C6 and BPC6. With both SVP and TZVP basis sets, the calculated binding energy of the Cs+ metal ion with BPC6 using B3LYP and TPSSH functional is nearly equal, however, the calculated binding energy using BP86 functional is lower by 2.9 kcal/mol than the value expected from B3LYP and TPSSH functional with SVP basis set. The binding energy from the SVP basis set for Na^+ ion with BPC6 is lower than that of Cs^+ ion, however, it was higher for Na^+ ion over Cs^+ ion with DCH18C6. The binding energy from the TZVP basis set for Na^+ with BPC6 demonstrates that Na^+ ion has a greater selectivity than Cs^+ ion.

Table 5. The binding energy of Na^+ and Cs^+ with DCH18C6 and BPC6 at various DFT functional and basis set [85]

	BE (kcal/mol)					
	SVP			TZVP		
System	B3LYP	BP86	TPSSH	B3LYP	BP86	TPSSH
DCH18C6						
Na^+	−62.9 (−96.4)	−56.1 (−92.1)	−61.8 (−96.6)	−88.4 (−95.8)	−82.8 (−95.5)	−87.2 (−95.8)
Cs^+	−55.8 (−67.2)	−52.5 (−62.9)	−55.9 (−67.3)	−55.8 (−67.2)	−54.2 (−67.6)	−55.9 (−67.3)
BPC6						
Na^+	−53.5 (−96.3)	−48.9 (−96.3)	−53.1 (−96.0)	−81.7 (−96.0)	−77.9 (−96.2)	−81.7 (−96.5)
Cs^+	−58.8 (−75.9)	−55.9 (−76.1)	−58.8 (−75.6)	−58.6 (−76.4)	−56.8 (−76.9)	−58.2 (−77.1)

Conclusion

The extraction of heavy metal ions from the aqueous phase using ILs is a subject of intense research with some intriguing prospects. ILs have a limited application in extraction procedures due to their relatively high cost and mostly unknown toxicity and long-term stability. Some newly designed ILs, such as [A336][TS] or [A336][Sal], have costs that are about an order of magnitude cheaper than some "traditional" RTILs, making their use more appealing [100-102].

Crown ether is widely used as an extractant agent in heavy metal ion extraction process. The dual cationic exchange was also identified by the DFT calculation of the free energy of transfer of the metal ion from the aqueous phase to the non-aqueous ILs solvent phase. The higher extraction of metal ions in IL is successfully modeled by the exchange of the metal ion with the cation of the IL. The fact that there is no complex metal ion hydration impact in the free energy difference between two competing ligands shows that BPC6 has a higher extraction capacity than DCH18C6. On the other hand, the varied water solubilities of ILs have a big impact on their stability as extracting agents, reusability, and time-of-life, along with their environmental impact [100].

References

[1] Wang, L. K.; Chen, J. P.; Hung, Y.-T.; Shammas, N. K. *Heavy Metals in the Environment*. CRC press, 2009.

[2] Khan, A. M.; Ganai, S. A. Removal and recovery of heavy metal ions using natural adsorbents. In *Modern Age Waste Water Problems*, Springer 2020, 251-260.

[3] Lawal, K. K.; Ekeleme, I. K.; Onuigbo, C. M.; Ikpeazu, V. O.; Obiekezie, S. O. A review on the public health implications of heavy metals. *World Journal of Advanced Research and Reviews* 2021, 10, 255-265.

[4] Ericson, B.; Hanrahan, D.; Kong, V. The world's worst pollution problems: the top ten of the toxic twenty. *Blacksmith Institute and Green Cross* 2008.

[5] Ahmad, W.; Alharthy, R. D.; Zubair, M.; Ahmed, M.; Hameed, A.; Rafique, S. Toxic and heavy metals contamination assessment in soil and water to evaluate human health risk. *Scientific Reports* 2021, 11, 1-12.

[6] Atangana, E.; Oberholster, P. J. Using heavy metal pollution indices to assess water quality of surface and groundwater on catchment levels in South Africa. *Journal of African Earth Sciences* 2021, 182, 104254.

[7] Kapoor, D.; Singh, M. P. Heavy metal contamination in water and its possible sources. In *Heavy Metals in the Environment* Elsevier: 2021; pp 179-189.

[8] Tchounwou, P. B.; Yedjou, C. G.; Patlolla, A. K.; Sutton, D. J. Heavy metal toxicity and the environment. *Molecular, Clinical and Environmental Toxicology* 2012, 133-164.

[9] Kentish, S.; Stevens, G. Innovations in separations technology for the recycling and re-use of liquid waste streams. *Chemical Engineering Journal* 2001, 84, 149-159.

[10] Fu, F.; Wang, Q. Removal of heavy metal ions from wastewaters: a review. *Journal of Environmental Management* 2011, 92, 407-418.

[11] Papadopoulos, A.; Fatta, D.; Parperis, K.; Mentzis, A.; Haralambous, K.-J.; Loizidou, M. Nickel uptake from a wastewater stream produced in a metal finishing industry by combination of ion-exchange and precipitation methods. *Separation and Purification Technology* 2004, 39, 181-188.

[12] Luo, H.; Dai, S.; Bonnesen, P. V.; Buchanan, A.; Holbrey, J. D.; Bridges, N. J.; Rogers, R. D. Extraction of cesium ions from aqueous solutions using calix [4] arene-bis (tert-octylbenzo-crown-6) in ionic liquids. *Analytical Chemistry* 2004, 76, 3078-3083.

[13] Chun, S.; Dzyuba, S. V.; Bartsch, R. A. Influence of structural variation in room-temperature ionic liquids on the selectivity and efficiency of competitive alkali metal salt extraction by a crown ether. *Analytical Chemistry* 2001, 73, 3737-3741.

[14] Visser, A. E.; Swatloski, R. P.; Reichert, W. M.; Griffin, S. T.; Rogers, R. D. Traditional extractants in nontraditional solvents: Groups 1 and 2 extraction by crown ethers in room-temperature ionic liquids. *Industrial & Engineering Chemistry Research* 2000, 39, 3596-3604.

[15] Arce, A.; Earle, M. J.; Rodríguez, H.; Seddon, K. R.; Soto, A. Bis {(trifluoromethyl) sulfonyl} amide ionic liquids as solvents for the extraction of aromatic hydrocarbons from their mixtures with alkanes: effect of the nature of the cation. *Green Chemistry* 2009, 11, 365-372.

[16] Nasirpour, N.; Mohammadpourfard, M.; Heris, S. Z. Ionic liquids: Promising compounds for sustainable chemical processes and applications. *Chemical Engineering Research and Design* 2020, 160, 264-300.

[17] Heintz, A. Recent developments in thermodynamics and thermophysics of non-aqueous mixtures containing ionic liquids. A review. *The Journal of Chemical Thermodynamics* 2005, 37, 525-535.

[18] Zhang, Q.; Shreeve, J. n. M. Energetic ionic liquids as explosives and propellant fuels: a new journey of ionic liquid chemistry. *Chemical Reviews* 2014, 114, 10527-10574.

[19] Lee, J. S.; Wang, X.; Luo, H.; Baker, G. A.; Dai, S. Facile ionothermal synthesis of microporous and mesoporous carbons from task specific ionic liquids. *Journal of the American Chemical Society* 2009, 131, 4596-4597.

[20] Singh, S. K.; Savoy, A. W. Ionic liquids synthesis and applications: An overview. *Journal of Molecular Liquids* 2020, 297, 112038.

[21] Kianfar, E.; Mafi, S. Ionic liquids: properties, application, and synthesis. *Fine Chemical Engineering* 2021, 22-31.

[22] Cocalia, V. A.; Gutowski, K. E.; Rogers, R. D. The coordination chemistry of actinides in ionic liquids: A review of experiment and simulation. *Coordination Chemistry Reviews* 2006, 250, 755-764.

[23] Vander Hoogerstraete, T.; Wellens, S.; Verachtert, K.; Binnemans, K. Removal of transition metals from rare earths by solvent extraction with an undiluted phosphonium ionic liquid: separations relevant to rare-earth magnet recycling. *Green Chemistry* 2013, 15, 919-927.

[24] Leyma, R.; Platzer, S.; Jirsa, F.; Kandioller, W.; Krachler, R.; Keppler, B. K. Novel thiosalicylate-based ionic liquids for heavy metal extractions. *Journal of Hazardous Materials* 2016, 314, 164-171.

[25] Yudaev, P. A.; Chistyakov, E. M. Ionic Liquids as Components of Systems for Metal Extraction. *ChemEngineering* 2022, 6, 6.

[26] Dietz, M. L.; Dzielawa, J. A. Ion-exchange as a mode of cation transfer into room-temperature ionic liquids containing crown ethers: implications for the 'greenness' of ionic liquids as diluents in liquid–liquid extraction. *Chemical Communications* 2001, 2124-2125.

[27] Casnati, A.; Pochini, A.; Ungaro, R.; Ugozzoli, F.; Arnaud, F.; Fanni, S.; Schwing, M.-J.; Egberink, R. J.; de Jong, F.; Reinhoudt, D. N. Synthesis, complexation, and membrane transport studies of 1, 3-alternate calix [4] arene-crown-6 conformers: a new class of cesium selective ionophores. *Journal of the American Chemical Society* 1995, 117, 2767-2777.

[28] Haverlock, T. J.; Bonnesen, P. V.; Sachleben, R. A.; Moyer, B. A. Analysis of equilibria in the extraction of cesium nitrate by calix [4] arene-bis (t-octylbenzo-crown-6) in 1, 2-dichloroethane. *Journal of Inclusion Phenomena and Macrocyclic Chemistry* 2000, 36, 21-37.

[29] Mudring, A. V.; Tang, S. Ionic liquids for lanthanide and actinide chemistry. *European Journal of Inorganic Chemistry* 2010, 2010, 2569-2581.

[30] Thasneema, K.; Dipin, T.; Thayyil, M. S.; Sahu, P. K.; Messali, M.; Rosalin, T.; Elyas, K.; Saharuba, P.; Anjitha, T.; Hadda, T. B. Removal of toxic heavy metals, phenolic compounds and textile dyes from industrial waste water using phosphonium based ionic liquids. *Journal of Molecular Liquids* 2021, 323, 114645.

[31] Dietz, M. L.; Jakab, S.; Yamato, K.; Bartsch, R. A. Stereochemical effects on the mode of facilitated ion transfer into room-temperature ionic liquids. *Green Chemistry* 2008, 10, 174-176.

[32] Luo, H.; Huang, J.-F.; Dai, S. Solvent extraction of Sr^{2+} and Cs^{+} using protic amide-based ionic liquids. *Separation Science and Technology*, 2010, 45, 1679-1688.

[33] Luo, H.; Dai, S.; Bonnesen, P. V. Solvent extraction of Sr^{2+} and Cs^{+} based on room-temperature ionic liquids containing monoaza-substituted crown ethers. *Analytical Chemistry* 2004, 76, 2773-2779.

[34] Xu, C.; Yuan, L.; Shen, X.; Zhai, M. Efficient removal of caesium ions from aqueous solution using a calix crown ether in ionic liquids: mechanism and radiation effect. *Dalton Transactions* 2010, 39, 3897-3902.

[35] Dietz, M. L. Ionic liquids as extraction solvents: where do we stand? *Separation Science and Technology* 2006, 41, 2047-2063.

[36] Nakashima, K.; Kubota, F.; Maruyama, T.; Goto, M. Feasibility of ionic liquids as alternative separation media for industrial solvent extraction processes. *Industrial & Engineering Chemistry Research* 2005, 44, 4368-4372.

[37] Li, H.; Ibrahim, M.; Agberemi, I.; Kobrak, M. N. The relationship between ionic structure and viscosity in room-temperature ionic liquids. *The Journal of Chemical Physics* 2008, 129, 124507.

[38] Katsyuba, S. A.; Zvereva, E. E.; Vidiš, A.; Dyson, P. J. Application of density functional theory and vibrational spectroscopy toward the rational design of ionic liquids. *The Journal of Physical Chemistry A* 2007, 111, 352-370.

[39] Sachleben, R. A.; Bryan, J. C.; Engle, N. L.; Haverlock, T. J.; Hay, B. P.; Urvoas, A.; Moyer, B. A. Rational Design of Cesium-Selective Ionophores: Dihydrocalix [4] arene Crown-6 Ethers. *European Journal of Organic Chemistry* 2003, 2003, 4862-4869.

[40] Shamova, L. I.; Shamov, G. A.; Antipin, I. S.; Konovalov, A. I. Modeling K^+ and Ag^+ complexation by thiacalix [4] arene amides using DFT: the role of intramolecular hydrogen bonding. *The Journal of Physical Chemistry A* 2009, 113, 5691-5699.

[41] Sieffert, N.; Wipff, G. Comparing an ionic liquid to a molecular solvent in the cesium cation extraction by a calixarene: a molecular dynamics study of the aqueous interfaces. *The Journal of Physical Chemistry B* 2006, 110, 19497-19506.

[42] Pouramini, Z.; Mohebbi, A.; Kowsari, M. H. The possibility of cadmium extraction to the ionic liquid 1-hexyl-3-methylimidazolium hexafluorophosphate in the presence of hydrochloric acid: a molecular dynamics study of the water–IL interface. *Theoretical Chemistry Accounts* 2019, 138, 1-14.

[43] Hunter, J. P.; Dolezalova, S.; Ngwenya, B. T.; Morrison, C. A.; Love, J. B. Understanding the recovery of rare-earth elements by ammonium salts. *Metals* 2018, 8, 465.

[44] Fujii, K.; Nonaka, T.; Akimoto, Y.; Umebayashi, Y.; Ishiguro, S.-i. Solvation structures of some transition metal (II) ions in a room-temperature ionic liquid, 1-ethyl-3-methylimidazolium bis (trifluoromethanesulfonyl) amide. *Analytical Sciences* 2008, 24, 1377-1380.

[45] Zuo, Y.; Liu, Y.; Chen, J.; Li, D. Q. The separation of cerium (IV) from nitric acid solutions containing thorium (IV) and lanthanides (III) using pure [C8mim] PF6 as extracting phase. *Industrial & Engineering Chemistry Research* 2008, 47, 2349-2355.

[46] Rajadurai, V.; Anguraj, B. L. Ionic liquids to remove toxic metal pollution. *Environmental Chemistry Letters* 2021, 19, 1173-1203.

[47] Pospiech, B.; Kujawski, W. Ionic liquids as selective extractants and ion carriers of heavy metal ions from aqueous solutions utilized in extraction and membrane separation. *Reviews in Chemical Engineering* 2015, 31, 179-191.

[48] Germani, R.; Mancini, M. V.; Savelli, G.; Spreti, N. Mercury extraction by ionic liquids: temperature and alkyl chain length effect. *Tetrahedron Letters* 2007, 48, 1767-1769.

[49] de los Ríos, A. P.; Hernandez-Fernandez, F.; Lozano, L.; Sanchez, S.; Moreno, J.; Godinez, C. Removal of metal ions from aqueous solutions by extraction with ionic liquids. *Journal of Chemical & Engineering Data* 2010, 55, 605-608.

[50] Lee, L. Y.; Morad, N.; Ismail, N.; Talebi, A.; Rafatullah, M. Optimization for liquid-liquid extraction of Cd (Ii) over Cu (ii) ions from aqueous solutions using ionic liquid aliquat 336 with tributyl phosphate. *International Journal of Molecular Sciences* 2020, 21, 6860.

[51] Messadi, A.; Mohamadou, A.; Boudesocque, S.; Dupont, L.; Guillon, E. Task-specific ionic liquid with coordinating anion for heavy metal ion extraction: Cation exchange versus ion-pair extraction. *Separation and Purification Technology* 2013, 107, 172-178.

[52] Swatloski, R. P.; Holbrey, J. D.; Rogers, R. D. Ionic liquids are not always green: hydrolysis of 1-butyl-3-methylimidazolium hexafluorophosphate. *Green Chemistry* 2003, 5, 361-363.

[53] Rout, A.; Binnemans, K. Liquid–liquid extraction of europium (III) and other trivalent rare-earth ions using a non-fluorinated functionalized ionic liquid. *Dalton Transactions* 2014, 43, 1862-1872.

[54] Seeley, F.; Crouse, D. Extraction of Metals from Chloride Solutions and Amines. *Journal of Chemical and Engineering Data* 1966, 11, 424-429.

[55] El-Wakil, A.; Farag, A.; Ez-Eldin, A. K. Liquid-liquid extraction of iron (III), cobalt (II), nickel (II) and cadmium (II) from aqueous halide media with Aliquat 336. *Fresenius' Zeitschrift für analytische Chemie* 1982, 311, 522-522.

[56] Lee, J.-Y.; Kumar, J. R.; Kim, J.-S.; Kim, D.-J.; Yoon, H.-S. Extraction and separation of Pt (IV)/Rh (III) from acidic chloride solutions using Aliquat 336. *Journal of Industrial and Engineering Chemistry* 2009, 15, 359-364.

[57] Nayl, A. Extraction and separation of Co (II) and Ni (II) from acidic sulfate solutions using Aliquat 336. *Journal of Hazardous Materials* 2010, 173, 223-230.

[58] Atanassova, M.; Dukov, I. L. Synergistic solvent extraction and separation of trivalent lanthanide metals with mixtures of 4-benzoyl-3-methyl-1-phenyl-2-pyrazolin-5-one and aliquat 336. *Separation and Purification Technology* 2004, 40, 171-176.

[59] Ali, A.; El-Nadi, Y.; Daoud, J.; Aly, H. Recovery of thorium (IV) from leached monazite solutions using counter-current extraction. *International Journal of Mineral Processing* 2007, 81, 217-223.

[60] Regel-Rosocka, M. Extractive removal of zinc (II) from chloride liquors with phosphonium ionic liquids/toluene mixtures as novel extractants. *Separation and Purification Technology* 2009, 66, 19-24.

[61] Pham, T. P. T.; Cho, C.-W.; Yun, Y.-S. Environmental fate and toxicity of ionic liquids: a review. *Water Research* 2010, 44, 352-372.

[62] Sun, X. Q.; Peng, B.; Chen, J.; Luo, F. An effective method for enhancing metal-ions' selectivity of ionic liquid-based extraction system: Adding water-soluble complexing agent. *Talanta* 2008, 74, 1071-1074.

[63] Lertlapwasin, R.; Bhawawet, N.; Imyim, A.; Fuangswasdi, S. Ionic liquid extraction of heavy metal ions by 2-aminothiophenol in 1-butyl-3-methylimidazolium hexafluorophosphate and their association constants. *Separation and Purification Technology* 2010, 72, 70-76.

[64] Domańska, U.; Rękawek, A. Extraction of metal ions from aqueous solutions using imidazolium based ionic liquids. *Journal of Solution Chemistry* 2009, 38, 739-751.

[65] Shen, Y.; Tan, X.; Wang, L.; Wu, W. Extraction of the uranyl ion from the aqueous phase into an ionic liquid by diglycolamide. *Separation and Purification Technology* 2011, 78, 298-302.

[66] Williams, D. B.; Stoll, M. E.; Scott, B. L.; Costa, D. A.; Oldham Jr, W. J. Coordination chemistry of the bis (trifluoromethylsulfonyl) imide anion: molecular interactions in room temperature ionic liquids. *Chemical Communications* 2005, 1438-1440.

[67] Dhumal, N. R.; Damodaran, K. Molecular interactions and normal vibrations of Fe–bis (trifluoromethanesulfonyl) imide and 1-ethyl-3-methylimidazolium–Fe–bis (trifluoromethane-sulfonyl) imide ionic liquids: a density functional study. *Journal of Molecular Structure: THEOCHEM* 2009, 906, 78-82.

[68] Dietz, M. L.; Stepinski, D. C. Anion concentration-dependent partitioning mechanism in the extraction of uranium into room-temperature ionic liquids. *Talanta* 2008, 75, 598-603.

[69] Kidani, K.; Hirayama, N.; Imura, H. Extraction behavior of divalent metal cations in ionic liquid chelate extraction systems using 1-alkyl-3-methylimidazolium bis (trifluoromethanesulfonyl) imides and thenoyltrifluoroacetone. *Analytical Sciences* 2008, 24, 1251-1254.

[70] Wei, G.-T.; Yang, Z.; Chen, C.-J. Room temperature ionic liquid as a novel medium for liquid/liquid extraction of metal ions. *Analytica Chimica Acta* 2003, 488, 183-192.

[71] Ajioka, T.; Oshima, S.; Hirayama, N. Use of 8-sulfonamidoquinoline derivatives as chelate extraction reagents in ionic liquid extraction system. *Talanta* 2008, 74, 903-908.

[72] Shimojo, K.; Goto, M. Solvent extraction and stripping of silver ions in room-temperature ionic liquids containing calixarenes. *Analytical chemistry* 2004, 76, 5039-5044.

[73] Visser, A. E.; Swatloski, R. P.; Griffin, S. T.; Hartman, D. H.; Rogers, R. D. Liquid/liquid extraction of metal ions in room temperature ionic liquids. *Separation Science and Technology* 2001, 36, 785-804.

[74] Hirayama, N.; Deguchi, M.; Kawasumi, H.; Honjo, T. Use of 1-alkyl-3-methylimidazolium hexafluorophosphate room temperature ionic liquids as chelate extraction solvent with 4, 4, 4-trifluoro-1-(2-thienyl)-1, 3-butanedione. *Talanta* 2005, 65, 255-260.

[75] Srncik, M.; Kogelnig, D.; Stojanovic, A.; Körner, W.; Krachler, R.; Wallner, G. Uranium extraction from aqueous solutions by ionic liquids. *Applied Radiation and Isotopes* 2009, 67, 2146-2149.

[76] Hoh, Y. C.; Chuang, W. S.; Yueh, P. S. The extraction of platinum from Pt (II) chloride solution by a quaternary ammonium compound. *Journal of Chemical Technology and Biotechnology. Chemical Technology* 1985, 35, 41-47.

[77] de Mendonça Fábrega, F.; Mansur, M. B. Liquid–liquid extraction of mercury (II) from hydrochloric acid solutions by Aliquat 336. *Hydrometallurgy* 2007, 87, 83-90.

[78] Cieszyńska, A.; Regel-Rosocka, M.; Wiśniewski, M. Extraction of palladium (II) ions from chloride solutions with phosphonium ionic liquid CyphosŽIL101. *Polish Journal of Chemical Technology* 2007, 9, 99-101.

[79] Alguacil, F. J.; Alonso, M.; Lopez, F. A.; Lopez-Delgado, A. Pseudo-emulsion membrane strip dispersion (PEMSD) pertraction of chromium (VI) using CYPHOS IL101 ionic liquid as carrier. *Environmental Science & Technology* 2010, 44, 7504-7508.

[80] Kogelnig, D.; Stojanovic, A.; Jirsa, F.; Körner, W.; Krachler, R.; Keppler, B. K. Transport and separation of iron (III) from nickel (II) with the ionic liquid trihexyl (tetradecyl) phosphonium chloride. *Separation and Purification Technology* 2010, 72, 56-60.

[81] Dai, S.; Ju, Y.; Barnes, C. Solvent extraction of strontium nitrate by a crown ether using room-temperature ionic liquids. *Journal of the Chemical Society, Dalton Transactions* 1999, 1201-1202.

[82] Yakshin, V.; Tsarenko, N.; Koshcheev, A.; Tananaev, I.; Myasoedov, B. Extraction of metal ions from aqueous solutions of mineral acids with crown ether-ionic liquid systems. *Radiochemistry* 2012, 54, 54-58.

[83] Yakshin, V.; Vilkova, O.; Tsarenko, N.; Tsivadze, A. Y.; Demin, S.; Zhilov, V. Control of the extractive power of crown ethers by alkyl substitution. *Radiochemistry* 2006, 48, 170-174.

[84] Yakshin, V.; Vilkova, O.; Laskorin, B. Extraction separation of radionuclides of cesium and strontium with-crown-ethers. *Doklady Akademii Nauk-Rossijskaya Akademiya Nauk* 1992, 325, 967-969.

[85] Ali, S. M.; Joshi, J.; Deb, A. S.; Boda, A.; Shenoy, K.; Ghosh, S. Dual mode of extraction for Cs^+ and Na^+ ions with dicyclohexano-18-crown-6 and bis (2-propyloxy) calix [4] crown-6 in ionic liquids: density functional theoretical investigation. *RSC Advances* 2014, 4, 22911-22925.

[86] Luo, H.; Dai, S.; Bonnesen, P. V.; Buchanan Iii, A. Separation of fission products based on ionic liquids: Task-specific ionic liquids containing an aza-crown ether fragment. *Journal of Alloys and Compounds* 2006, 418, 195-199.

[87] Luo, H.; Dai, S.; Bonnesen, P. V.; Haverlock, T.; Moyer, B. A.; Buchanan III, A. A striking effect of ionic-liquid anions in the extraction of Sr^{2+} and Cs^+ by dicyclohexano-18-crown-6. *Solvent Extraction and Ion Exchange* 2006, 24, 19-31.

[88] Vendilo, A. G.; Djigailo, D. I.; Smirnova, S. V.; Torocheshnikova, I. I.; Popov, K. I.; Krasovsky, V. G.; Pletnev, I. V. 18-Crown-6 and dibenzo-18-crown-6 assisted extraction of cesium from water into room temperature ionic liquids and its correlation with stability constants for cesium complexes. *Molecules* 2009, 14, 5001-5016.

[89] Vendilo, G.; Chistov, V. I.; Dikareva, J. M.; Popov, K. I. Crown ethers assisted cesium extraction from aqueous solutions into a hydrophobic room temperature ionic liquid 1-butyl-3-methylimidazolium bis [(trifluoromethyl) sulfonyl] imide. *Macroheterocycles 8: 181*, 2015, 184.

[90] Vendilo, A.; Chistov, V.; Privalov, V. Thermodynamics of complexation of cesium with dibenzo-21-crown-7 and dibenzo-24-crown-8 in 1-butyl-3-methylimidazolium bis ((trifluoromethyl) sulfonyl) imide. *Russian Journal of Inorganic Chemistry* 2012, 57, 629-633.

[91] Sun, X.; Luo, H.; Dai, S. Ionic liquids-based extraction: a promising strategy for the advanced nuclear fuel cycle. *Chemical Reviews* 2012, 112, 2100-2128.

[92] Visser, A. E.; Rogers, R. D. Room-temperature ionic liquids: new solvents for f-element separations and associated solution chemistry. *Journal of Solid State Chemistry* 2003, 171, 109-113.

[93] Vendilo, A. G.; Djigailo, D. I.; Rönkkömäki, H.; Lajunen, M.; Chernikova, E. A.; Lajunen, L. H.; Pletnev, I. V.; Popov, K. I. A correlation of caesium–18-crown-6 complex formation constants with the extraction capability for hydrophobic ionic liquids. *Mendeleev Communications* 2010, 2, 122-124.

[94] Sengupta, A.; Murali, M.; Mohapatra, P. A comparative study of the complexation of Am (III) and Eu (III) with TODGA in room temperature ionic liquid. *Journal of Radioanalytical and Nuclear Chemistry* 2013, 298, 405-412.

[95] Ohtsu, K.; Ozutsumi, K. Thermodynamics of solvation of 18-crown-6 and its alkali-metal complexes in various solvents. *Journal of Inclusion Phenomena and Macrocyclic Chemistry* 2003, 45, 217-224.

[96] Popov, K.; Vendilo, A.; Pletnev, I.; Lajunen, M.; Rönkkömäki, H.; Lajunen, L. H. *Formation of Complexes in RTIL and Ion Separations*. InTech: 2011.

[97] Arnaud-Neu, F.; Delgado, R.; Chaves, S. Critical evaluation of stability constants and thermodynamic functions of metal complexes of crown ethers (IUPAC Technical Report). *Pure and Applied Chemistry* 2003, 75, 71-102.

[98] Visser, A. E.; Swatloski, R. P.; Reichert, W. M.; Mayton, R.; Sheff, S.; Wierzbicki, A.; Davis Jr, J. H.; Rogers, R. D. Task-specific ionic liquids for the extraction of metal ions from aqueous solutions. *Chemical Communications* 2001, 135-136.

[99] Sieffert, N.; Wipff, G. Alkali cation extraction by calix [4] crown-6 to room-temperature ionic liquids. The effect of solvent anion and humidity investigated by molecular dynamics simulations. *The Journal of Physical Chemistry A* 2006, 110, 1106-1117.

[100] Stojanovic, A.; Keppler, B. K. Ionic liquids as extracting agents for heavy metals. *Separation Science and Technology* 2012, 47, 189-203.

[101] Villemin, D.; Didi, M. A. Extraction of rare earth and heavy metals, using ionic solvents as extraction medium (A Review). *Oriental Journal of Chemistry* 2013, 29, 1267-1284.

[102] Llaver, M.; Fiorentini, E. F.; Quintas, P. Y.; Oviedo, M. N.; Arenas, M. B. B.; Wuilloud, R. G. Task-specific ionic liquids: Applications in sample preparation and the chemistry behind their selectivity. *Advances in Sample Preparation* 2021, 1, 100004.

Chapter 10

The Roles of Ionic Liquids in Separation of Pollutants from Aqueous Solutions

Emmanuel A. Oke
and Sushma P. Ijardar*
Department of Chemistry,
Veer Narmad South Gujarat University, Gujarat, India

Abstract

Ionic liquids have emerged as potential substances for solving a few of the main problems confronting industrialized society regarding the use of hazardous volatile organic solvents in separation technology. Ionic liquids have been reported to be sustainable, effective, and environmentally benign by offering unique characteristic features such as low viscosity, low vapor pressure, high tunability, high thermal stability, and high electrical conductivity among others.

In recent times, contaminants are widely found in various environmental media owing to flourishing industrialization and other anthropogenic activities. People are now more concerned with the widespread of contaminants in the environment owing to the increasing progress in technology, scientific knowledge as well as socio-economic awareness. Contaminants pose threats and health risks to the ecosystem even at trace levels. Unfortunately, the available and traditional treatment methods currently have proved to either be inefficient, costly or not eco-friendly themselves.

Therefore, the utilization of ionic liquids in the removal of contaminants of high priority from various environmental matrices has surfaced recently. So, this chapter focuses majorly on the use of ionic liquids for the separation and extraction of an array of contaminants in our environment.

Keywords: ionic liquids, separation, dyes, pesticides, phenolic compounds, metal ions

Abbreviations

HPLC	High-performance liquid chromatography

* Corresponding Author's Email: sijardar@vnsgu.ac.in.

In: Properties and Applications of Ionic Liquids
Editors: Noureddine Abidi and Julia L. Shamshina
ISBN: 979-8-88697-635-9

RPLC-UV	Reversed-phase liquid chromatography-ultraviolet
HPLC-DAD	High-performance liquid chromatography- diode array detector
LOD	Limits of detection
DAD	Diode array detector
FT-IR	Fourier Transform Infrared
GC-MS	Gas chromatography-mass spectrometry
ILs	Ionic liquids
TSILs	Task-specific ionic liquids
[BMPYRR][NTf_2]	1-Butyl-1- methylpyrrolidinium bis(trifluoromethylsulfonyl)imide
[BMIM][PF_6]	1-Butyl-3-methylimidazolium hexafluorophosphate
[BMIM][Cl]	1-Butyl-3-methylimidazolium chloride
[HMIM][PF_6]	1-Hexyl-3-methylimidazolium hexafluorophosphate
[HMIM][BF_4]	1-Hexyl-3-methylimidazolium tetrafluoroborate
[OMIM][PF_6]	1-Octyl-3-methylimidazolium hexafluorophosphate
[OMIM][BF_4]	1-Octyl-3-methylimidazolium tetrafluoroborate,
[TCMA][SCN]	Tricaprylmethylammonium thiocyanate
[TCMA][DCA]	Tricaprylmethylammonium dicyanamide,
[TCMA][Bz]	Tricaprylmethylammonium benzoate
[TCMA][Hz]	Tricaprylmethylammonium hexanoate
[BBzIM][BF_4]	1-Butyl-3-benzylimidazolium tetrafluoroborate
[BBzIM][PF_6]	1-Butyl-3-benzylimidazolium hexafluorophosphate
[BBzIM][NTf2]	1-Butyl-3-benzylimidazolium bis(trifluoromethylsulfonyl)imide
[THTDP][Cl]	Trihexyltetradecylphosphonium chloride,
[THTDP][DCA]	Trihexyltetradecylphosphonium dicyanamide
[THTDP[NTf2]	Trihexyltetradecylphosphonium bis (trifluoromethyl sulfonyl) amide
[TBDDP][PF_6]	Tributyldodecylphosphonium hexafluorophosphate
[HMIM][Cl]	1-Hexyl-3-methylimidazolium chloride
[TBOP][Br]	Tributyloctylphosphonium bromide
[DMIM][BF_4]	1-Decyl-3-methylimidazolium tetrafluoroborate
[ODMIM][NTf2]	1-Octyl-2,3-dimethylimidazolium bis(trifluoromethylsulfonyl) imide
[DBIM][Br]	1,3-Dibutylimidazolium bromide
[MECM][NTf2]	3-Methyl-1-(ethoxycarbonylmethyl) imidazolium bis (trifluoromethylsulfonyl)imide
[THADH][SS]	Tetrahexylammonium dihexyl sulfosuccinate
[EMIM][NTf2]	1-Ethyl-3-methylimidazolium bis(fluorosulfonyl) imide
[BPYR][NTf2]	1-Butylpyridinium bis(trifluoromethylsulfonyl) imide,
[4BMPYR][NTf2]	4-Butyl-1-methylpyridinium bis(trifluoromethylsulfonyl) imide
[4BMPIP][NTf2]	4-Butyl-1-methylpiperidinium bis(trifluoromethylsulfonyl) imide
[4BMPRR][NTf2]	4-Butyl-1-methylpyrrolidinium bis(trifluoromethylsulfonyl)imide
[HMPRR][NTf2]	1-Hexyl-1-methylpyrrolidinium bis(trifluoromethylsulfonyl) imide
[THTDP][Dz]	Trihexyltetradecylphosphonium decanoate
[THTDP][TMPP]	Trihexyltetradecylphosphonium bis(2,4,4-trimethylpentyl) phosphinate
[TOMA][SC]	Trioctylmethylammonium salicylate

[DMIM][NTf2]	1-Decyl-3-methylimidazolium bis(trifluoromethanesulfonyl) imide
[Ch][NTf2]	Cholinium bis(trifluoromethanesulfonyl) imide
[BDMIM][PF_6]	1-Butyl-2,3-dimethylimidazolium hexafluorophosphate
[HMIM][NTf2]	1-Hexyl-3-methylimidazolium bis(trifluoro-methylsulfonyl)imide,
[HPIM][NTf2]	1-Hexyl-2-phenylimidazolium bis(trifluoromethylsulfonyl)imide
[BzMIM][NTf2]	1-Benzyl-3-methylimidazolium bis(trifluoromethylsulfonyl)imide
[TBP][OTF]	Tetrabutylphosphonium trifluoromethanesulfonate
[TBP][NTf2]	Tetrabutylphosphonium bis((trifluoromethylsulfonyl)imide
[TBP][PF_6]	Tetrabutylphosphonium hexafluorophosphate
[THMA][MC]	Trihexylmethylammonium methyl carbonate
[TOMA][MC]	Trioctylmethylammonium methyl carbonate
[THMA][AP]	Trihexylmethyl ammonium adipate
[THMA][SC]	Trihexylmethyl ammonium salicylate
[THMA][AL]	Trihexylmethyl ammonium azelate
[TOMA][DDD]	Trioctylmethyl ammonium dodecanedioate
[TOMA][OX]	Trioctylmethyl ammonium oxalate
[TOMA][AL]	Trioctylmethyl ammonium azelate
[TOMA][TSC]	Trioctylmethylammonium thiosalicylate
[TCMA][BzTBzt]	Tricaprylylmethylammonium 2-(benzylthio) benzoate,
[THTDP][BzTBzt]	Trihexyltetradecylphosphonium 2-(benzylthio) benzoate,
[TCMA][ETBzt]	Tricaprylylmethylammonium 2-(ethylthio) benzoate,
[THTDP][ETBzt]	Trihexyltetradecylphosphonium 2-(ethylthio) benzoate,
[TCMA][PTBzt]	Tricaprylylmethylammonium 2-(propylthio) benzoate
[THTDP][PTBzt]	Trihexyltetradecyl phosphonium 2-(propylthio) benzoate
[BPYR][NTf2]	3-Butylpyridinium bis(trifluoromethanesulfonyl)imide
[OMIM][BF_4]	1-Octyl-3-methylimidazolium tetrafluoroborate
[BIM][DEHP]	1-Butylimidazolium di(2-ethylhexyl) phosphate,
[MIM][DEHP]	1-Methylimidazolium di(2-ethylhexyl) phosphate
[BMMOR][BPI]	1-Butyl-1-methylmorpholinium butylphosphite,
[HMMOR][HPI]	1-Hexyl-1methylmorpholinium hexylphosphite
[OMMOR][OPI]	1-Octyl-1-methylmorpholium octylphosphite
[BzDMODA][TMPPA]	Benzyldimethyloctadecylammonium bis(2,4,4trimethylpentyl) phosphinate
[DDDMA][SAC]	Didecyldimethylammonium saccharinate
[DDDMA][ASU]	Didecyldimethylammonium acesulfamate
[TOA][OL]	Tetraoctylammonium oleate
[BMIM][DEHOX]	1-butyl-3-methylimidazolium-di(2-ethylhexyl)-oxamate,
[TOA][DODGLY]	Tetraoctylammonium dioctyl-diglycolamate
[TOA][DEHOX]	Tetraoctylammonium di(2-ethylhexyl)-oxamate
[MTOPPS][AC]	Methyltrioctylphosphonium pentylsulfanyl acetate
[MTOPHS][AC]	Methyltrioctylphosphonium hexylsulfanyl acetate
[MTOPBzS][AC]	Methyltrioctylphosphonium benzylsulfanyl acetate
[TBA][EHHEHP]	Tetrabutylammonium 2-ethylhexyl hydrogen-2-ethylhexyl phosphonate
[CMDMEA][NTf2]	*N*-(Carboxymethyl)-*N*,*N*-dimethylethanaminium bis-trifluoro methane-sulfonimide

[CETMA][NTf2]	*N*-(Carboxyethyl)-trimethylammonium bistrifluoromethane-sulfonimide
[BET][NTf2]	Betainium bis(trifluoromethylsulfonyl)imide

Introduction

The demand for efficient water treatment methodologies has risen in recent times owing to the burgeoning growth of the global population. The exponential growth in the global population has resulted in the scarcity of freshwater resources [1]. Therefore, more stringent regulation is being set by various environmental regulatory bodies for the sole purpose of maintaining the quality of water and protecting human health. The indiscriminate discharge of various pollutants from different anthropogenic activities into water bodies has reduced the quality of water. This causes potable water scarcity across the world. As a result of this challenge, the scientific community is now aggressively employing various approaches to remediate contaminated water bodies [2, 3]. Some of the pollutants identified in water and other aqueous matrices are heavy metals, organic contaminants, and nutrients (e.g., nitrogen, phosphorus, potassium, etc.). Copper, cadmium, lead, uranium, chromium, mercury, nickel, and zinc among others are the heavy metals that are being discharged by chemical companies into water bodies daily [4, 5]. Human health as well as aquatic life have been reported to be strongly affected by long-term exposure to toxic metals [6].

The common organic pollutants usually found in water and other aqueous media are synthetic halogenated substances, dyes, pesticides, phenolic compounds, pharmaceuticals, etc. In various industries, synthetic dyes are majorly utilized because they outperform the function of natural dyes. They have appealing colors and are recalcitrant to various environmental agents unlike their natural counterparts [7-9]. Another category of organic pollutants found in aqueous media is phenolic compounds which have been profiled by an environmental agency in the US (i.e., United States Environmental Protection Agency-USEPA) under Clean Water Act as priority pollutants [10, 11]. Traditional methods of separating these pollutants are not favorable due to their expensive nature and the fact that some of them are not eco-friendly [12].

Techniques such as separation involving the use of membranes, advanced oxidation, coagulation-flocculation, ion exchange, and solvent extraction have attracted the attention of many researchers for the remediation of water pollutants [13-18. However, these techniques are not cost-effective and some of them generate sludge/by-products and consequently create disposal/treatment challenges. They can as well cause secondary pollution owing to their large consumption of chemical substances [19]. Additionally, biological treatment methods generate biosolids and unsuitable discharge of nutrient-rich biosolids which could cause eutrophication thereby leading to the death of fish and other aquatic life [20, 21]. The use of greener substitutes in place of volatile organic solvents in analytical techniques adheres to the main principle of green chemistry. This is expected to contribute significantly to the minimization of the environmental effects of chemical technologies [22, 23]. So, in the last few decades, researchers have searched for environmentally benign solvents that could enhance the selectivity, sensitivity, and accuracy of analytical techniques. And of such greener solvents found are called ionic liquids, ILs [24].

ILs are simply molten salts constituted by large organic cations coupled with inorganic/organic anions with a melting point below 100^0C at room [23]. They have captivated the attention of researchers who are working on solvent-based extraction methods as a result of their unique physicochemical properties. Molecular design and preparation of novel ILs possessing diverse functionalities have widened the scope of ILs utilization [25]. The huge number of possible fusing of various organic cations and anions avails the chance to fabricate and tweak the physicochemical features of ILs by incorporating new structural motifs. Some of the structures of widely employed cations and anions during the preparation of ILs are shown in Figure 1. The properties of ILs depend on the appropriate selection of cation, and anion as well as the length of the substituent alkyl chain. ILs possess attention-catching properties like negligible vapor pressure at ambient conditions, low volatility, excellent thermal capacity, high stability, combustion resistance, low flammability, tunable viscosity, and wide electrochemical window compared to their neutral liquid counterparts [26, 27].

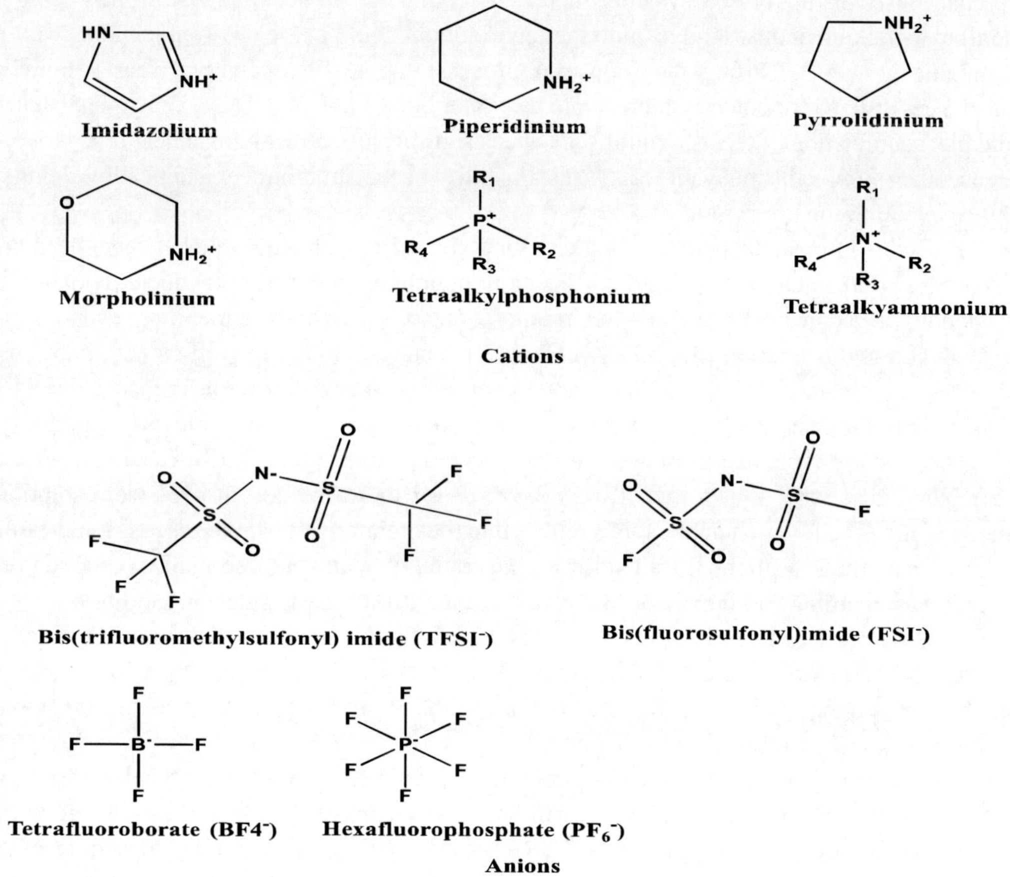

Figure 1. Chemical structures of commonly used cations and anions of ILs.

ILs have an outstanding capacity to associate with organic molecules or inorganic ions through various interactive forces such as π–π, dispersion, ion-exchange, electrostatic, hydrogen bonding, hydrophobic, and hydrophilic effects [28]. These forces are responsible for the separation of various analytes by ILs. Several research papers are discussed comprehensively outlining the successes of ILs as a separating tool in environmental science. So, this book chapter discusses the applications of ILs in the removal of dyes, pesticides, phenolic compounds, and toxic metals from various aqueous matrices. This article intends to provide relevant information expected to awake the consciousness of environmentalists in employing ILs for the decontamination of the environment.

Applications of ILs in the Separation of Various Pollutants

Separation of Dyes

Anthropogenic activities such as industrialization, and urbanization, among others, have led to the discharge of harmful chemical substances into the environment via aqueous wastes. These spontaneously disrupt our ecosystems, thereby affecting human health [29-32]. The release of non-biodegradable industrial dyes into the environment could cause environmental as well as economic concerns. During the industrial process, the textile industry releases inimical substances utilized for coloring fabrics into the water bodies [29, 30]. The toxicity, persistence, and bioaccumulation of these harmful substances significantly affect human health negatively. Hence, their removal is relevant for the sustainability of the environment and healthy living of the people inhibiting the world.

The conventional techniques for removing dyes from aqueous media have been ion exchange, photo catalytic degradation, filtration, membrane processes, chemical oxidation or reduction, biological, chemical precipitation, adsorption and extraction, evaporation, coagulation and ozonation [31, 33, 34]. However, challenges like low efficiency, high cost, slowness, and labor-intensive operation are preventing their adoption in the industry. For these reasons, the fabrication of cost-efficient and environmentally benign removal techniques is pertinent for the reclamation of wastewater containing dyes [18, 34-38]. Recently, several research studies have shown that ILs can be used for the extraction of dyes from solutions successfully. The research publications reported in the literature involve the use of liquid-liquid extraction methods whereby ILs are utilized. Interestingly, many of these studies reported great successes regarding the removal of various dyes from wastewater in addition to the regeneration/reuse of ILs with outstanding results. So, in this section, we described the use of ILs as environmentally friendly solvents for the extraction and removal of dyes from aqueous media. The summary of results reported regarding the removal of various dyes from aqueous matrices using ILs are presented in Table 1.

The extraction of two azo acid dyes known as Navy 5RE (acid blue dye) and acid red dye from water were investigated successfully by utilizing 1-butyl-1-methylpyrrolidinium bis(trifluoromethylsulfonyl)imide, [BMPYRR][NTf_2] [39]. A high extraction efficiency of 98% was obtained and the two analytes examined were recovered in addition to the recovery of the IL employed. Also, the partitioning of methyl orange, eosin yellow, and orange G anionic dyes were studied by utilizing various imidazolium-based ILs consisting of different alkyl chain length [32]. The partitioning of the investigated dyes was found to increase as the hydrophobicity of the IL increases i.e., 1-butyl-3-methylimidazolium hexafluorophosphate, [BMIM][PF_6]; <1-hexyl-3-methylimidazolium hexafluorophosphate, [HMIM][PF_6]; < 1-octyl-3-methylimidazolium hexafluorophosphate, [OMIM][PF_6]. The extraction of the investigated dyes was found to be strongly influenced by pH. After optimization, 85-99% of methyl orange, approximately 100% eosin yellows, and 69% of orange G extraction efficiencies were obtained from the analyzed water samples in one extraction step. By increasing the alkyl chain length of the cation of the ILs and temperature, the authors noticed a corresponding increase in the extraction rates of the dyes studied. Although, the introduction of potassium chloride to the aqueous solution containing the analyte did not have any significant effect on the extraction potential of the employed extraction systems. The thermodynamic studies conducted by the authors indicated that the extraction mechanism is driven by the hydrophobic interaction of the analytes and ILs.

Table 1. Summary of the removal of dyes by ILs

ILs	Dyes	Removal Efficiency (%)	References
[BMPYRR][NTf_2]	Acid blue & acid red	98	[39]
[BMIM][PF_6], [HMIM][PF_6] & [OMIM][PF_6]	Methyl orange eosin yellow & orange G	85–99 100 69	[32]
[BMIM][PF_6]	Acid yellow RN	81.31–99.98	[40]
[BMIM][PF_6]	Weak acid brilliant blue RAW	60.21	[41]
[BMIM][PF_6]	Weak acid blue 6B	95	[43]
[TCMA][SCN]	Methyl orange & methylene blue	89.09 64.14	[44]
[BBzIM][BF_4] [BBzIM][PF_6] & [BBzIM][NTf_2]	Rhodamime B	>88	[46]
[BMIM][PF_6]	Methylene blue	97.8	[47]
[BMIM][NTf_2], [HMIM][NTf_2], [OMIM][NTf_2], [HPIM][NTf_2] & [BzMIM][NTf_2]	Methylene blue	91.3 90.86 85.96 97.96 94.84	[48]
[TBP][OTF], [TBP][NTf_2] & [TBP][PF_6]	Methylene blue	84.95	[49]
[HMIM][BF_4] & [BMIM][PF_6]	Enriochrome black T & Crystal violet	99.65 98.3	[50]
[BMPYRR][NTf_2]	Rhodamine B	>98	[51]
[THTDP][Cl], [THTDP][DCA] & [THTDP[NTf_2]	Rhodamine B, methylene blue, methyl orange, malachite green, alizarin red S & Congo red	≈100	[30]

The extraction of acid yellow RN, acid brilliant red B, and weak acid brilliant blue RAW were examined by Li and collaborators from aqueous solutions [40, 41]. The investigators employed [BMIM][PF_6] in their studies. The removal percentages varying between 81.31% and 99.98% were successfully achieved for acid yellow RN. However, low removal percentage of 60.21% was obtained for weak acid brilliant blue RAW. The extraction activity of acid dyes is a form of anion exchange mechanism in which the soluble part of the IL participates as counter-ions. In another study, various imidazolium-based ILs were explored for the removal of 1-(phenylazo)-2-naphthol, 1-(2-pyridylazo)-2- naphthol, 4-(nitrophenylazo) resorcinol, 4-(2-pyridylazo) resorcinol, and 2-(4-dimethylaminophenylazo) benzoic acid from aqueous solution [42]. The distribution coefficient is strongly dependent on the pH of the water phase and the nature of the ILs employed. Generally, the distribution coefficient was found to increase with increasing alkyl chain length on the cations of the ILs. Also, the extraction rates using tetrafluoroborate-based ILs were found to be greater compared with that of hexafluorophosphate-based ILs.

[BMIM][PF_6] was also employed for the recovery of weak acid blue 6B from dyeing wastewater [43]. A high recovery efficiency of 95% was successfully achieved after the recovery process. The structure of acid blue 6B was not altered after the extraction as revealed by Fourier Transform Infrared (FT-IR) spectroscopic technique applied. The solubility of the weak acid blue 6B in the IL phase was reported to be higher than that of the aqueous rich phase. And this was the driving force responsible for the extraction. Additionally, the recovered IL and weak acid blue 6B were applied sequentially as retarding agents in the Tussah silk dyeing process.

Four quaternary ammonium-based ILs have been employed for the extraction of methyl orange and methylene blue from an aqueous solution [44]. The ILs employed for this study were: tricaprylmethylammonium thiocyanate, [TCMA][SCN]; tricaprylmethylammonium dicyanamide, [TCMA][DCA]; tricaprylmethylammonium benzoate, [TCMA][Bz]; and tricaprylmethylammonium hexanoate, [TCMA][Hz]. However, among these ILs, [TCMA] [SCN] demonstrated the best extraction efficiency i.e., 89.09% for methyl orange and 64.14% methylene blue. The extraction efficiency was improved significantly upon the inclusion of sodium chloride in the extraction system with the exception of [TCMA][SCN]. [TCMA][SCN] and [TCMA][DCA] resulted in 92.47% and 95.58% recoveries, respectively for methyl orange. Acetone, hexane, and chloroform were used to regenerate methylene blue from ILs. In this case, only chloroform was able to recover methylene blue from ILs. The removal rates of methylene blue obtained were 34.74% [TCMA][SCN] system and 45.22% for [TCMA][DCA] extraction platform in the first extraction cycle. The authors reported that the process reached 56.37% and 65.81% recoveries in the second extraction cycle.

Other types of ILs involving monoalkylguanidinium bis-trifluoromethane sulfonimides were utilized for the isolation of methyl orange from an aqueous solution [45]. Monoalkylguanidinium bis-trifluoromethane sulfonimides are regarded as functionalized hydrophobic ILs and demonstrated significant extraction efficiencies compared to traditional imidazolium-based ILs. The removal of rhodamine B has been examined from water through the utilization of benzyl-functionalized ILs [46]. The results obtained showed that benzyl functionalized ILs displayed higher extraction efficiencies (>88%) compared to dialkylimidazolium-based ILs. This study was carried out by varying the solution's pH from 2 to 12. The phase volume ratio and the addition of sodium chloride were noticed to influence the efficiency of the examined systems. The interaction between the IL and the analyte was found to involve π–π stacking, hydrophobicity, and electrostatic interactions. Nevertheless, hydrophobic interaction was regarded as the major driving force for the partitioning of rhodamine B from water into IL.

A nonionic surfactant-based ILs system has been employed for the separation of methylene blue from an aqueous solution [47]. In this investigation, [BMIM][PF_6] in combination with varying small amount of triton-x were employed. The extraction efficiency has the potential of reaching 97.8%, which indicates that the addition of triton-x surfactant improved the extraction of methylene blue significantly. The dye's concentration in the effluent indicated about 50-fold reduction under optimal conditions. Furthermore, the presence of potassium carbonate and low pH were found to hinder the extraction efficiency. Nevertheless, this could be useful for the regeneration of dye and surfactant- ILs.

Furthermore, five hydrophobic ILs were prepared and utilized for the extraction of methylene blue from an aqueous solution [48]. Several ILs based on [NTf_2] anion were used for the purpose of comparison. In this study, three of the ionic liquids were of commercial origin i.e., 1-butyl-3-methylimidazolium, [BMIM], 1-hexyl-3-methylimidazolium, [HMIM] and 1-octyl-3-methylimidazolium [OMIM] cations. The remaining two i.e., 1-hexyl-2-phenylimidazolium, [HPIM] and 1-benzyl-3-methylimidazolium, [BzMIM] were synthesized mainly for comparative study. The various parameters like length of the side alkyl chain, position and nature of the aromatic ring as well as the capacity to form hydrogen bonding on the extraction potentials of these ILs employed.

The results obtained indicated that the inclusion of the aromatic group led to outstanding dye removal compared to the commercially available ILs used (Figure 2 and Table 1). They also found out that it allowed easier recovery and subsequent reuse of the functionalized ILs. Similarly, the same dye examined by Kermanioryani et al. [48] was also examined by using phosphonium-based hydrophobic ILs with fluorous anions [49]. The phosphonium-based hydrophobic ILs employed demonstrated outstanding capacities and resulted in 84.95% methylene removal.

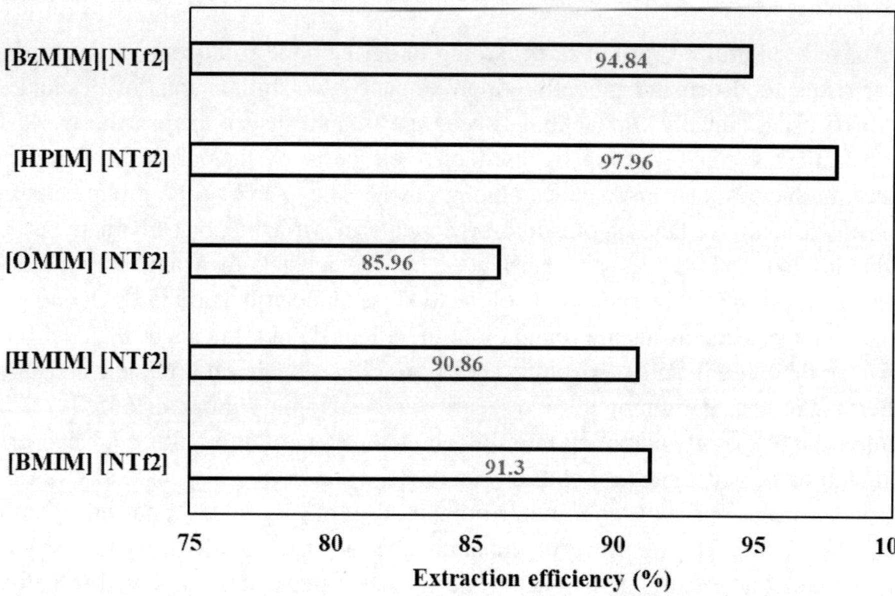

Figure 2. Extraction efficiencies of various ionic liquids employed for the extraction of methylene blue from water [48].

Three different imidazolium-based ILs were utilized for the separation of dyes that are recalcitrant in nature (i.e., eriochrome black T and crystal violet) from aqueous solutions [50]. The ILs employed for this study were: 1-hexyl-3-methylimidazolium tetrafluoroborate, [HMIM][BF_4]; [BMIM][PF_6] and 1-hexyl-3-methylimidazolium hexafluorophosphate, [HMIM][PF_6]. [HMIM][BF_4] IL displayed the highest removal rate (99.65%) for enriochrome black T whereas [BMIM][PF_6] displayed the best extraction efficiency (98.3%) for crystal violet. Additionally, maximum dye separation was achieved nearly at neutral pH. The exclusion of rhodamine B from aqueous solution has been systematically studied through the utilization of [BMPYRR][NTf_2] as an additive [51]. It was posited that imidazolium cation-based ILs such as [BMIM][BF_4] and [BMIM][PF_6] do not support the sequestration of rhodamine B dye from aqueous solutions. The rate of exclusion of compound of interest relies on the amount of [BMPYR][NTf_2] added to aqueous rhodamine B solution. The addition of a minute quantity of [BMPYR][NTf_2] was found to lead to the outstanding removal of dye (> 98%).

Recently, the removal of harmful textile dyes including rhodamine B, methylene blue, methyl orange, malachite green, alizarin red S, and congo red from their aqueous solutions were performed using trihexyltetradecylphosphonium chloride, [THTDP][Cl]; trihexyltetradecylphosphonium dicyanamide, [THTDP][DCA] and trihexyltetradecyl-phosphonium bis (trifluoromethyl sulfonyl) amide, [THTDP[NTf_2] ILs [30]. Under optimal conditions, the three ILs used were able to removal all the studied dyes completely (i.e., ≈100 removal efficiency). The toxicity of the investigated dyes and the ionic liquids employed were assessed using OSIRIS Software. The investigators found all the dyes to be toxic except methylene blue. Also, the ILs utilized were found to be non-toxic in nature.

Pesticides

Pesticides have been utilized for the safeguarding of agricultural produce in order to respond to the ever-rising food demand globally. Approximately two million tons of pesticides are reported to be used annually for tackling weeds, pests, and insects across the world [52]. Generally, pesticides can be grouped based on their pests of the target into fungicides, rodenticides, herbicides, and insecticides among others. They can also be grouped based on their chemical structure or heteroatoms into organochlorine, triazine, or carbamate pesticides [53]. Pesticides like Insecticides and herbicides are the most known and used pesticides, occupying 29.5% and 47.5%, respectively, of the total pesticide utilization [53]. Owing to their wide utilization, these chemicals are found often in various compartments of the environment [54]. However, they significantly constitute deleterious effects on the environment and humans, despite their importance pertaining to food security and economic significance [55].

Pollution can be majorly classified into soil, air, and water pollution [56-58]. Nevertheless, the contamination of water (water pollution) among them is a very big challenge worldwide and is usually occasioned by the indiscriminate use of pesticides and other inimical contaminants [59, 60]. The presence of contaminants in water modifies its features which makes water unfit for use [61]. This occurs mostly when pesticides employed in cultivated lands are leached apart by environmental agents such as rain, wind, and so on into bodies of water. The maintenance of tools and machines used to produce pesticides can also increase the concentration of pesticides in wastewater [62]. So, it is highly imperative to clean up and quantify pesticides in wastewater using relatively sustainable solvents and techniques. In most cases, high-performance liquid chromatography (HPLC) is normally employed for the quantification of pesticide residues after their extraction using ILs. The use of the gas chromatographic technique is usually hindered owing to the low volatile nature of most ILs [63]. Therefore, in this section, we elucidated the research findings whereby ILs were employed for the preconcentration and extraction of pesticides from different aqueous matrices.

Two organophosphorus pesticides (methylparathion and phoxim) from rain, groundwater, reservoir water, and river water were extracted by [HMIM][PF_6] [64]. In this study, the temperature-controlled IL dispersive liquid-phase microextraction method was employed prior to HPLC combined with ultraviolet detection. Factors influencing extraction performance such as the volume of IL, pH of solutions, extraction duration, centrifugation duration, dissoluble temperature, and salt effect were optimized. The compounds of interest showed a good linear relationship in the concentration range of 1–100 ng/L. The limits of detection (LOD) of methylparathion and phoxim were 0.17 ng/L and 0.29 ng/L, respectively. The recoveries of the analytes ranged between 88.2% and 103.6% were obtained. These outstanding results demonstrated that temperature-controlled IL dispersive liquid-phase microextraction could become a good tool in environmental science. In another study, a temperature-controlled IL dispersive liquid-phase microextraction has been utilized for the extraction of three triazine herbicides i.e., cyanazine, simazine, and atrazine from underground water, sewage treatment plant outlet drainage water as well as river water [65]. The analytes were extracted by [OMIM][PF_6] and determined by HPLC with an ultraviolet detector. LOD ranging between 0.05 μg/L and 0.06 μg/L were obtained by the researchers. Furthermore, high recoveries of 85.1% and 100% were successfully achieved after optimization.

Four pyrethroid insecticides (permethrin, cyhalothrin, fenpropathrin, and transfluthrin) were extracted from various water samples by tributyldodecylphosphonium hexafluorophosphate, [TBDDP][PF_6] IL [66]. In this study, the dispersion liquid-liquid microextraction method involving the solidification of sedimentary IL combined with HPLC was adopted. The effects of IL, ionic strength of the solution, and temperature of the sample solution as well as centrifugation time were investigated. These influencing factors were later optimized using a model called central composite design. The LOD for the pyrethroid insecticides ranged from 0.71μg/L to 1.54μg/L. Also, the recoveries for the pesticides examined ranged between 87.1 and 101.7% after optimization. Azoxystrobin, chlorothalonil, cyprodinil, and trifloxystrobin fungicides were recovered from various environmental water samples by utilizing [HMIM][PF_6] and 1-hexyl-3-methylimidazolium chloride, [HMIM][Cl] ILs as extractants [67]. This technique encompasses magnetic solid-phase microextraction and IL dispersive liquid–liquid microextraction in combination with HPLC. Under optimized conditions, between 0.02μg/L and 0.04μg/L LOD in addition to between 98% and 115% recoveries from real water samples were successfully obtained. Furthermore, enrichment factors ranging between 135 and 159 folds were also actualized.

In addition, pesticides based on phosphorus, including azinphos-methyl, chloropyriphos, parathion-methyl, diazinon, and phosalone have also been extracted by [BMIM][PF_6] IL [68]. In this investigation, a fast and overly sensitive ultrasound-assisted temperature-controlled IL dispersive liquid-phase microextraction coupled with reversed-phase liquid chromatography-ultraviolet (RPLC-UV) was employed. After optimization of all the extraction efficiency influencing factors, LOD ranging from 0.01ng/mL to 0.1ng/mL and extraction efficiencies ranging between 90 and 98% were achieved for the target compounds. A binary-solvent-based IL-assisted surfactant-enhanced emulsification microextraction technique was fabricated for the separation/preconcentration and determination of pyrimethanil, fludioxonil, cyprodynil, pyraclostrobin fungicides in apple juice and apple vinegar [69]. Amyl acetate (a non-chlorinated solvent) was employed as the primary extractant, and [HMIM][PF_6] IL as the secondary extracting agent. And the analytical technique employed for the quantification of analytes was HPLC. The LOD ranging from 0.5μg/L to 1.5 μg/L and recoveries between 51.9% and 107.8% were observed for the analytes under optimized conditions.

[BzMIM][NTf_2] IL was applied as an extractant in dispersive liquid-liquid microextraction for the enrichment and recovery of organophosphorus pesticides (i.e., phoxim, fenitrothion, chlorpyrifos, phorate, and parathion) in water samples [70]. In this study, HPLC was adopted for the quantification of the target compounds. Under optimized conditions, satisfactory recoveries ranging from 82.7%–118.3% and LOD between 0.01μg/L and 1.0 μg/L were reported. The availability of benzyl group in the IL employed was claimed to have elevated the extraction rate owing to the *π-π* interaction between the IL and the target compounds with a high enrichment factor of 339. In another study, an effervescence-assisted microextraction technique was utilized for the detection of azoxystrobin, triazolone, cyprodinil, and trifloxystrobin fungicides in water [71]. In this investigation, IL-based dispersive liquid–liquid microextraction was combined with the magnetic recuperation of the extraction solvent. The extractant dispersion and retrieval were achieved by employing a magnetic effervescent tablet constituted by Fe_3O_4 magnetic nanoparticles, sodium carbonate, sodium dihydrogen phosphate, and [HMIM][NTf_2]. HPLC coupled with a diode array detector (DAD) was the analytical technique employed for the quantification of analytes. The LOD obtained ranged from 0.02 μg/L and 0.10 μg/L and recoveries between 70.7% and 105% of analytes from real water samples were observed.

The isolation and preconcentration of organophosphorus- based pesticides (i.e., dimethoate, disulfoton, famphur, parathion, parathion-methyl, phorate, sulfotep, thionazin, and triethyl thiophosphate) from various water sources were carried out [72]. The IL employed in this study was 1-butyl-3-methylimidazolium chloride, [BMIM][Cl]. This preconcentration method was combined with gas chromatography-mass spectrometry (GC-MS) by employing the microbial insert thermal desorption technique. All extraction efficiency influencing parameters were optimized. Under these optimized states, the recoveries of the analytes were reported between 97% to 113%, and LOD ranged from 5 ng/L to 16 ng/L. Onsite IL dispersive liquid-liquid microextraction in combination with miniaturized Fe_3O_4 magnetic nanoparticles was constructed for the recovery and pretreatment of pyrethroid pesticides (deltamethrin, bifenthrin, cyhalothrin, and ethofenprox) in water samples [73]. The analytes were enriched by hydrophilic IL tributyloctylphosphonium bromide, [TBOP][Br] after the optimization of new anion-exchange reagents comprising sodium diethyldithiocarbamate and sodium dicyanamide. Ultra-small superparamagnetic iron Fe_3O_4 nanoparticles were utilized for the collection of IL and pyrethroids mixture followed by separation using acetonitrile. Good recoveries in the range of 80.20% to 117.31% with good enrichment factors (176–213) were noticed for the analyzed compounds.

The separation of aryloxyphenoxypropionate herbicides in aqueous were performed using IL-based air-assisted liquid–liquid microextraction in conjunction with HPLC-DAD [74]. In this analytical method, a small amount of [HMIM][PF_6] was utilized as extractant. Under optimized conditions, this technique demonstrated recoveries between 76% and 83% as well as enrichment factors ranging from 380 to 415. This method also showed low LOD between 0.16 ng/L and 0.31ng/L. Glyphosate and aminomethylphosphonic acid herbicides in various water sources were preconcentrated and determined by analytical temperature-assisted IL-based dispersive liquid–liquid microextraction [75]. The extracted target compounds were derivatized using 9-fluoroenylmethylchloroformate and quantified by liquid chromatography with fluorescence detection. Among the examined ILs, 1-decyl-3-methylimidazolium tetrafluoroborate, [DMIM][BF_4] displayed the best extraction performance. Eventually, the analytes recoveries and enrichment factors ranged between 75 and 99% and 57 and 76, respectively, after optimization. And the values obtained for LOD were 0.27ng/L for glyphosate and 0.22 ng/L for aminomethylphosphonic acid.

A microwave-assisted IL/IL dispersive liquid–liquid microextraction combined with HPLC was constructed for fast preconcentration and analysis of triazine herbicides in fruit juice

[76]. The triazine herbicides studied were desmetryn, atrazine, ametryn, propazine, terbuthylazine, prometryne, and dimethametryn. Rather than using harmful volatile organic solvents, two different categories of ILs, a hydrophobic IL ([HMIM][PF_6]), used as extraction solvent and a hydrophilic ionic liquid ([BMIM][BF_4]), utilized dispersion agent. Furthermore, an ion-pairing agent known as ammonium hexafluorophosphate was introduced to enhance the recovery of the IL phase.

The recoveries of 76.7% and 105.7% were obtained for the analytes in fruit juice after optimization. Diflubenzuron, triflumuron, hexaflumuron, flufenoxuron, lufenuron, diafenthiuron, transfluthrin, fenpropathrin, γ-cyhalothrin, and deltamethrin insecticides have also been preconcentrated and extracted from pear juice [77]. Ultrasound-assisted hybrid IL dispersive liquid-liquid microextraction technique coupled with HPLC (with a variable-wavelength detector) was used for the detection of the analytes. And a mixture of 1-octyl-2,3-dimethylimidazolium bis(trifluoromethylsulfonyl)imide, [ODMIM][NTf_2], and 1-hexyl-3-methylimidazolium bis(trifluoro-methylsulfonyl)imide, [HMIM][NTf_2] made up the extraction solvent. Recovery rates ranging from 88.9%–107% and enrichment factors within the range of 133 to 157 were obtained successfully after optimization for the analytes. The LOD for the compounds of interest as reported ranged between 0.19 μg/L and 0.69 μg/L.

Recently, triazole fungicides in water samples have been preconcentrated and determined by an onsite IL dispersive liquid–liquid microextraction coupled with miniature superparamagnetic Fe_3O_4 [78]. In this study, 1,3-dibutylimidazolium bromide, [DBIM][Br] was used as the extraction agent prior to HPLC. After optimization, high enrichment factors (187–323) were obtained for the target compounds. Also, the obtained recoveries of the analytes ranged between 70.1% and 115.0% with 0.74 μg/L to1.44 μg/L LOD ranges (Table 2). A low pretreatment time of fewer than 3 minutes as well as low consumption of extraction solvent were successfully actualized. The extraction of various fungicides (i.e., pyrimethanil, fludioxonil, azoxystrobin, cyprodinil, and pyraclostrobin) were performed from different fruit juices (grape, apple, and pear juices) using IL–based air-assisted liquid–liquid microextraction technique in combination with HPLC [79]. [HMIM][NTf_2] IL was used as the extraction solvent instead of the organic solvent. In this study, no organic dispersive solvent was required, and this makes the technique more environmentally benign. The authors found the LOD and recoveries of the analytes to range from 0.4 μg/L to 1.8 μg/L and 74.9% to 115.4%, respectively, under optimized conditions.

A functionalized IL 3-methyl-1-(ethoxycarbonylmethyl) imidazolium bis(trifluoromethyl-sulfonyl)- imide, [MECM][NTf_2] was explored for its separation power during the recovery of some organophosphorus pesticides (parathion, parathion-methyl, profenofos, chlorpyrifos, O-ethyl O-4-nitrophenyl phenylphosphonothioate) from river water samples [80]. In this examination, situ dispersive liquid-liquid microextraction in association with ultra-HPLC. π-π stacking interactions between the parent imidazolium cation core and the analytes were found to be responsible for the extraction process. During the liquid-liquid de-mixing, a turbid solution comprising fine drops of IL permitted for greater contact area between target compounds and IL. This allows the large transfer of the analyzed compounds into the ionic-rich phase. After optimization, greater than 400 enrichment factor and high recoveries within 105.5 to 114.2 were obtained successfully for the pesticides studied (Figure 3).

Table 2. Summary of the removal of pesticides by ILs

ILs	Pesticides	Matrix	LOD	Removal Efficiency (%)	References
[HMIM][PF_6]	Methylparathion & phoxim	Rain, river, reservoir & ground waters	0.17 ng/L & 0.29 ng/L	88.2 –103.6	[64]
[OMIM][PF_6]	Cyanazine, simazine & atrazine	Underground, river drainage & sewage treatment plant outlet waters	0.05 µg/L–0.06 µg/L	85.1–100	[65]
[TBDDP][PF_6]	Permethrin, cyhalothrin, cyprodinil & trifloxystrobin	Lake, river, paddy field & reservoir waters	0.71 µg/L–1.54 µg/L	87.1–101.7	[66]
[HMIM][PF_6] [HMIM][Cl]	Azoxystrobin, chlorothalonil cyprodinil & trifloxystrobin	Pool, stream, rain & river waters	0.02 µg/L–0.04 µg/L	98–115	[67]
[BMIM][PF_6]	Azinphos-methyl, chloropyriphos, parathion-methyl, diazinon & phosalone	Tap, well & lake waters	0.01 ng/mL–0.1 ng/mL	90–98	[68]
[BzMIM][NTf_2]	Pyrimethanil, fludioxonil, cyprodynil & pyraclostrobin vinegar	Apple juice & apple	0.5 µg/L–1.5 µg/L	51.9–107.8	[69]
[BzMIM][NTf_2]	Phoxim, fenitrothion, chlorpyrifos, phorate & parathion	Rain, tap & yellow river waters	0.01µg/L–1.0 µg/L	82.7–118.3	[70]
[HMIM][NTf_2]	Azoxystrobin, triazolone cyprodinil & trifloxystrobin	Lake waters	0.02 µg/L–0.10 µg/L	70.7–105	[71]
[BMIM][Cl]	Dimethoate, disulfoton, famphur, parathion, parathion- methyl, phorate, sulfotep, thionazin & triethyl thiophosphate	Environmental waters	5 ng/L–6 ng/L	97–113	[72]
[TBOP][Br]	Deltamethrin, bifenthrin, cyhalothrin & ethofenprox	Tap, pond, & river waters	0.16 µg/L–0.21 µg/L	80.2–117.3	[73]
[HMIM][PF_6]	Clodinafop-propargyl, haloxyfop-Rmethyl, fenoxap- rop-P-ethyl & diclofop-methyl	Surface, tap and well waters	0.16 ng/L–0.31 ng/L	76–83	[74]
[DMIM][BF_4]	Glyphosate & aminomethyl-phosphonic acid	Bottle, irrigation, tap & ground waters	0.27 ng/L–0.22 ng/L	75–99	[75]
[HMIM][PF_6]	Desmetryn, atrazine, ametryn, propazine, terbuthylazine, prometryne & dimethametryn	Fruit juice	1.01 µg/L–1.57 µg/L	76.7–105.7	[76]
ODMIM][NTf_2] & [HMIM][NTf_2]	Diflubenzuron, triflumuron hexaflumuron, flufenoxuron lufenuron, diafenthiuron transfluthrin, fenpropathrin γ-cyhalothrin & deltamethrin	Pear juice	0.19 µg/L–0.69 µg/L	88.9–107	[77]
[DBIM][Br]	Myclobutanil, tebuconazole, cyproconazole & prothioconazole	Lake, tap, river & pond waters	0.74 µg/L–1.44 µg/L	70.1–115	[78]
[HMIM][NTf_2]	Pyrimethanil, fludioxonil azoxystrobin, cyprodinil & pyraclostrobin	Grape, apple & pear juices	0.4 µg/L–1.8µg/L	74.9–115.4	[79]
[MECM][NTf_2]	Parathion, parathion-methyl imidazo profenofos, chlorpyrif-os, & O-ethyl O-4-nitrophenyl phenylphosphonothioate	River water	0.7 µg/L–2.7 µg/L	105.5–114.2	[80]
[BMIM][PF_6]	Imidacloprid, imidaclothiz, acetamiprid & thiacloprid	Wild & commercial honeys	0.30 µg/L–0.62 µg/L	81–103.4	[81]

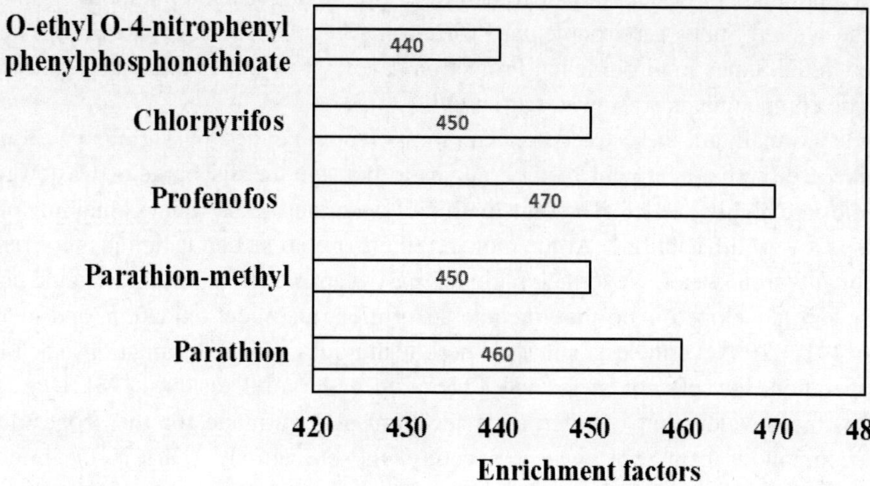

Figure 3. Enrichment factors of organophosphorus pesticides by a functionalized IL, [MECM][NTf_2] [80].

Neonicotinoid insecticides including imidacloprid, imidaclothiz, acetamiprid, and thiacloprid in wild and commercial honey samples were preconcentrated/pretreated by an in-situ IL dispersive liquid–liquid microextraction followed by ultra-HPLC [81]. This technique uses [BMIM][PF_6] as an extraction solvent. The IL was synthesized from the reaction between PF_6 and [BMIM][Br]. This technique does not require the use of the dispersive agent. The researchers obtained high enrichment factors ranging from 74 to 115 in addition to good recoveries (81% to 103.4%) for the analytes examined.

Phenolic Compounds

The mitigation of pollution caused by phenolic compounds is pertinent in ensuring the sustainability of our environment. This is so due to the fact that phenolic compounds are highly toxic to almost all living organisms in addition to their ability to persist and bio-accumulate in the environment [82, 83]. They have been found in wastewater generated from industrial chemical processes, including petrochemical, plastics, resins, pharmaceuticals, pulp and paper, textile as well as pesticide industries [84]. In these industries, high concentrations of phenolic compounds were detected ranging between 0.1 and 17,000 mg/L [85]. This poses a huge challenge for traditional treatment facilities. Even at phenol concentrations below 200 mg/L, these substances have been found to induce the growth of microorganisms in biological treatment facilities [86]. This threat has also been found in the environment. For instance, the dangerous effects of phenol on *Bufo arenarum* embryos were studied and it was observed that a concentration of 25 mg/L remarkably has a negative impact on this species [87].

Based on various environmental regulations, the permissible levels of phenols in water in order to guarantee the quality of freshwaters and prevent environmental effects are usually very low. For instance, pentachlorophenol, a very common phenol-derivative found in the environment has a maximum permissible concentration of 5 μg/L for potable water [88]. Also, phenolic compounds have been regarded as priority pollutants by international organizations such as the United States Environmental Protection Agency (USEPA) and European Union [89]. The United States Environmental Protection Agency has set the allowable concentration for phenolic compounds in wastewater at 1 mg/L [90].

In order to maintain these trace concentrations recommended by various environmental policymakers, effective treatment methodologies to manage the discharge of wastewater into water bodies are highly needed. The remediation of water and wastewater containing phenolic pollutants is a very difficult one. At the moment, there are no settled techniques for removing these pollutants from water. Nevertheless, in the past years, physical, biological, and chemical processes were developed. And they include adsorption, chemical oxidation, and membrane separation [91, 92]. Nevertheless, some of these techniques suffered limitations such as high cost of operation, low effectiveness, and generation of harmful residues [93]. Due to these drawbacks, the development of alternative techniques is germane for the sequestration of phenolic compounds from wastewater. Recently, the utilization of ILs in the removal of phenolic pollutants has surfaced and has gained popularity among scientists globally. In light of this, we have discussed the progress made so far in this section.

The capacity of 1-octyl-3-methylimidazolium tetrafluoroborate, [OMIM][BF_4]; [DMIM][BF_4]; [BMIM][PF_6] and [OMIM][PF_6] for the separation of phenol, tyrosol and p-hydroxybenzoic acid from aqueous solution were studied [94]. The results obtained demonstrated that ILs possessing BF_4^- anion are more effective in the sequestration of all the phenolic compounds examined compared to those with PF_6^- anion. In a similar study, the extraction of phenol, 4-nitrophenol, 2,4-dinitrophenol, 2,6-dinitrophenol, 2,4,6-trinitrophenol, 1-naphthol, 2-naphthol, 4-chlorophenol, catechol and resorcinol from water were performed by utilizing [BMIM][PF_6] IL [95]. This study was carried out at 295K and more than 90% extraction efficiencies were obtained for nitrophenols, naphtols, and 4-chlorophenol. Nevertheless, the study involving the effect of pH on the extraction performance indicates that the pH of the solution needs to be less than the pKa before high recovery can be obtained. However, the extraction rate was found to be significant when the pH was greater than pKa for 1-naphthol and 4-chlorophenol. In addition, the removal rate was also found to be more than 90% in the pH range 1.5–12 for 2,4,6-trinitrophenol.

The significance of alkyl chain length of alkyl-imidazolium ILs consisting of PF_6^- and BF_4^- anions in the removal of phenol, pentachlorophenol, bisphenol A, 4-octylphenol and 4-nonylphenol at the temperature of 297 K was investigated [96]. The extraction efficiency was found to increase as the alkyl chain length on the cation increases for a particular analyte. Higher extraction efficiencies were observed for IL-containing BF_4^- anion similar to the report by Vidal et al. [94]. Irrespective of ILs employed, the extraction efficiency of the phenolic compounds investigated follows the trend: bisphenol A>4-nonylphenol>4-octylphenol > pentachlorophenol > phenol. The highest extraction efficiency obtained for bisphenol A can be attributed to its possession of phenolic hydroxyls and two aromatic rings within its structure. And this boosted its stronger interactions with the various ILs studied.

Furthermore, imidazolium, pyrrolidinium, quaternary ammonium, and choline-based ILs have also been tested for their capacity to extract phenolic compounds from aqueous matrices [97, 98]. Compounds like phenol, 4-nitrophenol, 2,4-dinitrophenol, 2,6-dinitrophenol, 2,4,6-trinitrophenol, 1-naphthol, and 2-naphthol were extracted from water using tetrahexyl-ammonium dihexyl sulfosuccinate, [THADH][SS] and trioctylmethylammonium salicylate, [TOMA][SC]. And the capacity of these two ILs to extract the given analytes was compared with that of [HMIM][NTf_2] and 1-decyl-3-methylimidazolium bis(trifluoromethanesulfonyl) imide, [DMIM][NTf_2]. The distribution ratios for the imidazolium-based ILs were observed to be lower than those of the ammonium-based ILs. At more than pKa, the extraction rate was found to be satisfactory owing to the exchange of ions between the phenolic compound and the IL i.e., the phenolate form of the analytes and the salicylate anion of the IL. The high solubility of cholinium bis(trifluoromethanesulfonyl) imide, [Ch][NTf_2] is the only demerit of using it for extracting phenolic compounds [98]. [Ch][NTf_2] attained a concentration of more than 0.5 molar fractions of water in the organic-rich phase.

1-Butyl-2,3-dimethylimidazolium hexafluorophosphate, [BDMIM][PF_6] was utilized for the isolation of pentachlorophenol from an aqueous medium at 298.15 K [99]. Many imidazolium-based ILs containing NTf_2^-, PF_6^- and BF_4^- anions were examined for the separation of phenol, resorcinol, p-nitrophenol, guaiacol, and o-cresol from water [100]. Among the investigated ILs, the ILs consisting of BF_4^- anion displayed the best extraction potential. This behavior can be explained by their stronger capacity to form hydrogen-bonding with the compounds of interest [101]. The presence of hydroxyl group on the cations of the ILs with NTf_2^- and PF_6^- anions was found to enhance the extraction yield compared with those with the same anions. This effect was found to be pronounced for resorcinol. However, for the remaining four analytes, their extraction efficiencies were found to be comparable to those of the dialkyl-based ILs carrying BF_4^- anion.

The ability of 1-ethyl-3-methylimidazolium bis(fluorosulfonyl) imide, [EMIM][NTf_2] to separate o-cresol, 2-chlorophenol, resorcinol, and phenol from water have also been evaluated [102]. After the process of sequestration, the tolerance limit of 1 mg/L required by law for industrial wastewater was attained for various water sources with an initial concentration of less than 10 mg/L for all the phenolic compounds examined, with the exception of resorcinol. Because phenolic compounds are aromatic in nature, the conjugation π-π ingrained between the two rings (i.e., IL and analytes rings) has no crucial significant in the extraction performance. The low removal rates obtained for resorcinol can be linked to the availability of two hydroxyl groups in its molecule, which increases resorcinol-water interaction. The trend of the extraction of the analytes is in accordance with their hydrophobicity trend and it follows the order: 2-chlorophenol > o-cresol > phenol > resorcinol (Figure 4). The IL used in this study was regenerated easily using an aqueous solution of sodium hydroxide. The regenerated IL could be again used in the subsequent extraction process without any great loss in its extraction power.

Phenol, chlorophenol, and o-cresol were extracted from aqueous wastewater streams using 1-butylpyridinium bis(trifluoromethylsulfonyl) imide, [BPYR][NTf_2]; 4-butyl-1-methylpyridinium bis(trifluoromethylsulfonyl) imide, [4BMPYR][NTf_2]; 4-butyl-1-methylpiperidinium bis(trifluoromethylsulfonyl) imide, [4BMPIP][NTf_2] and 4-butyl-1-methylpyrrolidinium bis(trifluoromethylsulfonyl)imide, [4BMPRR][NTf_2] ILs [103]. The extraction efficiencies in all cases were found to be more than 90%. However, the performance of the extraction reduces when the concentration of phenols in the aqueous phase is less, and no saturation occurred in the IL phase when the initial concentration of phenols in the aqueous phase reached 15000 mg/L. The extraction efficiencies for o-cresol and chlorophenol were found to be higher than the ones obtained for phenol. In another study, 1-hexyl-1-methylpyrrolidinium bis (trifluoromethylsulfonyl) imide, [HMPRR][NTf_2] and [HMIM][NTf_2] were employed for the removal of phenol, *o*-cresol and resorcinol [104]. The phenol extraction realized was from 90%

to 95% within a contact time of 30 minutes by utilizing a phase volume ratio (water to IL) of 1.5.

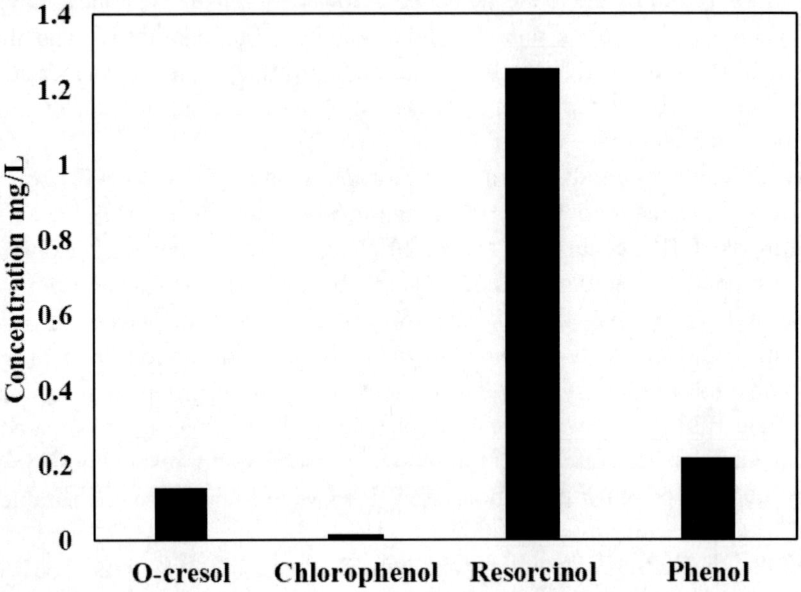

Figure 4. Removal of o-cresol, 2-chlorophenol, resorcinol, and phenol from water by [EMIM][NTf_2] IL [102].

A mixture constituted by [BMIM][BF_4] and tributylphosphate was utilized for the separation of phenol and 2,4-dichlorophenol from water [105]. Around 99.74% and 99.18% extraction rates were obtained for phenol and 2,4-dichlorophenol, respectively. Similarly, phenol and 2,4-dichlorophenol were removed from water by utilizing trihexyltetradecylphosphonium decanoate, [THTDP][Dz], and trihexyltetradecylphosphonium bis(2,4,4-trimethylpentyl)phosphinate, [THTDP][TMPP] ILs [106]. The ILs were observed to have the potential of extracting higher than 99% of 2,4-dichlorophenol and 89% of phenol within 10 min duration. The best extraction performance was obtained at lower temperatures and pH less than the pKa of these phenolic compounds.

Metal Ions Removal

Metal ions' existence in wastewater and other forms of aqueous media have been skyrocketing in recent times with industrialization and other anthropogenic activities, including the mining industry, plating and electroplating industry, textile industry, batteries, metal smelting, petrochemicals among others [107]. Wastewater contaminated by metal ions are discharged into the environment, consciously or unconsciously [108, 109]. They could threaten human well-being and the environment at large. Metal ions are non-biodegradable [110] and could be carcinogenic or teratogenic [111, 112]. So, their presence in water above the permissible limits has a tendency of posing critical health challenges to living things.

The most common metal ions in an aqueous environment are Pb^{2+}, Zn^{2+}, Hg^{2+}, Ni^{2+}, Cd^{2+}, Cu^{2+}, Cr^{2+}, and As^{2+} [107]. These metal ions need to be removed from aqueous matrices owing to their hazardous nature above their tolerable concentrations by using sustainable techniques and materials. Volatility, flammability, solvating ability, toxicity, and thermal stability are factors that are usually considered when choosing solvents for the removal of metal ions in separation technology [113]. However, traditional volatile organic solvents could lead to environmental and health safety consequences. Therefore, it is imperative to safeguard the environment via the elimination of volatile organic solvents. The core aim of sustainable chemistry is the avoidance of harmful solvents in chemical operations and waste reduction [114]. In this regard, ILs have enormously received attention from the scientific community regarding the recovery of different metal ions with outstanding results. Therefore, in this section, we have illuminated the use of ILs in the removal of various metal ions from aqueous matrices.

Scientists have proposed task-specific ionic liquids (TSILs) as futuristic extraction agents for the separation of metals a while ago. They are usually prepared via the introduction of functional groups such as phosphate, phospine, thioether, thiourea among others directly into the cations of ILs [115-117]. The utilization of TSILs have proven to increase the affinity of ILs for metal extraction, leading to a corresponding increase in extraction efficiency. The extraction of Cu^{2+}, Hg^{2+}, Pd^{2+}, and Ag^{+} ions were examined from the water some time ago [118]. In this study, it was found that ILs possessing disulfide functional group are more effective for the extraction of Cu^{2+} and Hg^{2+} from water (Table 3). And that ILs with nitrile functional group are highly powerful for Pd^{2+} and Ag^{+} ions extraction from water.

Table 3. Summary of the removal of toxic metals by ILs

ILs	Metals	Removal Efficiency (%)	References
[TOMA][SC]	Cu^{2+} & Fe^{2+}	89 for Cu^{2+} 99 for Fe^{2+}	[119]
[THMA][MC] [THMA][AP] [THMA][SC] [THMA][AL] [TOMA][DDD] [TOMA][OX] [TOMA][AL] & [TOMA][TSC]	Ni^{2+}, Cu^{2+}, Pb^{2+}, & Co^{2+}	4.4–96.4 for Ni^{2+} 47.2–99.7 for Co^{2+} 20.4–92.8 for Zn^{2+} 30.5–99.8 for Pb^{2+} 54.1–99.9 for Cu^{2+}	[120]
[TCMA][BzTBzt] [THTDP][BzTBzt] [TCMA][ETBzt] [THTDP][ETBzt] [TCMA][PTBzt]& [THTDP][PTBzt]	Cu^{2+}, Zn^{2+} & Cd^{2+}	≥ 94	[120]
[BPYR][NTf_2]	Cu^{2+}	82.6	[122]
[OMIM][BF_4] & [OMIM][NTf_2]	Hg^{2+}	≥ 90	[117]
[BIM][DEHP] & [MIM][DEHP]	Hg^{2+}	100	[123]
[BMMOR][BPI] [HMMOR][HPI] [OMMOR][OPI]	UO_2^{2+}	98.6 75.1 48.2	[124]

Table 3. (Continued)

ILs	Metals	Removal Efficiency (%)	References
[BzDMODA][TMPPA], [DDDMA][SAC] & [DDDMA][ASU]	Co^{2+} & Zn^{2+}	≥ 90	[125]
[TOA][DODGLY] & [TOA][DEHOX]	La^{3+}, Ce^{3+}, Nd^{3+}, Sm^{3+}, Gd^{3+}, Dy^{3+}, Er^{3+}, Yb^{3+}, La^{3+}, Sm^{3+} & Ho^{3+}	≈100	[126]
[TBA][EHHEHP]	Li^{+}	94.1	[127]
[MTOPPS][AC], [MTOPHS][AC] & [MTOPBzS][AC]	Cu^{2+} & Cd^{2+}	≥ 90	[128]
[CMDMEA][NTf_2] & [CETMA][NTf_2]	Eu^{3+}	≥ 90	[129]
[BET][NTf_2]	Sc^{3+}	≥ 95	[130]
[THTDP][Cl], [THTDP][DCA] & [THTDP[NTf_2]	As^{2+}, Cr^{2+}, Cd^{2+}, Cu^{2+}, Zn^{2+}, Pb^{2+} & Hg^{2+}	≈100	[30]

Basically, the TSILs that are utilized for the separation of metal ions from aqueous media are usually ammonium-based. The removal of Cu^{2+} and Fe^{2+} ions from the aqueous solution were performed with [TOMA][SC] [119]. The IL resulted in extraction efficiencies of 89.0% and 99.0% for Cu^{2+} and Fe^{2+} respectively. Also, ammonium-based TSILs– halogen-free water insoluble were investigated for the extraction of Ni^{2+}, Cu^{2+}, Pb^{2+}, Co^{2+}, and Zn^{2+} from aqueous solutions [120]. In this case, outstanding extraction rates were reported for the metal ions examined i.e., 4.4–96.4% for Ni^{2+}, 47.2–99.7% for Co^{2+}, 20.4–92.8% for Zn^{2+}, 30.5–99.8 % for Pb^{2+} and 54.1–99.9% for Cu^{2+}. Additionally, phosphonium and ammonium-based ILs containing thio and thioether functional groups were used for the extraction of Cu^{2+}, Zn^{2+}, and Cd^{2+} [121]. Although TSILs have proven to be efficient extractants, however, they have some demerits such as: not being easy to prepare, possessing high viscosities, and being too expensive among others [113].

The extraction of Cu^{2+} from an aqueous solution was successfully carried out by using 3-butylpyridinium bis(trifluoromethanesulfonyl)imide, [BPYR][NTf_2] without pH control and chelating agent employment [122]. An extraction efficiency of about 82.6% was obtained after this evaluation. Hydrophobic ILs involving 1-octyl-3-methylimidazolium tetrafluoroborate, [OMIM][BF_4] and [OMIM][NTf_2] were also utilized for the extraction of Hg^{2+} from an aqueous environment [117]. This investigation was done without the introduction of any chelating agent and greater than 90% extraction efficiencies were achieved. The addition of either sodium chloride or sodium bromide to the aqueous solution has the potential of removing Hg^{2+} totally within a very short time at a low temperature. This makes the system to be a very good one in the industry. In another study, 1-butylimidazoliumdi(2-ethylhexyl) phosphate, [BIM][DEHP] and 1-methylimidazolium di(2-ethylhexyl) phosphate, [MIM][DEHP] ILs were also utilized for the removal of Hg^{2+} [123]. In this study, complete removal (i.e., 100% extraction) was actualized.

The separation of uranium(VI) (UO_2^{2+}) was examined from the concentrated nitric acid solution containing colored metals otherwise known as transition elements [124]. ILs comprising of 1-butyl-1-methylmorpholinium butylphosphite, [BMMOR][BPI]; 1-hexyl-1methylmorpholinium hexylphosphite, [HMMOR][HPI] and 1-octyl-1-methylmorpholium octylphosphite, [OMMOR][OPI] were employed for this study. The results obtained demonstrated that [OMMOR][OPI] IL has the best separation efficiency for UO_2^{2+} from the sample matrix. The separation efficiencies of the three ILs employed in this study are illustrated graphically in Figure 5. The acidity of the aqueous phase, alkyl chain length of the ILs, molar quantity of ILs, the concentration of the aqueous feed, and separation duration were the factors upon which the separation of the analyte was found to rely.

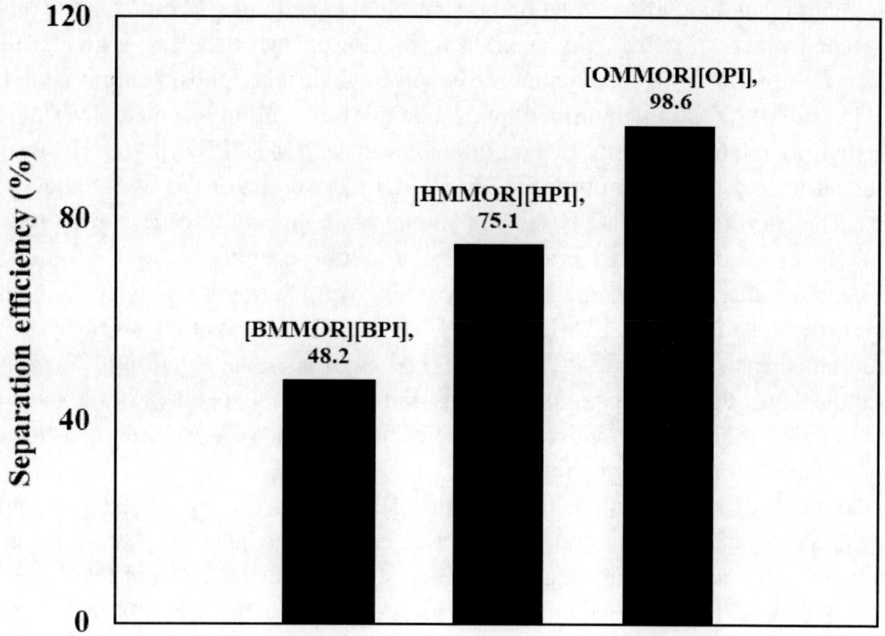

Figure 5. Separation capacities of ionic liquids based on H-phosphonate anions for uranium(IV) [124].

ILs involving benzyldimethyloctadecylammonium bis(2,4,4-trimethylpentyl)phosphinate, [BzDMODA][TMPPA]; didecyldimethylammonium saccharinate, [DDDMA][SAC] and didecyldimethylammonium acesulfamate, [DDDMA][ASU] were utilized for the sequestration of Co^{2+} and Zn^{2+} from aqueous solution [125]. The extraction was performed with the addition of chelating agents including 18-crown-6, 24-crown-8, n-tetracosane ($C_{24}H_{50}$) heptacosane ($C_{27}H_{56}$), Triton TM X-100, and L-histidine. The results obtained demonstrated that extraction performance of more than 90% can be achieved successfully within a very short time for the analytes investigated.

Tetraoctylammonium oleate, [TOA][OL]; 1-butyl-3-methylimidazolium-di(2-ethylhexyl)-oxamate, [BMIM][DEHOX]; tetraoctylammonium dioctyl-diglycolamate, [TOA][DODGLY] and tetraoctylammonium di(2-ethylhexyl)-oxamate, [TOA][DEHOX] were investigated for the extraction and selective separation of lanthanides from aqueous solutions [126]. The lanthanides extracted in this study were La^{3+}, Ce^{3+}, Nd^{3+}, Sm^{3+}, Gd^{3+}, Dy^{3+}, Er^{3+}, Yb^{3+}, La^{3+}, Sm^{3+} and Ho^{3+}. [TOA][DODGLY] and [TOA][DEHOX] ILs could result in approximately 100% extraction efficiency for all the lanthanides investigated under various experimental conditions. It was also found that [TOA][DEHOX] can be re-used without any loss in its extraction ability. Also, tetrabutylammonium 2-ethylhexyl hydrogen-2-ethylhexyl-phosphonate, [TBA][EHHEHP] IL was employed for the extraction of Li^+ from an aqueous solution [127]. The influencing factors such as IL concentration, aqueous acidity, and temperature were studied. The outstanding extraction efficiency of 94.1% was obtained at equilibrium. The extraction was performed to demonstrate the use of IL for high selectivity towards Li^{+} in the solution comprised of alkali metal ions.

ILs based on thioglycolate anion were prepared and purposely utilized for the isolation of Cu^{2+} and Cd^{2+} from water [128]. The extraction process of Cd^{2+} was found to be rapid and effective with more than 90% extraction percentage obtained after 30 min. On the other hand, the extraction process of Cu^{2+} was reported to be slower. Nevertheless, a greater than 90% extraction rate of Cu^{2+} was also actualized when methyltrioctylphosphonium pentylsulfanyl acetate, [MTOPPS][AC]; methyltrioctylphosphonium hexylsulfanyl acetate, [MTOPHS][AC] and methyltrioctylphosphonium benzylsulfanyl acetate, [MTOPBzS][AC] ILs were used. Aqueous nitric acid in addition to [MTOPHS][AC] were used for the back-extraction of the analytes. The alkyl thioglycolate based ILs employed in this study have a good prospect for application as heavy metal extraction agents. Furthermore, the extraction of europium ion (Eu^{3+}), a radioisotope was examine using *N*-(carboxymethyl)-*N*,*N*-dimethylethanaminium bis-trifluoromethane-sulfonimide, [CMDMEA][NTf_2] and *N*-(carboxyethyl)-trimethylammonium bistrifluoromethane-sulfonimide, [CETMA][NTf_2] from aqueous solution [129]. The results obtained displayed that the separation depends on pH but is independent of the ionic strength of the extracting solution. Extraction efficiencies higher than 90% were demonstrated by the two ILs employed for Eu^{3+} extraction.

Betainium bis(trifluoromethylsulfonyl)imide, [BET][NTf_2] IL was found to be appropriate for the extraction of Sc^{3+} from nitrate and chloride aqueous media [130]. At room temperature, the extraction process was demonstrated to be fast and efficient; and outstanding extraction efficiencies greater than 95% were achieved successfully. Phosphonium-based ILs below the pH of 7 and with a separation time of about 10 min were investigated for the extraction of As^{2+}, Cr^{2+}, Cd^{2+}, Cu^{2+}, Zn^{2+}, Pb^{2+}, and Hg^{2+} from their respective standard solutions [30]. Approximately 100% extraction rates for the metal ions were obtained.

Conclusion

In this book, we clearly revealed the utilization of various ILs in the separation of diverse emerging contaminants of high priority from water and wastewater based on the available literature. The ILs were tuned and enhanced for different applications by modifying their cation, anion as well as alkyl chain length accordingly by the researchers. Separation technology involving the utilization of ILs in place of toxic/volatile organic solvents for the exclusion of both organic and inorganic contaminants from various environmental compartments had been examined by many scientists. The main drawbacks encountered in ionic liquid solvent-based extraction were the need for a purification step, its expensive nature, the sophisticated synthetic route of ionic liquids, and low recyclability after extraction.

Nevertheless, sophisticated methodologies that are available could be employed to solve the issue of low recyclability of ionic liquids. Interestingly, some of these contaminants could be separated directly using ionic liquid and this gives an edge thereby making the process advantageous. Also, many of the results obtained using ionic liquids demonstrated great success in removing metals, dyes, pesticides, and phenolic compounds as described in this chapter.

References

[1] Gang, D.; Ahmad, Z. U.; Lian, Q.; Yao, L.; Zappi, M. E. A review of adsorptive remediation of environmental pollutants from aqueous phase by ordered mesoporous carbon. *Chem. Eng. J.* 2021, 403, 126286.

[2] Wang, S.; Sun, H.; Ang, H. M.; Tadé, M. O. Adsorptive remediation of environmental pollutants using novel graphene-based nanomaterials. *Chem. Eng. J.* 2013, 226, 336−347.

[3] Choudhury, M. R.; Anwar, N.; Jassby, D.; Rahaman, M. S. Fouling and wetting in the membrane distillation driven wastewater reclamation process–A review. *Adv. Colloid Interface Sci.* 2019, 269, 370−399.

[4] Gupta, V. K.; Kumar, R.; Nayak, A.; Saleh, T. A.; Barakat, M. A. Adsorptive removal of dyes from aqueous solution onto carbon nanotubes: a review. *Adv. Colloid Interface Sci.* 2013,193, 24−34.

[5] Lian, Q.; Ahmad, Z. U.; Gang, D. D.; Zappi, M. E.; Fortela, D. L. B.; Hernandez, R. The effects of carbon disulfide driven functionalization on graphene oxide for enhanced Pb (II) adsorption: Investigation of adsorption mechanism. *Chemosphere* 2020, 248, 126078.

[6] Ngah. W. S. W.; Teong. L. C.; Hanafiah, M. A. K. M. Adsorption of dyes and heavy metal ions by chitosan composites: A review. *Carbohydr. Polym.* 2011, 83,1446−1456.

[7] Yagub, M. T.; Sen, T. K.; Afroze, S.; Ang, H. M. Dye and its removal from aqueous solution by adsorption: a review. Adv. *Colloid Interface Sci.* 2014, 209, 172−184.

[8] Rafatullah, M.; Sulaiman, O.; Hashim, R.; Ahmad, A. Adsorption of methylene blue on low-cost adsorbents: a review. *J. Hazard. Mater.* 2010, 177, 70−80.

[9] Salleh, M. A. M.; Mahmoud, D. K.; Karim, W. A. W. A.; Idris, A. Cationic and anionic dye adsorption by agricultural solid wastes: a comprehensive review. *Desalination* 2011, 280, 1−13.

[10] Soto, M. L.; Moure, A.; Domínguez, H.; Parajó, J. C. Recovery, concentration and purification of phenolic compounds by adsorption: A review. *J. Food Eng.* 2011, 105, 1−27.

[11] Dąbrowski, A.; Podkościelny, P.; Hubicki, Z.; Barczak, M. Adsorption of phenolic compounds by activated carbon—a critical review. *Chemosphere* 2005, 58, 1049−1070.

[12] Tran, H. N.; You, S. J.; Hosseini-Bandegharaei, A.; Chao, H. P. Mistakes and inconsistencies regarding adsorption of contaminants from aqueous solutions: a critical review. *Water Res.* 2017,120, 88−116.

[13] Fu, F.; Wang, Q. Removal of heavy metal ions from wastewaters: a review. *J. Environ. Manage.* 2011, 92, 407−418.

[14] Da'na, E. Adsorption of heavy metals on functionalized-mesoporous silica: A review. *Microporous and Mesoporous Mater.* 2017, 247, 145−157.

[15] Shou, W.; Chao, B.; Ahmad, Z. U.; Gang, D. D. Ordered mesoporous carbon preparation by the in situ radical polymerization of acrylamide and its application for resorcinol removal. *J. Appl. Poly. Sci.* 2016, 133, 43426.

[16] Sultana, S.; Choudhury, M. R.; Bakr, A. R.; Anwar, N.; Rahaman, M. Effectiveness of electro-oxidation and electro-Fenton processes in removal of organic matter from high-strength brewery wastewater. *J. Appl. Electrochem.* 2018, 48, 519−528.

[17] Guieysse, B.; Norvill, Z. N. Sequential chemical–biological processes for the treatment of industrial wastewaters: review of recent progresses and critical assessment. *J. Hazard. Mater.* 2014, 267, 142−152.

[18] Isosaari, P.; Srivastava, V.; Sillanpää, M. Ionic liquid-based water treatment technologies for organic pollutants: Current status and future prospects of ionic liquid mediated technologies. *Sci. Total Environ.* 2019, 690, 604−619.

[19] Shamraiz, U.; Hussain, R. A.; Badshah, A.; Raza, B.; Saba, S. Functional metal sulfides and selenides for the removal of hazardous dyes from Water. *J. Photochem. Photobio. B: Bio.* 2016,159, 33−41.

[20] Rahman, M. Y. A.; Nachabe, M. H.; Ergas, S. J. Biochar amendment of stormwater bioretention systems for nitrogen and Escherichia coli removal: Effect of hydraulic loading rates and antecedent dry periods. *Bioresour. Technol.* 2020, 310, 123428.

[21] Ahmed, M.; Lin, O.; Saup, C. M.; Wilkins, M. J.; Lin, L. S. Effects of Fe/S ratio on the kinetics and microbial ecology of an Fe (III)-dosed anaerobic wastewater treatment system. *J. Hazard. Mater.* 2019, 369, 593−600.

[22] Anastas, P.; Eghbali, N. Green chemistry: principles and practice. *Chem. Soc. Rev.* 2010, 39, 301−312.

[23] Yavir, K.; Konieczna, K.; Marcinkowski, Ł.; Kloskowski, A. Ionic liquids in the microextraction techniques: The influence of ILs structure and properties. *Trends Anal. Chem.* 2020, 130, 115994.

[24] Rogers, R. D.; Seddon, K. R. Ionic liquids--solvents of the future? *Science* 2003, *302*, 792−793.

[25] Paucar, N. E.; Kiggins, P.; Blad, B.; De Jesus, K.; Afrin, F.; Pashikanti, S.; Sharma, K. Ionic liquids for the removal of sulfur and nitrogen compounds in fuels: a review. *Environ. Chem. Lett.* 2021, 19, 1205−1228.

[26] Maton, C.; De Vos, N.; Stevens, C. V. (2013). Ionic liquid thermal stabilities: decomposition mechanisms and analysis tools. *Chem. Soc. Rev.* 2013, 42, 5963−5977.

[27] Ventura, S. P.; e Silva, F. A.; Quental, M. V.; Mondal, D.; Freire, M. G.; Coutinho, J. A. Ionic-liquid-mediated extraction and separation processes for bioactive compounds: past, present, and future trends. *Chem. Rev.* 2017, 117, 6984−7052.

[28] Feng, J.; Loussala, H. M.; Han, S.; Ji, X.; Li, C.; Sun, M. Recent advances of ionic liquids in sample preparation. *Trends Anal. Chem.* 2020, 125, 115833.

[29] Oke, E. A.; Ijardar, S. P. Insights into the separation of metals, dyes and pesticides using ionic liquid based aqueous biphasic systems. *J. Mol. Liq.* 2021, 334, 116027.

[30] Thasneema, K. K.; Dipin, T.; Thayyil, M. S.; Sahu, P. K.; Messali, M.; Rosalin, T., Hadda, T. B. Removal of toxic heavy metals, phenolic compounds and textile dyes from industrial wastewater using phosphonium based ionic liquids. *J. Mol. Liq.* 2021, 323, 114645.

[31] Ferreira, A. M.; Coutinho, J. A.; Fernandes, A. M.; Freire, M. G. Complete removal of textile dyes from aqueous media using ionic-liquid-based aqueous two-phase systems. Sep. *Purif. Technol.* 2014, 128, 58−66.

[32] Li, Y. C.; Wang, J. J.; Xuan, X. P.; Fan, J.; Fan, M. Factors affecting ionic liquids based removal of anionic dyes from water. *Environ. Sci. Technol.* 2007, *41*, 5090−5095.

[33] Oke, E. A.; Ijardar, S. P. Advances in the application of deep eutectic solvents based aqueous biphasic systems: An up-to-date review. *Biochem. Eng. J.* 2021, 176, 108211.

[34] Archana, V.; Begum, K. M. S.; Anantharaman, N. (2016). "Studies on removal of phenol using ionic liquid immobilized polymeric micro-capsules. *Arab. J. Chem.* 2016, *9,* 371−382.

[35] Hynes, N. R. J.; Kumar, J. S.; Kamyab, H.; Sujana, J. A. J.; Al-Khashman, O. A.; Kuslu, Y.; Ene, A. Kumar, B. S. Modern enabling techniques and adsorbents based dye removal with sustainability concerns in textile industrial sector-A comprehensive review. *J. Cleaner Prod.* 2020, 272, 122636.

[36] Meksi, N.; Moussa, A. A review of progress in the ecological application of ionic liquids in textile processes. *J. Cleaner Prod.* 2017, *161*, 105−126.

[37] Chen, A. H.; Chen, S. M. Biosorption of azo dyes from aqueous solution by glutaraldehyde-crosslinked chitosans. *J. Hazard. Mater.* 2009, *172*, 1111−1121.

[38] Robinson, T.; McMullan, G.; Marchant, R.; Nigam, P. Remediation of dyes in textile effluent: a critical review on current treatment technologies with a proposed alternative. *Bioresour. Technol.* 2001, 77, 247−255.

[39] Vijayaraghavan, R.; Vedaraman, N.; Surianarayanan, M.; Macfarlane, D. R. Extraction and recovery of azo dyes into an ionic liquid. *Talanta* 2006, *69*, 1059−1062.

[40] Li, C.; Xin, B.; Xu, W.; Zhang, Q. Study on the extraction of dyes into a room-temperature ionic liquid and their mechanisms. *J. Chem. Technol. Biotechnol.* 2007, *82*, 196−204.

[41] Li, C.; Xin, B. Extraction and mechanisms of acid dyes into a room temperature ionic liquid. In May, *2008, 2nd International Conference on Bioinformatics and Biomedical Engineering,* pp. 3519−3522. IEEE.

[42] Fan, J.; Fan, Y.; Zhang, S.; Wang, J. (2011). Extraction of azo dyes from aqueous solutions with room temperature ionic liquids. *Sep. Sci. Technol. 2011*, *46*, 1172−1177.

[43] Lin, J.; Teng, Y.; Lu, Y.; Lu, S.; Hao, X.; Cheng, D. Usage of hydrophobic ionic liquid [BMIM][PF6] for recovery of acid dye from wastewater and sequential application in Tussah silk dyeing. *CLEAN–Soil, Air, Water* 2014, 42, 799−803.

[44] Chen, X.; Li, F.; Asumana, C.; Yu, G. Extraction of soluble dyes from aqueous solutions with quaternary ammonium-based ionic liquids. *Sep. Purif. Technol*. 2013, *106*, 105−109.

[45] Bouchal, R.; Prelot, B.; Hesemann, P. Alkylguanidinium based ionic liquids in a screening study for the removal of anionic pollutants from aqueous solution. *RSC Adv. 2016*, *6*, 39125−39130.

[46] Fan, Y.; Dong, X.; Li, Y.; Zhong, Y.; Miao, J.; Hua, S. Removal of Rhodamine B from Water by Benzyl-Functionalized Ionic Liquids. *CLEAN–Soil, Air, Water* 2016, 44, 1106−1112.

[47] Talbi, Z.; Haddou, B.; Ghouas, H.; Kameche, M.; Derriche, Z.; Gourdon, C. Cationic dye removal from aqueous solutions using ionic liquid and nonionic surfactant-ionic liquid systems: a comparative study based upon experimental design. *Chem. Eng. Commun.* 2014, 201, 41−52.

[48] Kermanioryani, M.; Mutalib, M. A.; Gonfa, G.; El-Harbawi, M.; Mazlan, F. A.; Lethesh, K. C.; Leveque, J. M. Using tunability of ionic liquids to remove methylene blue from aqueous solution. *J. Environ. Chem. Eng.* 2016, 4, 2327−2332.

[49] Ullah, Z.; Bustam, M. A.; Man, Z.; Khan, A. S. Phosphonium-based ionic liquids and their application in separation of dye from aqueous solution. *ARPN J. Eng. Appl. Sci*, 2016,11, 1653-1659.

[50] Mahajan, S.; Singh, N.; Kushwaha, J. P.; Rajor, A. Evaluation and mechanism of cationic/anionic dyes extraction from water by ionic liquids. *Chem. Eng. Commun.* 2019, 206, 697−707.

[51] Kumar, V.; Singh, M.; Behera, K.; Pandey, S. Ionic liquid induced removal of Rhodamine B from water. *J. Mol. Liq.* 2020, 319, 114195.

[52] De Gerónimo, E.; Aparicio, V. C.; Bárbaro, S.; Portocarrero, R.; Jaime, S.; Costa, J. L. Presence of pesticides in surface water from four sub-basins in Argentina. *Chemosphere* 2014, 107, 423−431.

[53] WHO. Pesticides (Training for the Health Sector), Biomarkers Hum. Biomonitoring. *World Health Organization* 2008, 1–35.

[54] Khademi, S. M. S.; Salemi, A.; Jochmann, M.; Joksimoski, S.; Telgheder, U. Development and comparison of direct immersion solid phase micro extraction Arrow-GC-MS for the determination of selected pesticides in water. *Microchem. J.* 2021, 164, 106006.

[55] Ye, M.; Beach, J.; Martin, J. W.; Senthilselvan, A. Pesticide exposures and respiratory health in general populations. *J. Environ. Sci.* 2017, 51, 361−370.

[56] Ates, B.; Koytepe, S.; Ulu, A.; Gurses, C.; Thakur, V. K. Chemistry, structures, and advanced applications of nanocomposites from biorenewable resources. *Chem. Rev.* 2020, 120, 9304−9362.

[57] Kumar, R.; Raizada, P.; Verma, N.; Hosseini-Bandegharaei, A.; Thakur, V. K.; Van Le, Q.; Singh, P. Recent advances on water disinfection using bismuth based modified photocatalysts: Strategies and challenges. *J. Clean. Prod.* 2021, 297, 126617.

[58] Rana, A. K.; Mishra, Y. K.; Gupta, V. K.; Thakur, V. K. Sustainable materials in the removal of pesticides from contaminated water: perspective on macro to nanoscale cellulose. *Sci. Total Environ.* 2021, 797, 149129.

[59] Sharma, B.; Thakur, S.; Mamba, G.; Gupta, R. K.; Gupta, V. K.; Thakur, V. K. Titania modified gum tragacanth based hydrogel nanocomposite for water remediation. *J. Environ. Chem. Eng.* 2021, 9, 104608.

[60] Thakur, V. K.; Voicu, S. I. Recent advances in cellulose and chitosan based membranes for water purification: A concise review. *Carbohydr. Polym.* 2016, 146, 148−165.

[61] Hasija, V.; Raizada, P.; Sudhaik, A.; Singh, P.; Thakur, V. K.; Khan, A. A. P. Fabrication of Ag/AgI/WO3 heterojunction anchored P and S co-doped graphitic carbon nitride as a dual Z scheme photocatalyst for efficient dye degradation. *Solid State Sci.* 2020, 100, 106095.

[62] Manasa, R. L.; Mehta, A. Wastewater: sources of pollutants and its remediation." In *Environmental Biotechnology* 2020, *Vol. 2*, pp. 197-219. Springer, Cham.

[63] Musarurwa, H.; Tavengwa, N. T. Emerging green solvents and their applications during pesticide analysis in food and environmental samples. *Talanta*, 2021, 223, 121507.

[64] Zhou, Q.; Bai, H.; Xie, G.; Xiao, J. Trace determination of organophosphorus pesticides in environmental samples by temperature-controlled ionic liquid dispersive liquid-phase microextraction. *J. Chromatogr. A,* 2008, 1188, 148−153.

[65] Zhou, Q. X.; Gao, Y. Y. Combination of ionic liquid dispersive liquid-phase microextraction and high performance liquid chromatography for the determination of triazine herbicides in water samples. *Chin. Chem. Lett.* 2014, 25, 745−748.

[66] Hu, L.; Zhang, P.; Shan, W.; Wang, X.; Li, S.; Zhou, W.; Gao, H. In situ metathesis reaction combined with liquid-phase microextraction based on the solidification of sedimentary ionic liquids for the determination of pyrethroid insecticides in water samples. *Talanta* 2015, 144, 98−104.

[67] Yang, M.; Xi, X.; Wu, X.; Lu, R.; Zhou, W.; Zhang, S.; Gao, H. Vortex-assisted magnetic β-cyclodextrin/attapulgite-linked ionic liquid dispersive liquid–liquid microextraction coupled with high-performance liquid chromatography for the fast determination of four fungicides in water samples. *J. Chromatogr. A* 2015, 1381, 37−47.

[68] Albishri, H. M.; Aldawsari, N. A.; El-Hady, D. A. Ultrasound-assisted temperature-controlled ionic liquid dispersive liquid-phase microextraction combined with reversed-phase liquid chromatography for determination of organophosphorus pesticides in water samples. *Electrophoresis* 2016, 37, 2462−2469.

[69] Chen, X.; Zhang, X.; Liu, F.; Hou, F. Binary–solvent–based ionic–liquid–assisted surfactant-enhanced emulsification microextraction for the determination of four fungicides in apple juice and apple vinegar. *J. Sep. Sci.* 2017, 40, 901−908.

[70] Ya-Li, W. A. N. G.; Li-Qin, Y. O. U.; Yu-Wen, M. E. I.; Jian-Ping, L. I. U.; Li-Jun, H. E. Benzyl functionalized ionic liquid as new extraction solvent of dispersive liquid-liquid microextraction for enrichment of organophosphorus pesticides and aromatic compounds. *Chin. J. Anal. Chem.* 2016, 44, 942−949.

[71] Yang, M.; Wu, X.; Jia, Y.; Xi, X.; Yang, X.; Lu, R.; Zhang, S., Gao, H., Zhou, W. Use of magnetic effervescent tablet-assisted ionic liquid dispersive liquid–liquid microextraction to extract fungicides from environmental waters with the aid of experimental design methodology. *Anal. Chim. Acta* 2016, 906, 118−127.

[72] Cacho, J. I.; Campillo, N.; Viñas, P.; Hernández-Córdoba, M. In situ ionic liquid dispersive liquid-liquid microextraction coupled to gas chromatography-mass spectrometry for the determination of organophosphorus pesticides. *J. Chromatogr. A* 2018, 1559, 95−101.

[73] Fan, C.; Liang, Y.; Dong, H.; Ding, G.; Zhang, W.; Tang, G.; Yang, J., Kong, D., Wang, D., Cao, Y. In-situ ionic liquid dispersive liquid-liquid microextraction using a new anion-exchange reagent combined Fe_3O_4 magnetic nanoparticles for determination of pyrethroid pesticides in water samples. *Anal. Chim. Acta* 2017, 975, 20−29.

[74] Farajzadeh, M. A.; Bamorowat, M.; Afshar Mogaddam, M. R. Ionic liquid-based air-assisted liquid–liquid microextraction for the extraction and preconcentration of aryloxyphenoxypropionate herbicides from aqueous and vegetable samples followed by HPLC-DAD. *Food Anal. Methods* 2017, 10, 749−758.

[75] Padilla-Alonso, D. J.; Garza-Tapia, M.; Chávez-Montes, A.; González-Horta, A.; Waksman de Torres; N. H.; Castro-Ríos, R. New temperature-assisted ionic liquid-based dispersive liquid–liquid microextraction method for the determination of glyphosate and aminomethylphosphonic acid in water samples. *J. Liq. Chromatogr. Relat. Technol.* 2017, 40, 147−155.

[76] Su, R.; Li, D.; Wu, L.; Han, J.; Lian, W.; Wang, K.; Yang, H. Determination of triazine herbicides in juice samples by microwave-assisted ionic liquid/ionic liquid dispersive liquid–liquid microextraction coupled with high-performance liquid chromatography. *J. Sep. Sci.* 2017, 40, 2950−2958.

[77] Zeng, H.; Yang, X.; Yang, M.; Wu, X.; Zhou, W.; Zhang, S.; Lu, R., Li, J., Gao, H. Ultrasound-assisted, hybrid ionic liquid, dispersive liquid–liquid microextraction for the determination of insecticides in fruit juices based on partition coefficients. *J. Sep. Sci.* 2017, 40, 3513−3521.

[78] Yang, J.; Fan, C.; Kong, D.; Tang, G.; Zhang, W.; Dong, H.; Cao, Y. Synthesis and application of imidazolium-based ionic liquids as extraction solvent for pretreatment of triazole fungicides in water samples. *Anal. Bioanal. Chem.* 2018, 410, 1647−1656.

[79] You, X.; Chen, X.; Liu, F.; Hou, F.; Li, Y. Ionic liquid-based air-assisted liquid–liquid microextraction followed by high performance liquid chromatography for the determination of five fungicides in juice samples. *Food Chem.* 2018, 239, 354−359.

[80] Lu, D.; Liu, C.; Deng, J.; Zhou, X.; Shi, G. Zhou, T. Rational design of an ionic liquid dispersive liquid–liquid micro-extraction method for the detection of organophosphorus pesticides. *Analyst* 2019, 144, 2166−2172.

[81] Yang, C.; Ran, L.; Xu, M.; Ren, D.; Yi, L. In situ ionic liquid dispersive liquid–liquid microextraction combined with ultra-high performance liquid chromatography for determination of neonicotinoid insecticides in honey samples. *J. Sep. Sci*. 2019, 42, 1930−1937.

[82] Michałowicz, J.; Duda, W. Phenols--Sources and Toxicity. *Pol. J. Environ. Stud*. 2007, 16, 347–362.

[83] Peyton, T. O. Biological disposal of hazardous waste. *Enzyme Microb. Technol*. 1984, *6*, 146−154.

[84] Saien, J.; Asrami, M. R.; Salehzadeh, S. Phase equilibrium measurements and thermodynamic modelling of {water+ phenol+[Hmim][NTf2]} ionic liquid system at several temperatures. *J. Chem. Thermodyn*. 2018, 119, 76−83.

[85] Mohammadi, S.; Kargari, A.; Sanaeepur, H.; Abbassian, K.; Najafi, A.; Mofarrah, E. Phenol removal from industrial wastewaters: a short review. *Desalination Water Treat*. 2015, 53 2215−2234.

[86] Marrot, B.; Barrios-Martinez, A.; Moulin, P.; Roche, N. Biodegradation of high phenol concentration by activated sludge in an immersed membrane bioreactor. *Biochem. Eng. J*. 2006, 30, 174−183.

[87] Paisio, C. E.; Agostini, E.; González, P. S.; Bertuzzi, M. L. (2009). Lethal and teratogenic effects of phenol on Bufo arenarum embryos. *J. Hazard. Mater*. 2009, 167, 64−68.

[88] WHO, G. Guidelines for drinking-water quality. *World Health Organization* 2011, 216, 303−304.

[89] Sas, O. G.; Domínguez, I.; González, B.; Domínguez, Á. (2018). Liquid-liquid extraction of phenolic compounds from water using ionic liquids: Literature review and new experimental data using [C2mim] FSI. *J. Environ. Manage*. 2018, 228, 475−482.

[90] Chen, Y.; Meng, Y.; Yang, J.; Li, H.; Liu, X. Phenol distribution behavior in aqueous biphasic systems composed of ionic liquids–carbohydrate–water. *J. Chem. Eng. Data* 2012, 57, 1910−1914.

[91] Ahmaruzzaman; M. Adsorption of phenolic compounds on low-cost adsorbents: a review. *Adv. Colloid Interface Sci. 2008*, 143, 48−67.

[92] Rodriguez-Narvaez, O. M.; Peralta-Hernandez, J. M.; Goonetilleke, A.; Bandala, E. R. "Treatment technologies for emerging contaminants in water: A review. *Chem. Eng. J*. 2017, 323, 361−380.

[93] Busca, G.; Berardinelli, S.; Resini, C.; Arrighi, L. "Technologies for the removal of phenol from fluid streams: a short review of recent developments. *J. Hazard. Mater*. 2008, 160(2-3), 265−288.

[94] Vidal, S. T.; Neiva Correia, M. J.; Marques, M. M.; Ismael, M. R.; Angelino Reis, M. T. Studies on the use of ionic liquids as potential extractants of phenolic compounds and metal ions. *Sep. Sci.Technol*. 2005, 39, 2155−2169.

[95] Khachatryan, K. S.; Smirnova, S. V.; Torocheshnikova, I. I.; Shvedene, N. V.; Formanovsky, A. A.; Pletnev, I. V. Solvent extraction and extraction–voltammetric determination of phenols using room temperature ionic liquid. Anal. *Bioanal Chem*. 2005, 381, 464–470.

[96] Fan, J.; Fan, Y.; Pei, Y.; Wu, K.; Wang, J.; Fan, M. Solvent extraction of selected endocrine-disrupting phenols using ionic liquids. *Sep. Purif. Technol*. 2008, 61, 324−331.

[97] Egorov, V. M.; Smirnova, S. V.; Pletnev, I. V. Highly efficient extraction of phenols and aromatic amines into novel ionic liquids incorporating quaternary ammonium cation. *Sep. Purif. Technol*. 2008, 63, 710−715.

[98] Cesari, L.; Canabady-Rochelle, L.; Mutelet, F. Extraction of phenolic compounds from aqueous solution using choline bis (trifluoromethylsulfonyl) imide. *Fluid Phase Equilibr*. 2017, 446, 28−35.

[99] Pilli, S. R.; Banerjee, T.; Mohanty, K. Liquid–liquid equilibrium (LLE) data for ternary mixtures of [C4DMIM]—[PF6]+[PCP]+[water] and [C4DMIM][PF6]+[PA]+[water] at T= 298.15 K and p= 1 atm. *Fluid Phase Equilibr*. 2014, 381, 12−19.

[100] Fan, Y.; Li, Y.; Dong, X.; Hu, G.; Hua, S.; Miao, J.; Zhou, D. Extraction of phenols from water with functionalized ionic liquids. *Ind. Eng. Chem. Res*. 2014, 53, 20024−20031.

[101] Lungwitz, R.; Strehmel, V.; Spange, S. The dipolarity/polarisability of 1-alkyl-3-methylimidazolium ionic liquids as function of anion structure and the alkyl chain length. *New J. Chem*. 2010, 34, 1135−1140.

[102] Sas, O. G.; Sanchez, P. B.; Gonzalez, B.; Dominguez, A. Removal of phenolic pollutants from wastewater streams using ionic liquids. *Sep. Purif. Technol*. 2020, 236, 116310.

[103] Sas, O. G.; Domínguez, I.; Domínguez, Á.; González, B. Using bis (trifluoromethylsulfonyl) imide based ionic liquids to extract phenolic compounds. *J. Chem. Thermodyn*. 2019, 131, 159−167.

[104] González, E. J.; Díaz, I.; Gonzalez-Miquel, M.; Rodríguez, M.; Sueiras, A. On the behavior of imidazolium versus pyrrolidinium ionic liquids as extractants of phenolic compounds from water: Experimental and computational analysis. *Sep. Purif. Technol.* 2018, 201, 214−222.

[105] Lakshmi, B. A.; Balasubramanian, A.; Venkatesan, S. J. C. S. Extraction of phenol and chlorophenols using ionic liquid [Bmim]+[BF4]−dissolved in tributyl phosphate. *CLEAN–Soil, Air, Water* 2013, 41, 349−355.

[106] Skoronski, E.; Fernandes, M.; Malaret, F. J.; Hallett, J. P. Use of phosphonium ionic liquids for highly efficient extraction of phenolic compounds from water. *Sep. Purif. Technol.* 2020, 248, 117069.

[107] Qasem, N. A. A.; Mohammed, R. H.; Lawal, D. U. Removal of heavy metal ions from wastewater: a comprehensive and critical review. *NPJ Clean Water* 2021, 4, 36.

[108] Oluyinka, O. A.; Oke, E. A.; Oyelude, E. O.; Abugri, J.; Raheem S. A. Recapitulating potential environmental and industrial applications of biomass wastes. *J. Mater. Cycl. Waste Manage.* 2022, 1−19.

[109] Oke, E. A.; Osibanjo, O., Raheem, S. A.; Oluyinka, O. A.; Ibe, K. K. Metals concentration levels in printed circuit boards of discarded cathode ray tube television: trends over the years. *Inter. J. Environ. Anal. Chem.* 2021, 1-12.

[110] El-Sherif, I. Y.; Tolani, S.; Ofosu, K.; Mohamed, O. A.; Wanekaya, A. K. Polymeric nanofibers for the removal of Cr (III) from tannery waste water. *J. Environ. Manage.* 2013, 129, 410−413.

[111] Zou, Y.; Wang, X.; Khan, A.; Wang, P.; Liu, Y.; Alsaedi, A.; Wang, X. Environmental remediation and application of nanoscale zero-valent iron and its composites for the removal of heavy metal ions: a review. *Environ. Sci. Technol.* 2016, 50, 7290−7304.

[112] Taseidifar, M.; Makavipour, F.; Pashley, R. M.; Rahman, A. M. Removal of heavy metal ions from water using ion flotation. *Environ. Technol. Innov.* 2017, 8, 182−190.

[113] Rajadurai, V.; Anguraj, B. L. Ionic liquids to remove toxic metal pollution. *Environ. Chem. Lett.* 2021, 19, 1173−1203.

[114] Krishnan, A.; Gopinath, K. P.; Vo, D. V. N.; Malolan, R.; Nagarajan, V. M.; Arun, J. Ionic liquids, deep eutectic solvents and liquid polymers as green solvents in carbon capture technologies: a review. *Environ. Chem. Lett.* 2020, 18, 2031−2054.

[115] Olivier, J. H.; Camerel, F.; Selb, J.; Retailleau, P.; & Ziessel, R. Terpyridine-functionalized imidazolium ionic liquids. *Chem. Commun.* 2009, 9, 1133−1135.

[116] Mehdi, H.; Binnemans, K.; Van Hecke, K.; Van Meervelt, L.; Nockemann, P. Hydrophobic ionic liquids with strongly coordinating anions. *Chem. Commun.* 2010, 46, 234−236.

[117] Mancini, M. V.; Spreti, N.; Di Profio, P.; Germani, R. Understanding mercury extraction mechanism in ionic liquids. *Sep. Purif. Technol.* 2013, 116, 294−299.

[118] Papaiconomou, N.; Lee, J. M.; Salminen, J.; Von Stosch, M., Prausnitz, J. M. Selective extraction of copper, mercury, silver, and palladium ions from water using hydrophobic ionic liquids. *Ind. Eng. Chem. Res.* 2008, 47, 5080−5086.

[119] Egorov, V. M.; Djigailo, D. I.; Momotenko, D. S.; Chernyshov, D. V.; Torocheshnikova, I. I.; Smirnova, S. V.; Pletnev, I. V. Task-specific ionic liquid trioctylmethylammonium salicylate as extraction solvent for transition metal ions. *Talanta* 2010, 80, 1177−1182.

[120] Vergara, M. A. V.; Lijanova, I. V.; Likhanova, N. V.; Vigueras, D. J.; Xometl, O. O. The removal of heavy metal cations from an aqueous solution using ionic liquids. *Can. J. Chem. Eng.* 2014, 92, 1875−1881.

[121] Leyma, R.; Platzer, S.; Jirsa, F.; Kandioller, W.; Krachler, R.; Keppler, B. K. Novel thiosalicylate-based ionic liquids for heavy metal extractions. *J. Hazard. Mater.* 2016, 314, 164−171.

[122] Reyna-González, J. M.; Galicia-Pérez, R.; Reyes-López, J. C.; Aguilar-Martínez, M. Extraction of copper (II) from aqueous solutions with the ionic liquid 3-butylpyridinium bis (trifluoromethanesulfonyl) imide. *Sep. Purif. Technol.* 2012, 89, 320−328.

[123] Guezzen, B.; Didi, M. A. Removal and analysis of mercury (II) from aqueous solution by ionic liquids. *J. Anal. Bioanal. Tech.* 2016, 7, 2.

[124] Zarrougui, R.; Mdimagh, R.; Raouafi, N. Highly efficient extraction and selective separation of uranium (VI) from transition metals using new class of undiluted ionic liquids based on H-phosphonate anions. *J. Hazard. Mater.* 2018, 342, 464−476.

[125] Łukomska, A., Wiśniewska, A., Dąbrowski, Z., & Domańska, U. Liquid-liquid extraction of cobalt (II) and zinc (II) from aqueous solutions using novel ionic liquids as an extractants. *J. Mol. Liq.* 2020, 307, 112955.

[126] Maria, L.; Cruz, A.; Carretas, J. M.; Monteiro, B.; Galinha, C.; Gomes, S. S.; Araujo, M. F., Paiva, I., Marcalo, J., Leal, J. P. Improving the selective extraction of lanthanides by using functionalised ionic liquids. *Sep. Purif. Technol.* 2020, 237, 116354.

[127] Shi, C.; Li, H.; Liu, B.; Qin, Y.; Song, G. Solvent extraction of lithium from aqueous solution using an ammonium ionic liquid. *J. Mol. Liq.* 2020, 304, 112756.

[128] Platzer, S.; Kar, M.; Leyma, R.; Chib, S.; Roller, A.; Jirsa, F., Krachler, R., MacFarlane, D.R., Kandioller, W., Keppler, B. K. Task-specific thioglycolate ionic liquids for heavy metal extraction: synthesis, extraction efficacies and recycling properties. *J. Hazard. Mater.* 2017, 324, 241−249.

[129] Dai, Y.; Cao, B.; Zhong, S.; Xie, G.; Wang, Y.; Liu, Y., Zhang, Zhibin, Liu, Y., Cao, X. Homogeneous liquid–liquid extraction of europium from aqueous solution with ionic liquids. *J. Radioanal. Nucl. Chem.* 2019, 319, 1219−1225.

[130] Onghena, B.; Binnemans, K. Recovery of scandium (III) from aqueous solutions by solvent extraction with the functionalized ionic liquid betainium bis (trifluoromethylsulfonyl) imide. *Ind. Eng. Chem. Res.* 2015, 54, 1887−1898.

Chapter 11

Insight into Cellulose Dissolution Mechanisms in Ionic Liquids/Cosolvent Systems

Vidura Thalangamaarachchige[1,*]
and Noureddine Abidi[2,†]
[1] Department of Chemistry and Physics,
McNeese State University, Lake Charles, Louisiana, USA
[2] Fiber and Biopolymer Research Institute,
Texas Tech University, Lubbock, Texas, USA

Abstract

The vast array of distinct applications of Ionic liquids (ILs) has recognized them as perhaps one of the most important class of solvents. During the last decade, ILs emerged as promising solvents for cellulose, which is not soluble in most common organic solvents due to extensive intra- and intermolecular hydrogen bonding.

There is a growing interest in this field of research to understand the mechanism of cellulose dissolution in ILs, and major breakthroughs appeared over the last few years. Recent work rather emphasizes the role of anion and cation of ILs in cellulose dissolution, which is still, however, under debate. Moreover, numerous research outcomes revealed that the addition of cosolvents to the IL enhances the dissolution process.

This chapter is focused on recent experimental discoveries on cellulose dissolution mechanism with diverse ILs and IL/cosolvent systems. In particular, imidazolium-based cations connected to various N-alkyl substituents and N-alkyl imidazole cosolvents will be discussed.

Keywords: ionic liquids, cellulose, dissolution, mechanism, co-solvent

Introduction

The world is now facing unprecedented challenges such as environmental protection, climate change, energy production, and sustainable economic growth. To address these global issues,

* Corresponding Author's Email: vthalangamaarac@mcneese.edu.
† Corresponding Author's Email: noureddine.abidi@ttu.edu.

In: Properties and Applications of Ionic Liquids
Editors: Noureddine Abidi and Julia L. Shamshina
ISBN: 979-8-88697-635-9

sustainable methodologies are required when designing processes for the manufacturing of products. Transferring society's dependence from petroleum to benign renewable feedstocks is vital for sustainable manufacturing. In this regard, lignocellulosic biomass is a perpetual and underutilized source of inexhaustible feedstock, which can help to reduce the global reliance on fossil fuel sources.

Currently, petroleum is being devoured 100,000 faster than nature is supplanting it [1]. Biomass is primarily made of cellulose and lignin. Cellulose is the most profuse biopolymer on the earth comprised of monomers of glucose linked via β-(1, 4) bonds (Figure 1). Above and beyond as a renewable feedstock for biofuels and chemicals, cellulose is utilized in various applications. For instance, cellulose and its derivatives exploited in the pharmaceutical industry as bio-adhesives for drug delivery systems, coatings of extended-release matrices, in semipermeable membranes (SPM), and in osmotic delivery systems [2-5]. Cellulose is predominently insoluble in most conevtional organic solvents due to its intra- and interchain hydrogen bonding (H-bonding) (Figure 1). Therefore, for cellulose to be dissolved in a solvent, the solvent must force itself between cellulose chains and disrupt the hydrogen bonding.

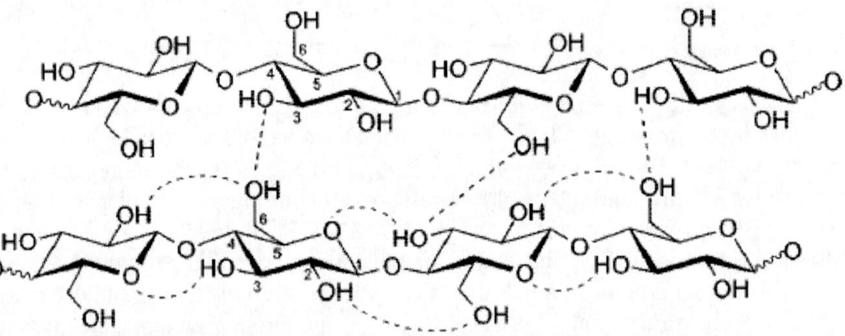

Figure 1. Structure of cellulose showing β-(1,4) linkages and intra- and interchain hydrogen bonds Reprinted with permission from Pinkert, A., Marsh, K. N., Pang, S., Staiger, M. P. Ionic Liquids and Their Interaction with Cellulose. *Chem. Rev.* 2009, *109*, 6712-6728. Copyright ©2009 American Chemical Society.

Even though the estimated annual production of cellulose is close to 40 billion tons, only a minimal amount (around 0.2 billion tons) is used as feedstock [6]. One of the main reasons for this underutilization is its inefficient and incomplete dissolution in most conventional organic solvents. N, N-dimethylacetamide (DMAc)/Lithium chloride (LiCl), aqueous sodium hydroxide (NaOH), and N-methylmorpholine-N-oxide (NMMO) are few examples for standard cellulose solvents. The drawback of using these solvents is mainly their toxicity to the environment and reduced solvation power [7-9]. Out of those solvents, NMMO is the most widely used solvent to dissolve cellulose in industrial Lyocell process. In general, the traditional cellulose dissolution processes require relatively harsh conditions [6-7]. Therefore, the quest for alternative solvents in this area of research has special interest. The need to improve the quality of cellulose-based materials (e.g., nanoparticles, aerogels, films, nanocomposites) is one of the main driving forces to invent new solvents [8].

Over the last 20 years, Ionic liquids (ILs) emerged as cellulose solvents attracting the scientific community as well as industry. ILs are salts composed of a cation and an anion, and have melting points at or below 100° C [10]. There is a great interest of using ILs as neoteric solvents in various reactions and chemical processes due to the ability of tuning the physico-chemical properties for instance wide liquid range, thermal stability, non-volatility, and non-flammability [11-12]. This unique feature of tuning the physical properties of ILs resulted in naming them as "designer" solvents [13]. Further, ILs also classified as "green" solvents which is useful in industrial scale [14] as well as laboratory scale [15]. Due to the versatility of these solvents, the field of ILs has come long way. They are not only useful as solvents, but also in other scientific and technological fields such as electrochemistry, biocatalysis, tribology, solar cells and ion-gel technology [14, 16-25].

During IL synthesis, its behavior can be controlled by either simply choosing different ions or by independently modifying the structure of cation or anions. The most common cations involved with ILs are imidazolium, pyridinium, pyrrolidinium, ammonium, phosphonium, piperidinium, thiazolium, pyrazolinium, and sulfonium. Out of these cations, imidazolium based ILs were greatly utilized in cellulose dissolution studies. The array of anions used in generating ILs for cellulose dissolution includes halides, acetates, formats, inorganic ions, and large organic ions [14]. Some ILs use as cellulose solvents are shown in Figure 2. After the dissolution of cellulose, it is regenerated by using anti solvents such as water or acetone to form various cellulose-based materials such as fibers, thin films etc. The regenerated cellulose is a combination of amorphous cellulose and cellulose II [26, 27].

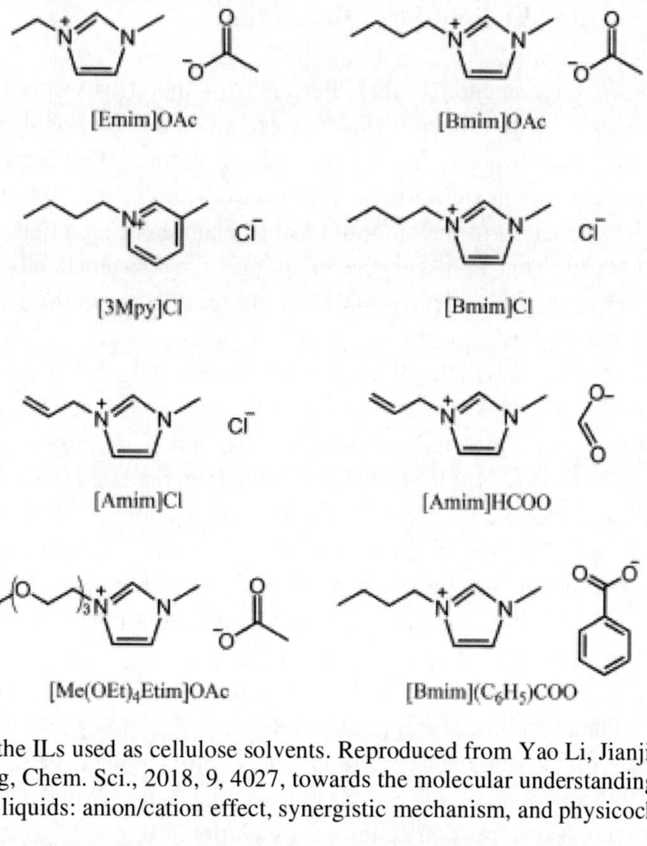

Figure 2. Some of the ILs used as cellulose solvents. Reproduced from Yao Li, Jianji Wang, Xiaomin Liu, Suojiang Zhang, Chem. Sci., 2018, 9, 4027, towards the molecular understanding of the cellulose dissolution in ionic liquids: anion/cation effect, synergistic mechanism, and physicochemical aspects (open access).

The breakthrough by Swatloski et al. in 2002 demonstrated that cellulose dissolution can be performed in certain ILs without pretreatment and the regeneration of cellulose can be accomplished by precipitation using water without degrading the cellulose. Of the ILs in this study, 1-butyl-3-methylimidazolium chloride produced the best results [28]. Subsequently, numerous studies have confirmed various ILs as promising cellulose solvents [6, 8, 29-36]. However, the first effort of using IL to dissolve cellulose was by Graenacher in 1934 using N-ethylpyridinium chloride IL in combination with N-containing bases [37]. Two-dimensional FTIR spectroscopy study by Kathirgamanathan et al. on cellulose dissolution in imidazolium-based ionic liquids confirms that the degree of hydrogen bonding between cellulose and IL ions was greater at elevated temperatures which leads to better dissolution [38]. More than a few review articles highlighted the fact that cellulose dissolution is ruled by numerous factors such as the kind of anion and cation in the IL, anion's basicity and H-bonding capability, cation's side chain length and the position [31, 33].

This chapter mainly focuses on recent experimental studies coupled with theoretical investigations on cellulose IL chemistry to figure out the underlying mechanism. To design effective ionic liquids as cellulose solvents, it is imperative to understand the cellulose dissolution mechanism in ILs. By looking at the rapid progress around IL-based cellulose and biomass processing, it is highly advantageous to have an overview about present understanding of cellulose dissolution in ILs from the described experimental and theoretical studies.

The Role of the Anion in Cellulose Dissolution

The widely accepted phenomenon is that the anion of the IL disrupts hydrogen bonds in cellulose and generates new hydrogen bonds between anion and hydroxyl protons in cellulose, leading to dissolution. This was confirmed from a study performed by Gupta et al. using ILs 1-Butyl-3-methylimidazolium hexafluorophosphate and 1-butyl-3-methylimidazolium acetate. During this study, a reduction in the amount of inter-chain hydrogen bonds on the surface of cellulose was observed. This indicates the breaking of hydrogen bonds of the surface, as recognized by the change in the orientation of OH groups due to cellulose –solvent interaction [39].

Swatloski et al. speculated that ILs with strong H-bond acceptor anions were highly effective in dissolving cellulose, whereas ILs with non-coordinating anions, for example hexafluorophosphate $[PF_6]^-$ and tetrafluoroborate $[BF_4]^-$ were ineffective in dissolving cellulose [28]. Multiple studies reported by Remsing and coworkers revealed that cellulose dissolution in ILs is dominated by anions. By using ^{13}C and $^{35/37}Cl$ NMR relaxation experiments on various carbohydrates in 1-butyl-3-methylimidazolium chloride, Remsing et al. examined the solvation of the solute in these model systems involves H-bonding with hydroxyl group protons in the carbohydrate and the chloride ions in a 1:1 ratio. In a follow up NMR study, Remsing and coworkers aimed at investigating the solvation of glucose and cellobiose in ILs 1-butyl-3-methylimidazolium chloride, and 1-butyl-3-methylimidazolium acetate. From the results, it was found that ^{13}C and $^{35/37}Cl$ relaxation rates of the acetate carbons and chloride ion, showed a strong dependence on the sugar concentration, confirming the impact of anions in dissolution whereas the ^{13}C relaxation rates of the carbons on the cations of the IL showed a weak dependence indicating less impact on cellulose dissolution [40, 41].

Youngs et al. utilized pulsed field gradient (PFG) and NOESY NMR, neutron diffraction (ND), and molecular dynamics (MD) simulations to elucidate the interactions between IL and glucose using 1-ethyl-3-methylimidazolium acetate. From experimental results, it was confirmed that the dominant glucose-IL interaction is H-bonding between the acetate oxygens and the glucose OH groups. Moreover, NOESY-NMR data revealed that hardly any interaction between the glucose OH groups and the acidic proton on 2-position (C2 proton) of the imidazolium cation [42]. This was found to be the case for experimental and simulation studies with D-glucose and 1,3-dimethylimidazolium chloride where solvation of carbohydrates in IL dominated by interactions with the anions presumably through hydrogen bonding, with weak or no interaction with the cation [43].

Moyna et al. utilized cellobiose (5 and 10 wt%) in 1-butyl-3-methylimidazolium chloride to calculate thermodynamic properties and compared NMR relaxation data with the H-bonding interactions between cellobiose and 1-butyl-3-methylimidazolium chloride. It was found that chloride anions interact with hydroxyl groups in an 1:1 ratio [44]. Singh and coworkers studied the dissolution of microcrystalline cellulose I_β in 1-Ethyl-3-methylimidazolium acetate [C_2mim][OAc]. Microfibrils were generated by using 3 layers of 3 cellulose chains and each layer has the I_β structure arrangement. They noticed that both the ions penetrated into cellulose while the anion formed strong hydrogen bonds with cellulose [45]. Casas et al. utilized COSMOS-RS method to examine cellulose solubility in more than 300 ILs. The excess enthalpy of cellulose/IL mixtures was determined to understand the affinity of various ILs toward cellulose. Their results suggested that the anion plays a prominent role in dissolution and the role of cation is negligible [46]. Yao et al. performed computational studies on the dissolution of cellulose in ILs 1-butyl-3-methylimidazolium chloride, 1-ethyl-3-methylimidazolium chloride [C_2C_1im][Cl] and 1-ethyl-3-methylimidazolium acetate [C_2C_1im][OAc] and their investigation was centered around the manner in which breakage of cellulose chain and ring opening utilizing cellobiose as a model particle. They found that alkyl chain of cations has minute impact on catalysis of the cleavage reaction. These outcomes showed that anions assumed a critical part in impacting the dissolution while cation effect is negligible [47].

The study by Ma et al. utilized cellulose with different degree of polymerization of $n = 1-12$ in the IL 1,3 dimethylimidazolium chloride to determine the intermolecular interaction via molecular dynamics (MD) simulations. They discovered that the electrostatic interactions between the cellulose and ionic liquid are substantially greater compared to dispersion forces. Additionally, they found that anion cellulose interaction energy is about 2.9 times relative to cellulose and cation interaction and concluded that the role of cation is negligible [48]. By using wide angle X-ray scattering (WAXS) and ^{13}C solid-state NMR spectroscopy, Endo et al. studied the interactions between [Emim][OAc] and cellulose. When cellulose concentration was in the range of 15–30 mol%, a periodic peak was observed in the WAXS pattern, which was attributed to cellulose chain alignment. At higher concentrations than 30 mol%, they suggested that because of cation anion intercalation, the structure could be altered to ordered layers. Further, this study revealed that at a lower concentration of cellulose, acetate anion interact with hydroxyl groups in cellulose at a 1 : 1 ratio, whereas at higher concentrations the acetate anion interacts with multiple hydroxyl groups to generate bridging structure to facilitate the dissolution (Figure 3) [49, 50].

To grasp whether there is an impact in carbene formation during the cellulose dissolution in ionic liquids, Holloczki´ et al. carried out a MD simulation study on a series of carbohydrate-ionic liquid systems in 1-ethyl-3-methylimidazolium acetate. The outcome of this study confirmed that solvation of the sugars is predominantly governed by the interactions with the anions, and that the ability of carbene formation has minimal impact on dissolution [51]. All the studies discussed so far claimed that the cation in the ILs has no impact or minimal impact on the cellulose dissolution. However, there are studies which showed that the cation does matter in cellulose dissolution.

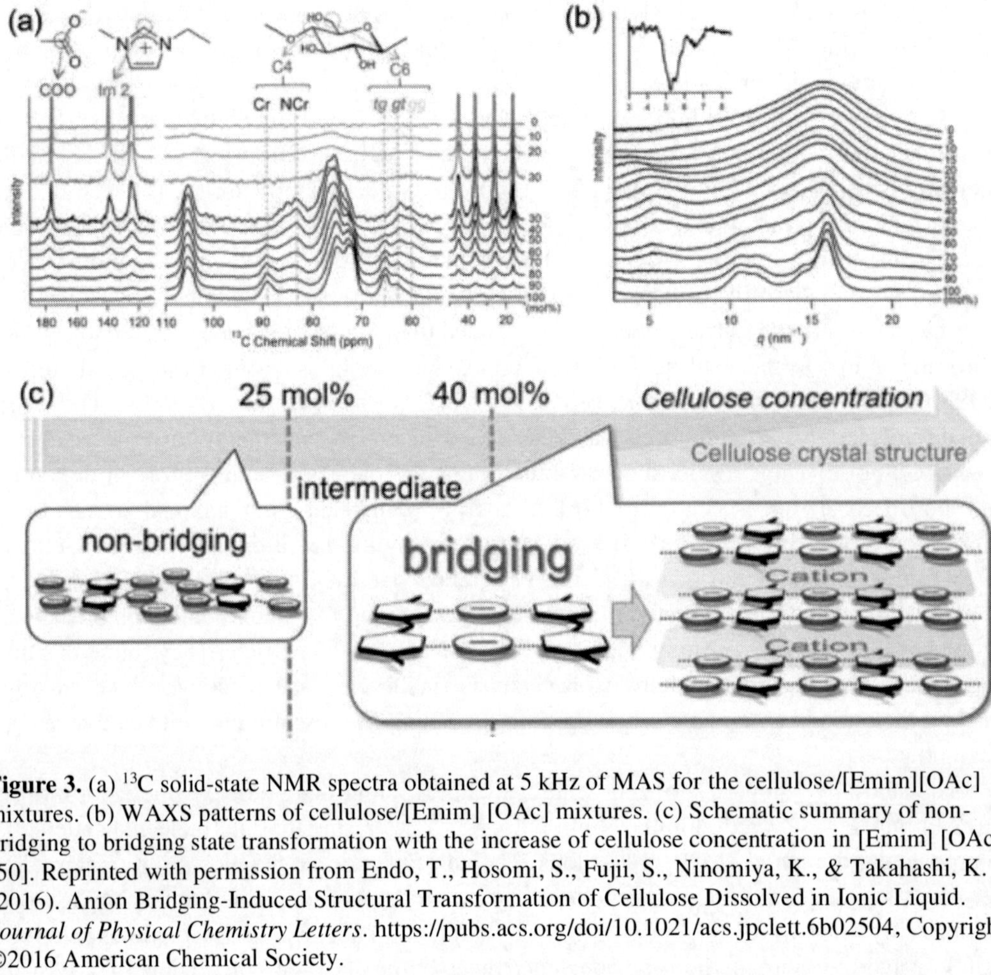

Figure 3. (a) ^{13}C solid-state NMR spectra obtained at 5 kHz of MAS for the cellulose/[Emim][OAc] mixtures. (b) WAXS patterns of cellulose/[Emim] [OAc] mixtures. (c) Schematic summary of non-bridging to bridging state transformation with the increase of cellulose concentration in [Emim] [OAc] [50]. Reprinted with permission from Endo, T., Hosomi, S., Fujii, S., Ninomiya, K., & Takahashi, K. (2016). Anion Bridging-Induced Structural Transformation of Cellulose Dissolved in Ionic Liquid. *Journal of Physical Chemistry Letters*. https://pubs.acs.org/doi/10.1021/acs.jpclett.6b02504, Copyright ©2016 American Chemical Society.

Synergistic Effect of Anion and Cation

In contrary to studies on anionic effect, numerous experimental studies on microcrystalline cellulose (MCC) in ILs indicated that the cation does play a vital role [29–34, 52, 53]. Zhang et al. using conventional and variable temperature ^{1}H NMR studied the solvation of the cellobiose in $[C_2C_1im][OAc]$ and found substantial broadening of the proton NMR peaks of the OH groups in cellobiose, which they ascribed to the exchangeable protons from cellulose OH groups with the acidic proton of $[C_2C_1im]^+$ cation [54]. Lu et al. examined the dissolution capability of cellulose in ILs with the imidazolium cation $[R_1R_2im]^+$ having fixed anion [OAc].

The result from this study is consistent with the acidic proton on 2-position of the imidazolium ring being involved in the dissolution of cellulose and these protons may be forming C-H---O H-bonds with OH groups and ether oxygen of cellulose. Additionally, these experiments indicated that the solubility of cellulose was decreased as a result of strong interactions of cation with the anion and also due to steric effect caused by the presence of bulky alkyl groups in the cation [52]. Insitu microscopy high throughput screening experiments revealed that the aromatic functionalized cations enable to alleviate the H-bonds amongst the cations and cellulose OH groups [55].

Youngs et al. performed molecular dynamic simulations (MD) using glucose to mimic cellulose. According to their study, anions are hydrogen bonded with glucose, while cations interacts weakly with glucose [56]. Singh and coworkers investigated the behavior of glucose oligomers in C_2C_1Im acetate to understand the behavior of cellulose. According to the results, oligomers hydroxyl group protons generated strong hydrogen bonds with anionic acetate and cations formed hydrophobic interactions with polysaccharides [57]. Zhang et al. examined the impact of the cation on cellulose dissolution by changing the alkyl chain length and the heterocyclic structure by using ILs [Cnmim]Cl (n = 2-10) and [C_4mpy] Cl. Young and coworkers affirmed that [C_4mpy] [Cl] is a better cellulose solvent than 1-butyl-3-methylimidazolium chloride and further stated that imidazolium cations with shorter alkyl chains exhibited better solubility. Supplementary studies confirmed that when electron withdrawing group is connected to the alkyl chain, it enhances the cation and cellulose interactions owing to inductive effect [45]. The extended study with the same cation [C2mim] with different anions such as chloride, acetate, and hexafluorophosphate demonstrated that the combination of factors such as alkyl chain length, electronegativity of H-bond acceptor, and electron-withdrawing group have impact on dissolution [58].

Mostofian et al. performed MD simulations to invetsigate the role of cations and anions in the same IL. The chloride anions interactions with hydroxyl groups of the surface led to disruption of H- bonds in the chains. Meanwhile, the cations experienced hydrophobic interactions with the surface stabilizing glucan chains [59]. Another interesting study by Chu and coworkers explored the thermodynamic driving force for cellulose dissolution. They considered two states, a crystalline microfibril and dissociated glucan chains for their study. From their results, it was determined that the solvent structure is perturbed by dissolved glucan chains and cation anion interactions with the glucan mainly determined the solubility [60]. A subsequent MD simulation study by Cho and coworkers studied cellulose deconstruction by using cellulose microfibril via shedding off an 11-residues glucan chain. They noticed in 1-butyl-3-methylimidazolium chloride, strong interactions of anion with hydroxyl groups, meanwhile peeled-off state's side chains and linker oxygens coupled with cations [61].

Another MD simulation study performed by Rabideau et al. investigated the breakup of cellulose bundles by using I_α and I_β structures in 1-butyl-3-methylimidazolium chloride, 1-ethyl-3-methylimidazolium acetate, and 1,3-Dimethylimidazolium dimethylphosphate. As expected, the three ILs selected could break up cellulose bundles. Anions bound strongly to –OH groups and led to the formation of negatively charged species. This weakened intra- strand hydrogen bonds, consequently reducing the strand rigidity. At the same time, cations coordinated with the complex, by pushing the chains apart initiating the breakup. Out of three ILs they used, [C_2mim][OAc] was the best IL because it has a high possibility to separate strands from the bundle. This detailed study further proved that the intra sheet H-bond disruption is more vital than the inter strand breakage to peel off strands. This study concluded that the role of cations is vital in the initial breakup of cellulose bundle in dissolution [62]. Computational study with pure 1-butyl-3-methylimidazolium chloride IL and three cellulose oligomer/IL systems revealed that 1-butyl-3-methylimidazolium chloride IL appeared to be an explicitly merged structure with alternative array of anions and cations. From gathered information on the oligomer/IL systems, it was noticed that the anionic chloride and the cationic imidazolium heterocyclic ring generate hydrogen bonds with the oligomer hydroxy functional groups. In light of the present outcomes, they suggested that the cellulose dissolution in 1-butyl-3-methylimidazolium chloride IL is a repercussion of the combined interactions of anions and cations with cellulose [63].

A systematic study by Cao et al. investigated IL- cellulose interactions using cellobiose to represent cellulose in the IL 1-butyl-3-methyl imidazolium acetate. This study showed that hydrogen bonds and van der Waals interactions between cellobiose and ILs, decides the dissolution capacity. Thermodynamic analysis proposed that increasing the temperature can unambiguously improves the dissolution of cellulose in [Bmim][OAc]. NMR studies suggested that acetate anion generate strong hydrogen bonding interactions with cellulose hydroxyl groups which can soften up the intrinsic hydrogen bonds in cellulose [64]. Baek et al. showed that oxygen-containing alkoxy alkyl groups corperated to C 2-position of 1,3-dialkyl-imidazolium cations in ILs enables to generate powerful solvents for cellulose. This is due to C2- alkoxy functionalized imidazolium ILs induced higher acidity at C4 - and C5 hydrogens in the imidazolium cations [65]. Erdmenger et al. explained the phenomena called the "odd–even" effect. According to that study, imidazoliuim cations having an even number of carbons containing alkyl chains are effective as cellulose solvents. A follow up study by Pinkert et al. elucidated the impact of the odd or even number of carbons in alkyl chain in the imidazolium cation on the polarization of the cation [31, 66]. Ebner et al. performed ^{13}C isotope labeling and fluorescence labeling experiments and found that C2 carbon atom in the imidazolium cation reacts with the reducing end of polysaccharide, in a basic medium [67]. In a later ^{1}H NMR study, Clough et al. observed the interaction between the cation and the acetal group of the glucose in the IL [Emim][OAc] [68].

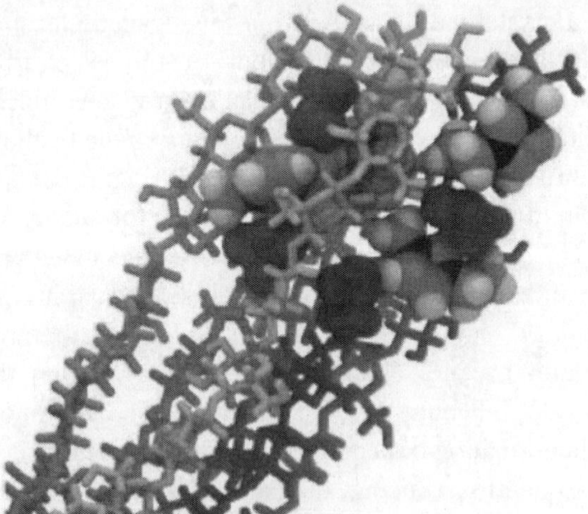

Figure 4. MD simulation of the structure of a I_α bundle in [C_2C_1im][OAc] showing initial separation of a strand (yellow) from the bundle. Anions (magenta) bind to hydroxyls of cellulose, as cations (cyan & white) force themselves between separating strands Rabideau, B. D., Agarwal, A., Ismail, A. E. *J. Phys. Chem. B* 2013, *117*, 3469-3479. Copyright ©2013 American Chemical Society. Permisions related to the material excerpted should be directed to the ACS. https://pubs.acs.org/doi/10.1021/jp310225t.

Two recent molecular dynamics (MD) simulations offer molecular-level information about interactions of the ions in ILs with cellulose [69, 70]. These investigations revealed that the mechanism for cellulose dissolution is much more complicated and that there are multiple effects impacting the dissolution other than just H-bonding effects. Rabideau et al. carried out cellulose dissolution MD simulation studies in three ILs - $[C_2C_1im][OAc]$, 1-butyl-3-methylimidazolium chloride, and $[C_2C_2im][DMP]$ ($[DMP]^-$ = dimethyl phosphate). The scrambling of small bundles of cellulose I_α and I_β in those ILs was examined [69]. It was noted that initially, the anion strongly interacted with the exterior strands' hydroxyl groups which led to the formation of a negatively charged complex. This weakens the inter and intra molecular hydrogen bonds in cellulose strands. As a result of the negatively charged complex formation, a charge imbalance occurs, and it was nullified by the intercalation of the cation to cellulose strands, eventually leading to the separation of strands. They also noticed when individual strand peeled off from the main bunch in the IL. As a result, glucan chain was separated from the main bundle (Figure 4).

In another similar MD simulation study, Li et al. used cellulose bundle with 7 strands in $[C_2C_1im][OAc]$, $[C_2C_1im][Cl]$, and 1-butyl-3-methylimidazolium chloride. They observed similar behavior to Rabideau et al. in MD simulations where 7 strands completely break up in $[C_2C_1im][OAc]$, but observed slight variation for $[C_2C_1im][Cl]$ and $[C_4C_1im][Cl]$ in the same time period. Further, they noticed that $[OAc]^-$ ions bind firmly to the hydroxyl groups, loosening the connectivity in adjacent strands. The space generated between strands permits excess ions to move into strands. Additionally, it was shown that cations interact via strong van der Waals forces with cellulose [70].

Rabideau and coworkers found support for van der Waals interactions in a MD study of a single cellulose chain in $[C_nC_1im][X]$, where they systematically looked at the effect of alkyl chain length n and of anion $[X]^-$ on interaction energies, H-bonding and solvation structure. It was elaborated that hydrophobic domain in cellulose interacts with the cation via dispersion forces and the anion interacts through electrostatic interactions with hydrophilic domains in cellulose. Additionally, this simulation study showed cations and anions act synergistically in the dissolution process. Cations interact with the six membered rings of cellulose by forming Vander Waals interactions while anions break the intra and intermolecular hydrogen bonds in cellulose. As anion concentration increased, cations intercalate between cellulose chains. Results showed that the acetate anion was able to form four different hydrogen bonds within cellulose chains, creating space for cations to enter the chains. On the contrary, chloride anion cannot create such space for cations. Therefore, EmimAc dissolves cellulose more rapidly than EmimCl [62].

A detailed study by Dissanayake et al. revealed that alkyl substituents in the imidazolium cation have a great impact on the cellulose dissolution behavior in ionic liquids by using a series of imidazolium based ILs, (see Table 1) [71].

Cations 1-3 in Table 1 allowed them to study how the dissolution is affected by chain length, while cations 3-5 provided information on how the dissolution is impacted by having linear alkyl versus cyclic alkyl, and aromatic versus nonaromatic substituents. During their study, 1-hepyl-3-methylimidazolium acetate, 1-(cyclohexylmethyl)-3-methylimidazolium acetate, 1,3-dibenzylimidazolium acetate, and 1-(2-napthylmethyl)-3-methylimidazolium acetate showed relatively low dissolution after microwaving whereas noticeable cellulose dissolution was observed in 1-benzyl-3-methylimidazolium acetate after microwaving.

Figure 5 represents PLM images of the state of the cellulose (5% w/w) dissolved in the above mentioned ILs separately and the images are captured at various stages of the cellulose dissolution. The PLM images showed that stirring at 90ºC for 4 h led to complete dissolution in 1-benzyl-3-methylimidazolium acetate. FTIR spectrum of regenerated cellulose in 1-benzyl-3-methylimidazolium acetate displayed significant changes in its vibrations at 3482, 3439, 3330, 3289, 1645, 1428, 1315, 1159, 1104,1052, 995, 895, 710, and 666 cm^{-1} respectively as compared to the spectrum of the starting material, cotton cellulose. These data are consistent with the values reported in the literature [72–77].

Table 1. Structure and size of the studied IL cations

cation	cation structure	vdW Volume[a] (Å^3)	abbreviation	formula
1-hepyl-3-methylimidazolium	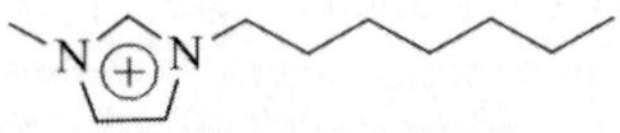	261.3	$[C_7C_1im]^+$	$C_{11}H_{21}N_2^+$
1-(cyclohexylmethyl)-3-methylimidazolium	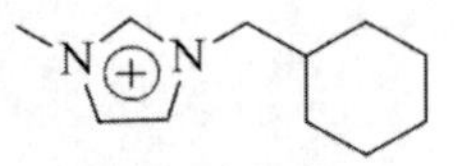	232.0	$[CyhmC_1im]^+$	$C_{11}H_{19}N_2^+$
1-benzyl-3-methylimidazolium	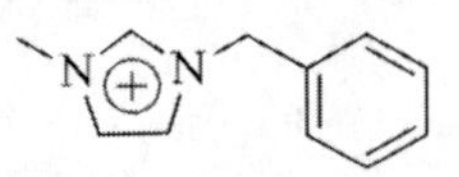	222.9	$[BnzC_1im]^+$	$C_{11}H_{13}N_2^+$
1,3-dibenzylimidazolium		303.6	$[(Bnz)_2im]^+$	$C_{17}H_{17}N_2^+$
1-(2-napthylmethyl)-3-methylimidazolium	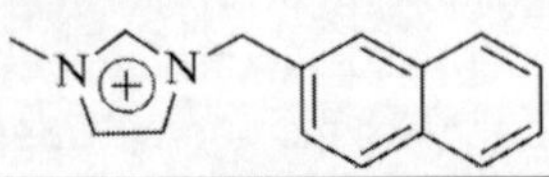	269.3	$[NapmC_1im]^+$	$C_{15}H_{15}N_2^+$

Adopted by permission from Springer, *Cellulose*, "Substituent effects on cellulose dissolution in imidazolium- based ionic liquids," N. Dissanayeke, V.D. Thalangamaarachchige, S. Troxell, E.L. Quitevis, N. Abidi, 2018.

According to this study, it was found that the degree to which the dissolution has occurred follows the trend 1-benzyl-3-methylimidazolium acetate >> 1-hepyl-3-methylimidazolium acetate > 1-(cyclohexylmethyl)-3-methylimidazolium acetate. These differences in the cellulose dissolution behavior in aforementioned ILs is explained using the size of the cation, the orientation of the substituents connected to nitrogen in imidazolium cation and the shape of the cation. Previously reported simulation studies confirmed that due to coulombic attraction, the cation is pulled by the anion into the cellulose fiber interior resulting in the separation of strands from the main bundle [69, 70]. From this study, it was evident that the intercalation of the cation between the strands, and the separation of cellulose strands contingent on the size, shape, and the orientation of space of the cation.

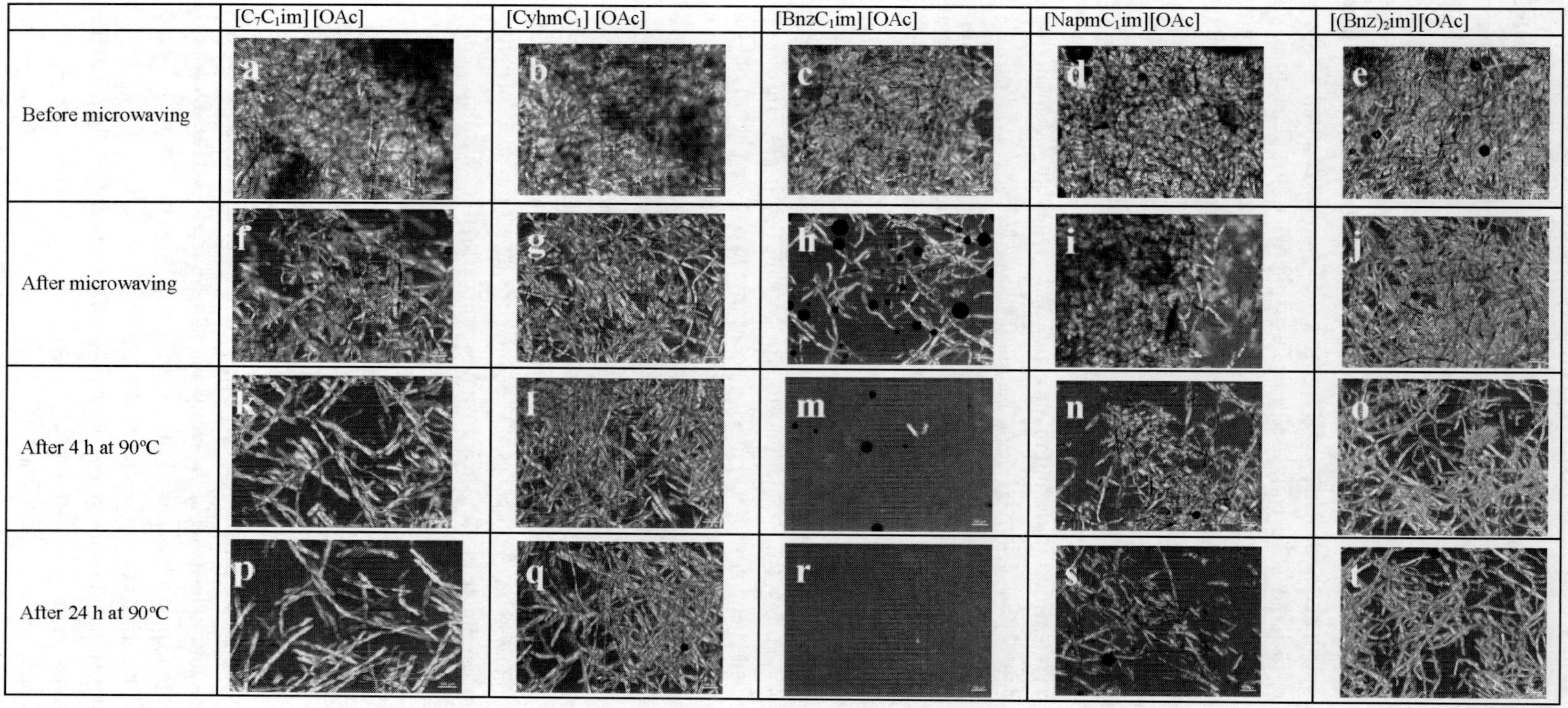

Figure 5. PLM images at various stages of dissolution in ILs with acetate anion and different alkyl substituents at the imidazolium cation. Reprinted by permission from Springer, *Cellulose*, "Substituent effects on cellulose dissolution in imidazolium- based ionic liquids," N. Dissanayeke, V.D. Thalangamaarachchige, S. Troxell, E.L. Quitevis, N. Abidi, Copyright ©2018.

The observed results are consistent with this mechanism where 1-(cyclohexylmethyl)-3-methylimidazolium acetate showed comparatively insignificant dissolution after microwaving while a noticeable cellulose dissolution was observed in the IL 1-benzyl-3-methylimidazolium acetate at the same stage. This is not mainly due to the size effect as the calculated Van Der Waals volumes suggested because the size difference is not significant between 1-(cyclohexylmethyl)-3-methylimidazolium cation and the 1-benzyl-3-methylimidazolium cation. This may be attributed to the shape effect where the cyclohexyl group which introduces steric hindrance due to larger size compared to planar phenyl rings. Therefore, cyclohexyl group hinders the intercalation and minimized the penetration of the cation into cellulose strands. Additionally noted that the nonplanar nature of cyclohexyl group where it exists boat or chair confirmation impact adversely during the intercalation.

From this study, it was revealed that 1-benzyl-3-methylimidazolium acetate was better cellulose solvent than 1,3-dibenzylimidazolium acetate. This was mainly attributed to the difficulty of intercalating between strands when two phenyl groups present in the cation. When considering 1-benzyl-3-methylimidazolium acetate and 1-naphthyl-3-methylimidazolium acetate Ionic liquids we expect 1-benzyl-3-methylimidazolium acetate to be more effective and it was explained via size effect. If the same rationale continues, 1-(2-napthylmethyl)-3-methylimidazolium acetate must be more effective cellulose solvent than 1,3-dibenzylimidazolium acetate where the calculated van der Waals volume 1-(2-napthylmethyl)-3-methylimidazolium cation being 11% smaller than 1,3-dibenzylimidazo-lium cation. But experimentally observed that 1,3-dibenzylimidazolium acetate is more effective IL than 1-(2-napthylmethyl)-3-methylimidazolium acetate. This was explained based on space-filling models which suggested in 1,3-dibenzylimidazolium acetate, two phenyl rings in two benzyl groups were twisted and lied in different planes. In contrary, two fused phenyl rings in naphthyl group in 1-(2-napthylmethyl)-3-methylimidazolium acetate lied in the same plane which may facilitate better interacalation between the cellulose strands. The variation in the dissolution behavior in ILs 1-hepyl-3-methylimidazolium acetate and 1-benzyl-3-methylimidazolium acetate was explained based on size effect where 1-hepyl-3-methylimidazolium cation is 17% larger than the 1-benzyl-3-methylimidazolium cation.

Nevertheless, 1-hepyl-3-methylimidazolium acetate appears to be an incongruity because 1-hepyl-3-methylimidazolium cation is 3% smaller than the 1-(2-napthylmethyl)-3-methylimidazolium cation. It is expected to have more effective dissolution in 1-hepyl-3-methylimidazolium acetate than in 1-(2-napthylmethyl)-3-methylimidazolium acetate, contradicting the experimental results. This study concluded that 3% size difference is not significant enough and proposed that multiple factors must have affected cellulose dissolution in these two ILs. One explanation was that heptyl group hold more confirmational freedom compared to Naphthyl group which could hinder the intercalation of cation between cellulose strands. When comparing $[BnzC_1im]^+$, $[(Bnz)_2im]^+$, and $[NapmC_1im]^+$ cations planarity of phenyl groups facilitate the intercalation between cellulose strands. This study concludes that size, shape, and orientation in space of the cation has a great impact on the dissolution capacity. This study clearly indicates the impact of the cation in the dissolution process. A recent review article summarized mechanistic studies in cellulose dissolution in ILs and further discussed the importance of both anion and cation in cellulose dissolution where the anion and the cation are synergistically impacting the dissolution [78].

Few recent studies also proved that both the cation and the anion are involved in the dissolution. Xin et al. developed novel ionic liquids with superbase cations and carboxylate anions and investigated the cellulose dissolution. From this study it was confirmed that the super base cations with larger ring facilitate the dissolution while electron withdrawing groups, short alkyl chains, less steric hindrance in the anion promote faster dissolution. Ishida performed computational study to understand the dissolution process using a cellulose I_{β} fiber in 1-ethyl-3-methylimidazoliuim acetate. The study confirmed that anions disrupt the hydrogen bonds and the cation stabilize detached cellulose chains via van der Waals interactions. Both studies revealed that both cation and anion are synergistically involved in the dissolution process [79, 80].

Cosolvent Effect on Cellulose Dissolution

A number of studies revealed that cosolvents, such as dimethyl sulfoxide (DMSO), and Dimethyl formamide (DMF) facilitate cellulose dissolution in ILs. The investigations on the role of the co-solvents (DMSO, DMF, and DMA) on the solubility of cellulose by Xu and coworkers stated that the solubility was enhanced as a result of solvation of the cation [81]. Several review articles highlighted the fact that cellulose dissolution is governed by multiple factors such as the nature of the anion and cation, anion basicity and its ability to form H-bonding with cellulose, and the position and alkyl chain length of the cation [31, 52].

Despite the effectiveness of ILs as cellulose solvents, the dissolution in the neat ILs is time consuming partly due to the high viscosity of the IL /cellulose solutions. Rinaldi et al. demonstrated that mixing a minor molar fraction of IL (χ_{IL}) with aprotic-cosolvents such as dimethylsulfoxide (DMSO), 1,3-dimethyl-2-imidazolidinone (DMI) and N, N-dimethyl-acetamide (DMA), can promptly dissolve large amounts of MCC. For example, immediate dissolution of 10 wt% MCC at 100°C was observed in [C_2C_1im][OAc]/[DMSO] at $\chi_{IL} = 0.08$ whereas it was observed in C_2C_1im][OAc]/[DMA] ($\chi_{IL} = 0.27$), which indicates DMSO is more effective cosolvent than DMA in cellulose dissolution [82].

By using the KAT HBA acidity α, the H-bond donor (HBD) basicity β, and dipolar/polarizability index π^* to characterize IL/cosolvent systems it was shown that both the IL and the cosolvent should have high β to lower the χ_{IL} needed to dissolve cellulose instantaneously and that the cosolvent should have shown no H-bond acidity ($\alpha = 0$) to evade competition between the cosolvent and cellulose for the H-bond basicity of the IL [83–85]. A study by Andanson et al. utilized MCC in 1-butyl-3-methyl imidazolium acetate mixtures and the dissolution was observed using polarized light microscopy (PLM). This study concluded that faster dissolution achieved due to a decrease in the viscosity [86].

To recognize the cosolvent effect, Velioglu et al. performed MD simulations utilizing microfibril with 36 strands in the neat IL 1-butyl-3-methylimidazolium acetate and equilmolar mixture of [C_4C_1im][OAc]/DMSO. From their study, it was confirmed that even with the high polarity of DMSO, it showed no interaction with cellulose. Additionally, they claimed that DMSO does not exhibit any specific association with IL ions. The lowering of the viscosity by the cosolvent facilitates the mass transport which enhances the dissolution process. This conclusion contradicts Rinaldi's empirical explanation. Velioglu et al. indicated that the cosolvent has no impact on IL ions interactions with cellulose [87].

The experimental results of Andanson et al. and of simulation results of Velioglu et al. are, however, in conflict with an extensive experimental study by Minnick et al. of the interactions between $[C_2C_1im][DEP]$ ([DEP] = diethyl phosphate)/cosolvent mixtures and cellulose using KAT analysis and FTIR and NMR spectroscopy. According to their studies, the cosolvent preferentially solvated IL ions which affected its capacity to dissolve cellulose [88]. 1D and 2D NMR spectroscopy were used in determining the interactions of cation and anion of -ethyl-3-methylimidazolium acetate with cellulose in the presence of co-solvent 1,3-dimethyl-2-imidazolidinone. It was noted that greater interaction between the dialkylimida-zolium ring C 2–H substituent and cellulose was observed at elevated temperatures [89].

The study by Goud and coworkers reported the impacts of IL and IL/co-solvent mixtures for the cellulose dissolution by using QC (Quantum Calculations) and MD (Molecular Dyanamic) simulations. This work is further supported by experimental studies. The addition of co-solvent to IL, led to dissociation of ion pairs, increasing the free cations and anions from IL. The separated free anions would then interact with cellulose and enhance the dissolution process. The obtained results, such as interaction energies and coordination numbers of cellulose-IL/cosolvent align with the mechanism [90]. Xu and coworkers stated that cellulose solubility was enhanced because of solvation of the cation in IL 1-butyl-3-methylimidazolium acetate when polar aprotic solvents were used as co-solvents. This study concluded that the solvation of cation $[C_4mim]^+$ and the anion $[CH_3COO]^-$ by the aprotic solvents enhance the "free" $[CH_3COO]-$ anion concentration by disturbing the interactions of anion-cation pairs in IL, which facilitate the anion - cellulose interaction, leading to the intensification in cellulose solubility [81]. Phadagi et al. and Omar et al. recently investigated the impact of DMF and DMSO on cellulose dissolution as co-solvents [91, 92].

Olsson and coworkers showed that methylimidazole was an effective cosolvent [93]. A follow up detailed study by Dissanayake et al. using equimolar mixtures of N- alkyl imidazoles (N = methyl, ethyl, propyl, butyl) and 1-butyl-3-methylimidazolium methylphosphonate IL confirmed that alkyl imidazoles are effective cosolvents in cotton cellulose (5% w/w) dissolution. The progression of the dissolution was examined via polarized light microscopy (PLM) and images were taken at different intervals [94]. Based on PLM images, it is evident that the IL/MIM solvent system most effectively dissolves cellulose compared to other solvent systems. This cellulose dissolution behavior in different IL+ cosolvent systems was explained via physio-chemical properties such as viscosity, conductivity, Kalmet and Taft (KAT) analysis of solvation properties of IL/Cosolvent Mixtures. Previous studies showed almost quantitative correlation between KAT parameters of IL/cosolvent mixtures and cellulose dissolution [82–86].

In a study by Xu et al., a linear correlation between H-bond basicity of anions with the cellulose dissolution in ILs was reported [95]. Dissanayake et al. also performed KAT solvatochromatic analysis in their study and the β parameter of IL/cosolvent systems was determined (see Table 2). Based on a preliminary KAT analysis of $[C_4C_1im][(OMe)(H)PO_2]$/1-n-alkylmidazole mixtures (alkyl = C_nH_{2n+1} with n = 1-4), they found that the values of β at $\chi_{IL} = 0.5$ to be 1.02 ± 0.02, 1.04 ± 0.02, 1.05 ± 0.01 for the mixtures with methyl- 1-*n*-ethyl- and 1-*n*-butylimidazole as the cosolvents, respectively, compared to $\beta = 0.96 \pm 0.01$ for neat $[C_4C_1im][(OMe)(H)PO_2]$ (Table 2). The fact that the values of β for the mixtures are nearly equal to the value for the neat IL is consistent with a leveling effect observed by Rinaldi [82].

Table 2. Kamat Taft β parameter values for IL and IL/co-solvent systems

Solvent[a]	***β*-parameter**[b]
IL	0.961 ± 0.006
IL+MIM	1.02 ± 0.02
IL+EIM	1.04 ± 0.01
IL+PIM	1.04 ± 0.02
IL+BIM	1.05 ± 0.01

[a]IL = [C4C1im][(OMe)(H)PO2]; MIM = 1 methylimidazole; EIM = 1-ethylimidazole; PIM = 1-propylimidazole (PIM); and BIM = 1-butylimidazole. Reprinted from *Carbohydrate Polymers* 221(13), "Dissolution of cotton cellulose in 1:1 mextures of 1-butyl-3-methylimidazolium methylphosphonate and 1-alkylimidazole co-solvents," N. Dissanayeke, V.D. Thalangamaarachchige, M. Thakurathi, M. Knight, S. Troxell, E.L. Quitevis, N. Abidi, 63-72, Copyright (2019) with permission from Elsevier

In addition to β parameter measurements, Dissanayake et al. also measured the viscosity and conductivity data of IL/cosolvent mixtures. Figures 6A and 6B show data for the conductivity (κ) and viscosity (η) of $[C_4C_1im][(OMe)(H)PO_2]$ and the 1:1 mole fraction mixtures of $[C_4C_1im][(OMe)(H)PO_2]$ in alkyl imidazole cosolvents as a function of temperature. As expected, κ increases with increasing temperature due to increased mobility of the ions. Moreover, neat IL, IL/1-ethylimidazole, and IL/methylimidazole are consistent with high conductivity being associated with low viscosity. However, the case IL/1-butylimidazole appears to be anomalous in that κ (neat IL) ≈ κ (IL/1-butylimidazole) and η (neat IL) > η (IL/1-butylimidazole).

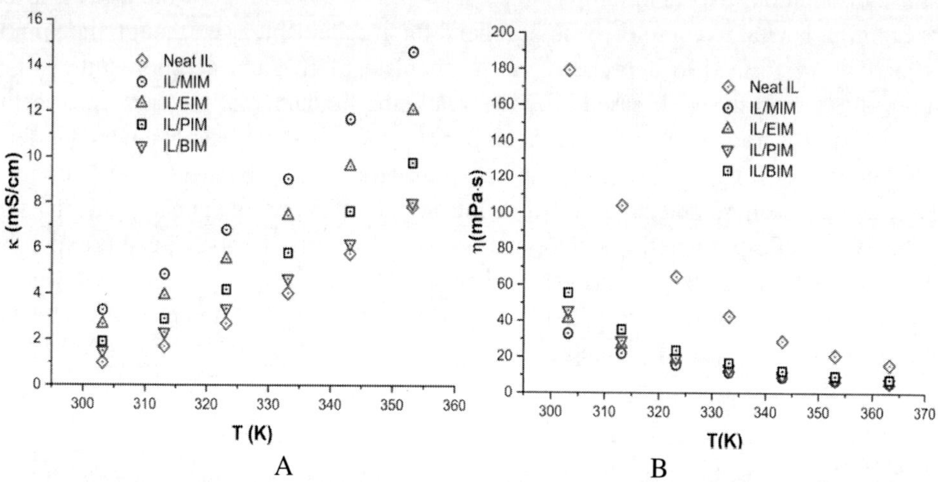

Figure 6. A) The specific conductivities of IL [C4C1im][(OMe)(H)(PO2)] and 1:1 mixtures of this IL with 1-alkylimidazoles. B) The viscosities of IL [C4C1im][(OMe)(H)(PO2)] and 1:1 mixtures of this IL with 1-alkylimidazoles. MIM = 1-methylimidazole, EIM = 1-ethylimidazole, PIM = 1-propylimidazole, BIM = 1-butylimidazole. Reprinted from *Carbohydrate Polymers* 221(13), "Dissolution of cotton cellulose in 1:1 mextures of 1-butyl-3-methyl imidazolium methyl phosphonate and 1-alkylimidazole co-solvents," N. Dissanayeke, V.D. Thalangamaarachchige, M. Thakurathi, M. Knight, S. Troxell, E.L. Quitevis, N. Abidi, 63-72, 2019 Copyright with permission from Elsevier.

From the rheological experiments, it was confirmed that the IL/MIM solvent system has the lowest viscosity, and it was found to be the most effective cellulose solvent system in this study. This is attributed to the increase in mass transport due to the lowering of viscosity in the IL/MIM mixture. The measured viscosity for solvent systems is as follows: Neat IL > IL/1-Butylimidazole > IL/1-Propylimidazole > IL/1-Ethylimidazole > IL/1-Methylimidazole (Figure 6B). From the conductivity measurements, it was noticed that the addition of cosolvents resulted in a noticeable increase in specific conductivity (κ).

As illustrated in Figure 6A, the conductivity improved and continued to trend upward when shortening the alkyl chain of imidazole. The specific conductivity of the data set is as follows: IL/1-Methylimidazole > 1-Ethylimidazole > IL/1-Propylimidazole > IL/1-Butylimidazole > Neat IL. From the experimental evidence it was concluded that higher conductivity in the mixture of IL-cosolvent leads to greater mobility of ions which enables ions to penetrate faster to cellulose chains. Unsurprisingly, the specific conductivity (κ) is inversely related to the viscosity. It was first thought that ILs were entirely comprised of dissociated ions.

As observed by MacFarlane et al. some ILs exhibit much lower ionic conductivity than others, even though their viscosities are comparable. The lower-than-expected conductivity suggested that viscosity is not the only property responsible for the conductivity. In particular, the formation of neutral ion-pairs or aggregates, if they are long-lived enough, can lower the ionic conductivity of the IL [96]. The anomalous behavior observed in Figures 6A and 6B for $[C_4C_1im][(OMe)(H)PO_2]$/1-butylimidazole suggests that the cosolvent 1-butylimidazole is promoting ion association.

Conclusion

Cellulose dissolution involves multiple steps. Detailed understanding of the aspects affecting the dissolution is vital to completely comprehend the mechanism. Even though the number of investigations performed to apprehend the mechanism of ILs and cellulose interactions in cellulose dissolution have increased over the years, the fundamental mechanism of cellulose dissolution in ILs remain vague and not conclusive as the key factors which effect the dissolution are not fully comprehended at a microscopic level. Advances have been made to investigate the dissolution mechanisms of cellulose in ILs, including the basic attributes of ILs that dissolve cellulose, though many details remain obscure. At present, some experimental results cannot be explained by described mechanism. Therefore, further extensive experimental and theoretical mechanistic studies should be performed in this field to elucidate the cellulose IL ions interactions and to fully comprehend the mechanism.

References

[1] Netravali, A. N., Chabba, S. Composites get greener. *Mater. Today* 2003, 6, 22–29.

[2] Lynd, L. R., Wyman, C. E., Gerngross, T. U. Biocommodity engineering. *Biotechnol. Prog.* 1999, 15, 777–793.

[3] Dufresne, A. Cellulose nanomaterial reinforced polymer nanocomposites. *Curr. Opin. Colloid Interface Sci.* 2017, 29, 1–8.

[4] Shokri, J., Adibki, K. Application of cellulose and cellulose derivatives in pharmaceutical industries. In *Cellulose - Medical, Pharmaceutical and Electronic Applications* 2013.

[5] Ragauskas, A. J., Williams, C. K., Davison, B. H., Britovsek, G., Cairney, J., Eckert, C. A., Frederick, W. J., Hallett, J. P., Leak, D. J., Liotta, C. L., Mielenz, J. R., Murphy, R., Templer, R., Tschaplinski, T. The path forward for biofuels and biomaterials. *Science.* 2006, 311, 484–489.
[6] Hermanutz, F., Gähr, F., Uerdingen, E., Meister, F., Kosan, B. New developments in dissolving and processing of cellulose in ionic liquids. *Macromol. Symp.* 2008, 262, 23–27.
[7] Heinze, T., Liebert, T. Unconventional methods in cellulose functionalization. *Prog. Polym. Sci.* 2001, 26, 1689–1762.
[8] Liebert, T. Cellulose solvents-remarkable history, bright future. In *ACS Symposium Series.* 2010, 3–54.
[9] Hu, Y., Thalangamaarachchige, V. D., Acharya, S., Abidi, N. Role of low-concentration acetic acid in promoting cellulose dissolution. *Cellulose* 2018, 25, 4389–4405.
[10] Holbrey, J. D., Rogers, R. D. Green Industrial applications of ionic liquids: Technology Review. 2002, 446–458.
[11] MacFarlane, D. R., Seddon, K. R. Ionic liquids-progress on the fundamental issues. *Aust. J. Chem.* 2007, 60, 3–5.
[12] Huddleston, J. G., Willauer, H. D., Swatloski, R. P., Visser, A. E., Rogers, R. D. Room temperature ionic liquids as novel media for "clean" liquid-liquid extraction. *Chem. Commun.* 1998, 1765–1766.
[13] Freemantle, M. Ionic liquids may boost clean technology development. *Chem. Eng. News* 1998.
[14] Plechkova, N. V., Seddon, K. R. Applications of ionic liquids in the chemical industry. *Chem. Soc. Rev.* 2008, 37, 123–150.
[15] Olivier-Bourbigou, H., Magna, L., Morvan, D. Ionic liquids and catalysis: Recent progress from knowledge to applications. *Appl. Catal. A Gen.* 2010, 373, 1–56.
[16] Wasserscheid, P., Welton, T. *Ionic Liquids in Synthesis: Second Edition*; 2008.
[17] Lee, J., He, M., Yeo, C. D., Kumar, G., Hu, Z., Quitevis, E. L., Thalangamaarachchige, V. D. Friction and wear of Pd-rich amorphous alloy (Pd43Cu27Ni10P20) under dry and ionic liquid (IL) lubricated conditions. *Wear* 2018, 408–409, 190–199.
[18] Cull, S. G., Holbrey, J. D., Vargas-Mora, V., Seddon, K. R., Lye, G. J. Room-temperature ionic liquids as replacements for organic solvents in multiphase bioprocess operations. *Biotechnol. Bioeng.* 2000, 69, 227–233.
[19] Sanes, J., Carrión, F. J., Bermúdez, M. D., Martínez-Nicolás, G. Ionic liquids as lubricants of polystyrene and polyamide 6-steel contacts. Preparation and properties of new polymer-ionic liquid dispersions. *Tribol. Lett.* 2006, 21, 121–133.
[20] Le Bideau, J., Gaveau, P., Bellayer, S., Néouze, M. A., Vioux, A. Effect of confinement on ionic liquids dynamics in monolithic silica ionogels: 1H NMR study. *Phys. Chem. Chem. Phys.* 2007, 9, 5419–5422.
[21] Seki, S., Ohno, Y., Kobayashi, Y., Miyashiro, H., Usami, A., Mita, Y., Tokuda, H., Watanabe, M., Hayamizu, K., Tsuzuki, S., Hattori, M., Terada, N. Imidazolium-based room-temperature ionic liquid for lithium secondary batteries - Effects of lithium salt concentration. *J. Electrochem. Soc.* 2007, 154, 173-177.
[22] Wang, P., Zakeeruddin, S. M., Exnarb, I., Grätzel, M. High efficiency dye-sensitized nanocrystalline solar cells based on ionic liquidpolymer gel electrolyte. *Chem. Commun.* 2002, 2972–2973.
[23] Gu, W., Chen, H., Yi-Chung Tung. Multiplexed hydraulic value actuation using ionic liquid filled soft channels and braille displays. *Appl. Phys. Lett.* 2007, 90.
[24] Thakurathi, M., Gurung, E., Cetin, M. M., Thalangamaarachchige, V. D., Mayer, M. F., Korzeniewski, C., Quitevis, E. L. The Stokes-Einstein equation and the diffusion of ferrocene in imidazolium-based ionic liquids studied by cyclic voltammetry: Effects of cation ion symmetry and alkyl chain length. *Electrochim. Acta* 2018, 259, 245–252.
[25] Lee, J., Yeo, C. D., Hu, Z., Thalangama-Arachchige, V. D., Kaur, J., Quitevis, E. L., Kumar, G., Koh, Y. P., Simon, S. Friction and wear of pd-rich amorphous alloy (Pd43cu27ni10p20) with ionic liquid (IL) as lubricant at high temperatures. *Metals (Basel).* 2019, 9.
[26] Cheng, G., Varanasi, P., Li, C., Liu, H., Melnichenko, Y. B., Simmons, B. A., Kent, M. S., Singh, S. Transition of cellulose crystalline structure and surface morphology of biomass as a function of ionic liquid pretreatment and its relation to enzymatic hydrolysis. *Biomacromolecules* 2011, 12, 933–941.
[27] Cheng, G., Varanasi, P., Arora, R., Stavila, V., Simmons, B. A., Kent, M. S., Singh, S. Impact of ionic liquid pretreatment conditions on cellulose crystalline structure using 1-ethyl-3-methylimidazolium acetate. *J. Phys. Chem. B* 2012, 116, 10049–10054.
[28] Swatloski, R. P., Spear, S. K., Holbrey, J. D., Rogers, R. D. Dissolution of cellose with ionic liquids. *J. Am. Chem. Soc.* 2002, 124, 4974–4975.
[29] Wu, J., Zhang, J., Zhang, H., He, J., Ren, Q., Guo, M. Homogeneous acetylation of cellulose in a new

ionic liquid. *Biomacromolecules* 2004, 5, 266–268.

[30] Zhang, H., Wu, J., Zhang, J., He, J. 1-allyl-3-methylimidazolium chloride room temperature ionic liquid: A new and powerful nonderivatizing solvent for cellulose. *Macromolecules* 2005, 38, 8272–8277.

[31] Pinkert, A., Marsh, K. N., Pang, S., Staiger, M. P. Ionic liquids and their interaction with cellulose. *Chem. Rev.* 2009, 109, 6712–6728.

[32] Sun, N., Rodríguez, H., Rahman, M., Rogers, R. D. Where are ionic liquid strategies most suited in the pursuit of chemicals and energy from lignocellulosic biomass? *Chem. Commun.* 2011, 47, 1405–1421.

[33] Wang, H., Gurau, G., Rogers, R. D. Ionic liquid processing of cellulose. *Chem. Soc. Rev.* 2012, 41, 1519–1537.

[34] Wang, H., Gurau, G., Rogers, R. D. Dissolution of biomass using ionic liquids. In *Structure and Bonding*, 2013, 79–105.

[35] Cruz, H., Fanselow, M., Holbrey, J. D., Seddon, K. R. Determining relative rates of cellulose dissolution in ionic liquids through in situ viscosity measurement. *Chem. Commun.* 2012, 48, 5620–5622.

[36] Cao, Y., Wu, J., Zhang, J., Li, H., Zhang, Y., He, J. Room temperature ionic liquids (RTILs): A new and versatile platform for cellulose processing and derivatization. *Chem. Eng. J.* 2009, 147, 13–21.

[37] Graenacher, C. Cellulose Solution, 1934.

[38] Kathirgamanathan, K., Grigsby, W. J., Al-Hakkak, J., Edmonds, N. R. Two-dimensional FTIR as a tool to study the chemical interactions within cellulose-ionic liquid solutions. *Int. J. Polym. Sci.* 2015, 2015, 1–9.

[39] Gupta, K. M., Hu, Z., Jiang, J. Mechanistic understanding of interactions between cellulose and ionic liquids: A molecular simulation study. *Polymer (Guildf).* 2011, 52, 5904–5911.

[40] Remsing, R. C., Swatloski, R. P., Rogers, R. D., Moyna, G. Mechanism of cellulose dissolution in the ionic liquid 1-n-butyl-3-methylimidazolium chloride: a 13C and 35/37Cl NMR relaxation study on model systems. *Chem. Commun.* 2006, 1271–1273.

[41] Remsing, R. C., Hernandez, G., Swatloski, R. P., Massefski, W. W., Rogers, R. D., Moyna, G. Solvation of carbohydrates in N,N'-dialkylimidazolium ionic liquids: A multinuclear NMR spectroscopy study. *J. Phys. Chem. B* 2008, 112, 11071–11078.

[42] Youngs, T. G. A., Holbrey, J. D., Mullan, C. L., Norman, S. E., Lagunas, M. C., D'Agostino, C., Mantle, M. D., Gladden, L. F., Bowron, D. T., Hardacre, C. Neutron diffraction, NMR and molecular dynamics study of glucose dissolved in the ionic liquid 1-ethyl-3-methylimidazolium acetate. *Chem. Sci.* 2011, 2, 1594–1605.

[43] Youngs, T. G. A., Hardacre, C., Holbrey, J. D. Glucose solvation by the ionic liquid 1,3-dimethylimidazolium chloride: A simulation study. *J. Phys. Chem. B* 2007, 111, 13765–13774.

[44] Liu, Z., Remsing, R. C., Moore, P. B., Moyna, G. Molecular dynamics study of the mechanism of cellulose dissolution in the ionic liquid 1-n-butyl-3-methylimidazolium chloride. In *ACS Symposium Series*, 335–350.

[45] Zhao, Y., Liu, X., Wang, J., Zhang, S. Effects of cationic structure on cellulose dissolution in ionic liquids: A molecular dynamics study. *ChemPhysChem* 2012, 13, 3126–3133.

[46] Casas, A., Palomar, J., Alonso, M. V., Oliet, M., Omar, S., Rodriguez, F. Comparison of lignin and cellulose solubilities in ionic liquids by COSMO-RS analysis and experimental validation. *Ind. Crops Prod.* 2012, 37, 155–163.

[47] Yao, Y., Li, Y., Liu, X., Zhang, X., Wang, J., Yao, X., Zhang, S. Mechanistic study on the cellulose dissolution in ionic liquids by density functional theory. *Chinese J. Chem. Eng.* 2015, 23, 1894–1906.

[48] Kan, Z., Zhu, Q., Yang, L., Huang, Z., Biaobing Jin; Ma, J. Polarization Effects on the cellulose dissolution in ionic liquids: Molecular dynamics simulations with polarization model and integrated tempering enhanced sampling method. *J. Phys. Chem. B* 2017, 121, 4319–4332.

[49] Endo, T., Hosomi, S., Fujii, S., Ninomiya, K., Takahashi, K. Nano-structural investigation on cellulose highly dissolved in ionic liquid: A small angle X-ray scattering study. *Molecules* 2017, 22.

[50] Endo, T., Hosomi, S., Fujii, S., Ninomiya, K., Takahashi, K. Anion Bridging-induced structural transformation of cellulose dissolved in ionic liquid. *J. Phys. Chem. Lett.* 2016, 7, 5156–5161.

[51] Gehrke, S., Schmitz, K., Hollóczki, O. Is carbene formation necessary for dissolving cellulose in ionic liquids? *J. Phys. Chem. B* 2017, 121, 4521–4529.

[52] Lu, B., Xu, A., Wang, J. Cation does matter: How cationic structure affects the dissolution of cellulose in ionic liquids. *Green Chem.* 2014, 16, 1326–1335.

[53] Remsing, R. C., Petrik, I. D., Liu, Z., Moyna, G. Comment on "NMR spectroscopic studies of cellobiose solvation in EmimAc aimed to understand the dissolution mechanism of cellulose in ionic

liquids" by J. Zhang, H. Zhang, J. Wu, J. Zhang, J. He and J. Xiang, Phys. Chem. Chem. Phys., 2010, 12, 1941. *Phys. Chem. Chem. Phys.* 2010, 12, 14827–14828.

[54] Zhang, J., Zhang, H., Wu, J., Zhang, J., He, J., Xiang, J. NMR spectroscopic studies of cellobiose solvation in EmimAc aimed to understand the dissolution mechanism of cellulose in ionic liquids. *Phys. Chem. Chem. Phys.* 2010, 12, 1941–1947.

[55] Zavrel, M., Bross, D., Funke, M., Büchs, J., Spiess, A. C. High-throughput screening for ionic liquids dissolving (ligno-)cellulose. *Bioresour. Technol.* 2009, 100, 2580–2587.

[56] Youngs, T. G. A., Holbrey, J. D., Deetlefs, M., Nieuwenhuyzen, M., Costa Gomes, M. F., Hardacre, C. A molecular dynamics study of glucose solvation in the ionic liquid 1,3-dimethylimidazolium chloride. *ChemPhysChem* 2006, 7, 2279–2281.

[57] Liu, H., Sale, K. L., Holmes, B. M., Simmons, B. A., Singh, S. Understanding the interactions of cellulose with ionic liquids: A molecular dynamics study. *J. Phys. Chem. B* 2010, 114, 4293–4301.

[58] Zhao, Y., Liu, X., Wang, J., Zhang, S. Effects of anionic structure on the dissolution of cellulose in ionic liquids revealed by molecular simulation. *Carbohydr. Polym.* 2013, 94, 723–730.

[59] Mostofian, B., Smith, J. C., Cheng, X. Simulation of a cellulose fiber in ionic liquid suggests a synergistic approach to dissolution. *Cellulose* 2014, 21, 983–997.

[60] Gross, A. S., Bell, A. T., Chu, J. W. Entropy of cellulose dissolution in water and in the ionic liquid 1-butyl-3-methylimidazolim chloride. *Phys. Chem. Chem. Phys.* 2012, 14, 8425–8430.

[61] Cho, H. M., Gross, A. S., Chu, J. W. Dissecting force interactions in cellulose deconstruction reveals the required solvent versatility for overcoming biomass recalcitrance. *J. Am. Chem. Soc.* 2011, 133, 14033–14041.

[62] Rabideau, B. D., Agarwal, A., Ismail, A. E. The role of the cation in the solvation of cellulose by imidazolium-based ionic liquids. *J. Phys. Chem. B* 2014, 118, 1621–1629.

[63] Xu, H., Pan, W., Wang, R., Zhang, D., Liu, C. Understanding the mechanism of cellulose dissolution in 1-butyl-3-methylimidazolium chloride ionic liquid via quantum chemistry calculations and molecular dynamics simulations. *J. Comput. Aided. Mol. Des.* 2012, 26, 329–337.

[64] Cao, B., Du, J., Du, D., Sun, H., Zhu, X., Hui Fu. Cellobiose as a model system to reveal cellulose dissolution mechanism in acetate-based ionic liquids: Density functional theory study substantiated by NMR spectra. *Carbohydr. Polym.* 2016, 149, 348–356.

[65] Baek, C. S., Lee, Y. J., Lee, S. J., Lee, S. G., Hyun-Chul Kim; Jeong, S. W. C2-Functionalized 1, 3-dialkylimidazolium ionic liquids for efficient cellulose dissolution. *J. Mol. Liq.* 2017, 234, 111–116.

[66] Erdmenger, T., Haensch, C., Hoogenboom, R., Schubert, U. S. Homogeneous tritylation of cellulose in 1-butyl-3-methylimidazolium chloride. *Macromol. Biosci.* 2007, 7, 440–445.

[67] Ebner, G., Schiehser, S., Potthast, A., Rosenau, T. Side reaction of cellulose with common 1-alkyl-3-methylimidazolium-based ionic liquids. *Tetrahedron Lett.* 2008, 49, 7322–7324.

[68] Clough, M. T., Geyer, K., Hunt, P. A., Son, S., Vagt, U., Welton, T. Ionic liquids: Not always innocent solvents for cellulose. *Green Chem.* 2015, 17, 231–243.

[69] Rabideau, B. D., Agarwal, A., Ismail, A. E. Observed mechanism for the breakup of small bundles of cellulose I-alpha and I-beta in ionic liquids from molecular dynamics simulations. *J. Phys. Chem. B* 2013, 117, 3469–3479.

[70] Li, Y., Liu, X., Zhang, S., Yao, Y., Yao, X., Xu, J., Lu, X. Dissolving process of a cellulose bunch in ionic liquids: A molecular dynamics study. *Phys. Chem. Chem. Phys.* 2015, 17, 17894–17905.

[71] Dissanayake, N., Thalangamaarachchige, V. D., Troxell, S., Quitevis, E. L., Abidi, N. Substituent effects on cellulose dissolution in imidazolium-based ionic liquids. *Cellulose* 2018, 25, 6887–6900.

[72] Abidi, N., Hequet, E., Cabrales, L., Gannaway, J., Wilkins, T., Wells, L. W. Evaluating cell wall structure and composition of developing cotton fibers using fourier transform infrared spectroscopy and thermogravimetric analysis. *J. Appl. Polym. Sci.* 2008, 107, 476–486.

[73] Oh, S. Y., Yoo, D. Il; Shin, Y., Seo, G. FTIR analysis of cellulose treated with sodium hydroxide and carbon dioxide. *Carbohydr. Res.* 2005, 340, 417–428.

[74] Schwanninger, M., Rodrigues, J. C., Pereira, H., Hinterstoisser, B. Effects of short-time vibratory ball milling on the shape of FT-IR spectra of wood and cellulose. *Vib. Spectrosc.* 2004, 36, 23–40.

[75] Salmén, L., Bergström, E. Cellulose structural arrangement in relation to spectral changes in tensile loading FTIR. *Cellulose* 2009, 16, 975–982.

[76] Ilharco, L. M., Garcia, A. R., Lopes Da Silva, J., Vieira Ferreira, L. F. Infrared approach to the study of adsorption on cellulose: influence of cellulose crystallinity on the adsorption of benzophenone. *Langmuir* 1997, 13, 4126–4132.

[77] Lan, W., Chuan-Fu Liu; Yue, F.-X., Sun, R.-C., John F.Kennedy. Ultrasound-assisted dissolution of cellulose in ionic liquid. *Carbohydr. Polym.* 2011, 86, 672–677.

[78] Li, Y., Wang, J., Liu, X., Zhang, S. Towards a molecular understanding of cellulose dissolution in

ionic liquids: anion/cation effect, synergistic mechanism and physicochemical aspects. *Chem. Sci.* 2018, 9, 4027–4043.

[79] Li, X., Li, H., Ling, Z., Xu, D., You, T., Wu, Y. Y., Xu, F. Room-Temperature superbase-derived ionic liquids with facile synthesis and low viscosity: Powerful solvents for cellulose dissolution by destroying the cellulose aggregate structure. *Macromolecules* 2020, 53, 3284–3295.

[80] Ishida, T. Theoretical Investigation of Dissolution and Decomposition mechanisms of a cellulose fiber in ionic liquids. *J. Phys. Chem. B* 2020, 124, 3090–3102.

[81] Xu, A., Zhang, Y., Zhao, Y., Wang, J. Cellulose dissolution at ambient temperature: Role of preferential solvation of cations of ionic liquids by a cosolvent. *Carbohydr. Polym.* 2013, 92, 540–544.

[82] Rinaldi, R. Instantaneous dissolution of cellulose in organic electrolyte solutions. *Chem. Commun.* 2011, 47, 511–513.

[83] Kamlet, M. J., Abboud, J. L., Taft, R. W. The solvatochromic comparison method. 6. The Pi* scale of solvent polarities1. *J. Am. Chem. Soc.* 1977, 99, 6027–6038.

[84] Laurence, C., Nicolet, P., Dalati, M. T., Abboud, J. L. M., Notario, R. The empirical treatment of solvent-solute interactions: 15 Years of π. *J. Phys. Chem.* 1994, 98, 5807–5816.

[85] Abboud, J. L. M., Notario, R. Critical compilation of scales of solvent parameters. Part I. Pure, non-hydrogen bond donor solvents. *Pure Appl. Chem.* 1999, 71, 645–718.

[86] Andanson, J. M., Bordes, E., Devémy, J., Leroux, F., Pádua, A. A. H., Gomes, M. F. C. Understanding the role of co-solvents in the dissolution of cellulose in ionic liquids. *Green Chem.* 2014, 16, 2528–2538.

[87] Velioglu, S., Yao, X., Devémy, J., Ahunbay, M. G., Tantekin-Ersolmaz, S. B., Dequidt, A., Costa Gomes, M. F., Pádua, A. A. H. Solvation of a cellulose microfibril in imidazolium acetate ionic liquids: Effect of a cosolvent. *J. Phys. Chem. B* 2014, 118, 14860–14869.

[88] Minnick, D. L., Flores, R. A., Destefano, M. R., Scurto, A. M. Cellulose solubility in ionic liquid mixtures: Temperature, cosolvent, and antisolvent effects. *J. Phys. Chem. B* 2016, 120, 7906–7919.

[89] Clough, M. T., Farès, C., Rinaldi, R. 1D and 2D NMR spectroscopy of bonding interactions within stable and phase-separating organic electrolyte–cellulose solutions. *ChemSusChem* 2017, 10, 3452–3458.

[90] Mohan, M., Banerjee, T., Goud, V. V. Effect of protic and aprotic solvents on the mechanism of cellulose dissolution in ionic liquids: A combined molecular dynamics and experimental insight. *ChemistrySelect* 2016, 1, 4823–4832.

[91] Phadagi, R., Singh, S., Hashemi, H., Kaya, S., Venkatesu, P., Ramjugernath, D., Ebenso, E. E., Bahadur, I. Understanding the role of dimethylformamide as co-solvents in the dissolution of cellulose in ionic liquids: Experimental and theoretical approach. *J. Mol. Liq.* 2021, 328.

[92] El Seoud, O. A., Bioni, T. A., Dignani, M. T. Understanding cellulose dissolution in ionic liquid-dimethyl sulfoxide binary mixtures: Quantification of the relative importance of hydrogen bonding and hydrophobic interactions. *J. Mol. Liq.* 2021, 322.

[93] Olsson, C., Hedlund, A., Idström, A., Westman, G. Effect of methylimidazole on cellulose/ionic liquid solutions and regenerated material therefrom. *J. Mater. Sci.* 2014, 49, 3423–3433.

[94] Dissanayake, N., Thalangamaarachchige, V. D., Thakurathi, M., Knight, M., Quitevis, E. L., Abidi, N. Dissolution of cotton cellulose in 1:1 mixtures of 1-butyl-3-methylimidazolium methylphosphonate and 1-alkylimidazole co-solvents. *Carbohydr. Polym.* 2019, 221, 63–72.

[95] Xu, A., Wang, J., Wang, H. Effects of anionic structure and lithium salts addition on the dissolution of cellulose in 1-butyl-3-methylimidazolium-based ionic liquid solvent systems. *Green Chem.* 2010, 12, 268–275.

[96] MacFarlane, D. R., Forsyth, M., Izgorodina, E. I., Abbott, A. P., Annat, G., Fraser, K. On the concept of ionicity in ionic liquids. *Phys. Chem. Chem. Phys.* 2009, 11, 4962–4967.

Chapter 12

The State of Natural Fiber Welding

Christopher D. Stachurski
Nathaniel E. Larm
David P. Durkin
and Paul C. Trulove*
Department of Chemistry,
US Naval Academy, Annapolis, Maryland, USA

Abstract

The unique physical properties and designer nature of ionic liquids (ILs) have facilitated their use as solvents for naturally-occurring biopolymer materials. Though IL-based full dissolution was introduced in 2002, partial dissolution techniques such as Natural Fiber Welding (NFW) have distinct advantages over conventional techniques in biopolymer processing.

In NFW, the outer region of fibers of biopolymer materials harvested from natural sources (i.e., cellulose, cotton, chitin/chitosan, silk, lignocellulose) can be mobilized allowing for substantial manipulation or modifications. Upon addition of an antisolvent to the mobilized biopolymer system, precipitation of an amorphous region is observed which, if performed on fibers adjacent to one another, can create unified composite materials. The composites produced through NFW can exhibit enhanced properties compared to their native counterparts, including increased tensile strength, higher thermal stability, and altered specific surface areas depending on the applied treatment. NFW also opens the door to the functionalization of biomaterials to an extent not possible with conventional solvents alone.

This chapter looks at how NFW has advanced over the last 10 years as a process to engineer functional biocomposites for a wide variety of commercially relevant applications.

Keywords: ionic liquid, biopolymer, natural fiber welding, cellulose, nanomaterials

---

* Corresponding Author's Email: trulove@usna.edu.

In: Properties and Applications of Ionic Liquids
Editors: Noureddine Abidi and Julia L. Shamshina
ISBN: 979-8-88697-635-9

Abbreviations

NFW	Natural Fiber Welding.
IL	Ionic Liquid.
DMAc	N,N-dimethylacetamide.
EMIAc	1-ethyl-3-methylimidazolium acetate.
SEM	Scanning electron microscopy.
CFM	Confocal fluorescence microscopy.
BET	Brunauer-Emmett-Teller.
ATR-FTIR	Attenuated total internal reflectance – Fourier Transform infrared spectroscopy.
MOF	Metal-organic framework.
TFSI	bis(trifluoromethane)sulfonimide.
EDS	Energy dispersive X-ray spectroscopy.
BMIAc	1-butyl-3-methylimidazolium acetate.

Introduction

The concept of Natural Fiber Welding (NFW) is based upon the non-derivatizing and partial solvation of biopolymers, such as cellulose, using ionic liquids (ILs). The process emerged as a solution to a common challenge met by industry in the processing of cellulosic materials: controlled disruption of the intricate hydrogen bonding network inherent to natural biopolymers to achieve any amount of dissolution. NFW was introduced and patented in 2010 by Haverhals et al. [1] and has evolved significantly since that time. While a review on the groundwork discovery of NFW was published in 2012, recent advances necessitate a backward glance at the history of biopolymer treatments and the fundamentals of fiber welding to better understand the recent discoveries, current work, future endeavors, and potential hurdles which must be overcome to further industrialize the process.

Throughout the ages, research into the solvation of abundant ecological sources of cellulosic materials (e.g., cotton, hemp, wood) has demanded the use of harsh solvents. While disruption of the hydrogen bonding network is often cited as the critical factor in solubilizing biopolymers, their amphiphilic nature, the presence of formal charges throughout the structure, the rigidity of the polymer backbone, and the entropic contributions to the free energy of dissolution must also be considered [2]. As such, biopolymer dissolution should be framed as a complex and dynamic process encompassing more than just hydrogen bond disruption when looking to design optimal solvent systems for these materials.

Early strategies for biopolymer (primarily cellulose, depicted in Figure 1A) solvation sought to chemically modify the intermolecular-bonding functional groups into more inert groups, as is the case in nitrate-, acetate-, phosphate or carbanilate-derivatized cellulose [3-4]. These chemical alterations are known to greatly increase the solubility of cellulose in a variety of solvents, including water. Non-derivatizing solvation of cellulose was first achieved using highly ionic solutions, such as ~8 wt% lithium chloride in N,N-dimethylacetamide (DMAc) [5-8] which exploit high ionic strength (i.e., competing coordination) to overcome thermodynamic barriers to dissolution [9]. The LiCl/DMAc system can dissolve up to 15 wt% cellulose with moderate heat (ca. 150°C) and allows for bulk, solution-based processing of dissolved cellulose [10]. Low temperature (< 10°C) solvation reduces the risk of oxidative degradation of cellulose, and can be performed with cold aqueous alkali hydroxide (NaOH, LiOH) and urea solutions [11-12]. Dissolved biopolymer can then be regenerated through solvent rinses to remove the disrupting ions and reorganize the hydrogen bonding networks.

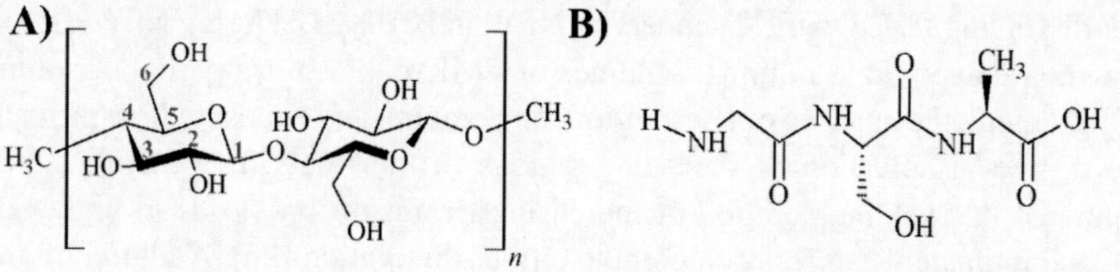

Figure 1. Native structures of (A) a cellulose monomer subunit with carbon numbering, and (B) a silk peptide comprising amino acids glycine-serine-alanine.

In the early 2000s, Swatloski et al. discovered that imidazolium-based ILs could also dissolve cellulose similar to earlier reported high ionic strength solutions [13]. Since this discovery, the most common ILs used for this application have been based on the dialkylimidazolium cation (Figure 2) paired with a chaotropic anion. The polar nature of ILs paired with the amphiphilic nature and hydrogen bond accepting capacity of the counter anions facilitates a similar hydrogen bond disruption as seen with highly ionic solutions [14]. In our lab, we see that while solution properties like viscosity, melting temperature, conductivity, and hydrophilicity are affected by the chemistries of both ions, the anion identity and hydrogen bonding character are key for biopolymer dissolution.

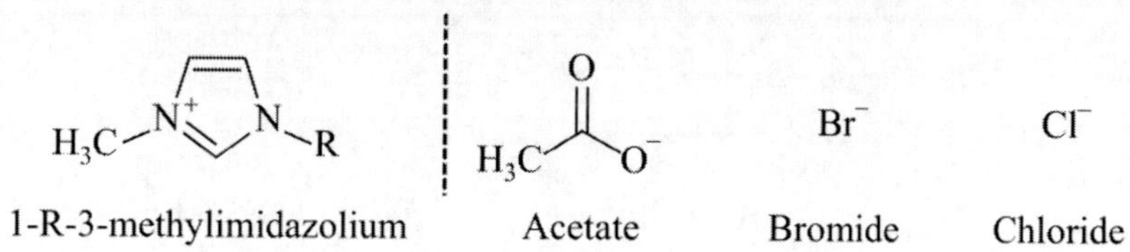

Figure 2. Chemical structure of a generic 1-R-3-methylimidazolium cation alongside common anion pairings, forming an IL suitable for biopolymer processing (R = alkyl substituent).

To this end, halides (particularly Cl^- and Br^-) and acetate show significant affinity towards cellulose solvation [15-16]. IL solutions of up to 10 wt% cellulose are attained with intensified conditions, such as mild heating (< 100°C), while non-conventional treatments methods, such as microwave heating, elevate the loading to above 25 wt% cellulose [13-14]. The diverse intra- and intermolecular interactions found in silk proteins (Figure 1B) are also susceptible to dissociation in 1-ethyl-3-methylimidazolium chloride, with up to 23 wt % solutions prepared at 100°C [17]. Interestingly, mixtures of ILs and hydrogen bonding co-solvents, such as water, express a competitive nature wherein the extra hydrogen bond source impedes biopolymer dissolution [18].

Alternatively, aprotic polar cosolvents (acetone, acetonitrile, dimethyl sulfoxide) can enhance biopolymer dissolution by reducing solution viscosity and improving mass transport

[5, 19-20]. Though excess dilution impedes solvation once solution ion concentrations decrease beyond a critical level [21]. In conclusion, IL dissolution of biopolymers is practical from the perspective of non-derivatization and solvation capacity, though complete dissolution will always suffer the loss of the admirable structural properties native to the neat crystalline material [22-23].

The discrete choice of 1-ethyl-3-methylimidazolium acetate (EMIAc) as a highly effective cellulose-dissolving IL led to the discovery of NFW in 2010 [24]. During NFW, fibrous natural biopolymers are exposed to minimal volumes of EMIAc to achieve partial dissolution of the fibers [25]. Tuning the quantity of IL, solution temperature, and exposure time provides control over the degree of solvation of the outer regions of the fibers. In general, increasing the temperature of the welding solution or increasing treatment time leads to higher degrees of welding, and ultimately can result in complete fiber dissolution [26]. Addition of anti-solvent to wash the fibrous material of IL initiates a collapse of the dissolved biopolymer onto the remaining undissolved fibers in a reorganized manner. The result of this fiber collapse effectively creates two distinct regions within treated fibers: a preserved native crystalline core and a distinct non-crystalline amorphous shell of regenerated biopolymer [26].

Beyond treatment of single fibers, NFW can be used to effectively "weld" adjacent fibers through careful leverage of partial dissolution as shown by the cartoons in Figure 3 which diagram the steps for the welding of two cotton fibers. Following initial NFW treatment, adjacent fibers (Figure 3A) undergo entanglement of their amorphous shells (Figure 3B).

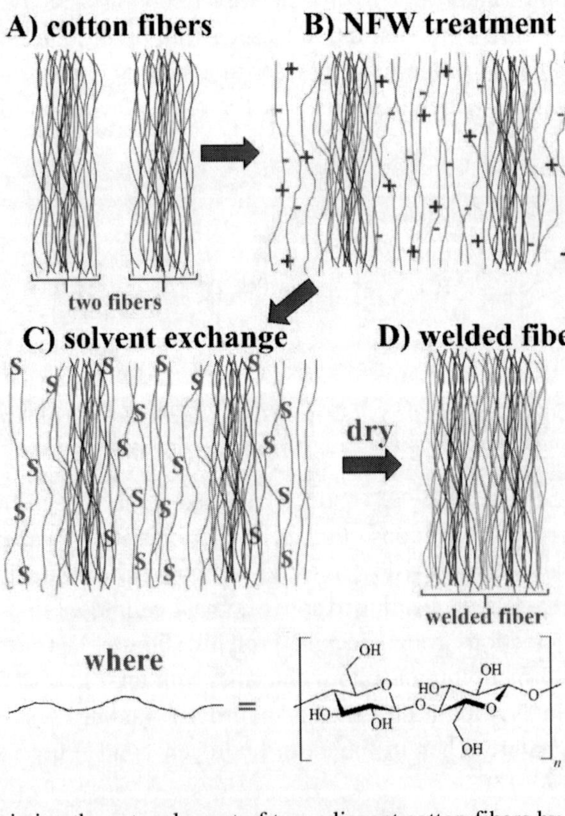

Figure 3. Cartoon depicting the entanglement of two adjacent cotton fibers by partial biopolymer (cellulose) dissolution and reconstitution using the NFW process.

Antisolvent is used to quench the cellulose dissolution process and remove the IL, "welding" the two fibers with the amorphous cellulose (Figure 3C). Once dried, this composite comprises crystalline cores from the original fibers surrounded by a shared amorphous shell (Figure 3D). Expanding this picture to a thread consisting of many fibers, the welding process results in a thread with an increased density as signified by a loss in loose periphery fibers and reduction of up to 50% from the original thread diameter [24, 27]. By design, NFW retains a portion of the crystalline cellulose core to prevent the loss of desirable structural properties, making it stand out among other solvation/regeneration techniques.

New textile applications of the NFW technique can exploit the hybrid nature of the product and its amorphous cellulosic network to produce welded composites and blended materials, as exhibited through scanning electron microscopy (SEM) and confocal fluorescence microscopy (CFM) imaging [24, 27]. Specifically, observations using SEM show the aforementioned reduction in loose periphery fibers of two crossed cotton (Figure 4A and B) or hemp (Figure 4C and D) threads before and after the NFW process, with the post-NFW threads being welded together by their amorphous shells [27]. The welding of these threads is also evident from tactile handling, with unraveling being unworkable and mechanical separation requiring the complete breakage of the composite. Images from CFM of welded fluorescently-labeled silk threads (Figure 4E) provide a view of the cross-sectional inter-fiber weaving accomplished by the NFW technique [27].

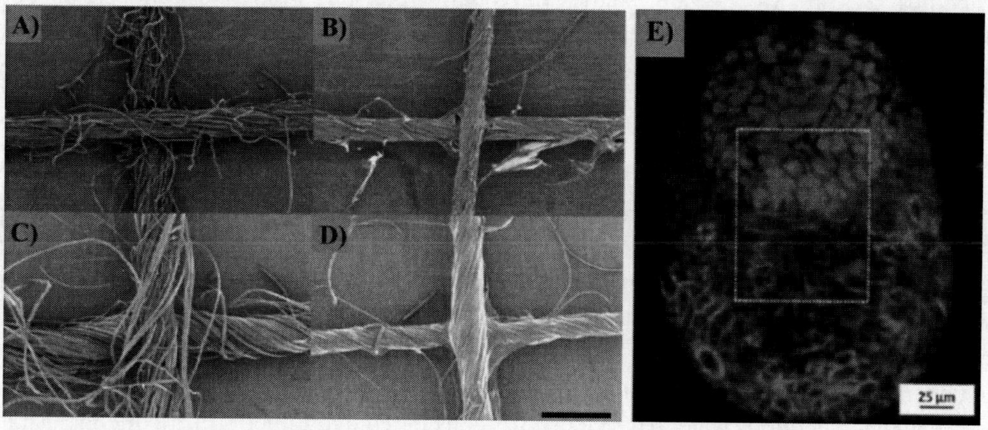

Figure 4. Two types of biopolymer subjected to NFW treatment. Native cotton thread before and after (panels A and B, respectively) welding in EMIAc at 60°C for 5 min. Hemp fibers before and after (panels C and D, respectively) welding in EMIAc at 60°C for 5 min. Scale bar = 500 µm. (E) Confocal microscopy images of adjacent silk yarn threads welded in EMIAc, depicting the welded interface. Reproduced with permission from reference 27.

As the NFW process expands into the industrial sphere [31], the benefits of the technique are clear. Cellulose is the most abundant natural biopolymer on Earth and is renowned for its structural properties, many of which are retained after the partial dissolution pathway afforded by NFW. The technique also allows for regeneration of used cellulosic materials via grafting during the welding process. Versatility in the final product can be greatly expanded by chemical modification of the amorphous shell, which is now much more accessible as shown by the higher surface area of the NFW product. In addition, attempts at up-scaling NFW are primarily limited by the cost of the IL (which can be salvaged, purified of dissolved cellulose, and recycled), and the properties of the products surpass those of many commercial competitors, particularly synthetic textiles. The extent of cellulosic research regarding ILs and NFW is too great to contain within these pages, and thus, in highlighting this emergent technique, this review will focus on advances (or state of the art) in two research areas: fundamental studies that elucidate the stages of the NFW process, and applied studies that demonstrate how NFW can be used to engineer novel, functional biocomposite materials.

Fundamental Studies on Ionic Liquid-Based Biopolymer Processing

At its core, NFW capitalizes on the solvating power of ILs towards biopolymers to provide a controllable route for processing and/or modifying bulk biopolymeric materials. Prior to an increase in commercial accessibility of ILs, alternative methods for dissolving and processing natural biopolymers like silk and cellulose relied on high ionic strength solutions, derivatizing strategies, or a combination of the two.

In order to take full advantage of the benefits associated with using ILs for biopolymer dissolution, it is critical to understand the processes at each stage of the NFW process. Past research has demonstrated the significant impact of welding parameters like treatment time, temperature, choice of IL, and target biopolymer on post-treatment physical and mechanical properties [32-34]. More recent studies have built upon this prior body of work to better understand topics such as how contaminants or additives impact welding efficiency, or how post-treatment solvent rinses impact the dynamic mesoporous shell developed post-welding.

Purity of the IL Welding Solution

At its core, NFW capitalizes on the ability of specific IL anions to disrupt inter-fiber hydrogen bonding to achieve the desired impact on biopolymers [24]. As such, the presence of any contaminant exhibiting hydrogen bonding capabilities, such as water or other polar protic solvents, diminishes the efficiency of welding through competitive bonding along the biopolymer chains [35]. Intrinsically, the vast majority of ILs are hygroscopic to some degree and contain adventitious water. Even those ILs considered hydrophobic due to the inclusion of aliphatic sidechains or highly fluorinated ions possess residential ion domains susceptible to adsorbing ambient moisture.

For comparison, commercial EMIAc typically comprises between 0.3 and 0.5 wt% water simply due to the synthesis process, handling, and storage [36]. Hydration can impact the NFW process, as adventitious water is known to decrease the solubility of cellulose in IL as well as protect the native structural properties of treated biopolymer from decrystallization and reconstitution [2, 36-37]. Furthermore, cellulose retains about 6.5–8.0 wt% water under ambient conditions due to the H-bonding networks. This implies that substrates should be dried prior to NFW, and any recycling of the IL should include reevaluation of residual water [16].

Additional contaminants, either foreign or native degradants of the IL, can also arise from excipients in the welding solution or during the welding process. For example, Durkin et al. reported commercially sourced EMIAc containing anywhere from 0.3 to 3.0 wt % acetic acid that led to preferential leaching of indium from alloyed palladium-indium catalysts fabricated through the NFW process [38].

Conversely, polar aprotic cosolvents (e.g., acetonitrile, N,N-dimethylformamide) have been shown to increase the efficiency of NFW processes when mixed with welding ILs in certain ratios [34, 38-39]. Enhanced biopolymer dissolution by these mixed solvents has been attributed to substantial decreases in viscosity of the welding medium. At lower viscosities, welding solutions can more effectively wet the interstitial spaces between fibers, allowing for more complete interaction between the IL and the biopolymer [34, 40]. In applications which warrant the use of dissolved materials in the welding solution, particularly extraneous biopolymer (which itself imparts an increase in solution viscosity with higher loadings), added cosolvent can be critical to enhancing NFW by lowering solution viscosity [34]. Advanced material processing techniques paired with NFW, such as electrospinning [40], in-line yarn modification [41], and 3-D printing, will benefit greatly by tuning this experimental parameter.

The Kinetics of Biopolymer Decrystallization

Early fiber-welding studies primarily used SEM and CFM to demonstrate that increasing welding time and/or temperature will promote higher degrees of biopolymer dissolution and improve NFW efficacy [24, 32]. As a process that requires mobilization of macroscopic material into a compatible solvent, it made sense that these process variables would positively impact the welded product. To quantify these effects, Cosby et al. used Raman mapping to study the kinetics of biopolymer decrystallization, capturing a clear progression of cellulose I to amorphous cellulose when exposed to systematically varied welding conditions (Figure 5) [26].

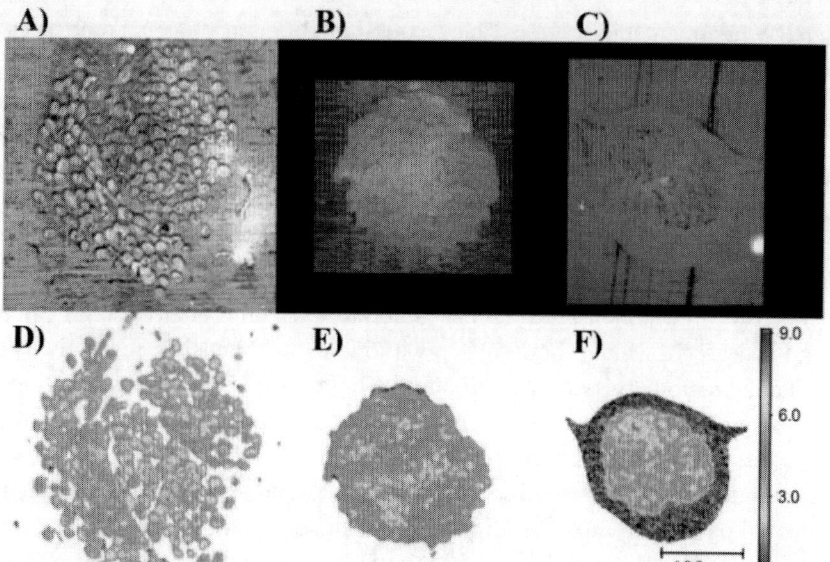

Figure 5. Optical images (A–C) and spatial Raman mapping of cellulose I signal (D–F) of cotton yarn unmodified (A, D) or subjected to NFW treatment for at 45°C for 5 min (B, E) or 70 min (C, F) in EMIAc. Reproduced with permission from reference 26.

2D Raman microscopy, affixed to modes associated with cellulose I crystallinity (the green shaded region of Figure 5D-F), was used to interrogate the collapse (loss of thread diameter) and gradual transition from cellulose I to amorphous cellulose (the blue shaded region of Figure 5D-F) in the peripheral regions of the thread upon NFW treatment. The heterogeneous loss of cellulose crystallinity occurs in two distinct rate-limited regimes, an initial rapid decrystallization at isolated regions along the perimeter of the thread, followed by a slow decrystallization which incrementally approaches the thread core [26].

Keeping all other processing variables consistent (time, weight ratio of IL to cellulose, post-treatment washing), an elevated welding temperature accelerates decrystallization rate, as noticed through abrupt physical changes to the welded substrate such as densification and increased rigidity. While this can readily be explained by thermodynamics (i.e., higher molecular kinetic energy at elevated temperature leading to more rapid disruption of H-bonds), it is likely that this trend can be systematically broken down into two underlying effects relevant to the IL system.

First, mobility and fluidity of the solute and solvent, respectively, both increase as temperature increases and both govern the dynamics of heterogeneous decrystallization of cellulose fibers. Experimentally, the rate of slow heterogeneous decrystallization increases nearly 8-fold from 0.008 min^{-1} at 30°C to 0.063 min^{-1} at 75°C [26], an effect likely attributed to these solvation kinetics [20, 42]. Secondly, elevated temperatures are known to improve ion mobility in ILs, leading to increased disruptions between the anion and the H-bonding network of cellulose. As such, NFW requires careful control of temperature, as modified materials can vary greatly with deviation in temperature over the duration of treatment.

Studying the Effect of Tension on NFW

Following NFW treatment, biopolymer fibers exhibit modified mechanical properties reflecting the conversion of crystalline cellulose I to amorphous cellulose as well as the densification of the fibers. More recent work, however, has focused on the impact of applied tension during the welding process on the physical properties of post-treatment of biopolymers [43-44]. Commercialization of NFW has relied on systems where threads are continuously pulled through sequential treatment environments, conditions in which constant tension is applied to the fibers as welding occurs.

Initial work by Fahey et al. with linen yarns revealed that applying tension along the length of the fiber led to higher measured Young's moduli over untreated material (increasing by factors of 1.4 and 1.7 for welded fiber treated without and with tension, respectively) [44]. The increase in Young's modulus was primarily due to higher loads to break, signifying a stronger material. Subsequent work expanded upon this idea to include additional temperatures, non-IL based treatments, and additional sources of biopolymer [43].

At elevated temperatures, the improvements in Young's modulus for linen fibers was further increased by roughly 30%, leading to stiffer materials following welding which could be favorable in select applications. The act of applying tension to a fiber during welding was confirmed to enhance the efficiency of welding by decreasing the cross-sectional surface area of the fiber and removing empty space between strands, effectively holding fibers in close proximity for the duration of treatment. Under these constrained conditions, mobilized biopolymer from adjacent strands can more easily homogenize, leading to greater inter-fiber hydrogen-bonding after the wash step. Interestingly, slight increases in measured tensile properties were seen in the absence of IL when treating fibers under tension with heated methanol or acetonitrile [43]. With no form of welding occurring, this increase is believed to be due to more efficient removal of organic contaminants or better hydration of the exposed biopolymer surface under taut conditions.

Post-Welding Solvent Rinses

Until recently, the most common way to halt the NFW process has been to quench with a suitable polar exchange solvent (i.e., water or methanol). This anti-solvent served to dilute the IL, stop the welding process and create a sol–gel where the amorphous cellulose regions remain exfoliated until resident solvent is removed by rinsing and drying. This process can be performed using multiple rinses of a single solvent (e.g., water or short-chain alcohols which are favorable for the IL) or iteratively by a sequential solvent exchange [24]. In traditional NFW, this approach typically led to a denser, fiber-welded structure as shown with cotton or hemp in Figure 4. In other NFW applications such as nanoparticle entrapment, solvent exchange may require a polar environment for solvation or less polar solvent to incorporate non-polar molecules within the NFW cellulose matrix. Recent unpublished work in our lab exploring the fate of NFW cellulose to single solvent systems suggests that the exfoliated cellulose does not collapse upon introduction of these non-polar solvents and remains swollen until drying, making the rinsing steps ideal for modification of the cellulose or inclusion of functional materials.

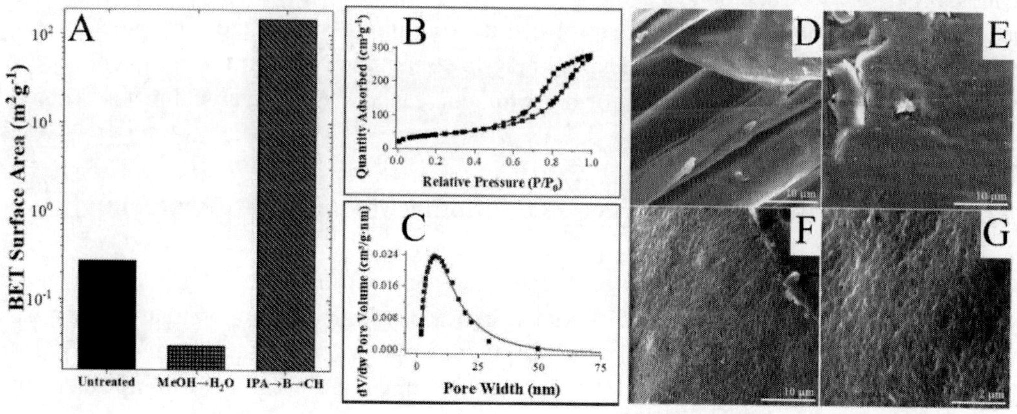

Figure 6. (A) Brunauer-Emmett-Teller (BET) surface areas for untreated cotton thread, welded cotton thread rinsed with polar solvents (MeOH and H_2O), and welded cotton thread rinsed with polar-to-nonpolar (isopropyl alcohol to 2-butanone to cyclohexane) iterative solvents. BET (B) N_2 physisorption isotherm and (C) pore size distribution for the polar-to-nonpolar rinse. SEM imaging of the (D) untreated cotton thread, (E) MeOH-H_2O rinse welded cotton thread, and (F) polar-to-nonpolar rinsed welded cotton thread. Note that (G) is a high-magnification image of (F). Reproduced with permission from reference 45.

Drying of NFW cellulose results in a collapse of the cellulose structure as solvent is removed. However, the extent of this collapse appears dependent upon the identity of the resident solvent, as investigated by Aiello et al. using cotton thread [45]. This recent publication demonstrates that drying from a polar solvent results in a fully collapsed structure with a Brunauer-Emmett-Teller (BET) surface area less than that of the native cotton thread (Figure 6A). However, drying from a non-polar solvent such as cyclohexane results in the formation of a xerogel, with a BET surface area several orders of magnitude greater than that of the native thread. This discovery is further probed in the next section, though it implies that the nascent porosity of the NFW cellulose can be tailored and controlled, perhaps even regenerated post-drying.

Commercialization of the NFW Process

The culmination of fundamental research into the process of NFW expands beyond the fundamental studies discussed in this review. In 2015, Natural Fiber Welding, Inc. was founded with the goal of scaling NFW to commercial levels in an economic and environmentally conscientious manner. While the process of NFW as a whole is considered "green" due to the recyclability of ILs, maintaining strong quality control over the composition of solutions used is crucial for reproducibility in commercial products.

A recent study by Horvath et al. addresses efforts to improve the way in which in-line solvent baths are monitored for contamination with IL [46]. Prior techniques involved a combination of conductivity and refractometry yet were severely limited in both range and selectivity towards EMIAc. Attenuated total internal reflectance–Fourier transform infrared spectroscopy (ATR-FTIR), however, proved to be effective in determining the quality of both welding medium and rinse baths throughout the NFW process. By fixing measurements to the characteristic C=O peak around 1176 cm^{-1}, diagnostic assessment of concentrations of welding medium in rinse solutions could be performed over an impressive range. Continued research towards these as well as related topics will help to streamline NFW in the commercial sector by highlighting the green application of ILs while increasing its financial viability.

Applying NFW to Engineer Novel, Functional Biocomposite Materials

The abundance of natural biopolymeric textiles imparts intrinsic advantages over synthetic materials, such as low cost, minimal toxicity, and accessibility to bioconjugation pathways. These traits are particularly attractive when applying biopolymers as supports and scaffolds for mixed organic-inorganic materials [47]. Indeed, nanomaterial-biopolymer composites show promise in the areas of sensing, filtration and water purification, UV protection, microbe inhibition, and catalysis [48]. However, the extent of biopolymer modification is limited to the exposed surfaces of textile threads (e.g., *in situ* synthesis [48], dip-coating [49], or printing [50]) and those locations within threads which can be reached by capillary action of the nanomaterial or its precursors [51]. The resulting hybrid materials retain the physical properties of the biopolymer, but the surface-bound nanomaterials are often at low loadings and are easy to remove by rinsing. The NFW process expands upon this scaffold application by exfoliating the biopolymer and exposing a higher surface area ripe for modification.

Preparing Cotton-Based Activated Carbon Electrodes through NFW

Traditionally, very few biopolymer materials have found use in electrochemical applications without prior chemical or physical modification. Most natural materials, while chemically stable, exhibit high electrical resistivities, barring their use as electrodes or circuit elements [52-53]. However, this is not to say that there is no interest in developing fabric-based electrodes, especially those consisting of green all-carbon materials. The prospect of wearable biosensors or electronics for real-time medical and decontaminative applications has drawn attention to modified fabrics or hybrid carbon fiber/biopolymer composites. Using NFW techniques, Jost et al. prepared electroactive materials suitable for applications as high energy capacitors. Micron-sized activated carbon was dispersed in welding solutions containing EMIAc, acetonitrile, and added cellulose binder, and then fiber-welded into bulk yarns (including cotton, linen, bamboo, and viscose) via NFW (Figure 7) [54].

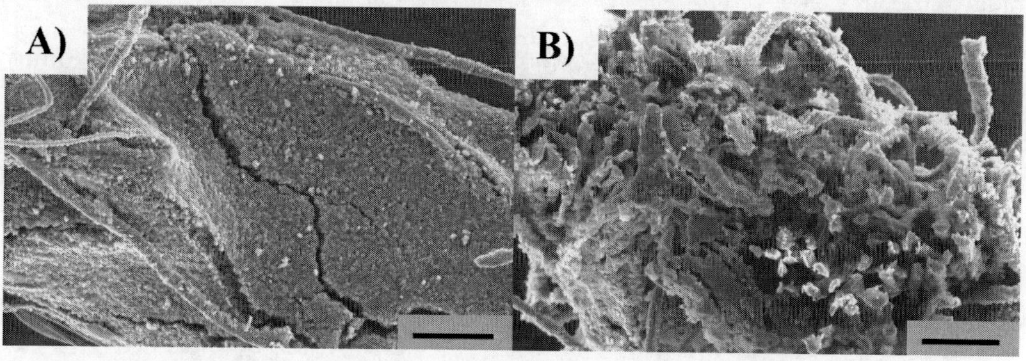

Figure 7. SEM micrographs of the (A) surface and (B) cross section of cotton yarn welded in the presence of 0.5 wt % binder composed of equal parts microcrystalline cellulose and milled cotton thread with added carbon black and activated carbon YP50. Scale bar = 100 μm.

The welding parameters were optimized to make functional energy storage yarns in bulk that could be knitted into flexible yarn supercapacitors which exhibited capacitances of up to 8 mF cm^{-2} with no additional electrode materials added, and up to 37 mF cm^{-2} when paired with a stainless-steel thread for enhanced conductivity and current collection. Furthermore, the welding apparatus used for fabric modification was scalable, capable of rapidly producing large quantities (>70 m) of material, limited only by the quantity of yarn and welding solution. The fabricated yarn supercapacitors demonstrated appreciable capacitances following thousands of cycles over three days, albeit with some gradual loss of performance, demonstrating strong potential for NFW to be applied in the field of wearable electronics.

Incorporating Metal–Organic Frameworks through NFW

Metal–organic frameworks, or MOFs, are compounds comprising a central metal ion and periphery organic linkers. These compounds are typically porous and are applied in the fields of molecular separation and storage, where their void spaces can be put to use, or in catalysis if the metal center is exploitable [55]. Storage of MOFs within a heterogeneous composite can help stabilize the structures, hasten their retrieval after application, and improve and expand upon applicability, particularly when the heterogeneous composite comprises natural fibrous materials used in textiles [56-57].

Standard procedures for the incorporation of MOFs within cloth threads usually requires *in situ* synthesis (particularly for MOFs which do not readily disperse in aqueous media), though a recent advancement by Bunge et al. demonstrates the use of NFW to add the water insoluble MOF UiO-66-NH_2 to pieces of cloth [58]. To accomplish this, the UiO-66-NH_2 is suspended within EMIAc IL welding solution, and then applied to pieces of fabric by either direct contact or using an airbrush. Welding at 80°C for variable lengths of time resulted in composites with encapsulated MOFs (Figure 8). X-ray diffraction data indicate little to no degradation of the UiO-66-NH_2 MOF under prolonged welding conditions in EMIAc, and the authors suggest that this may translate favorably to other MOF systems. The technique was further investigated for the modification of single cotton threads using pre-dissolved cellulose nanocrystals and a variety of EMIAc solvent ratios (Figure 8F), achieving BET surface areas of up to 100 $m^2\ g^{-1}$.

Figure 8. (A, B, and C) SEM images depicting unmodified cotton thread, a side view of cotton thread welded at 70°C in 30/70 volume mix of EMIAc and dimethyl sulfoxide comprising 10 wt% UiO-66-NH_2 MOF and 1.2 wt% cellulose nanocrystals, and a cross-section view of the same thread, respectively. (D, E, and F) Photographs of a 15.5 inch × 15.5 inch piece of unmodified cotton, the same fabric after welding in MOF EMIAc solution, and a spool of MOF-modified cotton thread, respectively. Reproduced with permission from reference 58.

Fabricating Free-Standing Catalytic Structures through NFW

Cellulose is a robust platform for nanomaterials, and the amorphous region of NFW composites offers an interesting opportunity for the stabilization of catalysts. For example, Durkin et al. demonstrated the capability of NFW supports for catalysts by first synthesizing monometallic Pd and bimetallic Pd-Cu catalysts on linen powders *in situ* using an incipient wetness technique and H_2 gas reduction [41]. These powders were then welded onto linen-scaffolds previously prepared via NFW, resulting in a composite lignocellulose free-standing structure containing fiber-welded Pd-based catalysts (Figure 9). The heterogeneous Pd-based catalyst was applied for nitrate reduction and its activity was unaffected by the NFW process, demonstrating robust stability on the macroscopic support as well as simple regeneration via thermal treatment.

A similar procedure was used by Durkin et al. to synthesize more robust and air-stable bimetallic Pd-In nanoparticle catalysts, where NFW yarn-coating procedures were applied to show scalability and control of catalysts deposition [47]. These studies demonstrated the versatility of NFW for applications in functional biocomposites that require uniform deposition of nanoparticles throughout a biopolymer matrix, and they paved the way for future studies where NFW catalysts can be considered an applied to both aqueous and gas-phase reactions.

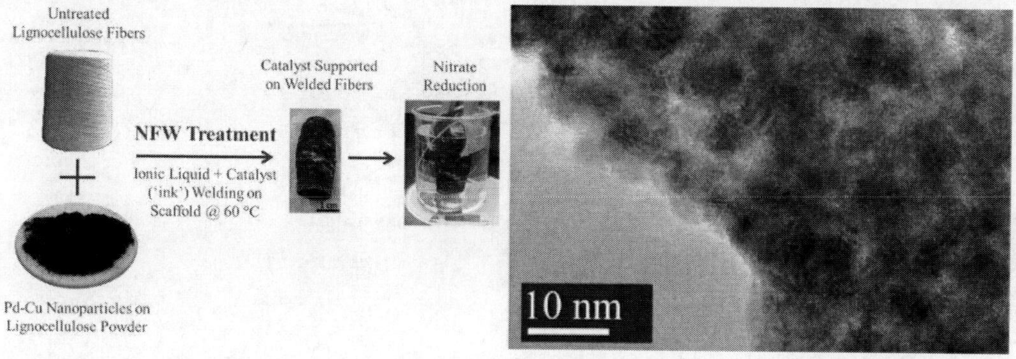

Figure 9. Schematic for the up-processing of linen thread to incorporate Pd-Cu bimetallic nanoparticles via an EMIAc welding procedure. Here, a pre-treatment of lignocellulose powder is performed to append Pd-Cu nanoparticles, followed by dissolution in EMIAc and NFW treatment of commercial linen threads. Application of the nanoparticles is demonstrated by nitrate reduction chemistry. Reproduced with permission from reference 41.

Tuning Textile Surface Properties through NFW

Silanization is a straightforward surface modification process for applying advantageous chemistries to cellulose fibers and can be summarized as: hydrolyzing trialkoxysilane at low pH, followed by condensation of the silane precursor under annealing temperatures by targeting cellulosic -OH groups (Figure 10). Variations of this process include solventless silanization via aerosol[59], aqueous "never-dry" silanization by removing the curing step [60], and silanization of modified cellulose, such as acetylated cellulose and nanocellulose [61-63].

The silanization process was first applied to NFW materials by Durkin et al. as a way to increase the thermal stability of lignocellulose [64]. Amino-, propyl-, or methyl-functionalized trimethoxysilanes were hydrolyzed and annealed in the presence of cellulose nanofibers, creating a silanol coating with a surface coverage dependent upon the wt% of silane to cellulose in solution [62]. Next, the modified cellulose nanofibers were dissolved in EMIAc welding solution and applied to the surface of linen yarns using an in-house yarn coating machine, where lignocellulose yarn were pulled through the welding solution containing the functionalized binders and onto a spool, where a heat treatment initiated the NFW process. The resulting products were rinsed and characterized, and found to possess improved thermal properties (compared to the native material), particularly regarding an increase in char yield when heating to 800°C under N_2 atmosphere (~8.5 wt% char for the control versus 56.7, 13.1, and 51.3 wt%

for methyl-, propyl-, and amino-functionalized NFW product, respectively, when using 10 wt% silane to cellulose nanofibers). Coupling the silanol coating and the NFW procedure results in inhibited low-temperature cellulose decomposition, a reduction in cellulose mass loss rates upon thermal treatment by ~25%, and an increase in thermal stability of the composite by ~17°C, all properties which are favorable in the creation of cellulose-based insulators.

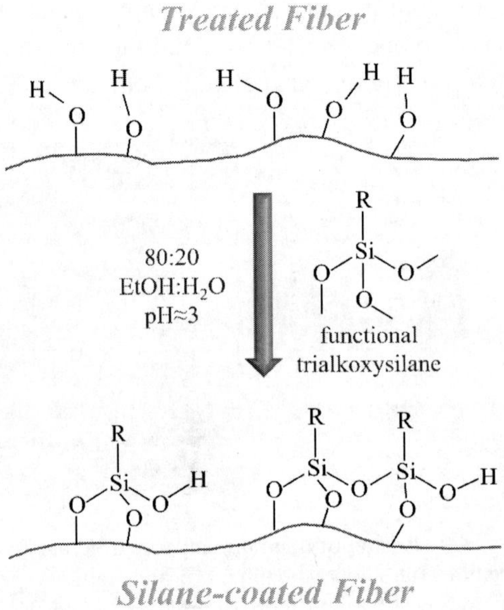

Figure 10. Illustration of the silanization process when using a functionalized trialkoxysilane. Silanization occurs at the cellulose -OH sites.

Future directions for silanol-coated NFW materials could exploit the higher (and more accessible) surface areas of a fiber-welded thread's amorphous shell to perform high-wt% silanization. This post-weld treatment gains the benefit of providing reinforcing cross-links via the silanol layer, an attribute which is absent when silanizing the pre-weld cellulose nanofibers, yet may increase the physical robustness of the product. Similarly, silanization using thiol or vinyl constituents could lead to an extensive and readily accessible click-chemistry network. The development of a "never-dry" post-weld silanization technique is ideal for these applications, as it promotes retention of the expanded hydrogen bonding network via the sol–gel structure.

Fiber-Welded Antimicrobial Cellulose Composites Containing Ag Nanoparticles

The application for antimicrobial or self-sterilizing cloth extends well into the realms of textiles, food packaging, and construction. Shen and coworkers have utilized the NFW process to produce antibacterial lignocellulose containing Ag nanoparticles without degrading the NFW biopolymer support [65]. The process for Ag impregnation is similar to incipient wetness, and a gas-phase reduction technique is used to achieve small, sub-10 nm particles. Impressively, the composites inhibit both the growth of the test bacterium and the formation of biofilm, a useful trait for the construction and longevity of buildings and marine platforms. The process also boasts impressive tunability (i.e., antimicrobial activity versus enhanced product physical properties) by controlling the metal identity and loading. NFW is an attractive engineering method for anti-microbial textiles for several reasons: it uses inexpensive, green support materials which in some cases can have natural anti-microbial properties (i.e., chitin or chitosan), and under certain circumstances, IL contaminants in welding solutions have been shown to leech certain metals from NFW hybrid materials, including the dealloying of In from Pd-In bimetallic nanoparticles [38]. This notable property might be exploited to further increase antimicrobial characteristics or catalytic effectiveness in a fiber-welded composite containing anti-microbial nanoparticles. Further, controlled leeching, can give rise to new nanostructures by way of templated and controlled porosity, a method toward materials which resist microbial adhesion.

Printing Functional NFW Composites

A novel approach toward re-defining custom cellulosic materials lies in the utilization of laser printers, materials printers, and a variety of additive manufacturing technologies. While printing of cellulose solutions is not necessarily novel, the application of NFW adds a new dimension to this phase space. Brown et al. demonstrated a spatially-specific, laser-induced NFW procedure wherein a CO_2 laser was used as a heat source to raster a pattern on EMIAc-coated cellulose paper [66]. The extent of the NFW process was controlled by IL concentration, laser power, raster time, and number of laser passes, and the method alludes to the use of lasers as heat sources for point-to-point NFW of cotton fabrics and other cellulosic materials.

A separate approach was demonstrated, by two additional reports from the Trulove group, wherein NFW was controlled using inkjet printing on cellulose paper as an IL deposition technique, followed by heating to induce the NFW process and rinsing to obtain an all-cellulose product [67-68]. More recently, Durkin et al. demonstrated how NFW could be used to print high resolution patterns with EMIAc-based inks containing Ag nanoparticle-laden lignocellulose fibers as binders [69]. These novel prints contained functional binders that were not only anti-microbial and anti-biofilm, but could also fluoresce (under UV irradiation) and demonstrated a surface-enhanced Raman effect due to the plasmon effect of the Ag nanoparticles (Figure 11).

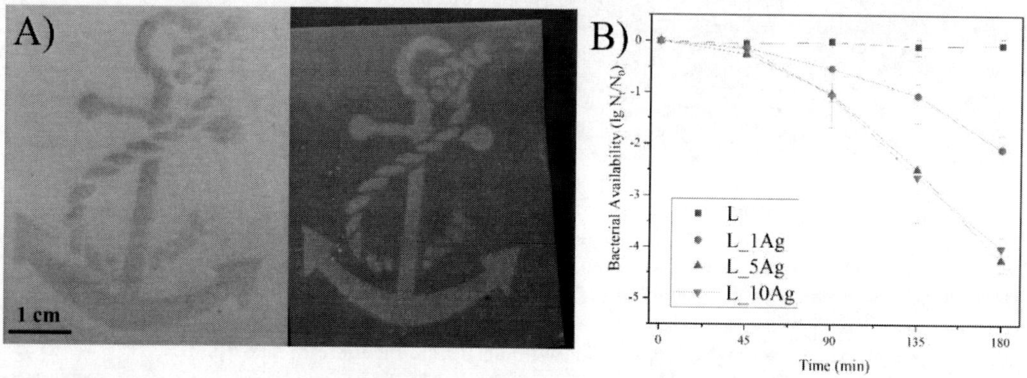

Figure 11. (A) Photograph comparing Ag nanoparticle-laden cellulose binder printed in controllable patterns on cellulose paper under either ambient light (left) or UV irradiation (right). (B) Plot showing the influence of exposure time of modified lignocellulose fibers on bacterial viability at different mass loadings of silver. Reproduced with permission from reference 65.

These studies report the formation of surface NFW structures with tight resolution control, indicating low IL mobility and overall impressive performance by inject and materials printers, and demonstrate the great potential for NFW to be applied to printing functional biocomposites. While additive manufactured technologies have not been demonstrated to any great degree yet, this application is likely on the horizon.

Fiber-Welded Polyionic Biocomposites

Classic NFW requires the complete removal of IL in order to stop the welding process, reconstitute the mobilized biopolymer fibers, and form the welded interface. A recent study by Durkin et al. eliminated this step by synthesizing two novel polymerizable ILs (Poly-ILs) that could be used as NFW solvents and then polymerized in-situ to create polyionic biocomposites [70]. First, the Poly-ILs (1-methyl-3-vinylimidazolium methylphosphonate and 1-ethyl-3-vinylimidazolium ethylphosphonate) were tested for their ability to dissolve cellulose, to evaluate their potential as an NFW solvent. Then, an NFW procedure was developed to maximally distribute the Poly-IL throughout the fibers before they were used to weld and polymerize *in situ*. The resultant polymer-cellulose composite (Figure 12) had polymerized Poly-IL uniformly distributed throughout the yarn bundles (green map in Figure 12A), was conductive enough to be imaged directly with SEM, and exhibited a massive increase in water uptake capacity (about 5 times increase, from ~8 wt % for neat cotton yarn to ~40 wt % for the polymer-cellulose composite).

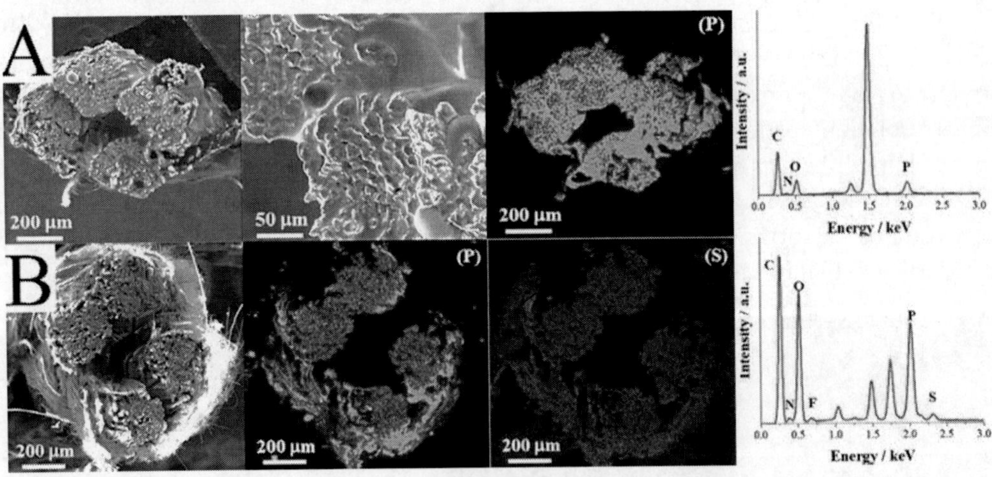

Figure 12. Scanning electron microscopy (SEM) and energy dispersive X-ray spectroscopy (EDS) for 4-ply cotton yarn welded with 1-methyl-3-vinylimidazolium methylphosphonate, where (A) depicts the welded biocomposite before anion exchange, including a higher magnification SEM image, an EDS map for elemental phosphorus (green) and resulting EDS spectrum; (B) is of the same material after exchanging ~5 mol% of the methylphosphonate for TFSI, and includes an EDS map for elemental sulfur from this exchanged anion. Reproduced with permission from reference 70.

After evaluating their water sorption properties, the composites were subjected to anion exchange (methylphosphonate for the more hydrophobic bis(trifluoromethane)sulfonimide (TFSI) anion). After a single pass anion exchange, the composites water sorption capacity decreased by ~15 wt% from the neat Poly-IL sample. The inclusion of TFSI was confirmed by energy dispersive X-ray spectroscopy (EDS; Figure 12), where regions of sulfur (purple map in Figure 12B) are observed. This exciting new study highlights how NFW can be tailored to make sustainable materials for new applications such as anion exchange and power and energy without the need to recycle the IL welding solvent.

NFW has developed considerably over the last ten years, and many studies have been reported that showcase the great potential of this engineering method. The future for NFW is bright, however there is much to consider as we chart the way forward. First, what methods exist for predicting the solvation of biopolymers in ILs when designing new welding solvents? Certainly, EMIAc is not the ideal IL for every NFW situation (i.e., recall the application of polymerizable ionic liquids). Perhaps other experimental goals will drive selection of a different welding solvent, such as thermal stability, nanoparticle dispersibility, hydrophobicity, etc.

A more targeted approach toward designing NFW solvents is desired, rather than simple trial and error. Additionally, how can NFW materials be applied most effectively, with special consideration to high-impact research or industrial applications? NFW has already made an impact on the textile industry through Natural Fiber Welding, Inc., a company dedicated to the up-processing of used cellulosic materials for improved fabric design. This type of value-added conversion of waste material is one impactful use of NFW. Our present research suggests that the ability to embed functional materials within the mesoporous cellulose shell of welded biopolymers can bring about the advent of a new generation of highly ordered textile materials.

Predicting Effective NFW Solvents Using Kamlet Taft Parameters

Future work towards developing NFW solvents (e.g., expanding the Poly-IL work described above) may benefit from a more targeted approach in predicting cellulose solvation. While the ability to completely dissolve cellulose is NOT a requirement for an NFW welding solvent, if a solvent can dissolve biopolymer then it will weld. Rather than determining this experimentally, trends associated with cellulose solubility in both high ionic strength solutions and ILs can be predicted using polarity parameters defined by Kamlet and Taft: hydrogen bond acidity (α) [71], hydrogen bond basicity (β) [72], and polarizability (π^*) [73]. These parameters are currently established for a range of solvent systems [74-77], and optimal cellulose dissolution has been shown to occur in solvents with an α value between 0.25 and 0.55 [78], a β value exceeding 0.8 [36, 78], and a π^* value between 0.85 and 1.15 [79].

The Sixta group takes these classifications a step further by analyzing correlations between β and "net basicity" (β-α), a value which accounts for the acidity of the cation in ionic mixtures [78]. The result is a solvation window where the (β-α) value should lie between 0.35 and 0.90, reflecting the need for balance between acidity and basicity. For ILs, the cation and anion identities are known to impart primary control over α and β, respectively, though changing either component will affect all solvent parameters to some degree. For example, the α, β, and π^* values for 1-butyl-3-methylimidazolium bistriflimide are 0.62, 0.24, and 0.98, respectively, while those for 1-butyl-1-methylpyrrolidinium bistriflimide are 0.43, 0.25, and 0.95, respectively [77]. Alkylation of the cation is also known to decrease π^* and α while increasing β [77, 80].

The Welton group has provided an excellent comparison of KT parameters for common ILs and conventional solvents [75], with another comparison containing temperature-dependent values for pyridinium ILs provided by Lee et al. [77], and yet another containing parameter values for 102 ILs provided by Spange et al. [80]. Unpublished work collected in our lab has shown direct correlations between KT parameters of welding solutions and their NFW efficacy. We provide a small compilation of a few common solvents, several solutions known to dissolve cellulose, and several key ILs in Table 1. Conscientious development of welding solutions based on predicted KT parameters may greatly benefit the advancement of NFW as new ILs are investigated [81].

Table 1. Kamlet-Taft parameters for several common imidazolium ILs, traditional solvents, and various solutions noted in this chapter for cellulose dissolution. For consistency, the nomenclature for imidazolium ILs is derived from the abbreviation for EMIAc (e.g., 1-butyl-3-methylimidazolium acetate is abbreviated BMIAc and 1-butyl-3-methylimidazolium chloride is abbreviated BMICl)

Solvent	α	β	π*	Ref.
EMIAc	0.40	0.95	1.09	[36]
EMIAc + 10 wt% H_2O	0.56	0.92	1.03	[36]
BMIAc	0.59	1.17	0.9	[36]
BMIAc + 10 wt% H_2O	0.54	0.99	0.97	[36]
BMICl	0.49	0.83	1.03	[82]
H_2O	1.17	0.15	1.09	[83]
4 mol% urea in H_2O	1.05	0.13	1.31	[84]
Methanol	0.93	0.62	0.60	[85]
2-Propanol	0.76	0.95	0.50	[85]
Acetonitrile	0.19	0.31	0.75	[85]
Acetone	0.08	0.48	0.71	[77, 86]
2-Butanone	0.06	0.48	0.67	[86]
N,N-dimethylformamide	0.00	0.69	0.88	[87]
N,N-dimethylacetamide	0.00	0.76	0.88	[86]
Cyclohexane	0.00	0.00	0.00	[86]
8 wt% LiCl in DMAc	0.65	1.06	0.87	[36]
4 mol% NaOH in H_2O	0.24	0.20	1.11	[84]
4 mol% NaOH + 4 mol% urea in H_2O	0.24	0.20	1.36	[84]
8 mol% NaOH in H_2O	0.04	0.39	1.22	[84]
4 mol% Triton B in H_2O	0.55	0.38	1.20	[84]
4 mol% Triton B + 4 mol% urea in H_2O	0.55	0.38	1.39	[84]
6.7 mol% Triton B in H_2O	0.57	0.47	1.24	[84]

Designing Functional Mesoporous Biocomposite Materials through NFW

The fiber-welded mesoporous biopolymer canvas that was first introduced by Aiello, et al. is an organic alternative to common inorganic materials such as mesoporous silica and aluminosilicates [88]. As previously stated, this scaffold can be coated with a variety of polymers (e.g., silanes, polymers) to impart changes on the overall composite, such as chemical functionality for surface tunability and modification, conductivity, water permeability and sorption, rigidity and enhanced physical properties, and many other properties for a variety of applications.

Adsorption of nanomaterials and MOFs add an entirely new dimension to the material in the realm of heterogeneous catalysis, separations, and magnetic susceptibility. These applications rely on the tunability of the mesoporous structure (size, shape, total surface area, and accessibility and depth of pores), with current research directions focusing in on BET surface area control by conscientious post-NFW solvent washing, pillaring and templating using embedded nanomaterials, reformed drying techniques, and general control over all of the variables surrounding the NFW process and cleanup. It is both enticing and poetic to consider that natural biopolymers are some of the first utilized materials by humans, and that the unlocked properties upon NFW treatment may allow them to rise in prominence and application over synthetic polymers in a time when petroleum and other limited resources become scarce.

Conclusion

Fiber welding creates a new type of composite unlike anything previously known, and further understanding of the fundamental processes of NFW can help fine-tune the produced structures and composites. While many studies to date have looked extensively at biopolymer dissolution, the implementation of welded composites into practical applications requires equal consideration towards post-welding solvent treatment, especially when considering post-treatment modification with functional materials (i.e., nanoparticles, molecular agents).

NFW materials dried from non-polar solvents have been shown to express high mesoporous surface areas, while a more collapsed shell is produced upon drying from polar conditions. A fundamental understanding of this collapse is necessary for long-term storage of NFW composites, particularly those which are hydroscopic with intended use in aqueous or environmental applications. Research in this area has focused on the retention of post-NFW mesoporosity, with little effort so far being directed towards regeneration of the desired high surface area state. It is likely that the collapse is not permanent and can be reversed, since the re-structured hydrogen bond network is unlikely to be as complete and extensive as it is for crystalline cellulose I warranting future investigation. This discovery may lead to a similar system to impart controllable and recyclable mesopores in other biopolymer systems.

In conclusion, NFW offers a unique approach to biomaterial processing not afforded by conventional dissolution techniques. The incomplete solvation of cellulosic materials allows for an unprecedented degree of biopolymer modification and functionalization, potentially increasing and expanding the impressive physical characteristics of the native material. Processing methods for NFW are always evolving, and the extent to which the resulting composites can be tailored is constantly expanding. There are a number of ways that NFW can be exploited to target specific applications, leaving much to yet be discovered. When such diverse chemistries, intriguing solvent interactions, and varied functionalities come together, the creativity of each researcher truly shines.

References

[1] DeLong, H. C., Trulove, P. C., Haverhals, L. M., Reichert, W. M., *US Patent* 8,202,379, 2012.

[2] Lindman, B., Karlström, G., Stigsson, L. On the mechanism of dissolution of cellulose. *J. Mol. Liq.* 2010, 156, 76-81.

[3] Saito, T., Kimura, S., Nishiyama, Y., Isogai, A. Cellulose nanofibers prepared by TEMPO-mediated oxidation of native cellulose. *Biomacromolecules.* 2007, 8, 2845-2491.

[4] Burchard, W. Solubility and Solution Structure of Cellulose Derivatives. *Cellulose.* 2003, 10, 213-225.
[5] Gindl, W., Keckes, J. All-cellulose nanocomposite. *Polymer.* 2005, 46, 10221-10225.
[6] Pillai, C. K. S., Paul, W., Sharma, C. P. Chitin and chitosan polymers: chemistry, solubility and fiber formation. *Prog. Polym. Sci.* 2009, 34, 641-678.
[7] Soykeabkaew, N., Sian, C., Gea, S., Nishino, T., Peijs, T. All-cellulose nanocomposites by surface selective dissolution of bacterial cellulose. *Cellulose.* 2009, 16, 435-444.
[8] McCormick, C. L., Callais, P. A., Hutchinson, B. H. Solution studies of cellulose in lithium chloride and N,N-dimethylacetamide. *Macromolecules.* 1985, 18, 2394-2401.
[9] Klemm, D., Heublein, B., Fink, H. P., Bohn, A. Cellulose: fascinating biopolymer and sustainable raw material. *Angew. Chem. Int. Ed.* 2005, 44, 3358 – 3393.
[10] Binder, J. B., Raines, R. T. Simple Chemical Transformation of Lignocellulosic Biomass into Furans for Fuels and Chemicals. *J. Am. Chem. Soc.* 2009, 131, 1979-1985.
[11] Cai, J., Zhang, L., Zhou, J., Li, H., Chen, H., Jin, H. Novel Fibers Prepared from Cellulose in NaOH/Urea Aqueous Solution. *Macromol. Rapid Commun.* 2004, 25, 1558-1562.
[12] Cai, J., Zhang, L. Rapid Dissolution of Cellulose in LiOH/Urea and NaOH/Urea Aqueous Solutions. *Macromolecular Bioscience.* 2005, 5, 539-548.
[13] Swatlotski, R. P., Spear, S. K., Holbrey, J. D., Rogers, R. D. Dissolution of cellulose [correction of cellose] with ionic liquids. *J. Am. Chem. Soc.* 2002, 124, 4974-4975.
[14] Kirchner, B. 2009. *Ionic Liquids*, 1 ed., Springer-Verlag Berlin Heidelberg.
[15] Tan, X., Li, X., Chen, L., Xie, F. Solubility of starch and microcrystalline cellulose in 1-ethyl-3-methylimidazolium acetate ionic liquid and solution rheological properties. *Phys. Chem. Chem. Phys.* 2016, 18, 27584-27593.
[16] Mantz, R. A., Fox, D. M., Green, J. M., Fylstra, P. A., De Long, H. C., Trulove, P. C. Dissolution of biopolymers using ionic liquids. *Z. Naturforsch., A: Phys. Sci.* 2007, 62, 275-280.
[17] Phillips, D. M., Drummy, L. F., Conrady, D. G., Fox, D. M., Naik, R. R., Stone, M. O., Trulove, P. C., De Long, H. C., Mantz, R. A. Dissolution and regeneration of bombyx mori silk fibroin using ionic liquids. *J. Am. Chem. Soc.* 2004, 126, 14350-14351.
[18] Fahey, P. J., Durkin, D. P., Fox, E. T., Gobet, M., Greenbaum, S., De Long, H. C., Trulove, P. C. Cellulose, cellobiose, and glucose cause similar decreases to molar conductivity and drastically different increases to dynamic viscosity of 1-ethyl-3-methylimidazoilum acetate based solvents. *Electrochem. Soc. Trans.* 2018, 86, 257-268.
[19] Andanson, J.-M., Bordes, E., Devemy, J., Leroux, F., Padua, A. A. H., Gomes, M. F. C. Understanding the role of co-solvents in the dissolution of cellulose in ionic liquids. *Green Chem.* 2014, 16, 2528-2538.
[20] Fahey, P. J., Fox, E. T., Scully, M. C., De Long, H. C., Trulove, P. C., Durkin, D. P. The apparent superionicity of ionic liquid solutions containing cellulose. *J. Electrochem. Soc.* 2019, 166, H140-H145.
[21] Viswanathan, G., Murugesan, S., Pushparaj, V., Nalamasu, O., Ajayan, P. M., Linhardt, R. J. Preparation of biopolymer fibers by electrospinning from room temperature ionic liquids. *Biomacromolecules.* 2006, 7, 415-418.
[22] O'Sullivan, A. C. Cellulose: the structure slowly unravels. *Cellulose.* 1997, 4, 173-207.
[23] Huber, T., Müssig, J., Curnow, O., Pang, S., Bickerton, S., Staiger, M. P. A critical review of all-cellulose composites. *J. Mater. Sci.* 2012, 47, 1171-1186.
[24] Haverhals, L. M., Reichert, W. M., De Long, H. C., Trulove, P. C. Natural fiber welding. *Macromol. Mater. Eng.* 2010, 295, 425-430.
[25] Mahmood, H., Moniruzzaman, M., Yusup, S., Welton, T. Ionic liquids assisted processing of renewable resources for the fabrication of biodegradable composite materials. *Green Chem.* 2017, 19, 2051-2075.
[26] Cosby, T., Aiello, A., Durkin, D. P., Trulove, P. C. Kinetics of ionic liquid-facilitated cellulose decrystallization by Raman spectral mapping. *Cellulose.* 2021, 28, 1321-1330.
[27] Haverhals, L. M., Nevin, L. M., Foley, M. P., Brown, E. K., De Long, H. C., Trulove, P. C. Fluorescence monitoring of ionic liquid-facilitated biopolymer mobilization and reorganization. *Chem. Commun.* 2012, 48, 6417-6419.
[28] Turner, M. B., Spear, S. K., Holbrey, J. D., Daly, D. T., Rogers, R. D. Ionic liquid-reconstituted cellulose composites as solid support matrices for biocatalyst immobilization. *Biomacromolecules.* 2005, 6, 2497-2502.

[29] Sun, N., Rahman, M., Qin, Y., Maxim, M. L., Rodríguez, H., Rogers, R. D. Complete dissolution and partial delignification of wood in the ionic liquid 1-ethyl-3-methylimidazolium acetate. *Green Chem.* 2009, 11, 646-655.

[30] Knill, C. J., Kennedy, J. F. Degradation of cellulose under alkaline conditions. *Carbohydr. Polym.* 2003, 51, 281-300.

[31] Haverhals, L. M., Durkin, D. P., Trulove, P. C., in *Commercial Applications of Ionic Liquids* (Ed.: M. B. Shiflett), Springer International Publishing, Cham, 2020, pp. 211-226.

[32] Haverhals, L. M., Sulpizio, H. M., Fayos, Z. A., Trulove, M. A., Reichert, W. M., Foley, M. P., De Long, H. C., Trulove, P. C. Process variables that control natural fiber welding: time, temperature, and amount of ionic liquid. *Cellulose.* 2012, 19, 13-22.

[33] Haverhals, L. M., Foley, M. P., Brown, E. K., Fox, D. M., De Long, H. C., Trulove, P. C., in *Ionic Liquids: Science and Applications, Vol. 1117*, American Chemical Society, 2012, pp. 145-166.

[34] Brown, E. K., Durkin, D. P., Fox, E. T., Haverhals, L. M., Sweely, K. D., De Long, H. C., Trulove, P. C. Natural fiber welded composite yarns. *Electrochem. Soc. Trans.* 2014, 64, 515-521.

[35] Haverhals, L. M., Foley, M. P., Brown, E. K., Nevin, L. M., Fox, D. M., De Long, H. C., Trulove, P. C. Ionic liquid-based solvents for natural fiber welding. *Electrochem. Soc. Trans.* 2012, 50, 603-613.

[36] Doherty, T. V., Mora-Pale, M., Foley, S. E., Linhardt, R. J., Dordick, J. S. Ionic liquid solvent properties as predictors of lignocellulose pretreatment efficacy. *Green Chem.* 2010, 12, 1967-1975.

[37] Mazza, M., Catana, D. A., Vaca-Garcia, C., Cecutti, C. Influence of water on the dissolution of cellulose in selected ionic liquids. *Cellulose.* 2009, 16, 207-215.

[38] Durkin, D. P., Ye, T., Chung, R. T., De Long, H. C., Shuai, D., Trulove, P. C. Preferential leaching of indium metal during room temperature ionic liquid processing of Pd–In nanoparticle-biopolymer composites. *Mater. Chem. Phys.* 2020, 249, 123179.

[39] Niu, X., Huan, S., Li, H., Pan, H., Rojas, O. J. Transparent films by ionic liquid welding of cellulose nanofibers and polylactide: Enhanced biodegradability in marine environments. *J. Hazard. Mater.* 2021, 402, 124073.

[40] Brown, E. K., Haverhals, L. M., Foley, M. P., De Long, H., Trulove, P. C. Electrospinning of Biopolymers from Ionic Liquid-Co-Solvent Systems. *Electrochem. Soc. Trans.* 2013, 50, 595.

[41] Durkin, D. P., Ye, T., Larson, E. G., Haverhals, L. M., Livi, K. J. T., De Long, H. C., Trulove, P. C., Fairbrother, D. H., Shuai, D. Lignocellulose fiber- and welded fiber- supports for palladium-based catalytic hydrogenation: a natural fiber welding application for water treatment. *ACS Sustain. Chem. Eng.* 2016, 4, 5511-5522.

[42] Endres, F., Zein El Abedin, S. Air and water stable ionic liquids in physical chemistry. *Phys. Chem. Chem. Phys.* 2006, 8, 2101-2116.

[43] Burn, A., Fahey, P. J., Durkin, D. P., De Long, H. C., Trulove, P. C. Evaluating the effect of applied tension during natural fiber welding of lignocellulose yarns. *Electrochem. Soc. Trans.* 2018, 86, 269-277.

[44] Fahey, P. J., Fox, E. T., Durkin, D. P., De Long, H. C., Trulove, P. C. Combining mechanical tension with natural fiber welding treatments of cellulose yarns. *Electrochem. Soc. Trans.* 2016, 75, 669-676.

[45] Aiello, A., Cosby, T., McFarland, J., Durkin, D. P., Trulove, P. C. Mesoporous xerogel cellulose composites from biorenewable natural cotton fibers. *Carbohydr. Polym.* 2022, 282, 119040.

[46] Horvath, A., Curry, J., Haverhals, L. M., Shaw, S. K. Practical Online Monitoring of Ionic Liquid Fiber Welding Solvent. *ACS Omega.* 2021, 6, 22367-22373.

[47] Durkin, D. P., Ye, T., Choi, J., Livi, K. J. T., Long, H. C. D., Trulove, P. C., Fairbrother, D. H., Haverhals, L. M., Shuai, D. Sustainable and scalable natural fiber welded palladium-indium catalysts for nitrate reduction. *Appl. Catal. B.* 2018, 221, 290-301.

[48] Velmurugan, P., Shim, J., Bang, K.-S., Oh, B.-T. Gold nanoparticles mediated coloring of fabrics and leather for antibacterial activity. *Journal of Photochemistry and Photobiology B: Biology.* 2016, 160, 102-109.

[49] Nguyen-Tri, P., Altiparmak, F., Nguyen, N., Tuduri, L., Ouellet-Plamondon, C. M., Prud'homme, R. E. Robust Superhydrophobic Cotton Fibers Prepared by Simple Dip-Coating Approach Using Chemical and Plasma-Etching Pretreatments. *ACS Omega.* 2019, 4, 7829-7837.

[50] Elmaaty, T. A., El-Nagare, K., Raouf, S., Abdelfattah, K., El-Kadi, S., Abdelaziz, E. One-step green approach for functional printing and finishing of textiles using silver and gold NPs. *RSC Adv.* 2018, 8, 25546-25557.

[51] Sivakavinesan, M., Vanaja, M., Annadurai, G. Dyeing of cotton fabric materials with biogenic gold nanoparticles. *Sci. Rep.* 2021, 11, 13249.

[52] Fujita, K., Murata, K., Masuda, M., Nakamura, N., Ohno, H. Ionic liquids designed for advanced applications in bioelectrochemistry. *RSC Adv.* 2012, 2, 4018-4030.

[53] Admassie, S., Ajjan, F. N., Elfwing, A., Inganäs, O. Biopolymer hybrid electrodes for scalable electricity storage. *Materials Horizons.* 2016, 3, 174-185.

[54] Jost, K., Durkin, D. P., Haverhals, L. M., Brown, E. K., Langenstein, M., De Long, H. C., Trulove, P. C., Gogotsi, Y., Dion, G. Natural fiber welded electrode yarns for knittable textile supercapacitors. *Adv. Energy Mater.* 2015, 5, 1-8.

[55] Morgan, S. E., Willis, M. L., Peterson, G. W., Mahle, J. J., Parsons, G. N. Green MOF-Fabrics: Benign, Scalable Sorption-Vapor Synthesis of Catalytic Composites to Protect against Phosphorus-Based Toxins. *ACS Sustain. Chem. Eng.* 2022, 10, 2699-2707.

[56] Couzon, N., Ferreira, M., Duval, S., El-Achari, A., Campagne, C., Loiseau, T., Volkringer, C. Microwave-Assisted Synthesis of Porous Composites MOF–Textile for the Protection against Chemical and Nuclear Hazards. *ACS Appl. Mater. Interfaces.* 2022, 14, 21497-21508.

[57] Balasubramanian, S., Kulandaisamy, A. J., Babu, K. J., Das, A., Balaguru Rayappan, J. B. Metal Organic Framework Functionalized Textiles as Protective Clothing for the Detection and Detoxification of Chemical Warfare Agents—A Review. *Ind. Eng. Chem. Res.* 2021, 60, 4218-4239.

[58] Bunge, M. A., Pasciak, E., Choi, J., Haverhals, L., Reichert, W. M., Glover, T. G. Ionic Liquid Welding of the UIO-66-NH2 MOF to Cotton Textiles. *Ind. Eng. Chem. Res.* 2020, 59, 19285-19298.

[59] Nguyen, S. T., Feng, J., Le, N. T., Le, A. T. T., Hoang, N., Tan, V. B. C., Duong, H. M. Cellulose Aerogel from Paper Waste for Crude Oil Spill Cleaning. *Ind. Eng. Chem. Res.* 2013, 52, 18386-18391.

[60] Beaumont, M., Bacher, M., Opietnik, M., Gindl-Altmutter, W., Potthast, A., Rosenau, T. A General Aqueous Silanization Protocol to Introduce Vinyl, Mercapto or Azido Functionalities onto Cellulose Fibers and Nanocelluloses. *Molecules (Basel, Switzerland).* 2018, 23, 1427.

[61] Lucia, A., Bacher, M., van Herwijnen, H. W. G., Rosenau, T. A Direct Silanization Protocol for Dialdehyde Cellulose. *Molecules (Basel, Switzerland).* 2020, 25, 2458.

[62] Frank, B. P., Durkin, D. P., Caudill, E. R., Zhu, L., White, D. H., Curry, M. L., Pedersen, J. A., Fairbrother, D. H. Impact of Silanization on the Structure, Dispersion Properties, and Biodegradability of Nanocellulose as a Nanocomposite Filler. *ACS Appl. Nano Mater.* 2018, 1, 7025-7038.

[63] Neves, R. M., Ornaghi, H. L., Zattera, A. J., Amico, S. C. The influence of silane surface modification on microcrystalline cellulose characteristics. *Carbohydr. Polym.* 2020, 230, 115595.

[64] Durkin, D. P., Frank, B. P., Haverhals, L. M., Howard Fairbrother, D., De Long, H. C., Trulove, P. C. Engineering lignocellulose fibers with higher thermal stability through natural fiber welding. *Macromol. Mater. Eng.* 2019, 0, 1900042.

[65] Shen, H., Gulbrandson, A. J., Park, S., Li, M., Shuai, D., Trulove, P. C., Durkin, D. P. Antimicrobial Biocomposites Fiber-Welded with Lignocellulose Containing Silver Nanoparticles. *Macromol. Mater. Eng.* 2022, 307, 2100872.

[66] Brown, E. K., Dennis, K., Haverhals, L. M., Sweely, K. D., Hugh, C., Trulove, P. C. Laser induced natural fiber welding of cellulosic substrates. *Electrochem. Soc. Trans.* 2014, 64, 507.

[67] Sweely, K. D., Fox, E. T., Brown, E. K., Haverhals, L. M., Hugh, C., Trulove, P. C. Inkjet printing ionic liquids for the fabrication of surface structures on biopolymer substrates. *Electrochem. Soc. Trans.* 2014, 64, 575.

[68] Haverhals, L. M., Brown, E. K., Foley, M. P., De Long, H. C., Trulove, P. C. Formation of surface structures on biopolymer substrates through inkjet printing of ionic liquids. *Electrochem. Soc. Trans.* 2012, 50, 615-621.

[69] Chen, G., Yan, L., Wan, X., Zhang, Q., Wang, Q. In situ synthesis of silver nanoparticles on cellulose fibers using D-glucuronic acid and its antibacterial application. *Materials.* 2019, 12, 3101.

[70] Durkin, D. P., Hoffman, C. E., Gulbrandson, A. J., Fessler, E. R., Chung, R. T., Aiello, A., Cosby, T., Woodcock, J. W., Trulove, P. C. Fiber-welded polyionic biocomposites using 1-alkyl-3-vinylimidazolium alkylphosphonate ionic liquids. *Journal of Ionic Liquids.* 2022, 2, 100024.

[71] Taft, R. W., Kamlet, M. J. The solvatochromic comparison method. 2. The .alpha.-scale of solvent hydrogen-bond donor (HBD) acidities. *J. Am. Chem. Soc.* 1976, 98, 2886-2894.

[72] Kamlet, M. J., Taft, R. W. The solvatochromic comparison method. I. The .beta.-scale of solvent hydrogen-bond acceptor (HBA) basicities. *J. Am. Chem. Soc.* 1976, 98, 377-383.

[73] Kamlet, M. J., Abboud, J. L., Taft, R. W. The solvatochromic comparison method. 6. The .pi.* scale of solvent polarities. *J. Am. Chem. Soc.* 1977, 99, 6027-6038.

[74] Muldoon, M. J., Gordon, C. M., Dunkin, I. R. Investigations of solvent–solute interactions in room temperature ionic liquids using solvatochromic dyes. *Journal of the Chemical Society, Perkin Transactions 2.* 2001, 433-435.

[75] Crowhurst, L., Mawdsley, P. R., Perez-Arlandis, J. M., Salter, P. A., Welton, T. Solvent–solute interactions in ionic liquids. *Phys. Chem. Chem. Phys.* 2003, 5, 2790-2794.

[76] Dzyuba, S. V., Bartsch, R. A. Expanding the polarity range of ionic liquids. *Tetrahedron Lett.* 2002, 43, 4657-4659.

[77] Lee, J.-M., Ruckes, S., Prausnitz, J. M. Solvent Polarities and Kamlet−Taft Parameters for Ionic Liquids Containing a Pyridinium Cation. *The Journal of Physical Chemistry B.* 2008, 112, 1473-1476.

[78] Hauru, L. K. J., Hummel, M., King, A. W. T., Kilpeläinen, I., Sixta, H. Role of Solvent Parameters in the Regeneration of Cellulose from Ionic Liquid Solutions. *Biomacromolecules.* 2012, 13, 2896-2905.

[79] Mäki-Arvela, P., Anugwom, I., Virtanen, P., Sjöholm, R., Mikkola, J. P. Dissolution of lignocellulosic materials and its constituents using ionic liquids—A review. *Ind. Crop. Prod.* 2010, 32, 175-201.

[80] Spange, S., Lungwitz, R., Schade, A. Correlation of molecular structure and polarity of ionic liquids. *J. Mol. Liq.* 2014, 192, 137-143.

[81] Lindman, B., Medronho, B., Alves, L., .Costa, C., Edlund, H., Norgren, M. The relevance of structural features of cellulose and its interactions to dissolution, regeneration, gelation and plasticization phenomena. *Phys Chem Chem Phys.* 2017, 19, 23704-23718.

[82] Brandt, A., Hallett, J. P., Leak, D. J., Murphy, R. J., Welton, T. The effect of the ionic liquid anion in the pretreatment of pine wood chips. *Green Chem.* 2010, 12, 672-679.

[83] Lu, J., Brown, J. S., Liotta, C. L., Eckert, C. A. Polarity and hydrogen-bonding of ambient to near-critical water: Kamlet–Taft solvent parameters. *Chemical Communications.* 2001, 665-666.

[84] Swensson, B., Larsson, A., Hasani, M. Probing Interactions in Combined Hydroxide Base Solvents for Improving Dissolution of Cellulose. *Polymers.* 2020, 12, 1310.

[85] Cheong, W. J., Carr, P. W. Kamlet-Taft .pi.* polarizability/dipolarity of mixtures of water with various organic solvents. *Anal. Chem.* 1988, 60, 820-826.

[86] Kamlet, M. J., Abboud, J. L. M., Abraham, M. H., Taft, R. W. Linear solvation energy relationships. 23. A comprehensive collection of the solvatochromic parameters, .pi.*, .alpha., and .beta., and some methods for simplifying the generalized solvatochromic equation. *JOC.* 1983, 48, 2877-2887.

[87] Marcus, Y. The properties of organic liquids that are relevant to their use as solvating solvents. *Chem. Soc. Rev.* 1993, 22, 409-416.

[88] Durkin, D. P., Hoffman, C. E., Cosby, T., Aiello, A., McFarland, J. A., De Long, H. C., Woodcock, J. W., Trulove, P. C. Developing polyionic biocomposite materials through natural fiber welding. *ECS Transactions.* 2020, 99, 99-112.

List of Contributing Authors

Abidi, Noureddine
Fiber and Biopolymer Research Institute,
Department of Plant and Soil Science,
Texas Tech University, Lubbock, TX, USA

Almeida, Jéssica S.
CICECO – Aveiro Institute of Materials,
Department of Chemistry, University of Aveiro, Aveiro, Portugal

Artemkina, Yulia M.
Department of General and Inorganic Chemistry,
Mendeleev University of Chemical Technology, Moscow, Russia.

Berton, Paula
Chemical and Petroleum Engineering Department,
University of Calgary, Calgary, AB, Canada.

Biswas, Rima
Process Simulation Research Group, School of Chemical Engineering,
Vellore Institute of Technology, Vellore, Tamil Nadu, India

Capela, Emanuel V.
CICECO – Aveiro Institute of Materials,
Department of Chemistry, University of Aveiro, Aveiro, Portugal

Durkin, David P.
Department of Chemistry,
US Naval Academy, Annapolis, MD, USA

Freire, Mara G.
CICECO – Aveiro Institute of Materials,
Department of Chemistry, University of Aveiro, Aveiro, Portugal

Ijardar, Sushma P.
Department of Chemistry,
Veer Narmad South Gujarat University, Surat, India.

Kharazi, Mona
Department of Applied Chemistry,
Bu-Ali Sina University, Hamedan, Iran

Larm, Nathaniel E.
Department of Chemistry,
US Naval Academy, Annapolis, MD, USA

Oke, Emmanuel
Department of Chemistry,
Veer Narmad South Gujarat University, Surat, India.

Plechkova, Natalia V.
Wellcome-Wolfson Institute for Experimental Medicine,
The Queen's University of Belfast, Belfast, UK

Saien, Javad
Department of Applied Chemistry,
Bu-Ali Sina University, Hamedan, Iran

Salma, Umme
Istituto di Nanotecnologia (NANOTEC) – CNR,
c/o Campus Ecotekne, Via Monteroni, Lecce, Italy

Shamshina, Julia L.
Fiber and Biopolymer Research Institute,
Department of Plant and Soil Science,
Texas Tech University, Lubbock, TX, USA

Shcherbakov, Vladimir V.
Department of General and Inorganic Chemistry,
Mendeleev University of Chemical Technology, Moscow, Russia

Stachurski, Christopher D.
Department of Chemistry,
US Naval Academy, Annapolis, MD, USA

Tamas, George
Bradley University, Peoria, Illinois, USA

Tavares, Ana P.M.
CICECO – Aveiro Institute of Materials,
Department of Chemistry, University of Aveiro, Aveiro, Portugal

Thalangamaarachchige, Vidura
Department of Chemistry and Physics,
McNeese State University, Lake Charles, LA, USA

Trulove, Paul C.
Department of Chemistry,
US Naval Academy, Annapolis, MD, USA

Index

B

C

D

K

L

M

Q

R

T

U

V

Editors' Contact Information

Dr. Noureddine Abidi, Ph.D.
Leidigh Endowed Professor,
Department of Plant and Soil Science
Director of Fiber and Polymer Research Institute
Texas Tech University
Lubbock, TX
noureddine.abidi@ttu.edu

Dr. Julia L. Shamshina
Research Assistant Professor,
Fiber and Biopolymer Research Institute
Texas Tech University
Lubbock, TX
jshamshi@ttu.edu